The Dynamics of Fields, Fluids, and Gauges

Book 2 of Physics from Maximal Information Emanation, a seven-book physics series.

ISBN 979-8-9888160-2-7

The Dynamics of Fields, Fluids, and Gauges

by

Stephen Winters-Hilt

ISBN 979-8-9888160-2-7

Golden Tao Publishing
Angel Fire, NM
USA

Dedication

This book is dedicated to my family that helped on this lengthy road of discovery: Cindy, Nathaniel, Zachary, Sybil, Eric, Joshua, Teresa, Steffen, Hannah, Anders, Angelo, John and Susan.

Contents

Preface to Physics Series on:

Physics from Maximal Information Emanation

"The Road goes ever on and on
Down from the door where it began.
Now far ahead the Road has gone,
And I must follow, if I can,
Pursuing it with eager feet,
Until it joins some larger way
Where many paths and errands meet.
And whither then? I cannot say"

– J.R.R. Tolkien, The Fellowship of the Ring

Variation, Propagation, and Emanation
This is a seven book Physics Series that starts with Classical Mechanics (Book 1 [1]), then Classical Field Theory, such as electromagnetism (Book 2 [116]), then Manifold Dynamics, such a General Relativity (Book 3 [5]). The switch to a quantum mechanics description is given in Book 4 [15], and to a quantum field theory, QED in particular, in Book 5 [12]. A 'quantum manifold theory' would be the obvious next step except it cannot be done (there is not a renormalizable Field theory for Gravitation). Instead a thermal quantum manifold theory is considered, as well as Black Hole thermodynamics in general, in Book 6 [6]. Book 7 [4] describes a new theory, Emanator Theory, that provides a deeper mathematical construct that undergirds Quantum theory, much like quantum theory can be shown to provide a deeper (complexified) mathematical construct based on the classical theory.

This is a modern exposition where subtleties of chaos theory are described in Book 1, of Lorentz Invariance in Book 2, of Covariant Derivatives (General Relativity) and Gauge Covariant Derivatives (Yang-Mills Field Theory) in Book3. Book 4 on Quantum Mechanics provides an extensive review of QM, then considers a full self-adjoint analysis on the full general relativistic solution to the spherical shell in-fall system (a result carried over from Book 3). Book 5 considers QFT basics in detail, along with alternate vacua in specific scenarios. Book 6 considers thermodynamics from the basics to the Hamiltonian thermodynamics of some Black Hole systems. Throughout, the odd recurrence of the alpha parameter is noted. In Book 7 we look to a deeper mathematical formulation from which the Quantum Path Integral formulation would result, as well as explaining the odd parameters and structures that have been discovered (such as alpha and Lorentz Invariance).

The physical description starts with the classic formulations of point particle motion. The first approach to doing this is using differential equations (Newton's 1st and 2nd Law); the second is using a variational function formulation to select the differential equation (Lagrangian variation); the third is using a variational functional formulation (Action formulation) to select the variational function formulation. Historically, it wasn't realized until much later that there are two domains for motion in many systems: non-chaotic; and chaotic.

In a description of particle motion, assuming not in a parameter domain with chaotic motion, several important limits are found to exist. Examples include: the universal constants from the aforementioned chaos phenomenon, that are still encountered in non-chaos regimes if driven "to the edge of chaos". Limits are found where scattering is defined in the asymptotic limit and perturbation theory is well-defined in the sense that it is convergent. Overall, if the evolution is described as a 'process' it is often a Martingale process, which has well-defined limits. So, we have descriptions for motion, typically reducible to an ordinary differential equation (ODE), and for which solutions (requiring limit-definitions) are typically found to exist.

The physical description then contends with field dynamics in 2D, 3D, and 4D (in Book 3 [5]). Two-dimensional ("2D") field dynamics can be described as a complex function (that maps complex numbers to complex numbers). A novelty of the 2D complex function is it also shows how to handle many types of singularities (the residue theorem), thus provides important information about fundamental structures in physics as well as fundamental mathematical techniques for solving many integrals. For the 3D field dynamics we do an analysis of the electromagnetic field in 3D. The level of coverage begins at an overview of electrostatics at the level of the graduate text Jackson [32]. Some problems from Jackson Ch's 1-3 are examined closely in developing the theory itself. For some this material (in Book 2 [116]) might provide a useful accompaniment to Jackson's text in a full course on EM (based from Jackson's text). A quick review of electrodynamics and electromagnetic wave phenomena is then given. In essence, we see many more examples of ODE problems with solutions, such as for the 3D Laplacian, usually involving separation of variables. We then review the famous transform, discovered by Lorentz in 1899 [37], that relates the EM field as seen by two observers differing by a relative velocity. With the existence of this transform, that brings in the time dimension along with the relative velocity, we effectively have a 4D theory.

From Lorentz Invariance we have, as a point transformation, rotational invariance under SO(3) or SU(2). If Lorentz Invariance is fundamental, then we should see both forms of rotation invariance, one of vector/tensor type from SO(3), and one of spinorial type from SU(2). This is the case, as gauge fields are vectorial and matter fields are spinorial. From Lorenz Invariance as a local invariance we have the Minkowski (flat) spacetime metric, which then generalizes to the Riemannian metric (in General Relativity).

As with the point particle dynamics, for the field dynamics we have three ways to formulate the behavior: (1) differential equation; (2) function variation (on Lagrangian); and (3) functional variation (on the Action). We will see similar limit phenomena as before, but also new phenomena, including (i) inevitable BH singularity formation (the Penrose singularity theorem); (ii) FRW Universe formation (from homogeneity and isotropy); (iii) the BH collapse singularity; (iv) the atomic collapse radiative 'singularity'.

Classical dynamics, thus, has two field-like formulations to describe the world: field and manifold. Such formulations can be interrelated mathematically, so what is happening is more a matter of physics emphasis and convenience. The emphasis on this difference, that appears to be no difference (mathematically), is that different physical phenomenologies are at play. Field descriptions appear to work for 'matter', where the fundamental elements are spinorial. Manifold descriptions appear to work best for geometrodynamics (GR), where the fundamental elements are vectorial (or tensorial, such as the metric). Matter fields are renormalizable, thus quantizable in the standard QFT formulation (to be described in Book 5 [12]), while gravitational manifolds are not renormalizable, and have constraints (weak energy condition and positive energy condition given the existence of spinor fields on the manifold).

The presentation in Books 1-3 [1,5,116], on 'classical' physics, is partly done to make the transition to quantum physics simple, obvious, and in some cases, trivial. Consider the functional variation (Action) formulation of the behavior (whether point-particle or field), this can be captured in integral form, as was done by D'Alembert very early [118] (then by Laplace [104]). Note the use of a large constant to effect a 'highly damped' integral for selection purposes (on variational extremum of the action). To transition to the quantum theory we also have the large constant from 1/h, and so the only difference is the introduction of a factor of 'i', to effect a 'highly oscillatory' integral for selection purposes.

After the transition to a quantum theory, for the point-particle descriptions, the classical collapse problem for atomic nuclei is eliminated. The spectral predictions have excellent agreement with theory, but there is still fine-structure in the spectra not fully explained. The theory is not relativistic and some initial corrections for this are possible (without going to a field-theory) and these indicate closer agreement and explain most of the fine-structure constant discrepancy (and reveal alpha in another place in the theory). It is shown in Book 3 [5] and Book 4 [15], that the GR singularity problem, however, remains unresolved (for the test case of spherical dust shell collapse, done in a full GR analysis, then quantized in a full self-adjoint quantization analysis [15]).

In Book 5 [12], the transition to quantum theory is continued to the field theory descriptions. A precise description/agreement of atomic nuclei is now possible with QED, and within the nuclei themselves (quark confinement) with QCD. The field theories have a small set of bothersome infinities, however, which is eventually solved by renormalization [12]. As mentioned, the quantization of manifold theories, such as GR, does not appear to be possible due to non-renormalizability. Not to be deterred, in Book 6 [6] we consider a Hamiltonian description of a GR system whose quantization would involve an energy spectrum based on that Hamiltonian, if we then use analytic continuation to take us to the thermal ensemble theory based on the partition function that results, we can consider the thermal quantum gravity (TQG) of such systems.

This last example (from Book 6), showing a consistent TQG theory if we use analyticity, is part of a long sequence of successful maneuvers involving analytic continuations in different settings. What is indicated is the presence of an actual complex structure to the stated theory. There is the trivial complex structure extension mentioned above that brought us from the standard classical physics theory to the standard path integral quantum theory. But we also see actual complex structure at the component level with time complexation (that ties to thermal version of the theory by defining the partition function), and we have complex structure as the dimension-level in the form of the successfully applied dimensional regularization procedure used in the renormalization program.

As well as covering the breadth of core physics topics at both undergraduate and graduate level (for courses taken at Caltech and Oxford), including extensive presentation of problems and their solutions, the Series also examines, in specific cases, the boundaries of the physical world "from the inside" (and then later "from the outside"). To this end exploration of spherical dust collapse to form a singularity is examined in a fully general relativistic formalism, and then carried-over to a quantum minisuperspace (quantum gravity) analysis (in Books 3 and 4 [5,15]). Also examined in-depth are the topics of black hole thermodynamics and quantum field theory with alternate vacua (part of Books 5 and 6 [6,12]). The in-depth material comprises the topics covered in my PhD dissertation [70], portions of which are published [71-74].

In recent work on machine learning, that includes statistical learning on neuromanifolds [13], we find a possible new source for a foundational element for statistical mechanics (entropy) via seeking a minimal learning process/path on a neuromanifold [13]. By the time the Series reaches thermodynamics in Book 6, therefore, the foundational thermodynamics elements have all been established from the physical descriptions discovered in Books 1-5, they just haven't been put together in a comprehensive analysis that gives us the fundamental constructs of thermodynamics and statistical mechanics. That said, it would seem that thermodynamics is, thus, entirely derivative from other, truly fundamental theories. Not so, in the joining of the parts to make thermodynamics we have something greater than the sum of the parts. In the 'system' descriptions we find that emergent phenomena exist. This, at least, is unique to thermodynamics, so it is fundamental in this "sum greater than the parts' aspect.

In Book 7 (the last) of the Series, we consider the standard physical world, described by modern physics, "from the outside." In doing this we've already eliminated part of the mystery of entropy by the geometric 'neuromanifold' description. If we can understand other oddities of the standard theory, and arrive at them naturally, then we might have an even deeper dive into modern physics, testing the limits of what is possible, and see possible future developments and unifications of the theory. This is what is described in papers [17,76-83], and organized along with current results into the final Book of the series.

Efforts in the last book of the Series involve choices and concepts identified in the prior six books of the Series, and theoretical maneuvers gleaned from the most advanced courses in physics and mathematical physics taken while at Caltech (as an undergraduate and then as a graduate) and the Oxford Mathematics Institute (as a graduate), and the University of Wisconsin at Milwaukee (as a graduate).

The broad range of topics covered in the Series is, initially, similar to the Landau & Lifshitz graduate textbook series (see [28,33,84]), with a similar exposition on classical mechanics at the start of Book 1. Even with well-established classical mechanics, however, there are significant, modern, updates, such as (modern) chaos theory. In the final two books of the Series (Books 6 and 7 [4,6]) we arrive at statistical mechanics and thermodynamics, together with modern topics such as black hole thermodynamics, thermal quantum gravity, and emanator theory.

Key constants and structures of physics, their discovery from the experimental data, and their theoretical placement in the "Grand Scheme," are emphasized throughout the Series. The constant alpha, a.k.a. the fine structure constant, appears in numerous settings so special note of the occurrence of alpha will be made in each chapter. This is the case even at the outset with Book 1, due to fundamental numerical constants appearing from chaos theory. In Book 7 we see the origin of alpha, as a maximal perturbation amount, appears naturally in a formalism for maximal information 'emanation'. But maximal perturbation in what space and in what manner? In Book 7 of the series [4] we will see a possible representation of such an information entity, and its space of existence, in terms of chiral trigintaduonions.

Thus, in the end, this is an effort to tell of a journey to a special place "where many paths and errands meet", giving rise to emanator theory and an answer to the mystery of alpha. Part of this journey is equivalent to 'finding the arkenstone' (alpha) in the most unlikely of places, the trigintaduonion emanation mathematics underpinning the emanator formalism (e.g., Smaug's Lair, described in Book 7 [4]). Why I should have wandered into such an odd place (mathematically speaking), and why I should posit a deeper form of quantum propagation using hypercomplex trigintaduonions, here called emanation, is why there is such extensive background on standard topics. This extensive background even impacts the classical mechanics description via its modern chaos theory material (due to a possible relation between C_∞ and alpha). The critical role of emergent phenomena is only understood at the end, including for manifolds in geometry and neuromanifolds in statistical mechanics, and leads to a Book 6 that goes from very basic (initial thermodynamics) to very advanced (emergent phenomena). Much is made clear with emanator theory, including how reality is both fractal and emergent. At this point in the journey, as with Tolkien, this much I can say: "The Road goes ever on and on … And whither then? I cannot say".

The seven books in the Series are as follows:

Book 1. Classical Mechanics and Chaos
Book 2. Classical Field Theory
Book 3. Classical Manifold Theory
Book 4. Quantum Mechanics and the Path Integral Foundation
Book 5. Quantum Field Theory and the Standard Model
Book 6. Thermal & Statistical Mechanics, and BH Thermo.
Book 7. Maximum Information Emanation and Emanator Theory

Overview of Book 1

Book 1 is a modern exposition of classical mechanics, including chaos theory, and including ties to later theoretical developments as well. The exposition consists, throughout, of the presentation of interesting problems with many solved, the others left for the reader. The problems are drawn from classical mechanics (CM) and mathematics courses taken at Caltech, Oxford, and the University of Wisconsin. The courses range from undergraduate level to advanced graduate level. The courses had a rich and sophisticated selection of textbook and reference material, as you might expect, and those reference texts are, similarly, drawn on here. Those classical mechanics texts, listed by author, include: Landau and Lifshitz [84]; Goldstein [85]; Fetter & Walecka [86]; Percival & Richards [87]; Arnold (ODE) [88]; Arnold (CM) [89]; Woodhouse [25]; and Bender & Orszag [43]. Notice how the first Arnold reference and the Bender and Orszag reference involve textbooks focused on ordinary differential equations (ODEs). Likewise, an analysis of the excellent, and rapid, exposition by Landau and Lifshitz, reveals that it partly progresses through the material by going through ODEs of increasing complexity (corresponding to more complicated pendulum motion, for example, such as by adding a frictional force). This strong alignment with the underlying mathematics of ODEs is continued in this exposition, so much so that an appendix is provided for a quick review of ODEs from the applied mathematics perspective.

Particle dynamics, with and without forces, are described, with all arriving at descriptions with chaotic motion, with chaos described in the latter half of Book 1 [1]. Universally it is found that systems transitioning to chaotic behavior do so with a remarkable period-doubling process and this will be described both mathematically and with computer results. In the analysis of such dynamical systems we will find that periodic physical systems can be described in terms of repeated "mappings", e.g., classic dynamic mappings [90], and when described in this way the transition to chaos is made much more mathematically evident (as will be shown). The familiar Mandelbrot set is generated by such a repeated mapping, where it's "edge of chaos" is defined by the fractal boundary of the classic Mandelbrot image.

Properties of the classic Mandelbrot set will be relevant to the physics discussed in Book 1 and Book 7, including the property that the fractal boundary has a fractal dimension of 2 (the fractal dimension of the boundary can be between 1 and 2, to get equal to 2 is special). With the Mandelbrot set we also recover the well-studied constants associated with the universal Feigenbaum constants [3]. In the Mandelbrot set we can clearly see the fundamental constant for maximum perturbation that is at maximum antiphase (negative) with magnitude C_∞, where the same results hold for a family of basic formulations (for a variety of Lagrangian formulations, for example).

From the Lagrangian variational formulation of 'action' for particle motion we will eventually define the path integral functional variational formulation involving that same Lagrangian to arrive at a quantum description for the non-relativistic quantum particle motion (described in detail in Book 4 [15], and relativistic in Book 5 [12]). From the quantum description we arrive at the propagator formalism for describing dynamics (this exists in the classical formulation too, but typically is not used much in that context). Complex propagators will then be found to have ties to statistical mechanics and thermodynamics properties (Book 6

[6]). The ties to statistical mechanics are further emphasized when at the "edge of chaos" but with the orbit motion still confined. This may be associated with an ergodic regime, thus an equilibrium and martingale regime, the existence of which can then be used at the start of Book 6 [6] statistical mechanics and thermodynamics derivations with the existence of equilibria established at the outset. The existence of the familiar entropy measures are already indicated in the neuromanifold description (Book 3 [5]), thus, together with equilibria, the Book 6 thermodynamics description is able to begin with a well-established foundation that is not claimed by fiat, rather claimed as a direct result of what has already been determined in the theory/experiment described in the previous books of the Series.

Overview of Books 2 & 3

When moving from a theory of point particles to a theory of fields, there's not much discussion in the core physics books on fields in a general sense, it usually just directly jumps to the main field of relevance, Electromagnetism (EM). If advanced, it may also cover General Relativity (GR), as with [38]. In what follows we will cover these topics, but we will also cover the more basic fields in 1, 2, and 3D (including fluid dynamics), as well as 4D Lorentzian Field formulations (for Special Relativity), the Gauge Field formulation (thus Yang Mills covered in a classical context), and the GR geometric and gauge formulations. This establishes the foundation for the standard forces, and upon quantization (Books 4 and 5 in the Series), lays the foundation for the standard renormalizable forces (all but gravitation).

The gravitational coupling constant 'G' is a dimensionful coupling (not like with alpha in EM), and gravitation with manifold construct can be described as a gauge field construct, although not renormalizable. Gravitation, and associated geometry/manifolds, appears to relate to its own emergent structure, as will be discussed in Book 6. From the local Lorentzian geometry and Lorentzian field descriptions we also see the first of many examples where there is system information in the complexification of some parameter, here the time component. If the Lorentzian is shifted to complex time, this shifts it to being a Euclidean field, with formally well-defined convergence properties (as occurs in statistical mechanics). Complex time also shows deep connections between classical motion and associated Brownian motion (where random walk reveals pi). Thus, it should not be surprising that an emergent manifold may have complex structure such that there is also an emergent 'thermal' manifold, possibly the neuromanifold described in Book 3 and the related partition functions examined in Book 6. Just like locally flat space-time is a natural construct in GR, so too are optimization "learning" steps on a neuromanifold such that relative entropy is selected as a preferred measure, and from it Shannon entropy and Boltzmann's statistical entropy. Thus, the manifold construct appearing at Book 3 has far reaching impact into the foundations of the thermodynamic and statistical mechanical theory described in Book 6.

Before we even get to the manifold/geometry complexities of GR, however, we have already established much with the EM field part of the theory: (i) from 'free' EM without matter we get the speed of light c, Lorentz invariance, and from that special relativity and locally flat space-time; (ii) from EM with matter we get the dimensionless coupling constant alpha.

In going over field theories to describe matter, force fields, and radiation we first describe the classical field theories (CFTs) of fluid mechanics, EM, and General Relativity, with many examples shown. This is then carried over to the quantum field theory (QFT) description in Book 5. A review of the core mathematical constructs employed in CFT and QFT is given in the Appendix. Even as the mathematical physics approach grows in sophistication, we still obtain solutions via variational extrema. Thus, determining the evolution of the system from its variational optimum now becomes the focus of the effort. System 'propagation' from one time to a later time can be described by a propagator. Although a 'propagator' formulation is possible mathematically in classical mechanics (CM) and classical field theory (CF), which are shown, this is usually not done, in favor of simpler representations for the experimental application at hand. As we move to descriptions in the quantum realm, however, the use of the propagator formalism becomes typical, and when used in the path integral formulations we arrive at a compact formulation describing both the evolution and stationary-phase solution at once.

In Book 2 the focus is on classical field theory in a fixed geometry, the main physical example is EM. In this setting alpha appears, for example, in the description of an electron-positron pair: $F = e^2/(4\pi\varepsilon a^2)$ for electron-positron distance 'a' apart, where alpha appears as the coupling constant. Later, in quantum mechanics (QM), both modern and in the early Bohr model, we have that alpha = $[e^2/(4\pi\varepsilon)]/(c\hbar)$. The appearance of alpha in these situations is occurring in bound systems. If we examine EM interactions that are unbound, on the other hand, such as with the Lorentz Force $F = q(E \times v)$, here there arises no alpha parameter, nor with the early quantum mechanical analysis of such systems such as with Compton scattering. Thus, we see an early role for alpha, but only in bound systems, thus only in systems with (convergent) perturbative expansions in system variables.

In Book 3, classical field theory with *dynamic* geometry, i.e. GR, we don't see alpha at all. Instead we see manifold constructs and the mathematics of differential geometry (and to some extent differential topology and algebraic topology). Manifold constructs are entirely encapsulated in the math background given in Book 3 and the Appendix there. An application in the area of neuromanifolds (see [13]), shows the equivalent of a geodesic path in this setting is evolution involving minimum relative entropy steps. Similar to the description of a locally flat space-time we now have a description of 'entropy' increasing/evolving according to minimum relative entropy.

General relativity (GR) stands apart from the other force fields. All the other force fields are part of an adjoint representation of the standard model vis-à-vis the stability subgroup $U(1)xSU(2)_LxSU(3)$. The form of which is derivable from the chiral T one-sided products described in Book 7. The standard model is uniquely obtained in this process, and with no mention of GR. Keep in mind, however, that the adjoint representation has operation on some space (hyperspinorial in case of simple octonion right-products, for example). The 'force' due to gravity is that due to manifold curvature, where the manifold construct is possibly emergent on the space of operation. Thus, the origin of the GR force is entirely different, and it will not allow quantization like the other forces, nor will its singular solutions be resolvable via quantum physics alone, as with EM in Books 4&5, but will also need thermal physics (as will be described in Book 6).

The existence of singular GR solutions, outside of specially symmetric cases (the classic Black hole solutions), wasn't firmly established until the Penrose singularity theorem [16] (awarded Nobel prize in Physics for this in 2020). Some of this material is covered in Book 3 to show how the mathematical formalism shifts to differential topology methods to describe the singularities, with examples referencing the Hawking and Ellis classic [92] and using Penrose diagrams. This, in turn, will come in handy when describing the classic FRW cosmologies with radiation and matter dominated phases (using notes from Peebles [93], Peebles won the Nobel in Physics in 2019).

The GR development would be remiss if it didn't briefly delve into cosmological models, the classic FRW cosmologies in particular. With the GR tools developed, cosmological results are examined, starting with the entry of the cosmological constant into the formalism (a candidate for Dark energy). Various observational data on galaxy rotations and universe simulations of galaxy cluster formation both indicate the existence of Dark matter. This, then, means we have new matter, non-interacting except gravitationally, and this is actually consistent with the latest observational data on the muon g-2 value [94], where the discrepancy between theory and experiment has grown to 4.2 standard deviations, where an extension in the Standard Model appears to be in the works. This is convenient as Emanator theory (Book 7 [4]), predicts such an extension.

We can thus arrive at field equations for EM, GR, and Yang-Mills Gauge Fields (Strong and weak). We can obtain wave and vortex phenomena (as hinted in fluid dynamics). We show the classical instability for atomic matter (classical EM instability) and classical gravitational instability (leading to black hole formation with singularity). From Lagrangian formulations we can then arrive at a QFT formulation (Book 5). The QFT formulation completes the QM (Book 4) cure of "non-relativistic atomic instability" with the cure of the fully relativistic atomic description of the radiative-collapse instability. Introduction of QFT also leads to new instability or infinities, but these can be eliminated by renormalization for the EM and electroweak formulations, and the Yang-Mills strong formulation, but not the GR (gauge) formulation. The current theoretical formulation in modern physics has one glaring gap, therefore: a quantum theory of gravitation. Perhaps this is not a missing element, however, if geometry/GR is a derivative phenomenon, like the field of statistical mechanics and thermodynamics appeared as derivative phenomenon when the complexified quantum propagator gives rise to a real (quantum) partition function. The hint of a deeper emanator theory suggests emergent structures of geometry and thermodynamics are arrived at in the process of emanation, with the information emanated being that of the renormalizable quantum matter fields. In Book 7 [4] a precise mathematical meaning will be found for describing maximal information emanation.

Overview of Book 4

By 1834, with Hamilton's Principle, there was a strong foundation for what is now called classical mechanics. By 1905, with Einstein's publication on the photoelectric effect [46], the rules of classical mechanics were being superseded by the new rules of quantum mechanics. The earliest appearance of quantum mechanics, however, began with the various observations of quantization of light, starting with the strange occurrence of spectral lines for hydrogen. The hydrogen spectrum was made even stranger by a precise fit to a succinct empirical formula by Balmer in 1885 [95]. This is the beginning of an amazing period of discovery. The developments of QM from introductory to advanced roughly follows that history.

The early phase of discovery for quantum mechanics moved into the modern quantum mechanics formalism with the discovery of Heisenberg of the successful application of matrix mechanics and the resultant uncertainty principle (1925) [96]. In 1926, Schrodinger showed that the problem of finding a diagonal Hamiltonian matrix in the Heisenberg's mechanics is equivalent to finding wavefunction solutions to his wave equation [97]. An interpretation of the wavefunction was then clarified in 1927 by Born [98]. Dirac developed a manifestly relativistic formalism for the wavefunction and wave-equation for fermionic matter (1928) [99]. An axiomatic reformulation of quantum mechanics was then given by Dirac (1930) [100], laying the foundation for much of modern quantum notation and for critical issues such as self-adjointness. Dirac then described a formulation of a quantum propagation path, with quantum propagator having the familiar phase factor involving the action, in his paper "The Lagrangian in Quantum Mechanics" in 1933 [101]. In essence, Dirac had obtained a single path, in what would eventually be generalized by Feynman to all paths with the invention of the path integral formalism (1942 & 1948) [102,103]. The equivalence of a quantum mechanical formulation in terms of path integrals and the Schrodinger formalism was shown by Feynman in 1948 [103].

In a path integral description, the quantum mixture state, semiclassical physics, and classical trajectories are all given by the stationary phase dominated component. A stationary phase solution that is dominated by a single path is typical for a classical system. Thus, variational methods are fundamental to analysis of physical systems, whether it be in the form of Lagrangian and Hamiltonian analysis, or in various equivalent integral formulations.

Feynman's discovery of the path integral formalism wasn't solely based on the prior work of Dirac (1933) [101], although by appending that paper to his PhD thesis (1946) its importance was clearly emphasized. Feynman also benefited from work going as far back as Laplace [104] for selection process based on highly oscillatory integral constructions that self-select for their stationary phase component. This branch of mathematics eventually became associated with Laplace's method of steepest descents, then to the work of Stokes and Lord Kelvin, then to the work of Erdelyi (1953) [105,106].

Feynman and others then invented quantum field theory for electromagnetism (QED) during 1946-1949 (more on this later). Extension to electroweak occurred in 1959, and to QCD in 1973, and to the "Standard Model" in 1973-1975. Thus, the impact of the path integral revolution in quantum physics was felt well into the 1970's, but this was only the beginning. At their inception path integrals were examined by Norbert Wiener, with the introduction of the Wiener Integral, for solving problems in statistical mechanics in diffusion and Brownian motion. In the 1970's this led to what is now known as "the grand synthesis" which unified quantum field theory (QFT) and statistical field theory (SFT) of a fluctuating field near a second-order phase transition, and where use of renormalization group methods enabled significant advances from QFT to be carried over to SFT.

The grand synthesis is one of many instances to come where we see analytic continuation of a constant or a parameter giving rise to familiar physics in the thermodynamic and statistical mechanics domains, showing a deeper connection (still not fully understood, see Book 7). The Schrödinger equation, for example, can be seen to be a diffusion equation with an imaginary diffusion constant. Likewise, the path integral can be seen to be an analytic continuation of the method for summing up all possible random walks.

In Book 4 we also carefully examine the closest gravitational equivalent to the hydrogenic atom (dust shell collapse). What results is an incomplete formulation due to boundary conditions, where to get the time choice you must input that time choice. No specific choice of time is indicated to avoid infall-collapse. The results, however, can show stability and consistency in a "full" thermal quantum gravity description where analyticity is employed. Success in this way, and not others, suggests possible fundamental role of analyticity and thermality (Books 6&7) and also suggests that thermal quantum gravity TQG may 'exist' or be well-formulate-able, while quantum gravity QG generally might not 'exist'. These results, shown in Book 6, provide the lead-in to the Book 7 discussion on Emanator theory, where core concepts in Books 1-6 that tie to emanator theory are brought together in a new theoretical synthesis.

Overview of Book 5

In Book 5 we show QFT's in the gauge field representation, which clearly relates the choice of field theory to a choice of Lie algebra, which, in turn, can be related to a choice of group theory (such as U(1) and SU(3)). From this we can see that non-classical algebraic constructs are ubiquitous in QM and QFT, so a review of Group Theory and Lie Algebras is given in the Appendix, as well as a review of Grassman Algebras, and other special algebras needed in QM and QFT. Similarly, as regards choice of approach, we find that the Schrodinger and Heisenberg formulations often provide the only tractable way to get a solution for bound systems. In critical theoretical considerations, however, the path integral approach is best (as will be shown). In seeking a deeper theory, the more unified path integral (PI) approach provides important hints as to a deeper theory (see Book 7).

In Book 5 we get the highest precision result for the value of alpha, in its role as perturbation parameter. If a calculation of the electron magnetic moment parameter g-2 is performed, with all of the Feynman diagrams appropriate to expansions up to 5th order, we get a determination of alpha up to 14 digits, where 1/alpha=137.05999..... . This gives us one of the most precise measurements of alpha known. When a similar analysis is done for the muon g-2, given the much larger muon mass, particle production pairs of other particles have a measurable effect, and we are able to probe the lower masses of the standard model that are present. In doing this, in preliminary experiments, there is a discrepancy indicating more particles, e.g. the Standard Model will need to be extended (possibly with a type of 'sterile' neutrino). These missing particles could be the missing "Dark Matter". The prediction of such in Emanator Theory, and why there should be an imbalance between the left and right neutrinos (hint: maximum information transmission) is described in Book 7.

Part of the description of quantum field theory entails use of analyticity and other complex structures to encapsulate more of the physics in a complex extension to the space (or dimension). This often leads to formulations in terms of complex integration, with the choice of complex contour specified, such as with the Feynman propagator. One of the main renormalization methods, for example, is to use dimensional regularization, which entails analytically continuing expressions with dimensionality to dimensionality as a complex parameter. There is also the aforementioned shift to complex and to "Wick rotate" expressions with real time to expressions with pure complex time. In doing this the statistical mechanical partition function for the system is obtained, with well-defined summation. Thus, a connection between 'thermality' and complex structure, in the time dimension at least, is indicated.

The second part of Book 5 describes QFT on curved space-time (CST), where we arrive at an early analysis of Black Hole thermodynamics. Here we find that space-time curvature gives rise to thermality and particle production effects. Black Hole thermality was revealed in Hawking radiation [111], due to the causal boundary at the horizon. Such thermality is even seen in flat space-time (Book 5) if causal boundaries are induced, such as in the case of an accelerated observer [112].

QFT on CST has one further gift, critical to the statistical mechanics formalism to follow in Book 6, and that's the spin-statistics relation. This relation is usually assumed, along with other critical notions, such as entropy, and the relation between entropy and density of states. These are all shown, with the presentation path chosen in this Physics Series, to be fundamental or derivative to the formalism already established in Books 1-5 (to prepare for Book 6).

The choice of time is related to choice of vacuum, which is related to choice of field geometry or observer motion (such as constant acceleration or expansion). If you have flat spacetime QFT with a boundary, then you have thermodynamic effects (e.g., the Rindler observer). In this setting we can compare the Hawking derivation of Hawking Radiation using the Euclideanization 'trick' vs the Bogoliubov transformations of the field to the Rindler geometry from the Minkowski geometry (if chosen as the asymptotic vacuum reference). With QFT on CST we also arrive at spin-statistics as mentioned, and get the final extension of the theory by way of Grassman algebras, to arrive at thermodynamically consistent Bose and Fermi statistical descriptions on quantum matter.

Overview of Book 6

Thermodynamics is the oldest of the physics disciplines (fire), with unapologetic use of phenomenological arguments and mysterious thermodynamic potentials (entropy). Obviously, thermodynamics is still prevalent today, including in its more quantified form via statistical mechanics. How is this not a failure of the mechanistic description of the universe indicated by CM and even QM? Concepts that appeared in QM, such as probability, are now occurring again. Other new concepts appear as well, including: approximate statistical laws; equations of state; heat as a form of energy; entropy as a variable of state; existence of equilibria; ensembles/distributions; and existence of the partition function. Many of these concepts appear in the path integral descriptions with the analyticity methods/extensions mentioned previously, so there are hints of a deeper theory that arrives at much of thermodynamics/Statistical mechanics foundation from the existing quantum theory.

Book 6 has been placed after the other chapters to await identification of entropy as fundamental in that it can be identified as an intrinsic system function even before getting to thermodynamics. We also already have experience with many particle systems, via QFT (especially in CST where particle creation is almost unavoidable), without directly tackling that scenario (due to QFT effectively already being many-particle, with analytic determination of many-particle system functions, such as entropy). With entropy presented at the outset as an important system variable, the derivation of thermodynamic potentials is then a straightforward process, as will be shown. The standard SM connections to thermodynamics can then be given. Thus, in covering Thermodynamics and Statistical Mechanics we start with the foundations of the theory mostly established, such as entropy (also with equipartition equivalent to sum on paths with no weightings, etc.), with no assumptions. Everything follows directly from the theoretical discoveries outlined in the preceding books in the Series. We don't see new connections to alpha, but we do see new structures/effects, especially manifold constructs (as with GR, where we also saw no role for alpha).

The close ties between QM Complexified giving rise to a particle ensemble partition function, and QFT complexified and field ensemble partition function, is now simply a derivative aspect of the fundamental complexation posited. This complexation will be posed in Book 7 with emanation in a complexified perturbation space.

From Atomic Physics, described in Book 4, we also obtain the standard rules on electron shell completion (that is encoded in the periodic table). Similarly, we can also understand the origins of the intermolecular quantum chemistry rules. When taken to the statistical mechanics (SM) extreme we have thermodynamic equilibrium emergent from (the Law of Large Numbers (LLN) and reverse Martingale convergence. With completion of application to chemical processes we have clear phase-transition effects, as well as equilibrium and near-equilibrium effects. The familiar chemistry results, with phases of matter.

From chemical equilibrium and near-equilibrium, with 10^{23} elements that interact weakly or not at all, we have two generalizations. The first is to consider chemical near-equilibrium and directly obtain an emergent process at this level, this is the branch that gives us biology/life at its most primitive level. The second is to consider equilibrium and near-equilibrium in general when the elements interact strongly (with10^{10}elements, say), this is the branch that describes biology/life at its most advanced social level and economics. In classic shot noise, the granularity of low-current flow (due to discreteness off electron charge) leads to a noise effect. Thus, as we consider situations with fewer elements, there are more complications, not less, due to granularity noise effects, and we enter the realm of machine learning with sparse data. Noise effects can be significant in complex systems, especially in biology where it is part of what is selected (such as in hearing, for background noise cancellation).

The second part of Book 6 explores the role of thermodynamics in efforts to extend to TQFT and TQG. This is done by exploring Black Hole settings. The recognition of a role for complex structure on system variables becomes apparent in this process (on top of the generalization to non-trivial algebras as already revealed).

In Book 6, part 2, we examine the Hamiltonian thermodynamics of some black hole geometries with stabilizing boundary conditions. In this foray into directly exploring a thermal quantum gravity (TQG) solution we assume a path integral form for the GR problem and shift directly to a partition function (by 'Wick rotation' mentioned above). We see that TQG is possible, where positive heat capacity shows stability. Another encouraging result as to an eventual unifying theory comes from String theory via its explanation of BH thermodynamics and BH horizon effects with the BH fuzz solution (via use of the holographic hypothesis and the related AdS-CFT relation [113,114]).

In Book 6, part 2, we also examine the propagator to partition-function transformation upon complexation, which leads to a thermodynamic theory for some equilibrium formulation, with certain parameter settings required for stability (positive heat capacity). This is doable in a variety of settings, suggesting how such thermodynamically consistent boundary conditions may be what constrains the classical motion and BH singularity formulation by the effect of this stabilization manifesting for certain internal geometries. Successful TQG (Thermal Quantum Gravity) formulations, such as for RNadS and Lovelock spacetimes shown in Book 6, via reformulation using analyticity, and not via non-analytic approaches, suggests a possible fundamental role of analyticity once again and also suggest that TQG may 'exist' or be well-formulate-able, while QG generally might not 'exist'. These results, together with core concepts from Books 1-6 that tie to emanator theory, are brought together in a new theoretical synthesis in Book 7.

Overview of Book 7

In Books 4,5, and 6 of the Series, we explored examples of QM with imaginary time, QFT in CST, Thermal QFT, minisuperspace QG, and Thermal QG. In this effort we find the path integral, and PI propagator, to provide the most general representation. In seeking a deeper theory in Book 7 we build on the sum-on-paths with propagator formulation to arrive at a sum-on-emanations with emanator formulation.

Propagation in a complex Hilbert space, in a standard QM or QFT formulation, requires the propagator function to be a complex number (not real or quaternionic, etc., [24]). This prohibits what would otherwise be an obvious generalization to hypercomplex algebras. In order to achieve this generalization, we have to introduce a new layer to the theory, one with universal emanation involving hypercomplex algebras (trigintaduonions) that is hypothesized to project to the familiar complex Hilbert space propagation with associated fixed elements (e.g., the emanator formalism projects out the observed constants and group structure of the standard model). The 'projection' is an induced mathematical construct, like having SU(3) on products of octonions, but here it we be the standard model U(1)xSU(2)xSU(3) on products of emanator trigintaduonions. Thus, in Book 7 a unified variational formulation is posed, one that arrives at alpha as a natural structural element, among other things, uniquely specified by the condition of maximal information emanation.

In Book 7 we also make note of the implications of a fundamental mathematical operation on a space that is repeated or added. The non-GR forces are given by the form of the operation (the sequence forming an associative algebra), the GR forces are given indirectly by the form of the space, this leaves the aspect “repeated or added” to be considered with care. If a purely ‘repeated’ operation, or mapping, occurs we can return to the dynamical mapping discussion of Book 1, where chaos can occur and is ubiquitous. There, the primal ‘phase transition’, the transition to chaos, is evident. If an operation with addition is involved (in the statistical sense of multiple elements), along with repeated overall steps, we arrive at the general framework of statistical mechanics with effects from the Law of Large Numbers (LLN) and reverse Martingale convergence, among other things (Book 6). Most notable, however, is the prevalence of a new effect, that of phase transitions and the emergence of new structure (order from disorder), including the remarkable structures of chemistry and biology.

Why the recurring ‘Cabbalistic formula’? was a question even in the time of Sommerfeld [115]. Now, the numerological parallel is more exact than realized at that time, so is too much a coincidence to be by chance. The non-coincidence appears to be due to the maximal nature of information transmission in a variety of circumstances (in physics, biology, and even human communication with sufficient optimization) as well as with the fractal-like repetition of key parameter sets that occurs in these different settings {10,22,78,137≅1/alpha}. We see that 10 expresses the dimensionality of propagation (or nodes of connectivity), while 22 corresponds to the number of fixed parameters in the propagation (in Book 7 we explore propagation in a 10 dimensional subspace of the 32 dimensional trigintaduonion space, leaving 22 dimensions at fixed values that appear as parameters in the theory). We will see the number 78 relates to generators of the motion, and that there are 4 chiralities of motion (‘doubly chiral’). We will also see that 137 is simply the number of independent tri-octonionic product terms in the general chiral trigintaduonion ‘emanation’.

Synopsis – Frodo Lives
Tolkien wrote of eucatastrophes [119], perhaps he anticipated the constructive role of emergent phenomena in maximum information transmission.

Preface to Physics Series, Book #2, on:

The Dynamics of Fields, Fluids, and Gauges

This is a modern exposition of classical field theory, including the centrality of Lorentz Invariance and Gauge Invariance in the formalism. The exposition consists, throughout, of the presentation of interesting problems with many solved, the others left for the reader. The problems are drawn from classical field theory courses taken at Caltech, Oxford, and the University of Wisconsin. The courses, and related homework and exam problems, ranged from undergraduate level to advanced graduate level. The courses had a rich and sophisticated selection of textbook and reference material, as you might expect, and those reference texts are, similarly, drawn on here.

In this book the focus is on classical fields, beginning with simple notions of field and fluid in 2D and 3D (Ch. 2), then a brief examination of Fluid Dynamics up to Shock formation (Ch. 3). Chapters 4-7 cover electromagnetic (EM) fields: (Ch.4) Electrostatics; (Ch. 5) Laplace's Equation in 3D; (Ch. 6) Multipoles and Macroscopic Matter; (Ch. 7) Electro- and magneto dynamics. In the process of examining the properties of EM fields we will find that they are Lorentz invariant (Ch. 8) and Gauge invariant (Ch. 10). A generalization of the applicability of Lorentz Invariance to all 4-vectors gives rise to Special Relativity (Ch. 8) which then generalizes (non-locally) to give General Relativity (Ch. 9). Only brief mention of General Relativity and Manifold constructs are given here, in Book 2 of the Series, since General Relativity manifolds are the main focus of Book 3 of the Series, where manifolds with spinor fields are considered. The final chapter, Ch. 11, is on Lie Groups and Gauge Covariant derivatives, where we examine the mathematical underpinnings of the Lorentz transform more closely (among other things).

Particular attention is given to the two forms of Lorentz Invariant object: the 4-vector and the Lorentzian spinor (there are other forms of invariant object that will be mentioned as well). The invariance of 4-vectors under Lorentz Transform is SO(3), while that of the spinor is SU(2). We show the U(1) gauge theory of EM and then examine the Yang Mills family of

Gauge theories, SU(2) and SU(3) in particular, since this is the basis of the gauge theory indicated by the Standard model, a seemingly unattractive direct product of three gauge theories: U(1)xSU(2)xSU(3). The oddity of the direct product agglomeration that is the Standard Model suggests a larger gauge theory for 'unification' such as SU(5). In Book 7 of the Series, however, we will see that there is a simple relation between the trigintaduonion emanator 'propagation' and a field with group structure precisely U(1)xSU(2)xSU(3). The most complicated aspect of all the groups examined, however, is the relation SU(2)→SO(3), thus special attention is given to mathematical exploration of this mapping.

Since the focus is on classical field theory in a fixed geometry, the main physical example is EM. In this setting alpha appears as the coupling constant. After the extensive description of classical field theory in fixed geometry, the foundation will be established for the two major branches of physics that follow in Books 3 and 4 of the Series, on General Relativity and Quantum Mechanics.

Chapter 1. Introduction

In Book 1 of this series [1] we established the theory of motion for dynamical point-like matter. We now want to extend this, in Book 2, to include the concept of dynamical field-like matter and force. The role of fields was minimal in Book 1 [1], or non-dynamical. As with the point particle dynamics, for the field dynamics we have three ways to formulate the behavior: (1) differential equation; (2) function variation (on Lagrangian); and (3) functional variation (on the Action). We will see similar limit phenomena as before, but also new phenomena, including (i) inevitable BH singularity formation (the Penrose singularity theorem); (ii) FRW Universe formation (from homogeneity and isotropy); (iii) the BH collapse singularity; and (iv) the atomic collapse 'singularity'.

The classical field theory texts drawn on in what follows, listed by author, include: Jackson [32]; Landau and Lifshitz Fluids [28]; Ramo et al. [34]; MTW [38]; Adler [59]; Liepmann and Roshko [29]; Churchill and Brown [23]. Notice the inclusion of a text on complex functions by Churchill & Brown. This is included because a complex function gives a 2D field theory, one of the simplest field theories (and also because we will use complex analytics to manage singularities later). Thus a careful review of complex functions is given in Appendix A.

We will consider field dynamics in 2D, 3D, and 4D (and more 4D in Book 3 [5]). Two-dimensional ("2D") field dynamics can be described as a complex function (that maps complex numbers to complex numbers). A novelty of the 2D complex function is it also shows how to handle many types of singularities (the residue theorem), thus provides important information about fundamental structures in physics as well as fundamental mathematical techniques for solving many integrals. For the 3D field dynamics we do an analysis of the electromagnetic field in 3D. The level of coverage begins at an overview of electrostatics at the level of the graduate text Jackson [Jackson]. Some problems from Jackson Ch's 1-3 are examined closely in developing the theory itself. For some this material (in Book 2 [116]) might provide a useful accompaniment to Jackson's text in a full course on EM (based from Jackson's text). A quick review of electrodynamics and electromagnetic wave phenomena is then given. In essence, we see many more examples of ODE problems with solutions, such as for the 3D Laplacian, usually involving separation of variables. We then review the famous transform, discovered by Lorentz in 1899 [37], that relates the EM field as seen by two observers differing by a relative velocity. With the existence of this transform, that brings in the time dimension along with the relative velocity, we effectively have a 4D theory.

From Lorentz Invariance we have, as a point transformation, rotational invariance under SO(3) or SU(2). If Lorentz Invariance is fundamental, then we should see both forms of rotation invariance, one of vector/tensor type from SO(3), and one of spinorial type from SU(2). This is the case, as gauge fields are vectorial and matter fields are spinorial. From Lorenz Invariance as a local invariance we have the Minkowski (flat) spacetime metric, which then generalizes to the Riemannian metric (in General Relativity).

In this book, #2 of the Series, we mainly cover Electromagnetism, where we encounter α in a variety of ways: as the strength of the coupling of the photon to the electron or as a single fit parameter in molecular spectra. In Book 3 [5] (Manifolds, Geometrodynamics) and Book 6 [6] (Statistical Mechanics and Thermodynamics), however, α does not appear, instead there is the novel role of the manifold construct. So, we have the manifold (geometry) construct, and we have the α perturbative maximum construct, but not in the same phenomenological group: we have either field or manifold, and with no explanation of why α has exactly the value it has ($\approx 1/137$). Contrast this with $C_\infty = -1.401155189$, where the constant can be obtained for a large family of situations, in a convergent limit. So we have a process and a general theory to generate the number C_∞, but not α (in Book 7 [4] we show a possible process for computing α that has significant implications). Thus Books 3 and 6 of the Series bring forth nothing new in regards to the mystery of α, but they do begin to reveal the existence of an underlying manifold/geometry for the systems described. Furthermore, this ubiquitous aspect of geometry that presents itself is often endowed with further structure, such as when considering shifting from real time to complex time, or some other instance of complexification, if not outright analyticity, required of some parameter. In Book 6 [6] we will see some of the most concrete (and useful) examples of the role of complex structure.

Book 2 goes from point-particle Lagrangians with static fields described in Book 1 to dynamical field Lagrangians with point-like matter sources (possibly in a dynamical geometry). To manage this transition with maximal re-use of knowledge gained from Book 1, we first consider the usual chaining of masses with springs. For example, consider two masses joined by a spring, then a line of three masses connected by springs, and so on, to eventually have a line density of masses. A related formulation for a lattice (2D or 3D) then relates to a 2D or 3D scalar field. Next we consider that a 2D complex-valued scalar field is a description of a complex function.

Differentiability for a complex function leads to the Cauchy-Riemann equations, and the 2-D Laplace equation. Another implication of the Cauchy-Riemann equations is the equivalence to 2D irrotational flow (curl 0). The important role of complex functions indicated in exploring the concept of field motivates an early section with careful review of complex function properties (and more extensive review, critical also for Book 7, is placed in App. A). A complex function representation allows for a powerful mathematical/analytic process to be applied in situations that can be described in a 2D setting, such as 3D when there is an axial symmetry (reducing to being 2D) such as with microwave transmission pipes, or with 2D electrostatics problems or 2D incompressible flow problems. Also note, analysis of a complex scalar field in a 3D setting will reveal charge quantization phenomenon. Many mathematical oddities such as this will appear as we progress, in the 3D case the echo of quantization as mathematical artifact will, indeed, be the case, but from a deeper mathematical formulation.

At 1D we describe a string construct (using Book 1 dynamics of masses with springs). Then at 2D we explore complex functions, where analysis allows solutions for a variety of (non-source) field problems mentioned above. At 3D we go directly to a full description of the electromagnetic (EM) field -- the 'core' field construct to be described in this book (even more so than the gravitational field, which is mainly covered in Book 3 [5]). We will eventually show how the EM field can be recast as a gauge force field, and the nature of other gauge force fields will also be revealed in this process. It's actually quite a jump to start the main field description with the full EM vector field description… but in some sense this is what happened historically, when the excellent observations of Faraday [7] were directly lifted into a concise mathematical formulation by Maxwell [8]. Interestingly, Maxwell's equations in vector calculus form are commonly known (sometimes even worn imprinted on the T-shirts of students), but Maxwell's quaternionic formulation is not so well-known. What the vector calculus and quaternionic formulations both try to address is an oddity of EM interactions, the strange appearance of the "cross product" (allowing forces orthogonal to the line to the source).

In Book 1 [1] we established Newton's 3rd Law of action equals reaction in the "strong form" where the forces were directly on the line between the interacting particles. For EM, what Faraday observed was force action that was in a direction not parallel to the line connecting the interacting particles, but perpendicular to that line! Mathematically, we need something like the cross product (e.g., we need things like $\hat{\imath} \times \hat{\jmath} = \hat{k}$). It turns out that, in 3D, the cross product is like working with multiplication of pure imaginary quaternions. But this implies a historical chronology where we had cross product first, then quaternions, but this is not the case. In fact this was the problem. The only math language available to Maxwell to describe his equations initially involved quaternionic multiplication properties. Whether included in the math in a verbal form or in a more mathematically rigorous form, the quaternionic references were very off-putting to his contemporaries of the time, such as Gibbs, another genius, who promptly invented vector calculus (and the cross product) to have a more friendly mathematical system [9]. Given the nice mathematical properties of vector calculus, this has become the main way to represent EM. Notice, however, that the quaternion-based formulation is more succinct in the sense that the representation of the fields and the field interactions is entirely done in the form of quaternions and quaternion products (multiplications). The quaternionic form provides the first hint of a possible universal mathematical representation in terms of a (hypercomplex) numbers and their multiplications (see Book 7 [4]). It has been known since 1974 [10], for example, that multiplication of octonions provides a possible presentation for the strong force, since it has the necessary SU(3) group structure emerging as a subgroup of its automorphism group G2. This is explored further in Book 7 [4]. The appearance of the cross product, thus, is a possible indicator of a deeper hypercomplex (Cayley algebra [11]) representation, and it appears not only in force fields via Maxwell's equations, but also the forces on charged matter via the Lorentz force relation.

Maxwell's equations are invariant under Lorentz transformations and describe the speed of EM propagation that we denote as 'c'. If we merely extend this invariance to the other forces with cross-product, in particular the Lorentz force, we effectively arrive at invariance of all matter (and radiation) dynamics under Lorentz transformation, e.g., Einstein's special theory of relativity. Special Relativity introduces the concept of 4-vector and of flat space-time, or Minkowski space, which then begs the question of what about curved space-time. From a locally flat special relativity (SR) geometry we will then generalize to a global, general relativity (GR), geometric description, in Book 3 [5].

A description of the EM field as a gauge field is explored once the basic EM field is understood (by exploring numerous examples). We will find that the EM field can be associated with a fiber bundle on space that is U(1), equivalent to each point in 3D space having a U(1) phase factor (group theory notation, such as U(1), is explained in detail in Book 5 [12]). It is found that the standard model of particle physics has forces described by a gauge field with fibration by the product group U(1) x $SU(2)_L$ x SU(3), with particles indicated by that group structure three times over (e.g., with three copies or generations). The U(1) x $SU(2)_L$ x SU(3) gauge theory is renormalizable, which allows a complete theory *for matter and gauge fields* to be given by the time we reach Book 5 [12] on Quantum Field Theory. GR, on the other hand, doesn't work in this regard. Although GR admits a gauge field formulation, it is not one that is renormalizable. This is the first irrefutable example about how physical fields and physical geometry are two different beasts entirely (this partly motivates the separation of Books 2 [116] and 3 [5]). In Book 3 we will also show the various types of positive energy conditions that are placed on the geometric manifold once a non-trivial spinor-field is posited to exist on the manifold. Such conditions do not exist for the (renormalizable) classical fields, that are the focus of this book. The positive energy theorem is the second example of how matter fields are different, without constraint, while geometric 'fields' have the positive energy constraint.

We will find that field descriptions with invariances lead to conservation laws, which lead to gauge-field dynamics. Matter fields are described with and without relativistic effect. Note that matter, or fluid, flow can already be complicated in 1D, even when non-relativistic, due to non-linear interactions allowing for shockwaves, etc., see Ch. 3 for further details.

In this book a number of specific EM problems are considered at the outset, such that the importance of separation of variables for tractable solutions becomes apparent, as well as the often critical role of boundary conditions in determining solutions. Many of the exercise problems are drawn from preliminary exams (qualifying exams) given to physics PhD students (typically exams required before advancing to PhD candidacy). GR problems in this book and Book 3 are used to show that Einstein's GR is indicated for a number of reasons, both theoretical and experimental. As with EM, a number of GR problems are considered. The problems are either dominated by (i) the observation of an observer on a trajectory or (ii) the study of the overall space-time (universe) dynamics, sometimes with non-trivial space-time surface contributions to the geometrodynamics.

When we apply EM to describe unbound free particle motion or scattering, we get results in exact agreement with theory. When we try to apply EM to bound systems, such as atoms, we get a failure in the theory. The classical EM theory predicts radiative collapse of atoms and this (fortunately) is not what we see. We will find a temporary fix at the level of quantum mechanics in Book 4 [15] (from the old Bohr model to the current Schrodinger/Heisenberg formulation), and a more complete fix in Book 5 [12] at the level of quantum electrodynamics (by way of the aforementioned renormalizability).

For GR, we have a similar issue, but the collapse is even more severe, leading to a new singularity in the description of the theory. Initially this singularity was shown to occur in highly symmetric collapse scenarios so it was surmised that the singularity may not occur in a 'normal' collapse. Work by Penrose (Nobel Prize, [16]) eventually showed that the singular outcomes were a general result (and thus general problem requiring resolution).

Now to consider fields, beginning with the simplest, in the lowest dimensionality. From the outset we will discover the field concept is part pure math, like complex function mapping, and part is phenomenology, like fluid flow with a specified equation of state.

Chapter2. Primitive notions and field fundamentals

2.1 Static and stationary fields

If a field is unchanging over time it is static. A one dimensional static field is a real function, and a 2D static field will be shown to be equivalent to a specification of a complex function. If a field (or flow) is changing but does so at a fixed, or steady, rate, then it is said to be stationary. A 2D integrable flow can, again, be represented using complex functions. By starting with a field or fluid that is static or stationary we've, effectively, carried over the inertial concept from point-particle dynamics first. Recall that Newton's second law F=ma is simply F=0. So, no force if motion is static or stationary (a=0). If no force, the system is unchanged in its motion (fixed or linear). We will start in Sec. 2.2 with complex functions as fields to elaborate further.

2.2 Complex functions as fields

A complex function can be viewed as a mapping from a complex number (a point in 2D) to another complex number. For a well-defined limit process on such a mapping the components of the complex number must be potentials satisfying the Cauchy-Riemann equations, or related 2D Laplacian, with whatever boundary conditions are indicated.

One might ask can quaternions be used to map a point in 4D to another in 4D? There are a variety of constructs where this is possible, but even more interesting, *some are Lorentz Invariant*, and describe 4-vectors. This connection to Lorentz Invariance (the topic of Ch. 8) is easily overlooked as a mathematical oddity at this stage if there wasn't more to it. As described in App. A, the Cayley algebras consist of algebras starting at 1D (the real numbers) and doubling in dimensionality for each new algebra (in effect, due to an added layer of 'complex structure'). After the real numbers come the 2D complex numbers, then the 4D quaternions, then the 8D octonions, then the 16D sedenions, then the 32D trigintaduonions, etc. Thus, it would seem we can describe an infinite number of mapping processes from 2D→2D, 4D→4D, 8D→8D, etc. It turns out, however, that the mapping is assumed to have an inverse (or no zero divisors), thus must be what is known as a division algebra, and the highest order Cayley algebra that retains this property are the Octonions.

So let's go one step further than the aforementioned quaternions and ask is there an 8D (octonion) mapping that is also Lorentz invariant? The answer is yes [20,21].

Suppose we don't want to be stopped so easily by the aforementioned division algebra constraint, and consider a bi-quaternion instead (dimensionality like an octonion since 4+4=8) and add complex structure (equivalent to all components generalized from real to complex) such that we have an object with the dimensionality of a sedenion (16D), but with the switch from a Cayley Product rule to a 'split' product rule, the products devolves to two octonion-like products, each with inverse, thus with overall inverse [11]. Now, with this ridiculously large 16D object, that is 'split', can we again arrive at a Lorentz invariant construct? The answer, remarkably, is yes, and this has been known since 1917 [22]. Can we go to higher even higher order using split-Cayley algebra's? Evidently not, and this and other features are discussed in Book 7 [4] on Emanator theory, where such maximal mathematical objects are the basis for a theory of maximal information 'emanation'.

Well before getting to hypercomplex numbers and their potential applications, however, regular 2D complex numbers will already have earned their primary place, from efficient computational methods using complex functions and the Schwartz-Christoffel transform), to providing mathematical hints at how to deal with singular sources. Some of the key properties are outlined next, with further background in App. A.

Let's now consider functions of a complex variable. Implicitly this means we have a complex number that is the value of a function of complex number, i.e., we are mapping from a complex number to a complex number (from 2D to 2D). Due to the richness of the complex numbers mathematically, the variety of such mappings that are possible is its own area of study that will only be mentioned briefly (see App. A for further details and references). This is notably different from the Quaternions and higher Cayley algebra's where there is nothing similar, and at its root it is because the notion of derivative only extends to complex functions and in a special way (fully specified in the free parameters introduced) while extensions beyond (to quaternions and higher) fail due to over-specification of the free parameters, thus no non-trivial solutions possible. Let's examine what makes complex functions so special. Suppose we have $f: \mathbb{C} \rightarrow \mathbb{C}$:

$$w = f(z) = f(x + iy) = u(x, y) + iv(x, y).$$

(Eqn. 2-1)

The definition of limits and continuity are as expected and are given in detail at [23]. For a definition of derivative at a point we generalize from $x \in \mathbb{R}$ to $z \in \mathbb{C}$ with:

$$f'(z_0) = \lim_{\Delta z \to 0} \frac{f(z_0 + \Delta z) - f(z_0)}{\Delta z}.$$

(Eqn. 2-2)

For this to be meaningful, however, it should not depend on how the limit is performed. If we let $\Delta z = (\Delta x, \Delta y)$ tend to zero horizontally, $(\Delta x, 0)$ to $(0,0)$, we get for $f = u + iv$:

$$f'(z_0) = \frac{\partial u(z_0)}{\partial x} + i\frac{\partial v(z_0)}{\partial x} \rightarrow f' = u_x + iv_x$$

If we let $\Delta z = (\Delta x, \Delta y)$ tend to zero vertically, $(0, \Delta y)$ to $(0,0)$, we get:

$$f' = -i(u_y + iv_y).$$

These two limit expressions for derivative must be in agreement if the definition of derivative is to be well defined, thus for the complex derivative of a function to exist its components must satisfy:

$$u_x = v_y \; and \; u_y = -v_x \, ,$$

(Eqn. 2-3)

which are known as the Cauchy-Riemann equations [23]. As can be seen, the function constraint for a derivative to exist is significant. We will explore next what complex functions are differentiable, but before moving on note how the system could easily be over-constrained when going to quaternions and the higher order Cayley algebras. Thus, we lose the critical derivative (variational) mathematical tools when going to higher order Cayley algebras. This is consistent with the fact that only the complex propagators can satisfy the quantum deFinetti relation [24] (since the quaternion and higher order propagators would be overly constrained in that situation as well). We thus see maximal information flow in terms of *projections* of Cayley algebras, which must eventually reduce to a representation of propagators in a complex Hilbert space (with complex propagators). We will describe maximal information flow in Book.9.

Typically we want to work with functions that are not only differentiable at a point, but in the neighborhood of that point. In the context of the complex derivative this is a more involved mathematical construct since the Cauchy-Riemann equations must then be satisfied on the domain (neighborhood) indicated. When this is the case, the function is said to be analytic (holomorphic) on the indicated domain. If a function is analytic on the entire complex plane, it is called 'entire'. If a function is analytic on a domain, then rearrangement of the Cauchy-Riemann equations shows that the component functions are harmonic (satisfy Laplace's equation) in the domain:

$$u_{xx} + u_{yy} = 0 \ \ and \ v_{xx} + v_{yy} = 0 \,.$$

(Eqn. 2-4)

If a function is analytic everywhere except at an (isolated) point z_0, i.e. it is analytic in every neighborhood of z_0 but not at $z_0$itself, then z_0 is called a singularity of the function. An example of this is $f(z) = 1/z$. The management of singular items in the 2D complex mapping will provide critical developments for 3D field theories, where there are point particles with singular potentials.

Let's now turn to complex integration (or anti-derivative). As might be expected, the integration is no longer defined on an interval on the real line, it is now generalized to being defined on a curve in the complex plane, i.e., we have the line integral (the line integral concept is then developed further, in 3D, in vector calculus introduced with EM). Let's compute the line integral for polynomials for the closed curve shown in Fig. 3.1.

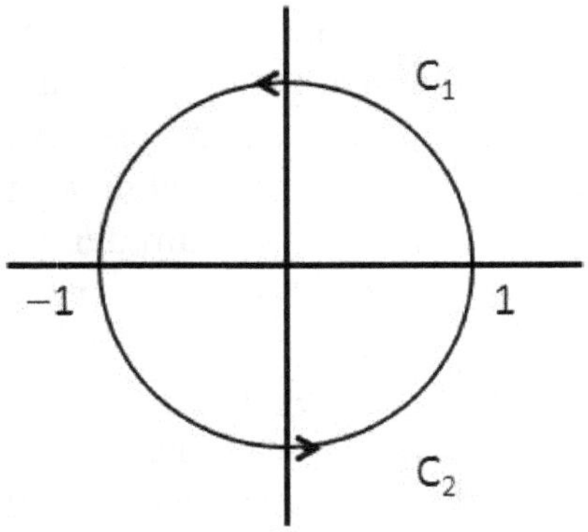

Fig. 3.1. Line integral for the unit circle.

The curves are specified by: $\mathcal{C}_1$: $z = e^{i\theta}$ for $0 \le \theta \le \pi$; and, $\mathcal{C}_2$: $z = e^{i\theta}$ for $\pi \le \theta \le 2\pi$. Let's now consider integrals of the form:

$$I = \oint_{\mathcal{C}} f(z)dz = \int_{\mathcal{C}_1} f(z)dz + \int_{\mathcal{C}_2} f(z)dz = I_1 + I_2.$$

Case 1, $f(z) = 1$:

$$I_1 = \int_{\mathcal{C}_1} dz = \int_0^{\pi} ie^{i\theta} d\theta = -2.$$

$$I_2 = \int_{\mathcal{C}_2} dz = \int_{\pi}^{2\pi} ie^{i\theta} d\theta = 2.$$

$$I = I_1 + I_2 = 0.$$

Case 2, $f(z) = z$:

$$I_1 = \int_{\mathcal{C}_1} zdz = \int_0^{\pi} ie^{i2\theta} d\theta = 0.$$

$$I_2 = \int_{\mathcal{C}_2} zdz = \int_{\pi}^{2\pi} ie^{i2\theta} d\theta = 0.$$

$$I = I_1 + I_2 = 0.$$

Case 3, $f(z) = z^2$:

$$I_1 = \int_{\mathcal{C}_1} z^2dz = \int_0^{\pi} ie^{i3\theta} d\theta = -2/3.$$

$$I_2 = \int_{\mathcal{C}_2} z^2dz = \int_{\pi}^{2\pi} ie^{i3\theta} d\theta = 2/3.$$

$$I = I_1 + I_2 = 0.$$

Clearly for $f(z) = z^n$ for $n > 2$ we will also have $I = I_1 + I_2 = 0$. For negative integers this relation will also hold, except for $n = -1$:

Case 4, $f(z) = z^{-1}$:

$$I_1 = \int_{\mathcal{C}_1} z^{-1}dz = \int_0^{\pi} id\theta = i\theta|_0^{\pi} = i\pi.$$

$$I_1 = \int_{\mathcal{C}_1} z^{-1}dz = \int_{\pi}^{2\pi} id\theta = i\theta|_{\pi}^{2\pi} = i\pi.$$

$$I = I_1 + I_2 = 2\pi i.$$

We have arrived at the residue theorem (see App. A for more detail) where we express any function as a Laurent series:

$$f(z) = \sum a_n z^n , \qquad n = 0, \pm 1, \pm 2, \pm 3, \ldots$$

The integration of that function, on any closed curve about the origin, is then: $I = \oint_C f(z)dz = 2\pi i a_{-1}$.
The coefficient of the 1/z term in the Laurent expansion, a_{-1}, is the only term with nonzero contribution, referred to as the residue. It appears that any complex function that is analytic throughout the region interior to a (closed) line integral will have result of zero for that integral. Let's show this following Cauchy:

$$\oint_C f(z)dz = \oint_C [u(x,y) + iv(x,y)][dx + idy]$$

$$= \oint_C udx - vdy + i\oint_C vdx + udy$$

Using Green's theorem (further details when we get to vector calculus):

$$\oint_C Pdx + Qdy = \iint_R (Q_x - P_y)dxdy$$

We get:

$$\oint_C f(z)dz = \iint_R (-v_x - u_y)dxdy + i\iint_R (u_x - v_y)dxdy$$

and substituting the Cauchy-Riemann relations we get:

$$\oint_C f(z)dz = 0.$$

When f is analytic in the region R, and the function derivative is continuous there (needed to use Green's Theorem) then the above equation holds. Later Goursat was able to show that the condition on the function derivative (to be continuous) is not needed. It is only required that the function be analytic interior to and on the boundary curve. From the above we can see that the integral of complex function on a closed contour will equal the sum of contributions from (1/z) terms only, the residues, each with a factor of $2\pi i$:

$$\oint_C f(z)dz = 2\pi i \sum residues.$$

(Eqn. 2-5)

2.3 Complex structure, field phenomenology, thermodynamics/analyticity

We saw the concept of 'complex structure' mentioned in Sec. 2.1, and how adding complex structure to a real function leads to a complex function (described in Sec. 2.2). Let's now formally define this construct.

Complex Structure [25]

A complex structure can be defined on a real vector space by adding the linear transformation: $J: V \to V$ such that $J^2 = -1$. (Note that complex structures only exist in spaces of even dimension since they entail a doubling of the dimensionality.) Use of complex scalars can then easily be absorbed into a vector space with complex structure and be replaced with a real scalar:

$$(x + iy)X = xX + yJX.$$

(Eqn. 2-6)

The quantum theory (Book 4 [15]) will have a sum on terms of the form e^{iEt}. Now suppose the time dimension (variable) is endowed with complex structure and (Wick rotate) to have the substitution $t = i/kT$, we then have a sum on terms of the form $e^{-E/kT}$, which is the classic form for the partition function of the theory. If it is the theory of an equilibrium system, then the partition function generates all of the thermodynamics properties of the system using standard methods (covered in Book 6 [6]). In Book 7 we will see that we need not posit the addition of such a complex structure as it is already present from the emanation process.

2.4 Field algebras and Cayley algebras

Field algebras and Cayley algebras can lose nice properties that we take for granted with the real (and complex) numbers, such as associativity. Recall that associative means that there isn't order dependence in computing 'abc': a(bc)=(ab)c, etc. As we go from real numbers to complex functions, aside from a trickier multiplication process, there hasn't been an order dependency on operations. As we generalize to hypercomplex (Cayley algebra) numbers or to "quantum numbers", various forms of order dependency begin to occur. Associativity is lost for Cayley algebras after the complex numbers, for example, but some partial associativity is still retained (alternativity).

Often when working with 'numbers', algebras, and operators, in quantum mechanics, it is useful to distinguish between C-numbers (classic numbers, with associativity) and Q-numbers (quantum numbers, often represented by a matrix). Using this terminology helps to clarify how different formations are possible, and equivalent, but with and without use of Q-numbers. When considering the quantum version of a classical system, e.g., 'quantizing ' the system, there is a standard process whereby C-numbers become Q-numbers (where the canonical commutator becomes nonzero, now having a magnitude Planck's constant in some systems). A less-well-known process is available by way of a path integral formulation [26] where there is no need to introduce Q-numbers, per se, but there is still a need for use of Grassmannian algebras when describing fermionic (anti-commutative) matter. By use of sum conventions over structured paths, it is even possible to eliminate all non-simple algebras in the Path Integral formulation (no need for Grassmannians) [26].

2.5 Dynamic field with finite propagation

For the static fields considered in Book 1 the matter of finite field propagation wasn't relevant. Once we allow field changes to occur the question naturally arises as to how fast those field changes propagate. For many situations, relative to the motions in the system, those speeds can be approximated as instantaneous. But for many other, simple, systems, it is easy to see the inertial properties of matter entering into the system to give finite propagation. One of the easiest ways to consider such systems is to take a masses connect by strings in a line and to "take the continuum limit" as this is turned into a 1D 'string'. To address this lets' consider some classic problems from Fetter and Walecka [27]:

Example 2.1 (Fetter 4.12)

Consider a linear chain of 2N masses, with springs alternating, connected to a wall (by a spring) at each end. Let the masses be the same, m, and the separations the same, a, but have alternating spring constants: k_1 and k_2. (a) Find the dispersion relation, examine limiting cases; (b) Apply periodic boundary conditions and find the allowed frequencies.

Solution

(a) The equations of motion for each mass depend on whether it is at index position that is even or odd, where it is assumed that indexing is from the leftmost mass and the leftmost spring has spring constant k_1, and the jth mass displacement is given by η_j:

$$m\ddot{\eta}_j = k_2(\eta_{j+1} - \eta_j) - k_1(\eta_j - \eta_{j-1}),\ j\ odd,$$

and

$$m\ddot{\eta}_j = k_1(\eta_{j+1} - \eta_j) - k_2(\eta_j - \eta_{j-1}),\ \ j\ even.$$

Using $\eta_j = \alpha e^{-i\omega t}\exp{(iqaj)}$ is used for $j\ even$ and $\eta_j = \beta e^{-i\omega t}\exp{(iqaj)}$ is used for $j\ odd$:

$$-\omega^2 m\alpha = -(k_1 + k_2)\alpha - (k_1 e^{iqa} + k_2 e^{-iqa})\beta,$$
$$-\omega^2 m\beta = -(k_1 + k_2)\beta - (k_2 e^{iqa} + k_1 e^{-iqa})\alpha.$$

These two equations can then be solved to obtain:

$$\omega^2 = \frac{(k_1 + k_2)}{m} \pm \left|\frac{k_1}{m} e^{iqa} + \frac{k_2}{m} e^{-iqa}\right|.$$

For $k_1 \rightarrow 0$ the system degenerates to masses separated by spring k_2: $\omega^2 = \frac{2k_2}{m}$ or 0.

For $k_1 = k_2$, $\omega^2 = \frac{2k}{m}(1 + \cos qa) = \frac{4k}{m}\sin^2{(\frac{qa}{2})}$.

Example 2.2 (Fetter 4.16)

Let's now consider a 2D version of the previous problem, but instead of springs let's have masses fixed on strings under tension, the strings forming a 2D grid, the masses at the intersections. Suppose $F = ka$ for springs is carried over to string under tension τ, then trivial that $k = \tau/a$ and we have:

$$L = \frac{1}{2}m\sum_{i,j=1}^{N} \dot{\mu}_{ij}{}^2 - \frac{1}{2}\frac{\tau}{a}\sum_{i,j=0}^{N}\left[(\mu_{i+1,j} - \mu_{i,j})^2 - (\mu_{i,j+1} - \mu_{i,j})^2\right],$$

where the mass displacements at grid position {i,j} is zero at the boundaries: $\mu_{0,j} = 0 = \mu_{N+1,j}$, etc. The equations of motion are:

$$m\ddot{\mu}_j = \frac{\tau}{a}[(\mu_{i+1,j} - \mu_{i,j}) - (\mu_{i,j} - \mu_{i-1,j}) + (\mu_{i,j+1} - \mu_{i,j}) - (\mu_{i,j} - \mu_{i,j-1})]$$

or

$$m\ddot{\mu}_{ij} = \frac{\tau}{a}[\mu_{i+1,j} + \mu_{i-1,j} + \mu_{i,j+1} + \mu_{i,j-1} - 4\mu_{i,j}]$$

To calculate the dispersion relation, let's use the form of solution:

$$\mu_{ij} = e^{i(k_x x_i + k_y y_j)}e^{-i\omega t}, \qquad where\ x_i = ia\ and\ y_j = ja.$$

We can then obtain:

$$\omega^2 = \frac{4\tau}{ma}\left[\sin^2\left(\frac{k_x a}{2}\right) + \sin^2{(\frac{k_y a}{2})}\right].$$

Let's take the limit of a continuous mass distribution:

$$\frac{N^2 m}{\left((N+1)a\right)^2} \to \frac{m}{a^2}\ as\ N \to \infty, so\ let\ \sigma = \frac{m}{a^2}.$$

and

$$m\ddot{\mu} = \tau\left[\frac{\mu(x+a,y)-\mu(x,y)}{a} - \frac{\mu(x,y)-\mu(x-a,y)}{a} + \cdots\right]$$

$$m\ddot{\mu} = \tau\left[\left(\frac{\partial\mu}{\partial x} + \frac{a}{2}\frac{\partial^2\mu}{\partial x^2}\right) - \left(\frac{\partial\mu}{\partial x} - \frac{a}{2}\frac{\partial^2\mu}{\partial y^2}\right)\cdots\right] = a\tau\left[\frac{\partial^2\mu}{\partial x^2} + \frac{\partial^2\mu}{\partial y^2}\right]$$

Thus,

$$\frac{\partial^2\mu}{\partial t^2} - c^2\left[\frac{\partial^2\mu}{\partial x^2} + \frac{\partial^2\mu}{\partial y^2}\right] = 0, \qquad where \quad c^2 = \tau/\sigma a.$$

Note that the dispersion relation for the discrete case is anisotropic in the x-y plane whereas that for the continuum is isotropic:

Discrete: $\omega^2 = \frac{4\tau}{ma}\left[\sin^2\left(\frac{k_x a}{2}\right) + \sin^2\left(\frac{k_y a}{2}\right)\right].$

Continuous: $\omega^2 = \frac{\tau a}{m}\left({k_x}^2 + {k_y}^2\right) = (ck)^2$ (independent of k).

Exact normal mode frequencies occur ($N < \infty$) at:

$$k_x = \frac{m\pi}{a(N+1)} \quad and \quad k_y = \frac{n\pi}{a(N+1)}$$

where:

$$\omega^2 = \frac{4\tau}{ma}\left[\sin^2\left(\frac{m\pi}{2(N+1)}\right) + \sin^2\left(\frac{n\pi}{2(N+1)}\right)\right]$$

For the continuous system, as a check, from L=T-V:

$$L = \frac{\sigma}{2}\int dxdy\left[\left(\frac{\partial\mu}{\partial t}\right)^2 - c^2\left\{\left(\frac{\partial\mu}{\partial x}\right)^2 + \left(\frac{\partial\mu}{\partial y}\right)^2\right\}\right],$$

for which the Euler-Lagrange equations give:

$$\frac{\partial^2\mu}{\partial t^2} - c^2\left[\frac{\partial^2\mu}{\partial x^2} + \frac{\partial^2\mu}{\partial y^2}\right] = 0,$$

in agreement.

Thus, in continuum limit, the differences can be related to differences in a scalar field:

$$\left(x_{j+1}(t) - x_j(t)\right) = \varphi\left(x_{j+1}, t\right) - \varphi\left(x_j, t\right),$$

and the kinetic terms have:

$$\dot{x}_j(t) = \dot{\varphi}(x_j, t).$$

2.6 Exercises

See Appendix A for exercises on Complex Numbers and Functions.

Chapter 3. Fluid Dynamics

A fluid supports no shear at rest. Fluids include: air, water, thixotropic gels, and amorphous solids.
The approach taken involves physics, thermodynamics, mathematics, and engineering (everything but religion, and even that if necessary). Fluids are sometimes described as a scalar field, if static, or as a vector field (with integrable ‘flow lines’).

There is interplay between the mechanical energy of fluid flow (thus an elaboration on what is known from classical mechanics [1]) and the energy associated with the thermodynamic state of the fluid. So we encounter a thermodynamic component to the theory that is foundational at the outset. Thus, ‘fluid flow’ has two phenomenologies: (i) classical mechanics with vector calculus; (ii) thermodynamics (using any pair of canonical variables, starting with standard density and pressure). From the classical mechanical we have the velocity field of the fluid (three components in 3D). From the thermodynamics we have pressure and density (later any two thermodynamics variable will suffice with known equation of state).

So, as much as multi-element and continuum formalisms have been deferred, such as statistical mechanics and thermodynamics, there is some minimal discussion of such needed at this juncture. But first let’s eliminate the complexity of multi-element by continuum limit, with a discussion of Loschmidt’s number, leaving an approximate continuum, thus a pure thermodynamic formalism, in what follows.

Loschmidt’s number indicates how many gas molecules are present in a cubic centimeter at the Earth’s surface (standard temperature and pressure):

$$N_L = \frac{N_A(Avogadro's\ \#)}{22.4 \times 10^3\ cm^3/mole} = 2.69 \times 10^{19}\ molecules/cm^3.$$

(Eqn. 3-1)

Even in a molecular-sized box of 500nm on a side, we have roughly 3,000 molecules. Clearly there is a typically lot of intermolecular interaction and thus rapid equilibration. (Not so for rarefied gases, a topic not discussed in what follows.)

3.1 Simple Fluid Dynamics: non-viscous; non-compressible; non-relativistic

In what follows a fluid is described, seemingly, at a point but this is actually taken to be a very small, almost infinitesimal, volume (recall a cm^3 has 10^{19} molecules, so we can go considerably smaller and still have 'law of large number' effects on the large number of molecules remaining). In this section the notation of Landau and Lifshitz is adopted from their Series book on Fluid Mechanics [28] and numerous problems are considered and solved from Gas dynamics by Roshko and Liepmann [29] and other sources.

3.1.1 Continuity Equation

Following the notation of L&L, consider a volume V_0, with fluid density ρ, we can write:

$$M = \int_{V_0} \rho\, dV,$$

(Eqn. 3-2)

and the exchange of mass at the surface is:

$$\delta M = \rho v \cdot df,$$

(Eqn. 3-3)

where v is the velocity vector field of the fluid, and ρ density as before, and df is a surface area oriented with outward normal. This convention then determines the sign conventions in what follows. Suppose for example, that we have a decrease in the mass given by:

$$-\frac{\partial}{\partial t}\int_{V_0} \rho\, dV.$$

(Eqn. 3-3)

From the surface exchange (assuming no creation or annihilation of mass inside the volume):

$$\oint \rho v \cdot df = \int_{V_0} \nabla \cdot (\rho v)\, dV.$$

Thus,

$$\int_{V_0} \left[\frac{\partial \rho}{\partial t} + \nabla \cdot (\rho v)\right] dV = 0,$$

and we get the continuity equation:

$$\frac{\partial \rho}{\partial t} + \nabla \cdot (\rho v) = 0.$$

(Eqn. 3-4)

3.1.2 Euler's Equation

So far we have considered surface flux of mass density (ρv). What of the other vectorial thermodynamic variable pressure? In examining this we arrive at Euler's equation (1755) [30]. Start with the total force acting on the volume element:

$$F_{V_0} = -\oint_{\partial V_0} p df = -\int_{V_0} \nabla \cdot p \, dV.$$

(Eqn. 3-5)

The element of force $-\nabla \cdot p$ indicated is acting according to the Newtonian $F = ma$ where a volume element dV has mass density ρ, and $a = \frac{dv}{dt}$, thus:

$$-\nabla \cdot p = \rho \frac{dv}{dt}.$$

(Eqn. 3-6)

There is a subtlety in this analysis regarding the differential notation. The dv/dt referenced above is not the standard derivative on time-shift alone:

$$\frac{dv}{dt} = \lim_{\delta t, \delta r \to 0} \left\{ \begin{array}{l} \frac{v(x+\delta x, y+\delta y, z+\delta z, t+\delta t) - v(x,y,z,t)}{\delta t} + \frac{v(x+\delta x, y+\delta y, z+\delta z, t+\delta t) - v(x,y,z,t)}{\delta x} \\ + \frac{v(x+\delta x, y+\delta y, z+\delta z, t+\delta t) - v(x,y,z,t)}{\delta y} + \frac{v(x+\delta x, y+\delta y, z+\delta z, t+\delta t) - v(x,y,z,t)}{\delta z} \end{array} \right\}$$

(Eqn. 3-7)

Thus,

$$\frac{dv}{dt} = \frac{\partial v}{\partial t} + (v \cdot \nabla)v.$$

(Eqn. 3-8)

This is known as the Eulerian derivative, where there is not only a time-shift on a streamline, but also a positional shift according to flow velocity vector field v. Accordingly the "Eulerian Derivative" for a vector field w is:

$$\frac{dw}{dt} = \frac{\partial w}{\partial t} + (v \cdot \nabla)w,$$

(Eqn. 3-9)

and for a scalar is:

$$\frac{ds}{dt} = \frac{\partial s}{\partial t} + v \cdot \nabla s.$$

(Eqn. 3-10)

Thus,

$$-\nabla \cdot p = \rho \frac{dv}{dt} = \rho \frac{\partial v}{\partial t} + \rho(v \cdot \nabla)v,$$

or

$$\frac{\partial v}{\partial t} + (v \cdot \nabla)v = \frac{-\nabla \cdot p}{\rho}.$$

(Eqn. 3-11)

If the fluid is in the Earth's gravitational field (at the surface), we get the extra force $\rho\vec{g}$, where $|\vec{g}| = g$, the acceleration due to gravity at the surface (directed "downwards"):

$$\frac{\partial v}{\partial t} + (v \cdot \nabla)v = \frac{-\nabla \cdot p}{\rho} + \vec{g}.$$

(Eqn. 3-12)

We will examine this last equation in the section on hydrostatics that follows.

A note on ordinary differential equations and generalized derivatives

The central role of ODE's in many physics descriptions has been described previously (and in Book 1 [1]), and an extensive appendix of introductory and advanced ODE examples/solutions is given in the appendices of Book 1 [1] (introductory ODEs) and this book (with Appendix D on Advanced ODEs). We have now seen two forms of derivative variants. One was in the context of complex functions in Ch. 2, where a well-defined derivative was obtained by adopting the derivative being the same at a point in the complex plane independent of direction of approach in the limit process. This generalization to the derivative for complex functions had many ramifications. Now we are seeing a new derivative function with reference to a vector (flow) field according to Euler. The rest of this chapter will explore many of the ramification of the Eulerian derivative. Looking ahead, however, it is natural to ask how such derivative generalizations will develop. Here is a list of derivative generalizations that will be considered (before arriving at Quantum Field Theory descriptions, where spinorial fermionic variables are needed as well):

(1) Complex derivative (Ch. 2).
(2) Eulerian derivative (Ch. 3 Fluid Mechanics);

(3) Gauge Covariant Derivative – Field Theory (Ch. 7,8,10,11);
(4) Covariant Derivative and Lie Derivative – Geometric Field Theory (Ch. 11 and Book 3)
(5) Derivative – Operator Theory in Quantum Mechanics (Book 4)

Adiabatic Flow

Adiabatic fluid flow occurs when there is "conservation of entropy", e.g., when the entropy per unit mass does not change in time:

$$\frac{ds}{dt} = 0 \rightarrow \frac{\partial s}{\partial t} + v \cdot \nabla \mathrm{s} = 0.$$

(Eqn. 3-13)

This is usually combined with mass continuity to arrive at an entropy flux continuity equation (when adiabatic flow):

$$\frac{\partial(\rho s)}{\partial t} + \nabla \cdot (\rho \mathrm{sv}) = 0.$$

(Eqn. 3-14)

Usually the behavior of the entropy density is quite simple, such as constant throughout the fluid (isentropic).

Isentropic Flow

Isentropic flow is when $\mathrm{s} = \mathrm{constant}$ throughout the fluid. Thermodynamics is the focus of Book 6 [6], but we will make use of minimal thermodynamic relations now to see some nice results. Consider for example the definition of enthalpy w per unit mass when $\mathrm{s} = \mathrm{const.}$:

$$dw = Tds + vdp \quad \rightarrow \quad dw = Vdp = \frac{dp}{\rho} \quad \rightarrow \quad \nabla w = \frac{\nabla \cdot p}{\rho}.$$

(Eqn. 3-15)

where $V = \frac{1}{\rho}$ is the specific volume, and T is temperature. We can, thus, write:

$$-\nabla w = \frac{\partial v}{\partial t} + (v \cdot \nabla)v.$$

(Eqn. 3-16)

A vector identity that is convenient (here an later when discussing streamlines) is:

$$\frac{1}{2}\nabla v^2 = v \times (\nabla \times v) + (v \cdot \nabla)v.$$

(Eqn. 3-17)

Using this we get:

$$\frac{\partial v}{\partial t} - v \times (\nabla \times v) = -\nabla\left(w + \frac{1}{2}v^2\right).$$

(Eqn. 3-18)

3.1.3 Hydrostatics

Starting with

$$\frac{\partial v}{\partial t} + (v \cdot \nabla)v = \frac{-\nabla \cdot p}{\rho} + \vec{g}.$$

(Eqn. 3-19)

Consider a fluid at rest, $v = 0$, thus $\nabla \cdot p = \rho\vec{g}$. Since $\vec{g} = -g\hat{z}$, we simply have $p = -\rho g z$ if the density is constant. If the density isn't constant, then we make use of a thermodynamic relation as with isentropic flow, this time the Gibbs Free energy:

$$d\Phi = -sdT + Vdp.$$

(Eqn. 3-20)

If we assume constant T:

$$d\Phi = Vdp = \frac{1}{\rho}dp \quad \rightarrow \quad \nabla\Phi = \frac{\nabla \cdot \mathrm{p}}{\rho} = \vec{g} \quad \rightarrow \quad \nabla(\Phi + \mathrm{gz}) = 0,$$

(Eqn. 3-21)

thus, $\Phi + \mathrm{gz} = \text{constant}$.

3.1.4 Mechanical equilibrium without thermal equilibrium (the absence of convection)

In the last example we were in both mechanical equilibrium (static, $v = 0$) and thermal equilibrium ($T = \text{constant}$). Is mechanical equilibrium possible when not in thermal equilibrium? Fluid motion to equalize temperature is known as convection. So, for equilibrium to be stable there must be no convection. Consider a specific volume $V(p, s)$ at a height z. Suppose an adiabatic displacement upwards by x is possible, with new $V(p', s)$, where p' is the pressure at height $z + x$. To have stability need: $V(p', s') > V(p', s)$, such that the displaced fluid volume is heavier (thus a restoring force). Since the difference is at fixed pressure and we have $\frac{s'-s}{x} \rightarrow \frac{ds}{dz}$ then:

$$\left(\frac{\partial V}{\partial s}\right)_p \frac{ds}{dz} > 0$$

(Eqn. 3-22)

Using the thermodynamic relation:

$$\left(\frac{\partial V}{\partial s}\right)_p = \frac{T}{c_p}\left(\frac{\partial V}{\partial T}\right)_p$$

(Eqn. 3-23)

this becomes:

$$\frac{T}{c_p}\left(\frac{\partial V}{\partial T}\right)_p \frac{ds}{dz} > 0\,,$$

(Eqn. 3-24)

and since $T > 0$, $c_p > 0$, and $\left(\frac{\partial V}{\partial T}\right)_p > 0$ for most substances (water being an exception), we must have:

$$\frac{ds}{dz} > 0,$$

(Eqn. 3-25)

in order for there to not be convection. Let's rewrite in terms of he temperature gradient:

$$\frac{ds}{dz} = \left(\frac{\partial s}{\partial T}\right)_p \frac{dT}{dz} + \left(\frac{\partial s}{\partial p}\right)_T \frac{dp}{dz} = \frac{c_p}{T}\frac{dT}{dz} - \left(\frac{\partial V}{\partial T}\right)_p \frac{dp}{dz} > 0,$$

(Eqn. 3-26)

and using $\frac{dp}{dz} = -\frac{g}{V}$:

$$-\frac{dT}{dz} < \frac{gT}{c_p}\left[\frac{1}{V}\left(\frac{\partial V}{\partial T}\right)_p\right].$$

(Eqn. 3-27)

Using the compressibility definition $\beta = \frac{1}{V}\left(\frac{\partial V}{\partial T}\right)_p$:

$$-\frac{dT}{dz} < \frac{g\beta T}{c_p}.$$

(Eqn. 3-28)

Condition for absence of convection

So for convection to not occur the temperature must decrease upwards with gradient less than $\frac{g\beta T}{c_p}$.

3.1.5 Bernoulli

Starting with

$$\frac{\partial v}{\partial t} - v \times (\nabla \times v) = -\nabla\left(w + \frac{1}{2}v^2\right).$$

(Eqn. 3-29)

let's now consider the special case where $\frac{\partial v}{\partial t} = 0$. For such there will be streamlines defined by:

$$\frac{dx}{v_x} = \frac{dy}{v_y} = \frac{dz}{v_z}.$$

(Eqn. 3-30)

Let's take the unit vector of the streamline to be $\vec{l}$, where we then have $\vec{l} \cdot \nabla w = \frac{\partial w}{\partial l}$. Streamlines are parallel to velocity, but $v \times (\nabla \times v)$ must be perpendicular to velocity, thus:

$$\vec{l} \cdot \left(\frac{1}{2} \nabla v^2 - v \times (\nabla \times v) \right) = \vec{l} \cdot (-\nabla w)$$

(Eqn. 3-31)

becomes:

$$\frac{\partial}{\partial l}\left(w + \frac{1}{2} v^2 \right) = 0 \rightarrow \; w + \frac{1}{2} v^2 = constant.$$

(Eqn. 3-32)

If there is Earth's gravitational field then:

$$w + \frac{1}{2} v^2 + gz = constant.$$

(Eqn. 3-33)

3.1.6 Energy Flux

Consider the energy of a unit volume of fluid:

$$\frac{1}{2} \rho v^2 + \rho \varepsilon,$$

(Eqn. 3-34)

where there is the usual kinetic term and then an internal energy term with ε the internal energy per unit mass. The change in this energy over time is given by

$$\frac{\partial}{\partial t}\left(\frac{1}{2} \rho v^2 + \rho \varepsilon \right).$$

(Eqn. 3-35)

Using the general adiabatic equation, the relation $d\varepsilon = T ds - p dV$, the equation of continuity, and Euler's equation, we can rewrite [28]:

$$\frac{\partial}{\partial t} \int \left(\frac{1}{2} \rho v^2 + \rho \varepsilon \right) dV = - \oint v \left(\frac{1}{2} \rho v^2 + \rho w \right) \cdot df$$
$$= - \oint v \left(\frac{1}{2} \rho v^2 + \rho \varepsilon \right) \cdot df - \oint p v \cdot df$$

(Eqn. 3-36)

where, $v\left(\frac{1}{2}\rho v^2 + \rho w\right)$ is the energy flux density vector; $-\oint v\left(\frac{1}{2}\rho v^2 + \rho\varepsilon\right)\cdot df$ is the energy transported; and $\oint pv \cdot df$ is the work done by the pressure forces. (Note convention that df is a surface area infinitesimal that is directed along the outward normal.)

3.1.7 Momentum Flux

For the momentum flux we want to consider $\partial(\rho v)/\partial t$ using tensor component notation [28]:

$$\frac{\partial(\rho v_i)}{\partial t} = \rho\frac{\partial v_i}{\partial t} + \frac{\partial \rho}{\partial t} v_i \, .$$

(Eqn. 3-37)

Use the equation of continuity:

$$\frac{\partial \rho}{\partial t} = -\frac{\partial(\rho v_i)}{\partial x_i},$$

and use Euler's equation:

$$\frac{\partial v_i}{\partial t} = -v_k\frac{\partial v_i}{\partial x_k} - \frac{1}{\rho}\frac{\partial p}{\partial x},$$

where Einstein's summation convention on repeated indices is in use, to get:

$$\frac{\partial(\rho v_i)}{\partial t} = -\frac{\partial[p\delta_{ik} + \rho v_i v_k]}{\partial x_k} = -\frac{\partial \Pi_{ik}}{\partial x_k},$$

(Eqn. 3-38)

where $\Pi_{ik} = p\delta_{ik} + \rho v_i v_k$. So,

$$\frac{\partial}{\partial t}\int \rho v_i dV = -\int \frac{\partial \Pi_{ik}}{\partial x_k} dV = -\oint \Pi_{ik} df_k \, ,$$

(Eqn. 3-39)

and using $df_k = n_k df$:

$$\Pi_{ik} n_k = p n_i + \rho v_i v_k n_k \, ,$$

(Eqn. 3-40)

where Π_{ik} is the momentum flux density tensor.

3.1.8 Conservation of Circulation

Consider the velocity circulation around a closed contour:

$$\Gamma = \oint v \cdot \delta r \, .$$

(Eqn. 3-41)

If we take the time derivative:

$$\frac{\mathrm{d}}{\mathrm{dt}}\Gamma = \oint \frac{\mathrm{dv}}{\mathrm{dt}} \cdot \delta r + \oint \mathrm{v} \cdot \frac{\mathrm{d}}{\mathrm{dt}}\delta r = \oint [-\nabla \mathrm{w}] \cdot \delta r + \oint \delta \left[\frac{\mathrm{d}}{\mathrm{dt}} v^2\right]$$
$$= \int \nabla \times (-\nabla \mathrm{w}) \cdot \delta f = 0,$$

(Eqn. 3-42)

where use is made of Stokes formula and the fact that the integral of a total derivative along a closed contour is zero. Thus we get:

$$\Gamma = \oint v \cdot \delta r = constant.$$

(Eqn. 3-43)

This is known as the law of conservation of circulation (aka, Kelvin's theorem [1869][31]). Using Stokes formula this can be written:

$$\Gamma = \int (\nabla \times v) \cdot df = constant,$$

(Eqn. 3-44)

which shows that the 'vorticity' $\nabla \times v$ "moves with the fluid" [28].

3.1.9 Potential Flow

Consider steady flow with streamline formation as described previously, and recall from the previous section:

$$\int (\nabla \times v) \cdot df = constant,$$

(Eqn. 3-45)

where df is a fluid surface element spanning a closed contour. Consider a closed contour containing a streamline that moves with the streamline. If $\nabla \times v = 0$ at a point on the streamline then it will remain so elsewhere on the streamline. Notably, if there is an asymptotic region of the flow with 'uniform flow' then we have $\nabla \times v = 0$, thus we would have $\nabla \times v = 0$ throughout. Such flow is known as potential flow since a vector filed having zero curl ($\nabla \times v = 0$) can be expressed as the gradient of a scalar:

$$\nabla \times v = 0 \rightarrow v = \nabla \varphi$$

(Eqn. 3-46)

since $\nabla \times \nabla \varphi = 0$ is an identity.

In practice the argument that $\nabla \times v = 0$ along a streamline will fail for the interesting boundary cases of streamlines that run adjacent to a surface. This is because the closed contour construct circling a point on the streamline will not be possible in such cases. It is typical to have $\nabla \times v \neq 0$ at surfaces. Moreover, at a surface the no viscosity assumptions will break down (the existence of boundary layer effects), so more information about surface behavior must generally be specified.

Away from surfaces, however, we can usually describe flow as potential flow, and substituting the potential form $v = \nabla\varphi$ involving the scalar field φ into Euler's equation we get:

$$\nabla\left(\frac{\partial\varphi}{\partial t} + w + \frac{1}{2}v^2\right) = 0 \quad \rightarrow \quad \frac{\partial\varphi}{\partial t} + w + \frac{1}{2}v^2 = f(t),$$

(Eqn. 3-47)

and for steady flow, $f(t) = constant$. This is known as Bernoulli's equation and, given the nature of potential flow, it is valid throughout the fluid (not just comparing on a streamline).

3.1.10 Incompressible Fluids and Complex Functions

Let's now consider incompressible fluids, specifically, and show the relation to complex functions. First recall the Euler equation with constant density:

$$\frac{\partial v}{\partial t} + (v \cdot \nabla)v = -\nabla \cdot (\frac{p}{\rho}) + \vec{g}.$$

(Eqn. 3-48)

The equation of continuity becomes a vector flow divergence zero relation: $\nabla \cdot v = 0$. If compared to the Bernoulli derivation we see we have w replaced by p/ρ. Thus, we have the relation:

$$\frac{p}{\rho} + \frac{1}{2}v^2 + gz = constant,$$

(Eqn. 3-49)

and also $\nabla \times v = 0$. Introducing $v = \nabla\varphi$ as with potential flow, we have:

$$\nabla^2\varphi = 0,$$

(Eqn. 3-50)

which is Laplace's equation. We will examine this equation for 3D extensively in later chapters describing electromagnetism. In 2D (due to symmetries or constraints) fluid flow (or electrostatics), however, Laplace's equation can be written in terms of a potential function like above and a second potential function related by way of the Cauchy-Riemann relations.

Since we have $\nabla \times v = 0$, we can write in terms of a potential function φ:

$$v = \nabla\varphi \quad \rightarrow \quad v_x\hat{x} + v_y\hat{y} = \frac{\partial\varphi}{\partial x}\hat{x} + \frac{\partial\varphi}{\partial y}\hat{y},$$

(Eqn. 3-51)

thus have:

$$v_x = \frac{\partial \varphi}{\partial x} \quad and \quad v_y = \frac{\partial \varphi}{\partial y}.$$

(Eqn. 3-52)

Since $\nabla \cdot v = 0$, we can write:

$$\nabla \cdot v = \frac{\partial}{\partial x} v_x + \frac{\partial}{\partial y} v_y.$$

(Eqn. 3-53)

If we introduce a second potential function ψ, then the above is satisfied if $v_x = \frac{\partial \psi}{\partial y}$ and $v_y = -\frac{\partial \psi}{\partial x}$ .The potentials are then related according to:

$$\frac{\partial \varphi}{\partial x} = \frac{\partial \psi}{\partial y} \quad and \quad \frac{\partial \varphi}{\partial y} = -\frac{\partial \psi}{\partial x}.$$

(Eqn. 3-54)

These are precisely the form of the Cauchy Riemann relations for a differentiable complex function:

$$w(z) = \varphi(z) + i\psi(z),$$

(Eqn. 3-55)

where $z = x + iy$ (see App. A for details, as well as Ch. 2). By connecting to complex function theory any problem can be easily solved, no matter the (2D) boundary, by way of the Schwartz-Christoffel transform (see App. A) and use of a computer (if still needed).

3.1.11 Viscous Fluids and Navier-Stokes Equation

Recall the momentum flux density when the viscosity is zero:

$$\frac{\partial(\rho v_i)}{\partial t} = -\frac{\partial \Pi_{ik}}{\partial x_k},$$

(Eqn. 3-56)

where $\Pi_{ik} = p\delta_{ik} + \rho v_i v_k$. Let's add a viscous tensorial element that will absorb the pressure term to arrive at the momentum flux density when viscosity is present:

$$\Pi_{ik} = -\sigma_{ik} + \rho v_i v_k,$$

(Eqn. 3-57)

where σ_{ik} is the viscous stress tensor.

Viscous stress occurs when there is relative motion between various parts of the fluid. Thus, σ_{ik} must depend on spatial derivatives of the velocity. If we consider small velocities, then we need only consider a theory to first order in $\frac{\partial v_i}{\partial x_k}$.

Consider fluid in uniform rotation $v = \Omega \times r$. Fluid with such motion will require first-order terms to cancel that have form:

$$\frac{\partial v_i}{\partial x_k} + \frac{\partial v_k}{\partial x_i}.$$

(Eqn. 3-58)

In general, for the non $p\delta_{ik}$ part of σ_{ik}, $\sigma_{ik}{}'$ we have:

$$\sigma'_{ik} = A\left(\frac{\partial v_i}{\partial x_k} + \frac{\partial v_k}{\partial x_i}\right) + B\frac{\partial v_i}{\partial x_k}.$$

Regrouping into terms that vanish upon [ik] contraction versus not, and introducing the standard viscosity coefficients $\{\eta, \zeta\}$:

$$\sigma'_{ik} = \eta\left(\frac{\partial v_i}{\partial x_k} + \frac{\partial v_k}{\partial x_i} - \frac{2}{3}\delta_{ik}\frac{\partial v_l}{\partial x_l}\right) + \zeta\delta_{ik}\frac{\partial v_l}{\partial x_l}.$$

Euler's equation with the above modifications becomes:

$$\rho\left(\frac{\partial v_i}{\partial t} + v_k\frac{\partial v_i}{\partial x_k}\right) = -\frac{\partial p}{\partial x_k} + \frac{\partial}{\partial x_k}\left\{\eta\left(\frac{\partial v_i}{\partial x_k} + \frac{\partial v_k}{\partial x_i} - \frac{2}{3}\delta_{ik}\frac{\partial v_l}{\partial x_l}\right)\right\} + \frac{\partial}{\partial x_k}\left\{\zeta\delta_{ik}\frac{\partial v_l}{\partial x_l}\right\}$$

Let's now assume that the viscosity coefficients $\{\eta, \zeta\}$ are mostly constant, and switch to vector form:

$$\rho\left[\frac{\partial v}{\partial t} + (v\cdot\nabla)v\right] = -\nabla\cdot p + \eta\nabla^2 v + \left(\frac{1}{3}\eta + \zeta\right)\nabla(\nabla\cdot v),$$

(Eqn. 3-59)

which is known as the Navier-Stokes Equation.

If incompressible, $\nabla \cdot v = 0$, and we get

$$\frac{\partial v}{\partial t} + (v\cdot\nabla)v = -\frac{1}{\rho}\nabla\cdot p + \frac{\eta}{\rho}\nabla^2 v,$$

(Eqn. 3-60)

where

$$\sigma_{ik} = -p\sigma_{ik} + \eta\left(\frac{\partial v_i}{\partial x_k} + \frac{\partial v_k}{\partial x_i}\right).$$

(Eqn. 3-61)

In the above, the coefficient η is known as dynamic viscosity and $\nu = \frac{\eta}{\rho}$ is known as kinematic viscosity.

3.1.12 Reynold's Number

From the form of the Navier-Stokes equation we only have a dependency on ν, v, and $\frac{p}{\rho}$, as well as boundary conditions for the flow that we will describe in terms of a characteristic length dimension l. Following the notation of [28] we let the velocity of the stream be u to avoid confusion. From dimensional analysis on: $l = cm$ and $u = cm/s$ and $v = cm^2/s$, we see the simplest dimensionless quantity is ul/v, which is known as the Reynold's number:

$$R = \frac{ul}{v}.$$

(Eqn. 3-62)

Consider general length r/l and general velocity v/u:

$$v = uf\left(\frac{r}{l}, R\right) \;\; and \;\; p = \rho u^2 f\left(\frac{r}{l}, R\right).$$

(Eqn. 3-63)

3.1.13 Transition to chaos (turbulence) by period doubling in periodic flow

In Book 1 [1] we examined the occurrence of chaos via period doubling for some simple systems. It was found that systems reducible to periodic mapping descriptions could then be directly compared and chaotic domains ascertained. Here we consider a flow that has periodicity because it is in a torus. In the general ODE stability analysis we consider the occurrence of a basic limit cycle in this flow (see App. A Book 1 for brief background on ODEs and App E of this book for advanced ODE methods). Consider the flux of path crossings near a point on the limit cycle that is cut orthogonally by surface σ. Each consecutive (periodic) crossing of the surface will satisfy a relation:

$$x_{j+1} = f(x_j, R),$$

(Eqn. 3-64)

where f depends on the Reynolds number. Such a mapping is known as a Poincare mapping or a sequence mapping. In L&L [28] a Poincare mapping is considered that has a fixed point (corresponding to periodicity) at $x = 0$:

$$x_{j+1} = -[1 + (R - R_1)]]x_j + (x_j)^2 + \beta(x_j)^3,$$

(Eqn. 3-65)

where $\beta > 0$. This mapping has a stable fixed point $x = 0$ for $R < R_1$, and unstable for $R > R_1$. The period doubling analysis requires iterating the map twice to get:

$$x_{j+2} = x_j + 2(R - R_1)x_j + 2(1+\beta)\left(x_j\right)^3.$$

(Eqn. 3-66)

For $R < R_1$ there is the stable fixed point as before. For $R > R_1$, the $x = 0$ pint is unstable as before, but now there is a pair of stable fixed points (corresponding to a stable limit cycle with a double period):

$$x_*^{(1),(2)} = \pm\sqrt{\frac{R - R_1}{1+\beta}}.$$

(Eqn. 3-67)

Repetition of the period doubling is a route to turbulence (chaos) that has universal properties [3].

3.1.14 Exercises

(Ex. 3.1) Derive Eqn. 3-8.
(Ex. 3.2) Derive Eqn. 3-11.
(Ex. 3.3) Derive Eqn. 3-14.
(Ex. 3.4) Derive Eqn. 3-21.
(Ex. 3.5) Derive Eqn. 3-28.
(Ex. 3.6) Derive Eqn. 3-43.
(Ex. 3.7) Derive Eqn. 3-59.
(Ex. 3.8) Derive Eqn. 3-65, 3-66, and 3-67.

3.2 Simple Fluid Thermodynamics

3.2.1 Variables of state; Extensive vs. Intensive; and the Carnot Cycle

Variables of state in fluid mechanics (and thermodynamics) are divided into two categories, extensive (dependent on the mass) and intensive (not dependent on the mass). Thus:

Extensive variables (depend on M)

$$\begin{matrix} E - \text{internal energy} \\ V - \text{volume} \\ S - \text{entropy} \end{matrix} \quad \begin{cases} e = E/_M \ \mathit{specific\ internal\ energy} \\ v = V/_M = \rho \ \mathit{specific\ volume\ (density)} \\ s = S/_M \ \mathit{specific\ entropy} \end{cases}$$

Intensive variables (don't depend on M)

P – pressure
T – temperature

If we place M in $V \Longrightarrow P(T)$ depends on T, need M, V, T to define P. Thus, we get an equation of state:

$$f(P,V,T) = 0$$

and a thermal equation of state:

$$P = P(V,T).$$

(Eqn. 3-68)

We have change in system energy in terms of a change in heat energy and a change in work energy (a form of the first law of thermodynamics):

$$dE = \delta Q + \delta W$$
$$de = \delta q + \delta w$$

(Eqn. 3-69)

The derivative notation:

d…→ path independent, exact differential

δ…→ path dependent

Note that the exact differential on E means that:

$$\oint dE = 0$$

For simple compressible substance: $\delta W = -pdV$, so $\delta Q = dE + pdV$. To understand δQ in terms of state variables in a classic analysis of what is known as a Carnot cycle, we use:

$$\oint \delta Q = Q_1 - Q_2, \quad \oint dW = -\oint pdV = W$$

(Eqn. 3-70)

Thus, the work done during one Carnot cycle (W) and efficiency are:

$$W + Q_1 - Q_2 = 0, \qquad \text{efficiency } \eta = \frac{work\ out}{work\ in} = \frac{Q_1 - Q_2}{Q_1} = 1 - \frac{Q_2}{Q_1}.$$

(Eqn. 3-71)

From this we see that the efficiency parameter for fluid A and fluid B must not be dependent on the fluid (or a violation of the second law can be obtained). Thus, the efficiency only depends on the temperatures (possibly only via their difference): $\eta = \eta(T_H, T_C)$.

Let's consider moving from temperature T_1 to T_0 to T_2, since:

$$\frac{Q_1}{Q_0} = f(T_1, T_0) \ \ and \ \ \frac{Q_0}{Q_2} = f(T_0, T_2)$$

(Eqn. 3-72)

We have

$$\frac{Q_1}{Q_2} = f(T_1, T_2) = f(T_1, T_0)f(T_0, T_2)$$

(Eqn. 3-73)

Note

$$f(T_0, T_2) = \frac{1}{f(T_2, T_0)}$$

Thus,

$$f(T_1, T_2) = \frac{f(T_1, T_0)}{f(T_2, T_0)} = \frac{\theta(T_1)}{\theta(T_2)}$$

where use of absolute temperatures is indicated, and with renaming temperature $T_1 = \theta(T_1)$ we then get

$$\frac{Q_1}{Q_2} = \frac{T_1}{T_2} \quad \rightarrow \quad \frac{\delta Q_1}{T_1} = \frac{\delta Q_2}{T_2}$$

(Eqn. 3-74)

In terms of the Carnot cycle above this becomes:

$$\oint \frac{\delta Q}{T} = 0,$$

(Eqn. 3-75)

which is a restatement of the 2nd law of thermodynamics. Let's introduce a new variable of state to account for this (entropy):

$$dS = \frac{dQ}{T}.$$

(Eqn. 3-76)

Now we can write

$$dE = TdS - pdV$$

(Eqn. 3-77)

or

$$de = Tds - pdv.$$

(Eqn. 3-78)

3.2.2 Equations of State; Legendre Transformations; and the Maxwell relations

Thus far we have a description using canonical conjugates paring intensive variables with extensive:

$$\text{Intensive} \begin{Bmatrix} p \leftrightarrow v \\ T \leftrightarrow s \end{Bmatrix} \text{Extensive}$$

If we work with specific energy in terms of entropy and work: $e = e(s, w)$, we have:

$$T = \left(\frac{\partial e}{\partial s}\right)_v; \quad p = -\left(\frac{\partial e}{\partial v}\right)_s$$

(Eqn. 3-79)

Let's transform the canonical equation of state to the caloric equation of state (desire e as a function of τ, v):

$$de = TdS - pdv = T\left\{\left(\frac{\partial s}{\partial v}\right)_T dV + \left(\frac{\partial S}{\partial T}\right)_V dT\right\} - pdV$$

$$= T\left(\frac{\partial S}{\partial T}\right)_V dT + \left\{T\left(\frac{\partial S}{\partial v}\right)_T - p\right\} dV$$

(Eqn. 3-80)

where $e = e(T, v)$ is the caloric equation of state, and $e = e(T)$ for perfect gas.

We can also move to alternative forms of the canonical equation by way of a Legendre transformation. Begin by noting that $d(pv) = pdv + vdp$ suggests a new energy function (enthalpy):

$$h = e + pv$$

(Eqn. 3-81)

and

$$dh = de + d(pv) = Tds + vdp = h(s,p).$$

(Eqn. 3-82)

Note that an energy plot in the h-s plane is known as a Mollier diagram (Carnot if p-v).

Canonical Eqns of State

$$e(s,v)\text{: } de = Tds - pdv; T = \left(\frac{\partial e}{\partial s}\right)_v, p = -\left(\frac{\partial e}{\partial v}\right)_s$$

(Eqn. 3-83)

$$h(s,p)\text{: } dh = Tds + vdp; T = \left(\frac{\partial h}{\partial s}\right)_p, v = \left(\frac{\partial h}{\partial p}\right)_s$$

(Eqn. 3-84)

Thus

$$f = e - Ts = f(T, v)$$

(Eqn. 3-85)

$$df = -sdT - pdv; s = -\left(\frac{\partial f}{\partial T}\right)_v, p$$

$$= -\left(\frac{\partial f}{\partial v}\right)_T$$

(Eqn. 3-86)

$$g = h - Ts = g(T, p)$$

(Eqn. 3-87)

$$dg = -sdT + vdp; s = -\left(\frac{\partial g}{\partial T}\right)_p, v = \left(\frac{\partial g}{\partial p}\right)_T$$

(Eqn. 3-88)

This indicates four thermodynamic potentials:
e – internal energy
h – enthalpy
f – free energy or helmholtz free energy
g – free enthalpy or gibbs free energy

Example 3.1. Maxwell relations:
(a) Show that for function, $f(x, y)$, which can be in expressed differential form as

$$df = a(x, y)dx + b(x, y)dy$$

the following partial derivatives must be equal:

$$\left(\frac{\partial a}{\partial y}\right)_x = \left(\frac{\partial b}{\partial x}\right)_y.$$

(b) Starting from the differential forms of the canonical equation of state, use part (a) to derive the four relations attributed to Maxwell.

Solution:

(a) $df = a(x, y)dx + b(x, y)dy \rightarrow \ a(x, y) = \frac{\partial f}{\partial x} \quad b(x, y) = \frac{\partial f}{\partial y}$

Thus,

$$\left(\frac{\partial a}{\partial y}\right)_x = \frac{\partial}{\partial y}\left(\frac{\partial f}{\partial x}\right) = \frac{\partial^2 f}{\partial y \partial x}; \left(\frac{\partial b}{\partial x}\right)_y = \frac{\partial}{\partial x}\left(\frac{\partial f}{\partial y}\right) = \frac{\partial^2 f}{\partial y \partial x} \rightarrow \ \frac{\partial^2 f}{\partial y \partial x} = \frac{\partial^2 f}{\partial x \partial y} \quad \text{so} \quad \left(\frac{\partial a}{\partial y}\right)_x = \left(\frac{\partial b}{\partial x}\right)_y$$

(b) $de = TdS - PdV, dF = -SdT - PdV, dG = -SdT + VdP, dH = TdS + VdP$

Thus,

$$\left(\frac{\partial T}{\partial V}\right)_S = -\left(\frac{\partial P}{\partial S}\right)_V, \left(\frac{\partial S}{\partial V}\right)_T = \left(\frac{\partial P}{\partial T}\right)_V, -\left(\frac{\partial S}{\partial P}\right)_T = \left(\frac{\partial V}{\partial T}\right)_P, \left(\frac{\partial T}{\partial P}\right)_S = \left(\frac{\partial V}{\partial S}\right)_P$$

3.2.3 Thermodynamic Derivatives: Specific Heats; Coefficients of Expansion/Compression

Let's now consider the Thermodynamic Derivatives, starting with the specific heats.

Specific Heats:

$$C_v = \left(\frac{\partial e}{\partial T}\right)_v = \left(\frac{\delta q}{\partial T}\right)_v = T\left(\frac{\partial S}{\partial T}\right)_v$$
$$= \left(\frac{\partial e}{\partial S}\right)_p \left(\frac{\partial S}{\partial T}\right)_v$$

(Eqn. 3-89)

$$C_p = \left(\frac{\delta q}{\partial T}\right)_p = T\left(\frac{\partial S}{\partial T}\right)_p = \left(\frac{\partial h}{\partial T}\right)_p$$

(Eqn. 3-90)

Writing $\delta q = de + pdv$:

$$\delta q = \left(\frac{\partial e}{\partial v}\right)_t dv + \left(\frac{\partial e}{\partial T}\right)_v dt + pdv = \left[\left(\frac{\partial e}{\partial v}\right)_T + p\right] dv + \left(\frac{\partial e}{\partial T}\right)_v dT$$

(Eqn. 3-91)

Thus

$$C_p - C_v = \left[\left(\frac{\partial e}{\partial v}\right)_T + p\right]\left(\frac{\partial v}{\partial T}\right)_p.$$

(Eqn. 3-92)

Example 3.2. Simplify the expression

$$c_p - c_v = \left[\left(\frac{\partial e}{\partial v}\right)_T + p\right]\left(\frac{\partial v}{\partial T}\right)_p$$

to one involving only *p*, *v*, and *T* by evaluating $(\partial e/\partial v)_T$ using the differential form of the canonical equation. Calculate this difference for a perfect gas. (The correct result is true for any thermally perfect gas whether or not it is calorically perfect.)

Solution

Have:

$$c_p - c_v = \left[\left(\frac{\partial e}{\partial v}\right)_T + p\right]\left(\frac{\partial v}{\partial T}\right)_p$$

Using $de = TdS - Pdv$ obtain:

$$\left(\frac{\partial e}{\partial v}\right)_T dv + \left(\frac{\partial e}{\partial T}\right)_v dT = T\left\{\left(\frac{\partial S}{\partial v}\right)_T dv + \left(\frac{\partial S}{\partial T}\right)_v dT\right\} - Pdv$$

Thus, $\left(\frac{\partial e}{\partial v}\right)_T = T\left(\frac{\partial S}{\partial v}\right)_T - P$ and from part (1) we have $\left(\frac{\partial S}{\partial V}\right)_T = \left(\frac{\partial P}{\partial T}\right)_V$, so:

$$c_p - c_v = T\left(\frac{\partial P}{\partial T}\right)_V \left(\frac{\partial v}{\partial T}\right)_P$$

For a perfect gas $Pv = RT, e \rightarrow e(\tau)$, thus:

$$\left(\frac{\partial P}{\partial T}\right)_V v = R \Rightarrow \left(\frac{\partial P}{\partial T}\right)_V = \frac{P}{T}; P\left(\frac{\partial v}{\partial T}\right)_P = R \Rightarrow \left(\frac{\partial v}{\partial T}\right)_P = \frac{R}{P},$$

and we get

$$c_p - c_v = R\,.$$

Some definitions relating to the log-derivatives:

Coefficient of thermal expansion:

$$\alpha = \frac{1}{V}\left(\frac{\partial V}{\partial T}\right)_p = -\frac{1}{\rho}\left(\frac{\partial \rho}{\partial T}\right)_p$$

(Eqn. 3-93)

Coefficient of isothermal compressibility:

$$k_T = -\frac{1}{V}\left(\frac{\partial V}{\partial p}\right)_T = \frac{1}{\rho}\left(\frac{\partial \rho}{\partial p}\right)_T$$

(Eqn. 3-94)

Coefficient of adiabatic compressibility:

$$k_s = \frac{1}{\rho}\left(\frac{\partial \rho}{\partial p}\right)_s$$

(Eqn. 3-95)

Example 3.3. Calculate the isothermal and isentropic compressibilities for a perfect gas. (What is their ratio?)

Solution

We have

$$K_T = -\frac{1}{V}\left(\frac{\partial V}{\partial P}\right)_T \quad and \quad K_S = -\frac{1}{V}\left(\frac{\partial V}{\partial P}\right)_S\,.$$

For adiabatic compression $PV^\gamma = P_0 V_0^\gamma$ for a perfect gas. Thus,

$$\left(\frac{\partial}{\partial P}\right)_S [PV^\gamma = P_0 V_0^\gamma] \Rightarrow V^\gamma + P\gamma V^{\gamma-1}\left(\frac{\partial V}{\partial P}\right)_S = 0 \Rightarrow V = -P\gamma\left(\frac{\partial V}{\partial P}\right)_S$$

$$\Rightarrow \left(\frac{\partial V}{\partial P}\right)_S = -\frac{V}{P\gamma} \Rightarrow K_S = \frac{1}{P\gamma}$$

For isothermal compression $PV = nRT$ for a perfect gas. Thus,

$$\left(\frac{\partial}{\partial P}\right)_T [PV = nRT] \Rightarrow V + P\left(\frac{\partial V}{\partial P}\right)_T = 0 \Rightarrow \left(\frac{\partial V}{\partial P}\right)_T = -\frac{V}{P} \Rightarrow K_T = \frac{1}{P}.$$

For the ratio:

$$K_S/K_T = {}^1\!/_\gamma .$$

$C_v = const$

Consider when we know (in the domain of interest) that $C_v = const$:

$$C_v = \left(\frac{\partial e}{\partial T}\right)_v \quad \rightarrow \quad e(T) = \int C_v \, dT + const$$

which can be solved to give:

$$e(t) = C_v T + e_0$$

which describes a calorically perfect gas.

From kinetic theory we know:

$C_v = {}^3\!/_2 R$ for perfect monatomic gas
$= {}^5\!/_2 R$ for perfect diatomic ($T > \theta_{rot}, T < \theta_{vib}$)
$= {}^7\!/_2 R$ for perfect a vibrating diatomic ($T > \theta_{vib}$)

$$C_v = \frac{(\#\ of\ quadratic\ terms\ in\ energy)}{2} R$$

3.2.4 Phase change processes; Van Der Waals Gases

For diatomic in center-of-mass, with spherical coordinates for rotational part, have for energy:

$$E = \frac{P_x^2}{2M} + \frac{P_y^2}{2M} + \frac{P_z^2}{2M} + A\dot{\theta}^2 + B\dot{\phi}^2 + \frac{P_r^2}{2M} + \frac{1}{2}Mr\omega^2.$$

(Eqn. 3-96)

Initially there are only the {x,y,z} motions, like monatomic, so 3 degrees of freedom. Then add the angular motions (r-fixed) to have 5 active terms. Lastly, have full vibrational, thus 7 degrees of freedom.

Interestingly, if we consider the plots of the pressure-volume curves for a gas at various temperatures we have a clear physical property when the curves are flat for any stretch of volume, i.e., when $\left(\frac{\partial p}{\partial v}\right) = 0$ on a constant temperature curve for a region of volumes. The formalism is already indicating a gas-liquid bi-phase in this constant region (condensation occurring). The first such region encountered as curves with lower temperatures are considered will not only have $\left(\frac{\partial p}{\partial v}\right) = 0$ but also $\left(\frac{\partial^2 p}{\partial v^2}\right) = 0$ and the constant region will be reduced to a point, known as the critical point. The temperature at the critical point then defines a clear separation in thermodynamic phases:

$T < T_c$ always a biphase region – condensation,
$T > T_c$ no biphase – no condensation,
$T = T_c$ where $\left(\frac{\partial p}{\partial v}\right) = 0$ and $\left(\frac{\partial^2 p}{\partial v^2}\right) = 0.$

If $T \gg T_c$ and $P \ll P_c$ then have thermally perfect gas. T can increase until dissociation or ionization occurs, $T_c \ll T < T_d$ or T_i .

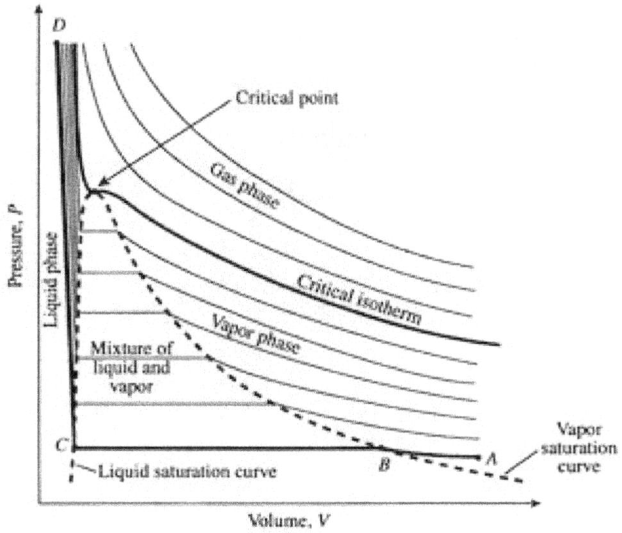

More detailed derivations of thermodynamic properties are given in [6], and we arrive at:

$$pV = NK_BT = nRT \quad or \quad pv = RT.$$

(Eqn. 3-97)

Example 3.4. The van der Waals equation of state is an analytic expression that captures in a general way some aspects of real gas effects (and has historical significance). Given the van der Waals expression

$$\left(p + \frac{a}{v^2}\right)(v - b) = RT$$

evaluate a and b in terms of p_c and T_c by using the fact that this thermal equation of state must demonstrate the correct physical *p*, *v*, *T* behavior near the critical point.

Solution

Starting with $\left(p + \frac{a}{v^2}\right)(v - b) = RT$, near the critical point:

$$\left(\frac{\partial P}{\partial V}\right)_T\bigg|_{\substack{T=T_c\\P=P_c}} = 0 \quad and \quad \left(\frac{\partial^2 P}{\partial V^2}\right)_T\bigg|_{\substack{T=T_c\\P=P_c}} = 0$$

Evaluating:

$$\left(\frac{\partial P}{\partial V}\right) = \frac{-RT}{(v-b)^2} + \frac{2a}{v^3} \quad and \quad \left(\frac{\partial^2 P}{\partial V^2}\right) \equiv \frac{2RT}{(v-b)^2} - \frac{6a}{v^4}$$

We have two equations and two unknowns:

$$\frac{-RT_c}{(v_c-b)^2} + \frac{2a}{v_c^3} = 0 \quad and \quad \frac{2RT_c}{(v_c-b)^3} - \frac{6a}{v_c^4} = 0$$

Solving, we get:

$$b = \frac{1}{3}v_c \quad and \quad a = \frac{9}{8}RT_Cv_c\,.$$

Thus,

$$\left(P + \frac{\left[\frac{9}{8}RT_Cv_c\right]}{v^2}\right)\left(v - \frac{1}{3}v_c\right) = RT$$

Since we need *a* and *b* in terms of P_c and T_c, evaluate at critical point:

$$\left(P_c+\frac{9}{8}\frac{RT_C}{v_c}\right)\left(\frac{2}{3}v_c\right)=RT_C \rightarrow \quad P_c=\frac{3RT_C}{2v_c}-\frac{9}{8}\frac{RT_C}{v_c}=\frac{3}{8}\frac{RT_C}{v_c} \rightarrow \quad V_c =\frac{3}{8}\frac{RT_C}{P_c}$$

Now we have:

$$a=\frac{1}{8}\left(\frac{3}{2}\right)^3\frac{(RT_C)^2}{P_c} \quad and \quad b=\frac{1}{8}\frac{RT_C}{P_c},$$

to get:

$$\left(P+\frac{1}{8}\left(\frac{3}{2}\right)^3\frac{1}{P_c}\left(\frac{RT_C}{V}\right)^3\right)\left(V-\frac{1}{8}\frac{RT_C}{P_c}\right)=RT$$

Example 3.5. Expand the van der Waals equation as a virial expansion and evaluate $B(T)$.

Solution

The virial expansion goes as: $\frac{pV}{RT}=1+\frac{B(T)}{V}+O(V^{-2})$. Using the form of the Vander Waals equation given above:

$$\left(P+\frac{a}{V^3}\right)(V-b)=RT$$

we get

$$\frac{PV}{RT}=1+\frac{Pb}{RT}-\frac{a}{RTV}+\frac{ab}{RTV^2}$$

where Pb/RT must be expanded by way of Taylor series with focus on terms of order $1/V$. The term $-a/RTV$ is explicitly of order $1/V$. Thus Start with

$$\left(\frac{b}{R}\right)\left(\frac{P}{T}\right)=\left(\frac{b}{R}\right)\frac{\left(\frac{RT}{(V-b)}-\frac{a}{V^2}\right)}{R}=\left(\frac{b}{R}\right)\left(\frac{R}{(V-b)}-\frac{a}{TV^2}\right)$$

$$=\left(\frac{1}{\left(V/_b\right)-1}\right)-\frac{ab}{RTV^2}$$

Thus, we need to Taylor expand $\left(\frac{1}{(V/_b)-1}\right)$ and pick off the order $\frac{1}{V}$ term:

For $V<b$: $\frac{1}{(V/_b)-1}\Rightarrow -\left(1-\frac{V}{b}+\left(\frac{V}{b}\right)^2\cdots\right)$, in which case there is no extra $\frac{1}{V}$ term and $B(T)=\left(-\frac{a}{RT}\right)$ for $V<\frac{1}{3}V_C$ (since $b=\frac{1}{3}V_C$) This case essentially corresponds to condensation, where the equation isn't as accurate.

For $V > b$: $\left(\frac{b}{V}\right)\frac{1}{1-(b/V)} \Rightarrow \left(\frac{b}{V}\right)\left(1-\left(\frac{b}{V}\right)+\left(\frac{Vb}{V}\right)^2 \cdots\right)$ and we have an additional $1/V$ term, thus
$B(T) = \left(b - \frac{a}{RT}\right)$ for $V > \frac{1}{3}V_C$.

3.2.5 Isentropic Change; Adiabatic Change; and Joule-Thomson Expansion

Let's consider the Perfect Gas undergoing Isentropic changes. Start with a perfect gas in variables exhibiting entropy:

$$de = Tds - pdv\ , \qquad ds = \frac{de}{T} + \frac{p}{T}dv$$

(Eqn. 3-98)

For thermally perfect gas: $de = C_v dT\ ,\ \frac{p}{T} = \frac{R}{v}$, so

$$ds = C_v\frac{dT}{T} + R\frac{dv}{v}$$

(Eqn. 3-99)

and we get:

$$S - S_0 = \int C_v\frac{dT}{T} + R\int\frac{dv}{v} \rightarrow \quad S - S_0 = C_v \ln\left(\frac{T}{T_0}\right) + R\ln\left(\frac{V}{V_0}\right)$$

(Eqn. 3-100)

Define $\equiv \frac{C_p}{C_v}$, in general $\gamma = \gamma(t)$ is a monotonically decreasing function, and we have:

$$C_p - C_v = R \quad ; \quad \frac{C_v}{R} = \frac{1}{\gamma - 1}\ , \qquad \frac{C_p}{R} = \frac{\gamma}{\gamma - 1}\ .$$

(Eqn. 3-101)

Now, specify isentropic conditions

$$\frac{C_V}{R}\frac{dT}{T} = -\frac{dv}{v}.$$

$$\int\frac{1}{\gamma - 1}\frac{dT}{T} = -\int\frac{dv}{v} \quad --> \quad \left(\frac{T}{T_0}\right)^{-\frac{1}{\gamma-1}} = \frac{v}{v_0}$$

(Eqn. 3-102)

If $\gamma = const$ (if calorically perfect) then:

$$v = (const)T^{-\frac{1}{\gamma-1}}$$

(Eqn. 3-103)

$$\rho = (const)T^{1/(\gamma-1)}, \quad and \text{ using } p = \rho RT$$

(Eqn. 3-104)

$$P = const\ T^{(\gamma/\gamma-1)} = const\ v^{-\gamma} = const\ \rho^{\gamma}$$

(Eqn. 3-105)

Example 3.6. Consider compressing a gas isentropically from $P_0 = 0.05MPa$ and $v_0 = 1L$ to $v_1 = 0.1L$. So, how much work is done? What is the final P?

Solution

Recall:

$$P_0 V_0^{\gamma} = P_1 V_1^{\gamma} \rightarrow P_1 = \frac{P_0 V_0^{\gamma}}{V_0^{\gamma}}$$

The work done on the gas $= W_{\Delta S=0} = \int p\, dV = -P_0 V_0^{\gamma} \int_{V_0}^{V_1} \frac{dV_1}{V_1^{\gamma}} = -\frac{P_0 V_0^{\gamma}}{\gamma - 1}\left[\frac{1}{V_1^{\gamma-1}} - \frac{1}{V_0^{\gamma-1}}\right]$, where, for air $\gamma = 1.4$. Thus,

$$W_{\Delta S=0,air} = \left(.05 \times 10^6 \frac{N}{m^2}\right)(10^{-3}m^3)\,[10^{.4} - 1]/(.4) = 190J.$$

and,

$$P = P_0 \left(\frac{V_0}{V}\right)^{\gamma} = (.05MP_a)(10)^{1.4} = 1.26MP_a$$

Now do same thing isothermally:

$$W_{\Delta T=0} = -\int_{V_0}^{V_1} p\, dV \quad with \quad p = \frac{RT}{V} = \frac{P_0 V_0}{V}.$$

Thus

$$W_{\Delta T=0} = -P_0 V_0 \ln(10) = 50J,$$

so it is less to compress isothermally than isentropically.

Example 3.7. Helium at 200 °C enters a system at 2.0 (standard) atm (1 atm = 0.1013 MPa), goes through a Joule-Thompson section dropping the pressure to 1.5 atm, and finally is expanded adiabatically to the environment (at 0.95 atm). What is the temperature change through this system? What is the specific entropy change?

If expanded isentropically:
Assume that we can consider Helium a perfect gas: (monatomic)

T = 200°C $\xrightarrow{J-T}$ T = 200°C $\xrightarrow{\text{isentropic}}$ T = ?

P = 2atm P = 1.5atm P = .95atm

$$\left(\frac{\partial T}{\partial P}\right)_H = -\frac{(\partial H/\partial P)_T}{(\partial H/\partial \tau)_P} = -\frac{1}{C_P}\left[V - T\left(\frac{\partial V}{\partial \tau}\right)_P\right]$$

For a perfect gas $PV = nRT$

$$\left(\frac{\partial V}{\partial \tau}\right)_P = \frac{V}{T}$$

Thus, $\left(\frac{\partial T}{\partial P}\right)_H = 0$ during the Joule Thompson section. During the isentropic expansion we use $P_1V_1^\gamma = P_0V_0^\gamma$ or $T_1V_1^{\gamma-1} = T_0V_0^{\gamma-1}$ where $\gamma = {}^5/_3$. Thus

$$T_1 = T_0\left(\frac{V_0}{V_1}\right)^{\gamma-1} \quad V = nR\frac{T}{P}$$

and

$$T_1 = T_0\left(\frac{P_1}{P_0}\right)^{2/5}$$

Consider the entropy change during Joule-Thompson expansion:

$$\left(\frac{\partial S}{\partial P}\right)_T = \frac{1}{T}\left[\left(\frac{\partial H}{\partial P}\right)_T - V\right] = -\left(\frac{\partial V}{\partial T}\right)_P = -\frac{V}{T}$$

for a perfect gas, thus

$$dS = \left(\frac{\partial S}{\partial T}\right)_P dT + \left(\frac{\partial S}{\partial P}\right)_T dP = \left(\frac{C_P}{T}\right)dT + \left(-\frac{nR}{P}\right)dP$$

Since temperature remains constant in the J-T expression this simplifies to:

$$dS = -nR\frac{dP}{P} \rightarrow \Delta S = nR\ln\left(\frac{P_0}{P_1}\right)$$

For the specific entropy change we have $\frac{\Delta S}{M} = \Delta S = \frac{nR}{M}\ln\left(\frac{P_0}{P_1}\right)$.

Example 3.8. Two rigid containers (volumes V_1 and V_2 of a perfect diatomic gas initially at pressures p_1 and p_2 and temperatures T_1 and T_2 are brought into thermal contact. Express the change in specific entropy of the entire system (i.e., both containers of gas) in terms of $\hat{T} = T_2/T_1$, $\hat{p} = p_2/p_1$, and $\hat{V} = V_2/V_1$. This entropy change must be consistent with the second part of the Second Law. Show this for the special case that the masses of gas within the two containers are equal.

Solution

For a perfect gas $u \to u(T)$. Since $C_V = \left(\frac{\partial u}{\partial \tau}\right)_V = \frac{du}{dT} = \frac{5}{2}nR$ for a perfect diatomic gas, we have:

$u = \frac{5}{2}nRT$, also, $PV = nRT$. Consider $TdS = du + PdV$:

$$dS = \frac{5}{2}\left({}^{nR}/_{T}\right)dT + \left({}^{nR}/_{V}\right)dV$$

So,

$$\Delta S = nR\ln\left[(T/T_0)^{5/2}(V/V_0)\right]$$

We're discussing the specific entropy so this is

$$\Delta S = \frac{n}{M}R\ln\left[(T/T_0)^{5/2}(V/V_0)\right]$$

If the systems are brought into thermal contact, eventually $T_{1f} = T_{2f} = T_f$, V_1 and V_2 remain the same, and P_1 and P_2 change according to the final temperature:

$$\Delta S_1 = \frac{n_1}{M_1}R\ln\left(T_f/T_1\right)^{5/2} \quad ; \quad \Delta S_2 = \frac{n_2}{M_2}R\ln\left(T_f/T_2\right)^{5/2}$$

Thus

$$\Delta S_T = R\ln\left(T_f/T_1\right)^{5/2 \cdot {}^{n_1}/_{M_1}} + R\ln\left(T_f/T_2\right)^{5/2 \cdot {}^{n_2}/_{M_2}}$$

and

$$\Delta S_T = \frac{5}{2}R\left\{\frac{n_1}{M_1}\ln\left(T_f/T_1\right) + \frac{n_2}{M_2}\ln\left(T_f/T_2\right)\right\}$$

Now, from conservation of energy: $\frac{5}{2}n_1RT_1 + \frac{5}{2}n_2RT_2 = \frac{5}{2}n_1RT_f + \frac{5}{2}n_2RT_f$, thus:

$$T_f = \frac{n_1T_1 + n_2T_2}{(n_1+n_2)}.$$

Substituting this back:

$$\Delta S_T = \frac{5}{2} R \left\{ \left(\frac{n_1}{M} + \frac{n_2}{M} \right) \ln \left[\frac{(1 + \hat{P}\hat{V})}{(1 + \hat{P}\,\hat{V}/\hat{T})} \right] - \frac{n_2}{M} \ln \hat{T} \right\}.$$

If the masses of gas are equal $n_1 = n_2 = n$ and we get:

$$\Delta S_T = \frac{5}{2} R \frac{n}{M} \left\{ \ln \hat{T} - 2 \ln \left[\frac{(\hat{T} + \hat{P}\hat{V})}{(1 + \hat{P}\hat{V})} \right] \right\}$$

The second part of the second law basically states that entropy is maximized during any reaction, thus $\frac{\partial \Delta S_T}{\partial \hat{T}} = 0$ if true, let's find out:

$$\frac{\partial \Delta S_T}{\partial \hat{T}} = \frac{5}{2} R \frac{n}{M} \left\{ \frac{1}{\hat{T}} - \frac{2}{\hat{T} + \hat{P}\hat{V}} \right\}$$

Together with $\hat{P}\hat{V} - \hat{T} = 0$ we get $\frac{\partial \Delta S_T}{\partial \hat{T}} = 0$, and it checks.

Example 3.9. Recall that for blackbody radiation $F = -pV$ and

$$p = A \frac{k_B^4}{(\hbar c)^3} T^4.$$

For an isentropic process, use the T, V behavior of this thermodynamic system to calculate an effective γ for a photon gas in analogy to an isentropic perfect gas.

Solution

$$E = 3pV = 3 \left[A \frac{K_\beta^4}{(\hbar C)^3} T^4 \right] V$$

$dS = \left(\frac{\partial S}{\partial T} \right)_V dT + \left(\frac{\partial S}{\partial V} \right)_T dV$ and using the Maxwell relations:

$$dS = \frac{C_V}{T} dT + \frac{1}{T} \left[\left(\frac{\partial E}{\partial V} \right)_T + P \right] dV.$$

Since

$$\left(\frac{\partial E}{\partial V} \right)_T = 3P \quad and \quad C_V = 12 \frac{pV}{T}$$

we get:

$$dS = 0 = 12pV \frac{dT}{T} + 4pdV \Rightarrow TV^{\frac{1}{3}} = \text{const.}$$

This corresponds to a perfect gas with $1/3 = \gamma - 1 \Rightarrow \gamma = 4/3$.

3.2.6 Exercises

(Ex. 3.9) Derive Eqn. 3-74.
(Ex. 3.10) Derive Eqn. 3-92.
(Ex. 3.11) Derive Eqn. 3-105.

3.3 Fluid Dynamics

Overall Flow Dynamics

System:
(1) A definite mass of fluid, a "particle", interacts with surroundings at boundaries.
(2) A definite volume in space through which fluid flows.

Variables:
(1) Thermodynamics state variables: E, S, V, (T, P)
(2) Mechanical state variables: velocity, kinetic energy= $\frac{1}{2}mv^2$, pressure P (note that pressure plays a both a thermodynamic and mechanical role).
(3) Specific values: e, s, v=$1/\rho$, $u^2/2$.

Changes of State:
(1) Thermodynamics equation of state (specifies fluid)
(2) Energy equation (includes $u^2/2$), a generalization of the 1st Law.
(3) Momentum equation
(4) Continuity equation

Reference System:
(1) Lagrangian system -- follows all of the fluid particles: $x = x(t, x_0)$.
(2) Eulerian system – specifies a velocity field $u(x, t)$, where a particle in the field at (x, t) moves with the instantaneous velocity given by the velocity field at that point. Note, for steady flow the velocity field is no longer time-dependent: $u = u(x)$, while such a simplification does not occur in the Lagrangian framework.

3.3.1 Quasi 1D flow

Quasi 1D flow occurs with flow in a channel, tube, or nozzle (where cross-sectional area depends on 1D only, e.g., with cross-sectional area

$A(x)$, for example). Quasi 1D flow is designed to exist in a jet engine nozzle, for example, and requires continuity and momentum flux, etc., equations specialized to Quasi 1D flow. Even with specialization to Quasi 1D flow, the description can quickly devolve to requiring a computer to handle computations. The focus here will be to carry through the analysis to the realization of a new feature akin to the emergence of a phase transition as seen in the P-V curves mentioned in Sec. X, where here that new feature is shock formation in the flow. Shocks are a remarkable and ubiquitous feature of Quasi 1D flow (such as jet-engine nozzle flow), an example of one of the emergent structural aspects of statistical mechanics that is discussed at length in Book 6 [6].

Using the notation introduced above, let's consider continuity (quasi 1D with only x-dim free):

$$M = \int_{x_1}^{x_2} \rho dV = \int_{x_1}^{x_2} \rho A(x) dx$$

(Eqn. 3-106)

Thus

$$\frac{dM}{dt} = \frac{d}{dt}\int_{x_1}^{x_2} \rho A(x) dx = \int_{x_1}^{x_2} \frac{\partial}{\partial t}(\rho A) dx + (\rho u A)_2 - (\rho u A)_1 .$$

(Eqn. 3-107)

Conservation of mass then becomes:

$$\int_{x_1}^{x_2} \frac{\partial}{\partial t}(\rho A) dx + (\rho u A)_2 - (\rho u A)_1 = 0.$$

(Eqn. 3-108)

Now, even though the flow profiles clearly show $A(x)$ is not constant (u(x) flow vectors not parallel), in the differential limit we can still take $\frac{\partial}{\partial t}(\rho A)dx \to \frac{\partial \rho}{\partial t} A(x) dx \to \frac{\partial \rho}{\partial t} dV$, thus:

$$\int_{x_1}^{x_2} \frac{\partial \rho}{\partial t} dV + (\rho u A)_2 - (\rho u A)_1 = 0.$$

(Eqn. 3-109)

The integral term is the time-dependent part (if present), while the boundary terms are the 'convective' part of the flow. If there is steady flow, then we simply have:

$$(\rho u A)_2 = (\rho u A)_1 .$$

(Eqn. 3-110)

Quasi 1D Continuity

Let's focus more specifically on continuity in the Quasi 1D case. If steady flow we have from above:

$$(\rho u A)_2 = (\rho u A)_1 \quad \rightarrow \quad \frac{d}{dx}(\rho u A) = 0,$$

where we now assume $\rho u A$ is differentiable. If flow is non-steady this generalizes to:

$$\frac{\partial}{\partial x}(\rho u A) + \frac{\partial}{\partial t}(\rho A) = 0.$$

(Eqn. 3-111)

If we compare to the equation (Z.1), for flow with unit cross-section:

$$\frac{\partial}{\partial x}(\rho u) + \frac{\partial}{\partial t}(\rho) = 0,$$

we see there is agreement.

Quasi 1D Bernoulli (with incompressible flow)

If flow in a pipe is steady, then equation (Z.2) indicated the Bernoulli result:

$$\nabla\left(w + \frac{1}{2}v^2\right) = 0 \rightarrow \ w + \frac{1}{2}v^2 = constant.$$

(Eqn. 3-112)

At this juncture it is convenient to shift from the notation adopted by L&L to the notation of Liepmann & Roshko [29], where an extensive analysis of shock formation and dynamics is discussed. In what follows we will now reprise the fluid dynamics of Sec. 3.1 for Quasi 1D, in the notation of R&L, and then extend to describe shock formation. This is done to demonstrate the existence of such phenomenon, so more complex dynamical aspects are left to the complete exposition given in R&L.

In the notation of Liepmann & Roshko [29] we have for enthalpy h not w, and for flow velocity we use u not v, and external heat added (if present) is denoted by q. Thus, for steady flow that is adiabatic ($q = 0$):

$$h_1 + \frac{1}{2}{u_1}^2 = h_2 + \frac{1}{2}{u_2}^2.$$

(Eqn. 3-113)

If external heat q is added:

$$q = \left(h_2 + \frac{1}{2}{u_2}^2\right) - \left(h_1 + \frac{1}{2}{u_1}^2\right),$$

(Eqn. 3-114)

where u_2 is at the cross-section downstream (after heat added).

Note the following definitions:

Thermally Perfect Gas: a gas is thermally perfect if $c_p dT + udu = 0$.
Calorically Perfect Gas: a thermally perfect gas where c_p is constant: $c_p T + \frac{1}{2}u^2 = constant$.

Quasi 1D Euler and Bernoulli (with compressible flow)
Previously we obtained for Euler's equation (eqn. Z.3):

$$\frac{\partial v}{\partial t} + (v \cdot \nabla)v = \frac{-\nabla \cdot p}{\rho},$$

(Eqn. 3-115)

and now specialized to Quasi 1D (Q1D):

$$\frac{\partial u}{\partial t} + u\frac{\partial u}{\partial x} = -\frac{1}{\rho}\frac{\partial p}{\partial x}.$$

(Eqn. 3-116)

For steady flow in Q1D we, thus, have:

$$u\frac{\partial u}{\partial x} = -\frac{1}{\rho}\frac{\partial p}{\partial x} \quad \rightarrow \quad udu = -\frac{dp}{\rho}.$$

(Eqn. 3-117)

Thus,

$$\frac{1}{2}u^2 + \int \frac{dp}{\rho} = constant$$

(Eqn. 3-118)

If incompressible this simplifies to

$$\frac{1}{2}u^2 + \frac{p}{\rho} = constant,$$

(Eqn. 3-119)

as indicated previously. The previous discussion, however, was more focused on liquids, with typical examples presented, while here we are mainly focused on gas dynamics, for which compressible flow must be considered.

The Q1D Momentum Equation
Combining Eqn. 3-116 and the continuity equation:

$$\frac{\partial}{\partial x}(\rho u^2 A) + \frac{\partial}{\partial t}(\rho u A) = -A\frac{\partial p}{\partial x}.$$

(Eqn. 3-120)

If we integrate with respect to x between two cross-sectional areas A_1 and A_2:

$$\frac{\partial}{\partial t}\int_1^2 \rho u A dx - (\rho_1 u_1^2 A_1 - \rho_2 u_2^2 A_2) = (p_1 A_1 - p_2 A_2) + \int_1^2 p\, dA$$

(Eqn. 3-121)

Some common definitions:

Isentropic Q1D flow (adiabatic, non-viscous, non-conducting flow is isentropic): A perfect gas with isentropic flow satisfies:

$$\left(\frac{p_2}{p_1}\right) = \left(\frac{\rho_2}{\rho_1}\right)^{\gamma} = \left(\frac{T_2}{T_1}\right)^{\gamma/(\gamma-1)} .$$

(Eqn. 3-122)

Compressibility: Compressibility of fluid flow is described by:

$$a^2 = \left(\frac{\partial p}{\partial \rho}\right)_s ,$$

(Eqn. 3-123)

where a is the speed of sound in the fluid. Using the above for a perfect isentropic gas in the form:

$$p = const\ \rho^{\gamma} \quad \rightarrow \quad a^2 = \frac{\gamma p}{\rho} = \gamma RT .$$

(Eqn. 3-124)

Mach Number (M):

The Mach number of the flow is a measure of speed in units of the speed of sound:

$$M = \frac{u}{a} .$$

(Eqn. 3-125)

If the flow speed has $M > 1$ it is supersonic. If $M < 1$ it is subsonic.

The Area-Velocity Relation

Recall that $\rho u A = const.$ along streamline tube, or:

$$\frac{d\rho}{\rho} + \frac{du}{u} + \frac{dA}{A} = 0 ,$$

(Eqn. 3-126)

for steady adiabatic flow. Recall that Euler's equation for steady flow was:

$$u du + \frac{dp}{\rho} = 0 .$$

Thus,

$$udu = -\frac{dp}{\rho} = -\frac{dp}{d\rho}\frac{d\rho}{\rho} = -a^2\frac{d\rho}{\rho},$$

or, using the Mach number notation with $a = u/M$:

$$\frac{du}{u} = -\frac{\frac{dA}{A}}{1 - M^2}.$$

(Eqn. 3-127)

From the above form we can immediately ascertain the following:
(1) If $M = 0$, then decrease in area is proportional to increase in velocity.
(2) If $M < 0$, subsonic, then as with $M = 0$, except greater change for velocity.
(3) If $M > 0$, supersonic, then increase in area is proportional to increase in velocity.
(4) If $M = 1$, then must have $dA = 0$, e.g., must be at throat in the flow tube (throat has $dA = 0$). Near $M = 1$, have extreme sensitivity to dA (if not zero).

Flow at a throat
There can be more than one throat constriction in a flow, but only one will be considered in the flow examples to follow. To have condition for throat $dA = 0$, we must have either $M = 1$ and $du \neq 0$ or $M \neq 1$ and $du = 0$. Furthermore, this indicates a maximum velocity at the throat for subsonic flow, and a minimum velocity for supersonic flow.

When $M = 1$, sonic flow, let's denote other variables with an asterisk. Just as a thermal reservoir or asymptotic flow might be a useful reference, so to with the sonic flow variables at a throat. Since $M = u/a$ in general, we have $u^* = a^*$. Let's now write flow relations with reference to reservoir values and the throat values. Consider the calorically perfect gas: $c_pT + \frac{1}{2}u^2 = const = c_pT_0$, where reference is first made to a reservoir at temperature T_0 (where $u = 0$). Since we have $a^2 = \gamma RT$:

$$\frac{u^2}{2} + \frac{a^2}{\gamma - 1} = \frac{a_0{}^2}{\gamma - 1} \rightarrow \frac{T_0}{T} = \left(1 + \frac{\gamma - 1}{2}M^2\right).$$

(Eqn. 3-128)

The isentropic relations at the throat (with reference to the reservoir) then become:

$$\frac{p_0}{p} = \left(1 + \frac{\gamma - 1}{2} M^2\right)^{\gamma/(\gamma-1)},$$

(Eqn. 3-129)

and

$$\frac{\rho_0}{\rho} = \left(1 + \frac{\gamma - 1}{2} M^2\right)^{1/(\gamma-1)}.$$

(Eqn. 3-130)

For air we have $\gamma = 1.40$.

Now let's specialize he above form to evaluation at the throat:

$$\frac{(a^*)^2}{a_0^2} = \frac{T^*}{T_0} = \frac{2}{\gamma + 1}.$$

(Eqn. 3-131)

For air: $\frac{T^*}{T_0} = 0.833$; $\frac{p^*}{p_0} = 0.528$; and $\frac{\rho^*}{\rho_0} = 0.634$. (Note that a throat need not exist for the sonic value to be used as a reference.)

According to the notation, $M^* = 1$, trivially, but it is convention to interpret

$$M^* = \frac{u}{a^*},$$

(Eqn. 3-132)

where it is understood that u need not be u^*. Using this (confusing) convention we can then write:

$$(M^*)^2 = \frac{\gamma + 1}{\frac{2}{M^2} + \gamma - 1},$$

(Eqn. 3-133)

or, inverting,

$$M^2 = \frac{2}{\frac{\gamma + 1}{(M^*)^2} - \gamma + 1}.$$

(Eqn. 3-134)

Note the critical behavior that if $M < 1$, then $M^* < 1$, and if $M^* > 1$, then $M > 1$.

Compressible Bernoulli Flow

Consider the calorically perfect gas, $c_p T + \frac{1}{2}u^2 = const = c_p T_0$, as before. Using the ideal gas law, $p = \rho RT$, with adiabatic flow, and we have the relation:

$$\frac{1}{2}u^2 + \frac{\gamma}{\gamma - 1}\left(\frac{p}{\rho}\right) = \frac{\gamma}{\gamma - 1}\left(\frac{p_0}{\rho_0}\right).$$

(Eqn. 3-135)

If the flow is also isentropic:

$$\frac{1}{2}u^2 + \frac{\gamma}{\gamma - 1}\left(\frac{p_0}{\rho_0}\right)\left(\frac{p}{p_0}\right)^{(\gamma-1)/\gamma} = \frac{\gamma}{\gamma - 1}\left(\frac{p_0}{\rho_0}\right).$$

(Eqn. 3-136)

Now, we can write

$$\frac{1}{2}\rho u^2 = \frac{1}{2}\gamma p M^2.$$

(Eqn. 3-137)

We can use a particular u as reference, and the associated M, from the above equation, will be denoted by M_1 (similarly for p_1, following the notation of R&L). We can then write:

$$c_p = \frac{2}{\gamma (M_1)^2}\left(\frac{p}{p_1} - 1\right).$$

(Eqn. 3-138)

If flow is also isentropic:

$$c_p = \frac{2}{\gamma (M_1)^2}\left(\left[\frac{2 + (M_1)^2(\gamma - 1)}{2 + M^2(\gamma - 1)}\right]^{\gamma/(\gamma-1)} - 1\right).$$

(Eqn. 3-139)

Example 3.10. Consider Quasi one dimensional flow, $A = A(x)$, where the momentum equation is

$$\frac{\partial}{\partial t}\int_1^2 \rho u A dx = \rho_1 u_1^2 A_1 - \rho_2 u_2^2 A_2 + p_1 A_1 + \int_1^2 p\frac{dA}{dx}dx - p_2 A_2 + \int_1^2 \rho f\, A dx$$

Let's write $x_2 = x + \Delta x$. Show that in the limit $\Delta x \to 0$ the equation reduces to

$$\rho\frac{\partial u}{\partial t} = -\rho u\frac{\partial u}{\partial x} - \frac{\partial p}{\partial x} + \rho f$$

(Thus, in the differential equations for quasi one dimensional flow in a channel of slowly varying cross section area $A(x)$, the latter appears only in the continuity equation, not in the momentum and energy equations.)

Solution

$$\int_1^2 \left\{\rho\frac{\partial u}{\partial t}A + \frac{u\partial\rho}{\partial x}A + \frac{\partial p}{\partial x}A - \rho f A\right\}dx = \rho_1 u_1^2 A_1 - \rho_2 u_2^2 A_2$$

So,

$$\left\{\rho\frac{\partial u}{\partial t} + \frac{\partial \rho}{\partial x} - \rho f\right\}A\Delta x = -\rho A u\frac{\partial u}{\partial x}\Delta x$$

and we get:

$$\rho\frac{\partial u}{\partial t} = -\rho u\frac{\partial u}{\partial t} - \frac{\partial \rho}{\partial x} + \rho f$$

Example 3.11. Consider flow through a constant duct area, with steady flow, and with a heat exchanger that is adjusted such that the input pressure is the same as the output pressure ($\dot{Q}$ is adjusted so that $p_2 = p_1$). Thus, at input ("1") we have: u_1, p_1, ρ_1, h_1, and with steady flow, $\dot{m} = \rho u A = const.$ Denote the drag force on the heat exchanger by D.

(a) Write down the conservation equations for the conditions given. Write these with the unknown conditions (2) on the left hand side of the equations, assuming that conditions (1) are known.

(b) Solve these equations for a perfect gas (for which $\frac{h_2}{h_1} = \frac{T_2}{T_1} = \frac{p_2}{p_1}\frac{\rho_1}{\rho_2}$). It is useful to write the equations in dimensionless form. Hints: Find $\frac{p_2}{p_1} = \left(1 - \frac{D}{\rho_1 u_1^2 A}\right)^{-1}$ and Find $\frac{\dot{Q}}{\dot{m}h_1}$ in terms of the dimensionless parameters $\frac{D}{\rho_1 u_1^2 A}$ and $\frac{u_1^2}{h_1}$.

(c) Is the required heat transfer into or out of the flow, $\dot{Q} \gtrless 0$?

(d) Is the entropy change $s_2 - s_1 \gtrless 0$?

Solution

From conservation of mass: $\rho_2 u_2 A_2 = \rho_1 u_1 A_1$, and for steady flow, A = const., so:

(i) $\rho_2 u_2 = \rho_1 u_1$

From conservation of energy: q + work done = increase of energy. The work done = $p_1 v_1 - p_2 v_2$ and the increase of energy = $\left(e_2 + \frac{1}{2}u_2^2\right) - \left(e_1 + \frac{1}{2}u_1^2\right)$. Thus

$$q + p_1 v_1 - p_2 v_2 = \left(e_2 + \frac{1}{2}u_2^2\right) - \left(e_1 + \frac{1}{2}u_1^2\right)$$

and using $h = e + pv$:

$$q = h_2 - h_1 + \frac{1}{2}u_2^2 - \frac{1}{2}u_1^2$$

Define our unit of heat such that we can let $q = \frac{\dot{Q}}{\dot{m}}$. Thus

(ii) $h_2 + \frac{1}{2}u_2^2 = h_1 + \frac{1}{2}u_1^2 + \frac{\dot{Q}}{\dot{m}}$

From conservation of momentum:$\Sigma F = \dot{p}$. Accounting for drag:

$$\Sigma F = P \cdot A = (P_1 - drag - P_2)A$$

Thus,

$$(P_1 - d - P_2)A = (\rho_2 u_2 A)u_2 - (\rho_1 u_1 A)u.$$

or

$$\rho_2 u_2^2 + P_2 = \rho_1 u_1^2 + P_1 - d$$

Furthermore, $\dot{Q}$ is adjusted so that $p_2 = p_1$, thus

(iii) $\rho_2 u_2^2 = \rho_1 u_1^2 - D/A$.

(b) So, we have $\rho_2 u_2 = \rho_1 u_1; h_2 + \frac{1}{2}u_2^2 = h_1 + \frac{1}{2}u_1^2 + \frac{\dot{Q}}{\dot{m}}; \rho_2 u_2^2 = \rho_1 u_1^2 - D/A; h_2 = h_1\left(\frac{\rho_1}{\rho_2}\right)$

Thus, $\frac{\rho_1}{\rho_2} = \left(1 - \frac{D}{\rho_1 u_1^2 A}\right)$ and

$$\frac{\dot{Q}}{\dot{m}h_1} = \frac{1}{2}\frac{u_1^2}{h_1}\left(\frac{D}{\rho_1 u_1^2 A}\right)^2 - \left(\frac{D}{\rho_1 u_1^2 A}\right)\left[1 + \frac{u_1^2}{h_1}\right]$$

$$u_2 = \left(\frac{\rho_1}{\rho_2}\right)u_2 = \left(1 - \frac{D}{\rho_1 u_1^2 A}\right)u_1$$

and

$$h_2 = h_1\left(1 - \frac{D}{\rho_1 u_1^2 A}\right).$$

(c) We have

$$\frac{\dot{Q}}{\dot{m}h_1} = \left(\frac{D}{\rho_1 u_1^2 A}\right)\left\{\frac{1}{2}\frac{D}{h_1 \rho_1 A} - \left(1 + \frac{u_1^2}{h_1}\right)\right\} = \left(\frac{D}{\rho_1 u_1^2 A}\right)X$$

If $D = 0$, then $\frac{\dot{Q}}{\dot{m}h_1} = 0$. If $D = AP_1$(its max): $\rho_2 u_2^2 = \rho_1 u_1^2 - P_1$

$$X = \frac{1}{2}\frac{D}{h_1\rho_1} - \left(1 + \frac{u_1^2}{h_1}\right) = \frac{1}{2}\frac{\rho_1 u_1^2 - \rho_2 u_2^2}{h_1\rho_1} - 1 - \frac{u_1^2}{h_1}$$
$$= \frac{1}{2}\frac{\rho_2 u_2^2}{\rho_1 h_1} - 1 - \frac{1}{2}\frac{u_1^2}{h_1} < 0$$

Thus, $\frac{\dot{Q}}{\dot{m}h_1} < 0$.

(d) The entropy of a closed system, one that exchanges neither heat nor work with its surroundings, always increases during any change of that system. The entropy change "$S_2 - S_1$" is not for a closed system, since there is a heat exchange $\dot{Q} \neq 0$ not accounted for in S_1 or S_2. Thus the entropy may even decrease locally since this isn't a closed system – this is actually the case as you will see.

$ds = \frac{1}{T}\left(dh - \frac{1}{\rho}dp\right)$ where $\dot{Q}$ is adjusted such that $P_2 = P_1$, which is a big entropy increase in heat reservoir that is unaccounted for is the expression $S_2 - S_1$; $ds = \frac{1}{T}dh$; and for a perfect gas:

$$S_2 - S_1 = \frac{C_P \ln T_2}{T_1} = -C_P \ln\left(\frac{h_1}{h_2}\right) = -C_P \ln\left[\frac{1}{1 - \frac{D}{\rho_1 u_1^2 A}}\right].$$

Since ln(>1)>0 and $C_P > 0$ we have:

$$S_2 - S_1 < 0\,.$$

3.3.2 Shock formation (RL notes, especially section 2.13)

Shock formation (emergent structure, see again in Book3 [5] chapter on Cosmology)

Flow in constant Areal cross-section

Consider the Q1D flow equations when the area is constant:

Continuity: $\rho_1 u_1 = \rho_2 u_2$.
Conservation of Momentum: $p_1 + \rho_1 {u_1}^2 = p_2 + \rho_2 {u_2}^2$
Conservation of Energy: $h_1 + \frac{1}{2}{u_1}^2 = h_2 + \frac{1}{2}{u_2}^2$.

Normal shock relations

Even if area is constant, a normal shock wave can occur. Let's consider the gas to be calorically perfect with isentropic flow before and after the shock:

$$u_1 - u_2 = \frac{p_2}{\rho_2 u_2} - \frac{p_1}{\rho_1 u_1} = \frac{(a_2)^2}{\gamma u_2} - \frac{(a_1)^2}{\gamma u_1}.$$

(Eqn. 3-140)

If we consider the reference at sonic flow we see that:

$$\frac{(u_1)^2}{2} + \frac{(a_1)^2}{\gamma - 1} = \frac{(u_2)^2}{2} + \frac{(a_2)^2}{\gamma - 1} = \frac{1}{2}\frac{\gamma + 1}{\gamma - 1}(a^*)^2,$$

(Eqn. 3-141)

which can be solved to give the Prandtl (or Meyer) relation:

$$u_1 u_2 = (a^*)^2.$$

(Eqn. 3-142)

Or, given $M^* = \frac{u}{a^*}$, this is simply:

$$M_1{}^* = \frac{1}{M_2{}^*}.$$

Recall that that if $M < 1$, then $M^* < 1$, and if $M^* > 1$, then $M > 1$. Thus, the flow across a normal shock must switch between subsonic and supersonic. In what follows we will see that only the switch from supersonic to subsonic is actually possible.

By using the relations$(M^*)^2 = \frac{\gamma+1}{\frac{2}{M^2}+\gamma-1}$ and $M^2 = \frac{2}{\frac{\gamma+1}{(M^*)^2}-\gamma+1}$, but now with $M_1{}^* = \frac{1}{M_2{}^*}$, and we can relate M_1 and M_2:

$$(M_2)^2 = \frac{1 + \left(\frac{\gamma - 1}{2}\right)(M_1)^2}{\gamma (M_1)^2 - \left(\frac{\gamma - 1}{2}\right)}.$$

(Eqn. 3-143)

We also have

$$\frac{u_1}{u_2} = \frac{(u_1)^2}{(a^*)^2} = (M_1{}^*)^2,$$

thus

$$\frac{\rho_2}{\rho_1} = \frac{u_1}{u_2} = \frac{(u_1)^2}{(a^*)^2} = \frac{(\gamma + 1)(M_1)^2}{(\gamma - 1)(M_1)^2 + 2}.$$

(Eqn. 3-144)

Similarly, shock strength is given by:

$$\frac{p_2 - p_1}{p_2} = \frac{2\gamma}{\gamma + 1}((M_1)^2 - 1).$$

(Eqn. 3-145)

Similarly,

$$\frac{T_2}{T_1} = \left(\frac{p_2}{p_1}\right)\left(\frac{\rho_1}{\rho_2}\right) = 1 + \frac{2(\gamma - 1)}{(\gamma + 1)^2}\frac{(\gamma(M_1)^2 + 1)}{(M_1)^2}((M_1)^2 - 1).$$

(Eqn. 3-146)

Let's now compute the change in entropy across the shock. Recall:

$$\frac{s_2 - s_1}{R} = \ln\left[\left(\frac{p_2}{p_1}\right)^{\frac{1}{\gamma-1}}\left(\frac{\rho_2}{\rho_1}\right)^{\frac{-\gamma}{\gamma-1}}\right]$$

(Eqn. 3-147)

using the above relations on the ratios and using $m = (M_1)^2 - 1$:

$$\frac{s_2 - s_1}{R} = \ln\left[\left(1 + \frac{2\gamma}{\gamma + 1}m\right)^{\frac{1}{\gamma-1}}(1 + m)^{\frac{-\gamma}{\gamma-1}}\left(\frac{\gamma - 1}{\gamma + 1}m + 1\right)^{\frac{\gamma}{\gamma-1}}\right].$$

(Eqn. 3-148)

If $M_1 \cong 1$ then $m \cong 0$, and to lowest order:

$$\frac{s_2 - s_1}{R} \cong \frac{2\gamma}{(\gamma + 1)^2}\frac{((M_1)^2 - 1)^3}{3}.$$

(Eqn. 3-149)

Since $s_2 - s_1 > 0$, must have $(M_1)^2 > 1$, thus $M_1 > 1$, confirming that only supersonic to subsonic jump allowed.

Example 3.12.

(a) Show that in perfect gas the relation of Mach numbers downstream and upstream of a normal shock wave is given by

$$\frac{M_2^2}{M_1^2} = \frac{P_1}{P_2}\frac{\rho_1}{\rho_2}$$

And thus obtain M_2 as a function of M_1.

(b) Find the form of M_2 for $M_1 \gg 1$ (strong shocks)

(c) Find the form of M_2 for $M_1^2 - 1 \ll 1$ (weak shocks)

Hint: Let $M_1^2 \equiv 1 + \mu, M_2^2 = 1 + v$. Show that for $\mu \ll 1, v \doteq -\mu$. (Thus $1 - M_2 \doteq M_1 - 1$.)

Solution

(a) We have the relations derived in 2.45, 2.47, and 2.48 [29] to give:

$$\frac{P_2}{P_1} = 1 + \frac{2\gamma}{\gamma + 1}(M_1^2 - 1) \quad and \quad \frac{\rho_2}{\rho_1} = \frac{(\gamma + 1)M_1^2}{(\gamma - 1)M_1^2 + 2}$$

Thus,

$$\frac{P_2}{P_1}\frac{\rho_2}{\rho_1} = \frac{(\gamma + 1)M_1^2}{(\gamma - 1)M_1^2 + 2} + \frac{2\gamma(\gamma + 1)(M_1^2)}{(\gamma - 1)M_1^2 + 2} = M_1^2\left[\frac{\frac{2\gamma}{(\gamma - 1)}M_1^2 - 1}{M_1^2 + \frac{2}{(\gamma - 1)}}\right]$$

$$= M_1^2\left[\frac{\gamma M_1^2 - \frac{\gamma - 1}{2}}{1 + \frac{\gamma - 1}{2}M_1^2}\right] = \frac{M_1^2}{M_2^2}$$

Thus, $\frac{M_2^2}{M_1^2} = \frac{P_1}{P_2}\frac{\rho_1}{\rho_2}$.

(b) If $M_1 \gg 1$, and have $M_2^2 = \left[\frac{1+\frac{\gamma-1}{2}M_1^2}{\gamma M_1^2 - \frac{\gamma-1}{2}}\right]$:

$$\lim_{M_1 \to \alpha} M_2^2 = \frac{((\gamma - 1)/2)}{\gamma} = \left(\frac{\gamma - 1}{2\gamma}\right) \quad and \quad M_2 \sim \left[\frac{\gamma - 1}{2\gamma}\right]^{1/2}$$

(c) For $M_1^2 - 1 \ll 1$, let $M_1^2 \equiv 1 + \mu$ and $M_2^2 + 1 + v$ as recommended, then have

$$(1 + \mu) - 1 \ll 1 \text{ or } \mu \ll 1$$

and

$$(1 + v) = \left[\frac{1 + \frac{\gamma - 1}{2}(1 + \mu)}{\gamma(1 + \mu) - \frac{\gamma - 1}{2}}\right] \quad \rightarrow \quad v = -\mu$$

Example 3.13. (Caltech Ae/APh 101a; Midterm Exam; November, 1987)

Consider airflow in a nozzle with a throat. Consider two cross-sections to the right (downstream) of the throat. Have $M_1 = 1.5$, $A_2 = 3A_1$, and $\gamma = 1.4$. What are the values of M_2 and p_2/p_1?

Solution

$M_1 = 1.5 \rightarrow {A^*}/{A_1} = 0.8502,\quad {P_1}/{P_0} = .2724$ per Table. Thus

$$A^*/A_2 = \frac{A^*}{A_1}\frac{A_1}{A_2} = (.8502)\frac{1}{3} = 0.2834$$

$$\therefore M_2 = 2.81 \ , \ P_2/P_0 = .3629 \rightarrow \frac{P_2}{P_1} = \frac{.03629}{.2724} = .133\ .$$

Example 3.14. (Caltech Ae/APh 101a; Midterm Exam; November, 1987)
A nozzle discharges into a large reservoir where the pressure is P_E and where nozzle exit area is 4 times the throat area. A shockwave has formed for some P_0, how should the (input) reservoir pressure be adjusted to move the shock wave to the nozzle exit?

Solution
$A^*/A_\epsilon = \frac{1}{4} \rightarrow M = 2.94$ for a shock at $A_\epsilon \therefore P_E/P_1 = 9.918$. So $P_1/P_0 = .02980$, and

$$\frac{P_0}{P_E} = \frac{P_0}{P_1}\frac{P_1}{P_E} = 3.38$$

3.3.3 Exercises
(Ex. 3.12) Air which has been shock heated and compressed to $T_0 = 8000°K$ and $p_0 = 4.8$ atm is expanded (isentropically) in the nozzle to a pressure of 0.01 atm.
(a) What are the values of *h*, *T*, ρ and *u* in the nozzle?
Notes:

(i) Use the handout of pages from NACA TN 4265
(ii) The reservoir condition corresponds to $\frac{s_0}{R} = 40.0$
(iii) The "compressibility factor" $z = p/R\rho T$ in the plots can be used to obtain ρ (Problem 1.18, p 34, in Liepmann & Roshko [29])

(b) What values of *h*, *T*, ρ and *u* would be obtained if perfect-gas relations were used?

(Ex. 3.13) Consider:
(a) The mass flux density ρu, the dynamic pressure $\frac{1}{2}\rho u^2$ (or the momentum flux density ρu^2) and the kinetic energy flux density $\frac{1}{2}\rho u^3$ are important flow variables. We have shown that, in isentropic flow, ρu has a maximum value, at $M = 1$. It is clear that $\frac{1}{2}\rho u^2$ and $\frac{1}{2}\rho u^3$ also have maxima. At what values of M do they occur?

(b) The result in (a) is general. Obtain the dimensionless dependence on M. for a perfect gas, i.e., $\frac{\rho u}{\rho_0 a_0}$, $\frac{\frac{1}{2}\rho u^2}{\frac{1}{2}\rho_0 a_0^2}$ and $\frac{\frac{1}{2}\rho u^3}{\frac{1}{2}\rho_0 a_0^3}$ as functions of M. Sketch these (roughly)

(c) Also of interest are the ratios based on ρ^*, a^* which are related to the others by constants which depend on γ. Determine $\frac{\rho^* a^*}{\rho_0 a_0}$, $\frac{\rho^* a^{*2}}{\rho_0 a_0^2}$ and $\frac{\rho^* a^{*3}}{\rho_0 a_0^3}$ as functions of γ.

(d) In practice, it is easier to measure p_0 and T_0, then ρ_0 and a_0. Obtain $\rho_0 a_0, \frac{1}{2}\rho_0 a_0^2$ and $\frac{1}{2}\rho_0 a_0^3$ in terms of p_0 , T_0.

(Ex. 3.14) Consider inflow from a reservoir, where the velocity at a point downstream can be determined from the values of static pressure at that point as well as the pressure and temperature in the reservoir. Assuming perfect gas, u can be obtained from

$$p = p_0\left(1 - \frac{\gamma - 1}{2}\frac{u^2}{a_0^2}\right)^{\frac{\gamma}{\gamma-1}} \qquad (1)$$

How much error will there be if one uses instead the incompressible Bernoulli formula

$$p = p_0 - \frac{1}{2}\rho_0 u^2 \qquad (2)$$

Proceed as follows. Since $\frac{\gamma-1}{2}\frac{u^2}{a_0^2} < 1$, do a binomial expansion on the quantity in brackets in (1).

(a) Show that the first term of the expansion leads to the incompressible formula (2).

(b) Keeping the second term in the expansion, obtain a correct factor K such that

$$p_0 - p = \rho_0 u^2(1 + K)/2,$$

and $K = 1 + A\frac{u^2}{a_0^2}$. Determine A.

(c) If the gas is air and $T_0 = 300°K$, at what speed u does the error become 1%?

(d) Obtain the correction factor in terms of Mach number, $K = 1 + BM^2$. Determine B.

(e) For sonic speed, and $\gamma = 1.4$, show that the exact formula (1) gives $p/p_0 = 0.528$. What does the incompressible formula (2) give? The following formulas from Liepmann and Roshko [29], might be helpful:

$$\dot{m} = p_0 a_0 A * M\left(1+\frac{\gamma-1}{2}M^2\right)^{-1/2\frac{\gamma+1}{\gamma-1}}$$

$$\dot{m}_{max} = \dot{m}^* = p_0 a_0 A^* \left(\frac{2}{\gamma+1}\right)^{1/2\frac{\gamma+1}{\gamma-1}}$$

and

$$\frac{p_0}{p_E} = \left(1+\frac{\gamma-1}{2}M^2\right)^{\frac{\gamma}{\gamma-1}}\left(\frac{2\gamma}{\gamma+1}M^2 - \frac{\gamma-1}{\gamma+1}\right)^{-i}.$$

(Ex. 3.15) Consider a supersonic nozzle with normal shock wave formation downstream from the throat (but still in the nozzle). Suppose we have a perfect gas, $\gamma = 1.4$, and $A_1/A' = 2$ and $A_E/A' = 4$, where A' is the throat area, A_1 is the area at the shock, and A_E is the area at the exit of the nozzle.

(a) Calculate the following:

$M_1, M_2, M_E; \frac{P_1}{P_0}, \frac{P_2}{P_0}, \frac{P_E}{P_0}; \frac{\rho_1}{\rho_0}, \frac{\rho_2}{\rho_0}, \frac{\rho_E}{\rho_0}; \frac{T_1}{T_0}, \frac{T_2}{T_0}, \frac{T_E}{T_0}; \frac{u_1}{a_0}, \frac{u_2}{a_0}, \frac{u_E}{a_0};$

(b) The flow at the exit is now allowed to pass through a second converging-diverging nozzle, with throat area A". Calculate M" and p''/p_0 for $A''/A' = 2$ and for $A''/A' = 1.3$. (For the last case, the shock position will have to change).

Note: For efficient solution of these problems, the use of isentropic-flow and shock Tables is recommended.

Partial Solution

First, we need to find M_1:

($A^* = A'$ for supersonic flow)

$$\left(\frac{A}{A'}\right)^2 = \frac{1}{M^2}\left[\frac{2}{\gamma+1}\left(1+\frac{\gamma-1}{2}M^2\right)\right]^{(\gamma+1)/(\gamma-1)}$$

$\left(\frac{A_1}{A'}\right) = 2,$ so, $4 = \frac{1}{M^2}[(.833)(1+(.2)M^2)]^6$

Thus, need to solve

$$1.26M^{1/3} - .167M^2 = .833$$

Time to start guessing…

M	$1.26M^{1/3} - .167M^2$
1	1.093

2	.9195
3	.3142
2.5	.666
2.2	.831

So, $M_1 = 2.2$ (supersonic).Similarly, $M_2^2 = \frac{1+\frac{\gamma-1}{2}M_1^2}{\gamma M_1^2-\frac{\gamma-1}{2}} \rightarrow M_2 = .547$ (subsonic). So, across the shock we have: $P_2/P_1 = 5.48$; $\rho_2/\rho_1 = 2.951$; $T_2/T_1 = 1.857$; $a_2/a_1 = 1.363$; and $P_2^0/P_1^0 = .6281$.

(Ex. 3.16) A blunt nosed body is re-entering the atmosphere. At an altitude of approximately 100,000 ft, the atmospheric conditions are

$P_a = 0.01$ atm and $T_a = 402°R$ (233°K). Thus
$a_a = 1009$ ft/sec (306 m/sec). The re-entr speed is 20,000 ft/sec.

(a) What are the values of $\frac{\rho_2}{\rho_a}, \frac{h_2}{h_a}, \frac{P_2}{P_a}, \frac{T_2}{T_a}$ immediately behind the shock wave that forms?

(b) What are those values at the stagnation point (at the blunt 'tip')?
(c) What stagnation-point values would result from a perfect-gas calculation?

(Ex. 3.17) (Caltech Ae/APh 101a; Midterm Exam; November, 1988) Air ($\gamma = 1.4$) at a temperature of 0°C is flowing in a channel of cross-sectional area A_1 at a velocity of 130 m/sec. Assume that the flow is isentropic and the gas perfect.
(a) To double the speed, by narrowing the channel, how much should the channel be changed; i.e., what is A_2/A_1?
(b) What then is the pressure ratio p_2/p_1?

(Ex. 3.18) (Caltech Ae/APh 101a; Midterm Exam; November, 1988) Consider the flow:

→	u = 500 m/sec (= 1640 ft/sec) → T = 12000°K P = 10 atm

where air in a channel has been shock compressed to the conditions shown. If this air is now accelerated in a nozzle isentropically to a pressure of 1 atm, what will be the velocity?

(Ex. 3.18) (Caltech Ae/APh 101a; Midterm Exam; November, 1988)
If we write $\frac{dM}{M} = f(M)\frac{du}{u}$. Derive f(M) for isentropic flow of a perfect gas.

Chapter 4. Fields – Electrostatics

In Chapters 4 thru 7 classic electromagnetism (EM) is covered, with extensive coverage on the Lorentz transform (EM is Lorentz invariant) in Ch. 8, which then leads to special relativity and general relativity and other topics in later chapters. This chapter (Ch. 4) begins with electrostatics, and when the force is attractive the mathematics is the same as that already encountered for gravitational systems in Book 1 [1], for which the behavior is already well-understood.

Note on Examples and Exercises
The chapters on electromagnetism draw, foremost, from the excellent source, Classical Electrodynamics by J.D. Jackson [32], that was the required text for many years at Caltech. Many of the Jackson problems develop the physics theory further (e.g., Gauss's Theorem) or further develop the mathematical techniques (for ODEs, definition of delta function as limit, etc.). With my own notes from the courses at Caltech, and upon later preliminary exam review at the Univ. Wisconsin, the addition of the completed problems from various sources were acquired in a way that seemed to provide a 'mapping' of understanding. I wanted to convey that understanding here. Solutions to these key problems are carefully worked out in what follows, along with notes on the theory from Jackson as well as from [33-35]. Basic exercises are given to re-derive or verify the theory. The solutions to the exercises here, for the most part, are available via the aforementioned Jackson text (with the detailed derivations provided there). Many examples and related problems are done in order to cover material from basic undergraduate problems to advanced engineering (for antenna design, say), so the notes and problems also draw from undergraduate and graduate EM class notes/exams, and the other texts mentioned ([33-35]). The remarkable success and modern applications of EM will be made evident, but at the end of Ch. 7, when we carefully analyze the hydrogen atom, we will find that there is radiative collapse in the classical theory. This fundamental flaw in the theory, when dealing with even the simplest bound atomic system, will be corrected with the development of the quantum theory in Book 4 [15].

4.1 Early Electromagnetism: Faraday's Observations and Maxwell's Unification

The electromagnetism (EM) examples that follow are in Gaussian units. Many examples will be drawn from Jackson [32], the classic graduate-level EM text. For a more complete collection of undergraduate- and graduate-level examples and problems see that text. Here the focus is on the core ideas, with many ODE examples, many in the Appendix on ODE's, and many in Ch. 5 on solving the 3D Laplacian, as well as on numerous graduate-level (preliminary-exam) examples.

For an extensive history of electromagnetism see Whittaker's books [36]. There the amazing contribution of Faraday in describing the field behavior is recounted and how Maxwell, with excellent mathematical skill, arrived and recognized a mapping to known mathematical constructs and their operations (the quaternions) such that mathematical equations expressing the relations could be constructed. We will jump to Maxwell's mathematical encapsulation in terms of "Maxwell's Equations" at the outset.

A remarkable feature of classical EM is that it is already (implicitly) a relativistic theory. This follows from the theory being invariant under Lorentz transformations [37]. What the Lorentz transformation reveals, in fact, is a deeper theory and interplay between the electric and magnetic fields seen according to reference frame [35]. If you just say the Lorentz transformation applies to everything (in particular the Lorentz Force relation), e.g., not just force fields but also to equations governing particulate and ponderable media, you then have special relativity *a la* Einstein [38]. An excellent undergraduate-level text on EM with special relativity transformations is by Purcell [35].

4.2 Vector Calculus

Historically, Maxwell developed his equations and expressed operations like cross-product with reference to quaternions (that had similar mathematical behavior). This hints at a deeper implementation of the theory with a deeper, more unified, presentation. For practical applications, however, the quaternion math was not well known or 'friendly' so Gibbs [9] developed modern-day vector calculus to describe the same mathematics, and to do so in a manner more compatible with Stokes theorems and other integral forms of the theory (as we will see). Here is a brief recounting of some of the key vector calculus relations:

$$a \cdot (b \times c) = b \cdot (c \times a) = c \cdot (a \times b)$$

(Eqn. 4-1)

$$a \times (b \times c) = (a \cdot c)b - (a \cdot b)c$$

(Eqn. 4-2)

$$(a \times b) \cdot (c \times d) = (a \cdot c)(b \cdot d) - (a \cdot d)(b \cdot c)$$

(Eqn. 4-3)

$$\nabla \times \nabla \psi = 0$$

(Eqn. 4-4)

$$\nabla \cdot (\nabla \times \vec{a}) = 0$$

(Eqn. 4-5)

$$\nabla \times (\nabla \times \vec{a}) = \nabla(\nabla \cdot a) - \nabla^2 a$$

(Eqn. 4-6)

$$\nabla \cdot (\psi \vec{a}) = a \cdot \vec{\nabla} \psi + \psi \vec{\nabla} \cdot \vec{a}$$

(Eqn. 4-7)

$$\nabla \times (\psi \vec{a}) = \nabla \psi \times a + \psi \nabla \times a$$

(Eqn. 4-8)

$$\nabla(a \cdot b) = (a \cdot \nabla)b + (b \cdot \nabla)a + a \times (\nabla \times b) + b \times (\nabla \times a)$$

(Eqn. 4-9)

$$\nabla \cdot (a \times b) = b \cdot \nabla \times a - a \cdot (\nabla \times b)$$

(Eqn. 4-10)

$$\nabla \times (a \times b) = a(\nabla \times b) - b(\nabla \times a) + (b \cdot \nabla)a - (a \cdot \nabla)b$$

(Eqn. 4-11)

Vector operations in different coordinate systems when given a metric function:

If there is a metric function $g_{\mu\nu}$, then we can write distance in 3D as:

$$ds^2 = g_{11}dx_1{}^2 + g_{22}dx_2{}^2 + g_{33}dx_3{}^2.$$

(Eqn. 4-12)

For which the gradient operation is:

$$\vec{\nabla} = \left(\frac{1}{\sqrt{g_{11}}}\frac{\partial}{\partial x_1}\right)\hat{e}_1 + \left(\frac{1}{\sqrt{g_{22}}}\frac{\partial}{\partial x_2}\right)\hat{e}_2 + \left(\frac{1}{\sqrt{g_{33}}}\frac{\partial}{\partial x_3}\right)\hat{e}_3 .$$

(Eqn. 4-13)

Thus, for $\vec{\nabla}\Psi$, with Ψ a scalar field, the formula above is trivially applied in whatever coordinate system.

The general form of the Laplacian operator, $\nabla^2\Psi$ is also straightforward:

$$\nabla^2\Psi = \frac{1}{|g|^{\frac{1}{2}}}\,\partial_\mu\left(|g|^{\frac{1}{2}}g^{\mu\nu}\partial_\nu\Psi\right).$$

(Eqn. 4-14)

What is not straightforward is the expression for $\vec{\nabla} \cdot A$, where A is a vector field. The trick to getting the expression of $\vec{\nabla} \cdot A$ appropriate to the coordinate system chosen is to get it from the simpler $\nabla^2\Psi$ and $\vec{\nabla}\Psi$ relations. Let's consider the example of cylindrical coordinates:

$$ds^2 = d\rho^2 + \rho^2 d\theta^2 + dz^2 \to \quad g^{\rho\rho} = 1;\ g^{\theta\theta} = \rho^{-2};\ g^{zz} = 1;\ |g|^{\frac{1}{2}} = \rho.$$

(Eqn. 4-15)

Thus,

$$\vec{\nabla}\Psi = \hat{\rho}\frac{\partial\Psi}{\partial\rho} + \hat{\theta}\frac{1}{\rho}\frac{\partial\Psi}{\partial\theta} + \hat{z}\frac{\partial\Psi}{\partial z}$$

(Eqn. 4-16)

and

$$\nabla^2\Psi = \frac{1}{\rho}\frac{\partial}{\partial\rho}\left(\rho\frac{\partial\Psi}{\partial\rho}\right) + \frac{1}{\rho^2}\frac{\partial^2\Psi}{\partial\theta^2} + \frac{\partial^2\Psi}{\partial z^2}.$$

(Eqn. 4-17)

Matching these relations under $\nabla^2\Psi = \vec{\nabla} \cdot \vec{\nabla}\Psi$:

$$\nabla^2\Psi = \frac{1}{\rho}\frac{\partial}{\partial\rho}\left(\rho\left(\vec{\nabla}\Psi\right)_\rho\right) + \frac{1}{\rho}\frac{\partial}{\partial\theta}\left(\left(\vec{\nabla}\Psi\right)_\theta\right) + \frac{\partial}{\partial z}\left(\left(\vec{\nabla}\Psi\right)_z\right),$$

(Eqn. 4-18)

we then have that:

$$\vec{\nabla} \cdot A = \frac{1}{\rho}\frac{\partial}{\partial\rho}(\rho A_\rho) + \frac{1}{\rho}\frac{\partial}{\partial\theta}(A_\theta) + \frac{\partial}{\partial z}(A_z).$$

(Eqn. 4-19)

4.3 Maxwell equations in vacuum

4.3.1 Vacuum equations

Maxwell's equations govern electromagnetic phenomena, which for sources in vacuum satisfy the equations:

$$\nabla \cdot \vec{E} = 4\pi\rho$$

(Eqn. 4-20)

$$\nabla \times \vec{B} - \frac{1}{c}\frac{\partial\vec{E}}{\partial t} = \frac{4\pi}{c}\vec{J}$$

(Eqn. 4-21)

which combine to give $\frac{\partial\rho}{\partial t} + \nabla \cdot \vec{J} = 0$, the continuity equation, and for the relations independent of source in vacuum we have:

$$\nabla \times \vec{E} + \frac{1}{c}\frac{\partial \vec{B}}{\partial t} = 0$$

(Eqn. 4-22)

$$\nabla \cdot \vec{B} = 0.$$

(Eqn. 4-23)

where $\vec{E}$ is the electric field, a vector field, which acts on charged particles, likewise for $\vec{B}$, the magnetic field. The electric charge density is given by ρ, and the electric charge current by $\vec{J}$. There is no magnetic charge or magnetic charge current. (there is some mystery here. Note that the formalism can be redone to have the ratio of electric to magnetic charge be fixed and have the same result. In the end, the U(1) part of the standard model gives the freedom for only one (free) charge, where our convention is that it be the electric charge with the convention above. The equations shown are given in a modern form involving vector calculus (more details to follow).

The action of the $\vec{E}$ and $\vec{B}$ force fields on a charged particle is given by the Lorentz force equation:

$$\vec{F} = q\left(\vec{E} + \frac{\vec{v}}{c} \times \vec{B}\right).$$

(Eqn. 4-24)

In Gaussian (cgs) units, convenient for many examples in what follows, we have length and time defined separately such that c is measured empirically: $c = 299{,}792{,}456.2 \pm 1.1\ m/s$. In SI units (since 1983) we have c defined by convention: $c = 299{,}792{,}458 m/s$. For (electric) charged particles or bodies we find that the charge is restricted to discrete values (the first clear indication of 'quanta', here the quantum of charge). The electron, for example, has one such quantum of charge:

$$|q_e| = 1.6021917(70) \times 10^{-19} Coulomb.$$

(Eqn. 4-25)

Fundamental Linearity

Due to linearity we have superposition property thus arrive at simple (summation) descriptions of macroscopic bodies and media and interfaces between those media. To manage interfaces will make use of the pillbox formation of junction conditions.

Aside from accounting for dielectric constants, Maxwell's equations (vacuum version) are:

$$\text{(I)} \quad \nabla \cdot \vec{E} = 4\pi\rho$$

$$\text{(II)} \quad \nabla \times \vec{B} - \frac{1}{c}\frac{\partial \vec{E}}{\partial t} = \frac{4\pi}{c}\vec{J}$$

$$\text{(III)} \quad \nabla \times \vec{E} + \frac{1}{c}\frac{\partial \vec{B}}{\partial t} = 0$$

$$\text{(IV)} \quad \nabla \cdot \vec{B} = 0.$$

(Eqn.s 4-26)

We will begin examining the application of these equations with classic electrostatics. As the name suggests, there is no time dependence, so the derivative terms are zero. Likewise we assume that no currents are present (initially), thus, we have conservative fields $\nabla \cdot \vec{E} = 4\pi\rho$ and $\nabla \cdot \vec{B} = 0$. The latter differential relation is simply the homogeneous solution to the former, so just focusing on the former to start we ask how we arrived at the relation $\nabla \cdot \vec{E} = 4\pi\rho$.

Maxwell's first law, $\nabla \cdot \vec{E} = 4\pi\rho$, can be obtained from the vector calculus relations above and the static Lorentz force law. If static, velocity is zero in Lorentz's force relation, and we have:

$$\vec{F} = q\vec{E},$$

(Eqn. 4-27)

where the E-field for point charges is given by Coulomb's Law:

$\vec{F} = kq_1q_2\frac{(\vec{x}_1-\vec{x}_2)}{|x_1-x_2|^3}$ with corresponding field at q_2 from q_1 : $\vec{E} = kq_1\frac{(\vec{x}_1-\vec{x}_2)}{|x_1-x_2|^3}$.

(Eqn. 4-28)

As mentioned, we have superposition with multiple (n) sources, so we can simply write:

$$\vec{E}(\vec{x}) = \sum_{k=1}^{n} q_i\frac{(\vec{x}_1-\vec{x}_2)}{|x_1-x_2|^3} ,$$

(Eqn. 4-29)

or if the matter description is continuous rather than discrete:

$$\vec{E}(\vec{x}) = \int \rho(\vec{x})\frac{(\vec{x}_1-\vec{x}_2)}{|x_1-x_2|^3}d^3x.$$

(Eqn. 4-30)

From the perspective of flux through a surface, we arrive at Gauss's Law:

$$\vec{E} \cdot \vec{n}da = \frac{q}{r^2}\cos\theta da = qd\Omega .$$

(Eqn. 4-31)

The flux though any simple closed surface S (that can radially project down to a 2-sphere) is, thus,

$$\oint_S E \cdot nda = \begin{cases} 4\pi q & q\ inside\ S \\ 0 & q\ outside\ S \end{cases}$$

(Eqn. 4-32)

With linear superposition we have:

$$\oint_S E \cdot nda = 4\pi \sum_i q_i$$

(Eqn. 4-34)

$$\oint_S E \cdot nda = 4\pi \int_v \rho(\vec{x}) d^3x$$

(Eqn. 4-35)

Obviously we have the integral vector calculus form, but paired with such, according to vector calculus (Stoke's Law, etc.), is a differential calculus form. To obtain it we begin with use of the divergence theorem:

$$\oint_S \vec{A} \cdot \vec{n} = \int_v \vec{\nabla} \cdot \vec{A} d^3x \ ,$$

(Eqn. 4-36)

which yields:

$$\oint_S E \cdot nda = \int_V \nabla \cdot \vec{E} d^3x = 4\pi \int_V \rho(\vec{x}) d^3x$$

(Eqn. 4-37)

or

$$\nabla \cdot E = 4\pi\rho$$

which is Maxwell's first law.

The static fields are conservative, with line integrals independent of path

Since $\nabla \cdot E = 4\pi\rho$ and $\nabla \times \vec{E} = 0$, we have a conserved field, i.e., we can write the E vector field in terms of a scalar 'potential' field:

$$E(\vec{x}) = \int \rho(\vec{x}) \frac{(\vec{x}_1 - \vec{x}_2)}{|x_1 - x_2|^3} d^3x' = \int \rho(\vec{x}') \left[-\nabla \left(\frac{1}{\vec{x} - \vec{x}'} \right) \right] d^3x'$$
$$= -\nabla \int \frac{\rho(\vec{x}')}{|\vec{x} - \vec{x}'|} d^3x' = -\nabla\Phi,$$

(Eqn. 4-38)

where the potential field is given by:

$$\Phi = \int \frac{\rho(\vec{x}')}{|\vec{x} - \vec{x}'|} d^3x'.$$

(Eqn. 4-39)

Let's compute the work done on a charge q to move it from A to B in a potential Φ:

$$W = -\int_A^B \vec{F} \cdot \overrightarrow{dl} = -q \int_A^B \vec{E} \cdot \overrightarrow{dl} = q \int_A^B \vec{\nabla}\Phi \cdot \overrightarrow{dl}$$
$$= q \int_A^B d\Phi = q\,(\Phi_B - \Phi_A)$$

(Eqn. 4-40)

Clearly this result is not dependent on path. We can see how there is not a dependence on path in a general sense via use of Stoke's Theorem:

$$\oint_C \vec{A} \cdot \overrightarrow{dl} = \int_S (\vec{\nabla} \times \vec{A}) \cdot \vec{n} da\,.$$

(Eqn. 4-41)

For the E field this becomes:

$$\oint_C \vec{E} \cdot \overrightarrow{dl} = \int_S (\vec{\nabla} \times \vec{E}) \cdot \vec{n} da = 0$$

(Eqn. 4-42)

where $\oint_C \vec{E} \cdot \overrightarrow{dl} = 0$ means that the line integral between two point A an B will be independent of path.

Let's now use Gauss's Theorem to prove surface charge properties (related to Problem 1.1 from [32]).

Example 4.1. Use Gauss's Theorem to show that:
(a) Excess charge on a conductor must reside on its surface:
$\vec{E} = 0$ inside a conductor → $\nabla \cdot E = 4\pi q = 0$ for surface constructed just under conductive surface, thus all charge must reside at the surface.
(b) A closed, hollow, conductor shields its interior from fields due to charges outside, but not due to charges inside:
Again, construct a surface of integration just under the outer surface of the conductive shell. For charge outside, since inside the conductive shell

we have $\vec{E} = 0$ and on the inner conductive surface we have $\nabla \cdot E = 0$, there is $\vec{E} = 0$ inside the shell. For charge inside, we have $\nabla \cdot E = 4\pi\rho$.
(c) The electric field at the surface of a conductor is normal to the surface with magnitude $4\pi\sigma$.
From the Gauss Theorem application in the pillbox junction boundary problem we know that there's the relation: $n \cdot (\vec{E}_2 - \vec{E}_1) = 4\pi\sigma$. Since inside the conductor we have $\vec{E}_1 = 0$ this simply leaves $n \cdot \vec{E}_2 = 4\pi\sigma$. Since on the surface of the conductor we have zero tangential field, $n \times \vec{E}_2 = 0$, we have that $|\vec{E}_2| = 4\pi\sigma$ and is directed normal to the surface.

4.3.2 Junction conditions

Now that we know the Maxwell equations that govern in a given macroscopic body or medium we can speak of having different media and of the interfaces between those media. In other words, what are the junction conditions indicated by Maxwell's equations (being valid in both regions and at the interface of those regions). This sets up what is known as the 'pillbox' formulation of the boundary conditions, where we (i) go from differential to integral form in Maxwell's equations using vector integral calculus; and (ii) we set up a "pillbox" volume that straddles the two regions encompassing the interface between them with a simple normal to the surface comprising the vertical dimension of a (very short) pillbox at the interface between the regions.

Maxwell's equations in integral form:

$$\nabla \cdot \vec{D} = 4\pi\rho \quad \rightarrow \quad \oint_S \vec{D} \cdot \vec{n} da = 4\pi \int_V \rho d^3x$$

(Eqn. 4-43)

$$\nabla \cdot \vec{H} - \frac{1}{c}\frac{\partial \vec{\rho}}{\partial t} = \frac{4\pi}{c}\vec{J} \quad \rightarrow \quad \oint_C \vec{H} \cdot \vec{dl} = \int_{S'} \left[\frac{4\pi}{c}\vec{J} + \frac{1}{c}\frac{\partial \vec{D}}{\partial t}\right] \cdot \vec{n}' \cdot da$$

(Eqn. 4-44)

$$\nabla \cdot \vec{E} - \frac{1}{c}\frac{\partial \vec{B}}{\partial t} = 0 \quad \rightarrow \quad \oint_C \vec{E} \cdot \vec{dl} = -\frac{1}{c}\int_{S'} \frac{\partial \vec{B}}{\partial t} \cdot \vec{n}' \cdot da$$

(Eqn. 4-45)

$$\nabla \cdot \vec{B} = 0 \quad \rightarrow \quad \oint_S \vec{B} \cdot \vec{n} \cdot da = 0$$

(Eqn. 4-46)

The boundary conditions at interfaces between different media with appropriate choice of 'pillbox':

$$(i)\ (\vec{D}_2 - \vec{D}_1) \cdot \vec{n} = 4\pi\sigma$$
$$(ii)(\vec{B}_2 - \vec{B}_1) \cdot \vec{n} = 0$$
$$(iii)\ \vec{n} \times (\vec{E}_2 - \vec{E}_1) = 0$$
$$(iv)\ n \times (H_2 - H_1) = 4\pi/c\, \vec{k}$$

(Eqn.s 4-47)

where σ is surface density and $\vec{k}$ is surface current. A final complication is then for the case of a moving boundary, $\vec{v} = c\vec{\beta}$. This is described in P.D. Noerdlinger, Am. J. Physics 39, 191 (1971), where we find that (i) and (ii) hold without modification (where σ is surface density as observed in the laboratory). In determining (iii) and (iv) we can no longer ignore the time-derivative terms. So let's go to the comoving frame of the interface where we have:

$$\int \frac{1}{c}\frac{\partial D}{\partial t} \cdot \vec{t} da = 0$$

(Eqn. 4-48a)

Since the convective derivative is zero:

$$0 = \int \frac{1}{c}\frac{\partial \vec{D}}{\partial t} \cdot \vec{t} da + \int [(\vec{\beta} \cdot \vec{\nabla})] \cdot \vec{t} da$$

(Eqn. 4-48b)

Upon substitution, along with use of a vector calculus relation (from Sec. 4.2), we thus have:

$$\int \frac{1}{c}\frac{\partial \vec{D}}{\partial t} \cdot \vec{t} da = \int [\vec{\nabla} \times (\vec{\beta} \cdot \vec{D}) - \vec{\beta}\vec{\nabla} \cdot \vec{D}] \cdot \vec{t} da,$$

(Eqn. 4-49)

thus, we find after some reworking:

$$(iii')\ \vec{n} \times (\vec{E}_2 - \vec{E}_1) - \vec{n} \cdot \vec{\beta}(\vec{B}_2 - \vec{B}_1) = 0$$
$$(iv')\ \vec{n} \times (\vec{H}_2 - \vec{H}_1) + \vec{n} \cdot \vec{\beta}(\vec{D}_2 - \vec{D}_1) = 4\pi/c\, \vec{k}$$

(Eqn. 4-50)

In the simplest case where $\vec{D} = \vec{E}$ and $\vec{H} = \vec{B}$ we get:

$$(iv'')\ \left[1 - (\vec{n} \cdot \vec{\beta})^2\right] \vec{n} \times (\vec{B}_2 - \vec{B}_1) = \frac{4\pi}{c}\vec{k},$$

(Eqn. 4-51)

which is clearly exhibiting a correction at second order in $\vec{\beta}$. So, as long as we are considering boundaries moving at non-relativistic speeds only, then there is no correction to the simple pillbox boundary analysis.

Likewise, as long as we aren't considering exotic ferromagnetic matter, atomic matter, or neutrons stars (or any star), then we have simple diamagnetic and dielectric coefficients. This is the focus of the standard applications of Maxwell's equations in this chapter. This is the majority of applications that aren't of an astrophysics research nature. Relativistic magnetohydrodynamic considerations will be briefly considered in Book 3 [5], when much of the math and physics has been covered for other reasons.

Let's now turn to the junction condition issues that can specifically arise in the electrostatic case. For this we will concern ourselves with surface distributions of charges, dipoles, and the possibility of discontinuities in field and potential. This is the first encounter with special functions and their well-definedness by way of distribution theory[39-41] (not yet invented at the time Dirac introduced the delta function in what follows).

Consider a surface with a charge density σ:

$$\left(\vec{E}_2 - \vec{E}_1\right) \cdot \vec{n} = 4\pi\sigma \quad and \quad \Phi(\vec{x}) = \int_s \frac{\sigma(\vec{x}')}{|\vec{x} - \vec{x}'|} da'.$$

(Eqn. 4-52)

Let's now consider what's known as a dipole layer, where we have charge densities on parallel surfaces, d apart, and with surface charge $\pm\sigma$. What potential is produced by a dipole layer?

$$\Phi(\vec{x}) = \int_s \frac{\sigma(\vec{x}')}{|\vec{x} - \vec{x}'|} da' - \int_{s'} \frac{\sigma(\vec{x}')}{|\vec{x} - \vec{x}' + \vec{n}d|} da''$$

(Eqn. 4-53)

Since we are dealing with a distance d to be made arbitrarily small, we can make use of a Taylor expansion:

$$\frac{1}{|\vec{x} - \vec{a}|} = \frac{1}{x} + \vec{a} \cdot \nabla\left(\frac{1}{x}\right) + \cdots$$

(Eqn. 4-54)

to get at first order in d:

$$\Phi(x) = \int_S \frac{\sigma(\vec{x}')}{|\vec{x} - \vec{x}'|} da'$$

$$- \int_{S'} \sigma(\vec{x}) \left\{\frac{1}{|\vec{x} - \vec{x}'|} + \vec{n}d(\vec{x}) \cdot \nabla\left(\frac{1}{|\vec{x} - \vec{x}'|}\right)\right\} da''$$

(Eqn. 4-55)

Let's introduce surface dipole moment density $D(\vec{x}') = \sigma(\vec{x}')d(\vec{x}')$:

$$\Phi(x) = \int_s D(\vec{x}')\vec{n}\cdot\vec{\nabla}'\left(\frac{1}{|\vec{x}-\vec{x}'|}\right)da' \quad and \quad \vec{p} = \vec{n}Dda'.$$

(Eqn. 4-56)

Thus, the potential at $\vec{x}$ caused by a point dipole at $\vec{x}'$ is:

$$\Phi(x) = \frac{\vec{p}\cdot(\vec{x}-\vec{x}')}{|\vec{x}-\vec{x}'|^3}$$

(Eqn. 4-57)

Since

$$-\vec{n}\cdot\vec{\nabla}\left(\frac{1}{|\vec{x}-\vec{x}'|}\right)da' = -\frac{\cos\theta da'}{|\vec{x}-\vec{x}'|^2} = -d\Omega$$

(Eqn. 4-58)

we can also write:

$$\Phi(x) = -\int_S D(\vec{x}')d\Omega,$$

(Eqn. 4-59)

where $d\Omega$ has positive sign if the angle is acute, i.e., when the observation point views the inner (negatively charged) side of the dipole layer. For a constant surface dipole moment density D, the potential is the product of the moment and the solid-angle subtended at the observation point by the surface, regardless of shape. Consider what this means for the potential seen upon crossing a charged double layer. Just under the double layer we have inner surface subtending 2π solid angle (half), so potential is $-2\pi D$, on the outer side we have $2\pi D$, thus the potential step seen when crossing the surface is

$$\Phi_2 - \Phi_1 = 4\pi D \rightarrow \Phi = 4\pi\theta(x), \; where \; \theta(x) \; is \; the \; step \; function.$$

(Eqn. 4-60)

Let's now consider the Poisson equation (with source) and Laplace equation (no source) from this perspective. Thus, from $\nabla\cdot E = 4\pi p$ and $E = -\nabla\Phi$ we have $\nabla^2\Phi = 4\pi p$ (Poisson's equation), and if no charge is present, then $\nabla^2\Phi = 0$ (Laplace's equation). Notice the oddity that $\nabla^2\left(\frac{1}{r}\right) = 0$ for $r \neq 0$ while its volume integral is -4π, which was managed by Dirac with introduction of the Dirac delta function $\delta(\vec{x})$:

$$\nabla^2\left(\frac{1}{|\vec{x}-\vec{x}'|}\right) = -4\pi\delta(\vec{x}-\vec{x}').$$

(Eqn. 4-61)

Thus, we see the distribution theory [41] elements of the step function $\theta(x)$ and the delta function $\delta(\vec{x})$. In what follows, the differential expressions involving discontinuous functions can always be expressed in terms of their integral forms, where those discontinuities have well-defined evaluation. Thus, the integral transforms due to the divergence theorem and integration by parts are critical tools in applying the theory. To this end, early developments were done by Green, whose theorems [42], thus, follow as simple applications of the divergence theorem:

$$\int_v \vec{\nabla} \cdot \vec{A} d^3x = \oint_s \vec{A} \cdot \vec{n}\, ds,$$

(Eqn. 4-62)

where Green introduces:

$$\vec{A} = \varphi \nabla \psi \quad such\ that \quad \vec{\nabla} \cdot \vec{A} = \varphi \nabla^2 \psi + \varphi \nabla \cdot \nabla \psi \quad and \quad \vec{A} \cdot \vec{n} = \varphi \nabla \psi \cdot \vec{n} = \varphi \frac{\partial \psi}{\partial n}.$$

(Eqn. 4-63)

Green's first identity is then:

$$\int_v (\varphi \nabla^2 \psi + \varphi \nabla \cdot \vec{\nabla} \psi) d^3x = \oint_s \varphi \frac{\partial \psi}{\partial n} da,$$

(Eqn. 4-64)

and Green's second identity is:

$$\int_v (\varphi \nabla^2 \psi + \psi \nabla^2 \varphi) d^3x = \oint_s \left[\varphi \frac{\partial \psi}{\partial n} - \psi \frac{\partial \varphi}{\partial n}\right] da.$$

(Eqn. 4-65)

The general form of Green's function to be introduced is based on generalizing the interesting form that results when we use Green's second identity with $\frac{1}{R} = \frac{1}{|\vec{x} - \vec{x}'|}$, which gives $\nabla^2 \psi = -4\pi\delta$, where $\delta = \delta(\vec{x} - \vec{x}')$ is the delta function, and we also choose $\varphi = \Phi$ where we have $\nabla^2 \Phi = -4\pi\rho$:

$$\Phi(\vec{x}) = \int_V \frac{\rho(\vec{x}')}{R} d^3x' + \frac{1}{4\pi} \oint_S \left[\left(\frac{1}{R}\right) \frac{\partial \Phi}{\partial n'} - \Phi \frac{\partial}{\partial n'} \frac{1}{R}\right] da'$$

(Eqn. 4-66)

As $S \to \infty$ we get the usual result that $\Phi = \int_v \frac{\rho}{R} d^3x'$, but for the boundary surface not at infinity we have a form that makes clear that two types of terms exist at the surface boundary (that relate to Dirichlet and Neumann boundary conditions in what follows). Note, if the reference coordinate $\vec{x}$ is not inside the boundary surface (in the volume V) then we have:

$$0 = \int_V \frac{\rho(\vec{x}')}{R} d^3x' + \frac{1}{4\pi} \oint_S \left[\left(\frac{1}{R}\right) \frac{\partial \Phi}{\partial n'} - \Phi \frac{\partial}{\partial n'} \frac{1}{R} \right] da'.$$

(Eqn. 4-67)

Green's analysis then goes further to consider the general Green's function that has form:

$$\nabla'^2 G(\vec{x}, \vec{x}') = 4\pi \delta(\vec{x} - \vec{x}')$$

(Eqn. 4-68)

where

$$G(\vec{x}, \vec{x}') = \frac{1}{|\vec{x} - \vec{x}'|} + F(\vec{x}, \vec{x}') \quad \text{and} \quad \nabla'^2 F(\vec{x}, \vec{x}') = 0,$$

(Eqn. 4-69)

and we are free to choose $F(\vec{x}, \vec{x}')$ to eliminate one of the surface terms (e.g., use Dirichlet or Neumann boundary conditions). We know that we can write

$$\Phi(\vec{x}) = \int_v \rho(\vec{x}') G(\vec{x}, \vec{x}') d^3x'$$

$$+ \frac{1}{4\pi} \oint_S \left[G(\vec{x}, \vec{x}') \frac{\partial \Phi}{\partial n'} - \Phi(\vec{x}') \frac{\partial G(\vec{x}, \vec{x}')}{\partial n'} \right] da'$$

(Eqn. 4-70)

and with Dirichlet b.c.'s: $G_D(\vec{x}, \vec{x}') = 0$ for $\vec{x}'$ on S and we get the form:

$$\Phi(\vec{x}) = \int_v \rho(\vec{x}') G_D(\vec{x}, \vec{x}') d^3x' - \frac{1}{4\pi} \oint_S \left[\Phi(\vec{x}') \frac{\partial G_D}{\partial n'} da' \right].$$

(Eqn. 4-71)

For Neumann b.c.'s we can't similarly choose $\frac{\partial G_N}{\partial n} = 0$ as this is too restrictive. To find the simplest allowable b.c. on G_N start with the form $\nabla'^2 G_N(\vec{x}, \vec{x}') = 4\pi \delta(\vec{x} - \vec{x}')$ and use Gauss's theorem to get the relation:

$$\oint_S \frac{\partial G_N}{\partial n'} (\vec{x}, \vec{x}') = -4\pi.$$

(Eqn. 4-72)

The simplest b.c. is then:

$$\frac{\partial G_N}{\partial n'} = -\frac{4\pi}{|S|},$$

(Eqn. 4-73)

where $|S|$ is the total area of the boundary, and we get the relation:

$$\Phi(\vec{x}) = \langle\Phi\rangle_S + \int_v \rho(\vec{x}')G_N(\vec{x},\vec{x}')d^3x' + \frac{1}{4\pi}\oint_S \left[\frac{\partial\Phi}{\partial n'} G_N da'\right].$$

(Eqn. 4-74)

In this process, the freedom to choose $F(\vec{x},\vec{x}')$ to 'zero-out' boundary terms at a surface is equivalent to the freedom to choose a system of charges external to the volume to achieve that b.c. This is equivalent to using the method of images.

Note that the Green's function approach works with respect to ODE's using any *Division Algebra*, thus can extend to quaternion and octonion formulations (as well as complex).

Calculate the electrostatic potential energy for bringing a charge into a field

Let's now calculate the work done to move a charge from infinity to a point $\vec{x}_i$:

$$W_i = q_i\Phi(\vec{x}_i)$$

(Eqn. 4-75)

since $\Phi(\vec{x}_i) = \sum_{j=1}^{n-1} \frac{q_j}{|\vec{x}_i - \vec{x}_j|}$ we have $W_i = q_i \sum_{j=1}^{n-1} \frac{q_j}{|\vec{x}_i - \vec{x}_j|}$. The total potential energy is then:

$$W = \frac{1}{2}\sum_{ij} \frac{q_i q_j}{|\vec{x}_i - \vec{x}_j|} \quad (i \neq j)$$

(Eqn. 4-76)

or, for continuous source (charge density):

$$W = \frac{1}{2}\iint \frac{\rho(\vec{x})\rho(\vec{x}')}{|\vec{x} - \vec{x}'|} d^3x d^3x' = \frac{1}{2}\int \rho(x)\Phi(x)d^3x$$

(Eqn. 4-77)

From the last form, substituting Poisson's equation $\nabla^2\Phi = 4\pi\rho$:

$$W = -\frac{1}{8\pi}\int \Phi\nabla^2\Phi d^3x = \frac{1}{8\pi}\int |\Phi\nabla|^2 d^3x = \frac{1}{8\pi}\int \left|\vec{E}\right|^2 d^3x.$$

(Eqn. 4-78)

The energy density can be seen to be:

$$w = \frac{1}{8\pi}\left|\vec{E}\right|^2.$$

(Eqn. 4-79)

Capacitance
Suppose you have a system of n conductors, each with charge Q_i and at potential V_i, and arranged in some fixed geometry in otherwise empty space. We know from the linear superposition properties that we must be able to write the potential at conductor "i" in terms of a sum over all of the conductor charges, each with a fixed factor (dependent on geometry of conductors and their placement):

$$V_i = \sum_{j=1}^{n} p_{ij} Q_j,$$

(Eqn. 4-80)

which is inverted to have:

$$Q_i = \sum_{j=1}^{n} C_{ij} V_j.$$

(Eqn. 4-81)

The C_{ii} values are known as the capacitances while the C_{ij} with (i ≠ j) are known as coefficients of induction. To get the capacitance of an object, hold all other (nearby) conductors at zero potential (ground) and measure the charge present when the conductor is held at unit potential.

To determine the work required to charge a conductor (a capacitor) we have

$$W = \frac{1}{2}\sum_{i=1}^{n} Q_i V_i = \frac{1}{2}\sum_{i,j} C_{ij} V_i V_j\,.$$

(Eqn. 4-82)

Example 4.2. Consider the flux at an element of an irregular enclosing surface.

Solution
Since the flux at the surface element of each solid angle division of the surface may be given by $D_r = q/4\pi r^2$ and since the flux impinges upon a surface described by the surface $dS_\perp = r^2 d\Omega$, we have

$$\int D_r\, dS_\perp = \int \left(\frac{q}{4\pi r^2}\right) r^2 d\Omega = \frac{q}{4\pi}\cdot 4\pi = q$$

with 4π sterardians. We also have

$$\int D \cdot n\, dS = \int D_r\, dS_\perp$$

thus,

$$\int D \cdot n\, dS = q.$$

Example 4.3. Derive the continuity equation from the following Maxwell equations:

$$\nabla \cdot D = \rho \text{ and } \nabla \times H = J + \frac{\partial D}{\partial t}$$

Solution
We have

$$0 = \nabla \cdot (\nabla \times H) = \nabla \cdot J + \nabla \cdot \frac{\partial D}{\partial t}$$

Thus,

$$0 = \nabla \cdot J + \frac{\partial(\nabla \cdot D)}{\partial t} = \nabla \cdot J + \frac{\partial \rho}{\partial t}$$

Example 4.4. *Dirac delta function as (improper) limit of Gaussian function*

Solution
Consider the three-dimensional Gaussian:

$$D(\alpha; x, y, z) = (2\pi)^{-\frac{3}{2}} \alpha^{-3} exp\left[-\frac{1}{2\alpha^2}(x^2 + y^2 + z^2)\right]$$

Let's take the delta function to be the limit:

$$\delta(\vec{x} - \vec{x}') = \lim_{\alpha \to 0} D(\alpha; x, y, z)$$

In generalized coordinates du/u, dv/v, dw/w, as $\alpha \to 0$, we are only interested in the infinitesimal length element

$$ds^2 = \left(\frac{du}{u}\right)^2 + \left(\frac{dv}{v}\right)^2 + \left(\frac{dw}{w}\right)^2$$

and,

$$D(\alpha; x, y, z) = (2\pi)^{-\frac{3}{2}} \alpha^{-3} exp\left[-\frac{1}{2\alpha^2}(ds^2)\right]$$

$$= (2\pi)^{-\frac{3}{2}} \alpha^{-3} exp\left[-\frac{1}{2\alpha^2}\left(\left(\frac{du}{u}\right)^2 + \left(\frac{dv}{v}\right)^2 + \left(\frac{dw}{w}\right)^2\right)\right].$$

Thus,

$$D(\alpha; x, y, z) = \delta\left(\frac{u' - u}{u}\right) \delta\left(\frac{v' - v}{v}\right) \delta\left(\frac{w' - w}{w}\right) = \delta(\vec{x} - \vec{x}')$$

and

$$\delta(\vec{x}-\vec{x}') = uvw\delta(u'-u)\delta(v'-v)\delta(w'-w).$$

Example 4.5. *Let's now consider spherical and cylindrical delta functions:*

(a) For spherical coordinates we have the metric:

$$ds^2 = dr^2 + r^2(d\theta^2 + \sin^2\theta\, d\varphi^2).$$

Thus,

$$u = 1, v = \frac{1}{r}, w = \frac{1}{rsin\theta}$$

and we have:

$$\delta(\vec{x}-\vec{x}') = \frac{1}{r^2 sin\theta}\delta(r'-r)\delta(\theta'-\theta)\delta(\varphi'-\varphi).$$

Suppose there is charge Q over the spherical shell, then, in terms of delta functions:

$$\rho(x) = A\delta(r-R)$$

is the charge distribution, for which we have total charge $Q = \int \rho(x)$. Solving for the form of A:

$$\int \rho(x) = A\int (Rd\theta)(Rsin\theta d\varphi) = A4\pi R^2 = Q \Rightarrow A = \frac{Q}{4\pi R^2}$$

Thus,

$$\rho(x) = \frac{Q}{4\pi R}\delta(r-R).$$

(b) For cylindrical coordinates we have the metric:

$$ds^2 = dr^2 + r^2 d\theta^2 + dz^2.$$

Thus,

$$\delta(\vec{x}-\vec{x}') = \frac{1}{r}\delta(r'-r)\delta(\theta'-\theta)\delta(z'-z).$$

Suppose there is a charge per unit length, λ, on a long cylinder, radius b:

$$\rho(x) = A\delta(r-b) \rightarrow \int \rho(x) = A\int r\delta(r-b)drd\theta dz = Ab2\pi z = Q.$$

where,

$$\lambda = \frac{Q}{z} = Ab2\pi, \quad A = \frac{\lambda}{2\pi b}, \quad Q = \lambda z.$$

Thus,

$$\rho(x) = \frac{\lambda}{2\pi b}\delta(r - b).$$

Example 4.6. *Consider three spheres of radius 'a' with total charge Q:* (i) conducting; (ii) uniform charge density; and (iii) spherically symmetric charge density, varying radially as r^n where $n > -3$. Find the behavior of the electric fields inside and outside the sphere for (i), (ii), and (iii) with $n = \pm 2$.

Solution

(i) $\int (\nabla \cdot E) dv = \int \rho\, dv \rightarrow \oint_S E \cdot \vec{n} = 4\pi \int_V \rho dv \rightarrow E(4\pi r^2) = 4\pi Q \ outside.$

For the conducting sphere we know $\vec{E} = 0$ inside the conducting region $(r < a)$. While outside the conducting region (e.g., $r > a$) we have:

$$\vec{E} = \frac{Q}{r^2}\hat{r} .$$

Let's now consider $r = a$: we have the junction condition $\vec{n} \cdot (\vec{E}_1 - \vec{E}_2) = 4\pi\sigma$, with $E_2 = 0$ (on the inside). So, $\vec{E} = 4\pi\sigma\hat{r} = \frac{Q}{a^2}\hat{r}$ at the surface since $\vec{E}_1 \times \vec{n} = 0$ and $Q = 4\pi a^2 \sigma$.

(ii) $\oint_S \vec{E} \cdot \vec{n} = 4\pi \int_V \rho dv \xrightarrow{r<a} 4\pi r^2 E = 4\pi\left(\frac{4}{3}\pi r^3\right)\rho$

Inside the sphere we have $E = \left(\frac{r}{a^3}\right) Q\hat{r}$. Outside the sphere we have $E = \left(\frac{Q}{r^2}\right)\hat{r}$.

(iii) Need to get charge density, starting from the form: $\rho(x) = Ar^n$:

$$\int \rho(x) = Q = 4\pi A \int_0^a r^n r^2 dr = \frac{4\pi A}{(n+3)} a^{n+3}$$

$$A = \frac{n+3}{4\pi}\frac{Q}{a^{n+3}}$$

So, we have:

$$\rho(x) = \frac{(n+3)Q}{4\pi}\frac{r^n}{a^n}\frac{1}{a^3} \leftarrow r \leq a$$

Working with the usual:

$$\oint_S \vec{E} \cdot \vec{n} = 4\pi \int_V \rho dv$$

we get for $r \leq a$:

$$(4\pi r^2)|E| = 4\pi Q \frac{r^{n+3}}{a^{n+3}} \quad thus \quad |E| = Q\frac{r^{n+1}}{a^{n+3}}$$

While for $r > a$ we simply have: $|E| = \frac{Q}{r^2}$. Let's now consider the specific cases of $n = \pm 2$: both will have $E = \left(\frac{Q}{r^2}\right)\hat{r}$ for $r > a$, while for $r \leq a$ we have: $|E| = \frac{Q}{ar}$ for $n = -2$ and $|E| = \frac{Qr^3}{a^5}$ for $n = 2$.

Example 4.7. *Consider the neutral hydrogen atom:*
Estimate using the time-averaged potential:

$$\Phi = q\frac{e^{-\alpha r}}{r}\left(1 + \frac{\alpha r}{2}\right).$$

where q is the magnitude of the electronic charge, α^{-1} equals half the Bohr radius. Find the distribution of charge, both continuous and discrete.

Solution
First, to get the discrete distribution, notice that as $r \to 0$ we get $\lim_{r\to 0} \Phi = \frac{q}{r}$. From the definition of the Dirac Delta function, $\nabla^2\left(\frac{1}{|\vec{x}-\vec{x}'|}\right) = -4\pi\delta(\vec{x}-\vec{x}')$, we get $-4\pi\rho = \nabla^2\Phi = -4\pi q\delta(r) \rightarrow \rho = q\delta(r)$. Thus, we have a point charge of q at $r = 0$, which corresponds to the positively charged nucleus consisting of the point particle (at this energy scale) of the proton.

Let's now consider the continuous part, which has only a radial dependency, thus:

$$\nabla^2\Phi = \frac{1}{r^2}\frac{\partial}{\partial r}\left(\frac{r^2\partial\Phi}{\partial r}\right) = -\frac{q}{r^2}\frac{\partial}{\partial r}\left[e^{-\alpha r}\left(\alpha r + 1 + \frac{\alpha^2 r^2}{2}\right)\right] = \frac{\alpha^3 q e^{\alpha r}}{2}.$$

Thus,

$$\rho = -\frac{\alpha^3 q e^{\alpha r}}{8\pi}.$$

The total charge in the continuous distribution is:

$$Q = \int \rho dv = -\frac{\alpha^3 q}{8\pi}\int_0^\infty r^2 e^{\alpha r} dr = -\frac{q}{2}\Gamma(3) = -q.$$

The continuous distribution has total charge – q, the "electron cloud" with negative charge corresponding to one electron..

Example 4.8. *Calculating Capacitance for various geometries.*
Recall the capacitance relation: $Q = CV$, that results from the definion that two conductors with a potential difference of V and a charge of Q (on one, but matching opposite). Using Gauss's Law, compute the capacitance for the following scenarios:

(a) two large flat sheets of area A separate by d,
where $d \ll \sqrt{A}$, so we can ignore edge effects and/or approximate using an infinite plane:

The E field near an infinite plane is $4\pi\sigma$. The potential near an infinite plane, thus, only changes as you move away from the plane: $E = -\nabla\Phi \rightarrow \Phi = 4\pi\sigma x$, where the I've chosen x as the dimension separating the two planes. Thus, the potential difference between the two planes separated by $x = d$ will be: $\Phi = 4\pi\sigma d$. The charge on one plate is $A\sigma$, so the capacitance is:

$$C = \frac{A\sigma}{4\pi\sigma d} = \frac{A}{4\pi d}$$

(b) two concentric conducting spheres with radii a, b:
From previous results we know inside the innermost conducting sphere we have $E = 0$, and the same is true outside the outer sphere (since the charge enclosed by the enveloping sphere is zero). So, only between the spheres we have:

$$E = \frac{q}{r^2}, \quad \Phi = \frac{q}{r}, \quad and \quad Q = q$$

The potential difference is: $V = \frac{q}{a} - \frac{q}{b} = q\frac{(b-a)}{ab}$, while the charge is $Q = q$, so:

$$C = \frac{q}{q\frac{(b-a)}{ab}} = \frac{ab}{b-a}$$

(c) two concentric conducting cylinders, radii a, b, of length L,
with $L \gg b > a$, $a < r < b$:

$$\oint_S E\cdot\vec{n}da = 4\pi\int_V \rho dv \rightarrow E(2\pi r)z = 4\pi Q \rightarrow E = \frac{2\lambda}{r}, \text{ where } \lambda = \frac{Q}{z}.$$

$$E = \frac{2\lambda}{r} \quad \rightarrow \Phi = 2\lambda\ln\left(\frac{r}{r_0}\right)$$

The difference in potential between the cylinders is then:

$$V = 2\lambda\left\{\ln\left(\frac{q}{r_0}\right) - \ln\left(\frac{b}{r_0}\right)\right\} = 2\lambda\ln\left(\frac{a}{b}\right)$$

$$Q = \lambda L$$

and

$$C = \frac{Q}{V} = \frac{L}{2\ln\left(\frac{a}{b}\right)}.$$

Example 4.9. *Consider two long cylindrical conductors, with radii* $a, b,$ *separated by distance* $d \gg b > a$.

For each cylinder we have:

$$E = \frac{2\lambda}{r} \quad \rightarrow \quad \Phi = 2\lambda\ln\left(\frac{r}{r_0}\right).$$

Thus, the potential at each cylinder

$$V_1 = 2\lambda\ln\left(\frac{d}{a}\right)$$

$$V_2 = -2\lambda\ln\left(\frac{d}{b}\right)$$

$$V_{Total} = V_1 - V_2 = 2\lambda\left(\ln\left(\frac{d}{a}\right) + \ln\left(\frac{d}{b}\right)\right) = 4\lambda\ln\left(\frac{d}{\sqrt{ab}}\right).$$

The capacitance of the system is thus:

$$C = \frac{L}{4\ln\left(\frac{d}{\sqrt{ab}}\right)}.$$

Example 4.10. *For the capacitor geometries of (Example 4.8)(abc) get the total electrostatic energy and show* $W = \frac{1}{2}QV$.

Recall $W = \frac{1}{8\pi}\int\left|\vec{E}\right|^2 d^3x$.

(a) For parallel plates, $|\vec{E}| = 4\pi\sigma \Rightarrow W = \frac{(4\pi\sigma)^2}{8\pi}(Ad) = 2\pi Q^2 \frac{d}{A} = Q\left(\frac{2\pi Qd}{A}\right) = \frac{1}{2}\text{QV}$.

(b) For concentric spheres, $|\vec{E}| = \frac{Q}{r^2}$ and

$$W = \frac{1}{8\pi}\int \frac{Q^2}{r^4} 4\pi r^2 dr = \frac{Q^2}{2}\int \frac{1}{r^2} dr = \frac{Q^2}{2}\left(\frac{1}{b} - \frac{1}{a}\right) = \frac{1}{2}\left(\frac{b-a}{ab}\right) Q^2$$

and since

$$C = \frac{ab}{b-a}$$

we get

$$W = \frac{1}{2} QV.$$

(c) for concentric cylinders, $|\vec{E}| = \frac{2Q}{Lr}$ and

$$W = \frac{1}{8\pi}\int \frac{4Q^2}{L^2}(L2\pi r dr) = \frac{Q^2}{L}\int \frac{dr}{r} = \frac{Q^2}{L}\ln\left(\frac{b}{a}\right),$$

and since

$$\mathrm{C} = \frac{\mathrm{L}}{2\ln\left(\frac{\mathrm{a}}{\mathrm{b}}\right)}$$

we get

$$W = \frac{1}{2} QV.$$

Example 4.11. ***Calculate attractive force between conductors for (Example 4.8)(a), the parallel plates.***
(a) For parallel plates separated by distance 'x' we have:

$$W = \frac{2\pi Q^2 x}{A} \rightarrow F = \frac{dW}{dx} = \frac{2\pi Q^2}{A}.$$

Example 4.12. ***Prove the mean value theorem of electrostatics:***
The mean value theorem of electrostatics states that for charge-free space the value of the electrostatic potential at a point is the same as the average of the electrostatic potential on a sphere centered at that point.

We have from the divergence theorem:

$$\int_V \nabla \cdot A d^3x = \oint_S \vec{A} \cdot \vec{n} da$$

Let $\vec{A} = \varphi \nabla \psi$:

$$\nabla \cdot \vec{A} = \varphi \nabla^2 \psi + \nabla \varphi \cdot \nabla \psi$$

$$\vec{A} \cdot \vec{n} = \varphi \vec{\nabla} \psi \cdot \vec{n} = \varphi \frac{\partial \psi}{\partial n}$$

To arrive at Green's theorem

$$\int_V (\varphi \nabla^2 \psi + \nabla \varphi \cdot \nabla \psi) d^3x = \oint_S \varphi \frac{\partial \psi}{\partial n} da$$

and

$$\int_V (\varphi \nabla^2 \psi - \psi \nabla^2 \varphi) = \oint_S \left[\varphi \frac{\partial \psi}{\partial n} da - \psi \frac{\partial \varphi}{\partial n}\right] da$$

Let $\nabla^2 \psi = -4\pi \delta(\vec{x} - \vec{x}') = \nabla^2 \left(\frac{1}{R}\right)$, $\psi = \left(\frac{1}{R}\right)$ for a unit point charge and $\varphi = \Phi(\vec{x}')$ such that $\nabla^2 \varphi = -4\pi\rho(\vec{x}')$:

$$\int_V \left[-4\pi\Phi(\vec{x}')\delta(\vec{x} - \vec{x}') + \frac{4\pi}{R}\rho(\vec{x}')\right] d^3x'$$

$$= \oint_S \left[\Phi \frac{\partial}{\partial n'} \frac{1}{R} - \frac{1}{R} \frac{\partial \Phi}{\partial n'}\right] da'$$

or

$$\Phi(\vec{x}') = \int_V \frac{\rho(\vec{x}')}{R} d^3x' + \frac{1}{4\pi} \oint_S \left[\frac{1}{R} \frac{\partial \Phi}{\partial n'} - \Phi \frac{\partial}{\partial n'} \left(\frac{1}{R}\right)\right] da'$$

$$= \frac{1}{4\pi R} \int_v \nabla \cdot \vec{E} d^3x + \oint_S \frac{1}{4\pi R^2} \Phi d\vec{a}$$

$$\Phi(\vec{x}') = \frac{\oint_S \Phi d\vec{a}}{4\pi R^2}$$

Example 4.13. ***Use Gauss's Theorem*** to prove that at the surface of a charged conductor with principle radii of curvature R_1 and R_2 has normal derivative of its electric field given by

$$\frac{1}{E} \frac{\partial E}{\partial n} = -\left(\frac{1}{R_1} + \frac{1}{R_2}\right).$$

Starting with Gauss's Theorem:

$$\oint_S E\cdot\vec{n}da=\int_V \nabla\cdot Ed^3x=\int_V 4\pi\rho(x)d^3x=4\pi\int_V \rho(x)d^3x=4\pi Q,$$

and since $da = R_1R_2d\theta_1d\theta_2$

$$(E\cdot\vec{n})R_1R_2d\theta_1d\theta_2=4\pi dQ \;\rightarrow\; |\vec{E}|=\frac{4\pi dQ}{R_1R_2d\theta_1d\theta_2}$$

Note that:

$$\vec{n}=\frac{\vec{R}_1}{|R_1|}+\frac{\vec{R}_2}{|R_2|} \quad\Rightarrow\quad \frac{\partial E}{\partial n}=\frac{\partial E}{\partial R_1}+\frac{\partial E}{\partial R_2}$$

Thus,

$$\frac{1}{E}\frac{\partial E}{\partial n}=\frac{1}{\left(\frac{4\pi dQ}{R_1R_2d\theta_1d\theta_2}\right)}\left\{\frac{-4\pi dQ}{{R_1}^2R_2d\theta_1d\theta_2}+\frac{-4\pi dQ}{R_1{R_2}^2d\theta_1d\theta_2}\right\}$$

$$=-\left(\frac{1}{R_1}+\frac{1}{R_2}\right).$$

Example 4.14. ***Prove Green's reciprocation theorem:***

$$\int_V \rho\Phi' d^3x+\int_S \sigma\Phi' da=\int_V \rho'\Phi d^3x+\int_S \sigma'\Phi da$$

where we have two specifications of potential and charge on a given volume V with surface S.

Recall the second form of Green's Theorem:

$$\int_V \{\varphi\nabla^2\psi-\psi\nabla^2\varphi\}\,d^3x=\oint_S \left[\varphi\frac{\partial\psi}{\partial n}-\psi\frac{\partial\varphi}{\partial n}\right]da.$$

Let $\varphi=\Phi$ and $\psi=\Phi'$, then $\nabla^2\varphi=-4\pi\rho$ and $\nabla^2\psi=-4\pi\rho'$ and recall that $\sigma=\frac{1}{4\pi}\frac{\partial\Phi}{\partial n}$:

$$\int_V \{\Phi(-4\pi\rho')-\Phi'(-4\pi\rho)\}d^3x=\oint_S [\Phi(4\pi\sigma')-\Phi'(4\pi\sigma)]da$$

Thus we get the desired result:

$$\int_V \rho\Phi' d^3x+\int_S \sigma\Phi' da=\int_V \rho'\Phi d^3x+\int_S \sigma'\Phi da.$$

Example 4.15. ***Consider two infinite, grounded, conducting planes separated by d.*** A point charge q is placed between the planes, distance x' from one plate. Show induced charge is $-q\frac{x'}{d}$ at other plate.

Starting from Green's reciprocation theorem:

$$\int_V \rho\Phi' d^3x + \int_S \sigma\Phi' da = \int_V \rho'\Phi d^3x + \int_S \sigma'\Phi da$$

and choosing prime case with charge only on surface ($\Phi' = 4\pi\sigma' x'$ and $\rho' = 0$):

$$\int_V q\Phi' d^3x + \int_S \sigma\Phi' da = \int_S \sigma'\Phi da$$

and

$$q4\pi\sigma' x' + 4\pi\sigma' d\int_{x'=d} \sigma da = \sigma'\int_S \Phi da$$

Thus,

$$\int_{x'=d} \sigma da = -q\frac{x'}{d}.$$

Example 4.16. ***Prove Thomson's Theorem***

Thomson's Theorem: If a collection of surfaces are fixed in position and a given total charge is placed on each surface, then the electrostatic energy of the system is at a minimum when the charges adjust so that every surface is an equipotential, e.g., as if each surface were a conductor:

The electrostatic energy is given by:

$$U = \frac{1}{8\pi}\int_V |E|^2 dV,$$

where the volume is the space between the conductors. Thus,

$$\delta U = \frac{1}{4\pi}\int_V \vec{E}\cdot\delta\vec{E}\, dV = -\frac{1}{4\pi}\int_V \vec{\nabla}\Phi\cdot(\delta\vec{E})dV$$

$$= +\frac{1}{4\pi}\int_V \Phi\vec{\nabla}\cdot(\delta\vec{E})dV - \frac{1}{4\pi}\int_V \vec{\nabla}\cdot(\Phi\delta\vec{E})dV$$

$$\delta U = -\frac{1}{4\pi}\int_S (\Phi\delta\vec{E})\cdot\vec{n}\,dS = -\frac{1}{4\pi}\sum_i\left[\int_{S_i}\Phi_i\delta\vec{E}\cdot\vec{n}dS\right]$$

And, $\delta U = 0$ when each of the Φ_i is constant on the surface S_i because then we have

$$\delta U = -\frac{1}{4\pi}\sum_i\left[\Phi_i\int_{S_i}\delta\vec{E}\cdot\vec{n}dS\right] = 0$$

since $\delta\vec{E}$=0 at surfaces if Φ_i are constant on the surfaces.

4.4 Method of Images and introduction to Green' Functions (following conventions/notation of Jackson [32] Ch. 2)

When introducing Green's Functions, brief mention was made of the Method of Images. The basic idea is that by placing charges external to the region of interest we might elicit the electric field and boundary conditions on the actual region of interest. In particular, this then eliminates the need for a complicated surface charge description. Consider a point charge q distance d from a mirror (conducting surface), this produces the same equipotential surfaces and field lines as having an opposite charge distance d on the opposite side of the mirror (with mirror removed). Let's consider the slightly more complicated scenario of a point charge near a grounded conducting sphere. Let's take the origin of the coordinate system to be that of the sphere. Let's place the mirror charge on the same radial line, but inside the sphere.... Let's now show that if we have charge q at radial distance y, the mirror charge will be $-aq/y$ and at distance a^2/y.

Charge outside a conducting sphere: potential by method of images

To begin, let's consider a point charge q in the presence of a charged (Q), insulated, conducting sphere (radius a). Using superposition, let's establish the field exterior to the sphere due to the charge with conducting sphere boundary to be charges q and $q' = -\frac{a}{y}q$, and interior to the sphere, spherically distributed, the charge $(Q - q')$ exists, such that the potential in the region outside the sphere is:

$$\Phi(\vec{x}) = \frac{q}{|\vec{x}-\vec{y}|} - \frac{aq}{y\left|\vec{x}-\frac{a^2}{y^2}\vec{y}\right|} + \frac{Q+\frac{a}{y}q}{|\vec{x}|}$$

(Eqn. 4-83)

The radial force acting on the exterior charge is then:

$$\vec{F} = \frac{q}{y^2}\left[Q - \frac{qa^3(2y^2 - a^2)}{y(y^2 - a^2)^2}\right]\frac{\vec{y}}{|y|}$$

(Eqn. 4-84)

Note that this force is attractive at short distances even if the charges have the same sign. The work function of a metal, for an electron to be removed, is in large part the work done against this attractive image force.

Suppose the conducting sphere is at a fixed potential (instead of charged and insulated), such that the potential $\Phi(a) = V$ at the surface, then using mirror charge as before:

$$\Phi(\vec{x}) = \frac{q}{|\vec{x} - \vec{y}|} - \frac{aq}{y\left|\vec{x} - \frac{a^2}{y^2}\vec{y}\right|} + \frac{Va}{|\vec{x}|}$$

(Eqn. 4-85)

and force:

$$\vec{F} = \frac{q}{y^2}\left[Va - \frac{qay^3}{(y^2 - a^2)^2}\right]\frac{\vec{y}}{|y|}.$$

(Eqn. 4-86)

Consider next what it means to have a conducting sphere in a uniform electric field. A uniform electric field, E_0, can be represented, locally, by two charges equal and opposite, and equal and opposite in distance from the field position of interest. If we estimate the local field, by superposition to be

$$\frac{Q}{R^2} - \frac{(-Q)}{R^2} = \frac{2Q}{R^2} = E_0,$$

(Eqn. 4-87)

and allow $R \to \infty$ and $Q \to \infty$ such that $2Q/R^2 = E_0$, we get the uniform field. Return, for the moment, to R and Q finite, and consider the mirror charges (taking advantage of superposition) to establish neutralizing boundary conditions on a spherical boundary with radius a. The potential with placement of mirror charges is then, using the previous notation of reference point $\vec{x}$ and charge q location point $\vec{y}$ (and charge $-q$ location point $-\vec{y}$):

$$\Phi(\vec{x}) = \left(\frac{-Q}{|\vec{x}-\vec{y}|} + \frac{(a/y)Q}{|\vec{x}-(a^2/y^2)\vec{y}|}\right) + \left(\frac{Q}{|\vec{x}-(-\vec{y})|} + \frac{(-a/y)Q}{|\vec{x}-(a^2/y^2)(-\vec{y})|}\right).$$

(Eqn. 4-88)

A better notation for this would start from $\vec{x} = \vec{r}$ and $\vec{y} = \vec{R}$, such that

$$|\vec{x}-\vec{y}| = (r^2 + R^2 - 2rRcos\theta)^{\frac{1}{2}},$$

(Eqn. 4-89)

where θ is the angle between $\vec{r}$ and $\vec{R}$. Then we have:

$$\Phi = \frac{Q}{(r^2 + R^2 + 2rRcos\theta)^{\frac{1}{2}}} - \frac{Q}{(r^2 + R^2 - 2rRcos\theta)^{\frac{1}{2}}} - \frac{aQ/R}{\left(r^2 + \left(\frac{a^2}{R}\right)^2 + \frac{2a^2r}{R}Rcos\theta\right)^{\frac{1}{2}}} + \frac{Q}{\left(r^2 + \left(\frac{a^2}{R}\right)^2 - \frac{2a^2r}{R}Rcos\theta\right)^{\frac{1}{2}}}$$

(Eqn. 4-90)

which simplifies to:

$$\Phi = \frac{Q}{R}\left(-\frac{2r}{R}cos\theta\right) + \frac{aQ}{Rr}\left(\frac{2a^2}{Rr}cos\theta\right) = -E_0\left(r - \frac{a^3}{r^2}\right)cos\theta.$$

(Eqn. 4-91)

The surface charge induced on the shell is given by:

$$\sigma = -\frac{1}{4\pi}\frac{\partial\Phi}{\partial r}\bigg|_{r=a} = \frac{3}{4\pi}E_0 cos\theta.$$

(Eqn. 4-92)

Green's function for space with conducting sphere

Let's consider the Green's function for when there is a conducting sphere as described previously. Now we see this is like a Green's function due to a single point charge to now include one with an image charge:

$$G(\vec{x},\vec{x}') = \frac{1}{|\vec{x}-\vec{x}'|} - \frac{a}{x'\left|\vec{x} - \frac{a^2}{x'^2}\vec{x}'\right|}$$
$$= \frac{1}{(x^2 + x'^2 - 2xx'cos\gamma)^{\frac{1}{2}}} - \frac{1}{\left(\frac{x^2x'^2}{a^2} + a^2 - 2xx'cos\gamma\right)^{\frac{1}{2}}}$$

(Eqn. 4-93)

Since we are describing a conducting sphere with zero charge, we have Dirichlet boundary conditions, thus we have the formula:

$$\Phi(\vec{x}) = \int_V \rho(\vec{x}')G_D(\vec{x},\vec{x}')d^3x' - \frac{1}{4\pi}\oint_S \Phi(\vec{x})\frac{\partial G_D}{\partial n'}da'$$

(Eqn. 4-94)

where $da' = a^2 d\Omega'$ and:

$$\left.\frac{\partial G_D}{\partial n'}\right|_{x'=a} = -\left\{\frac{-(a - xcos\gamma) + \left(\frac{x^2}{a} - xcos\gamma\right)}{(x^2 + a^2 - 2axcos\gamma)^{\frac{3}{2}}}\right\}$$

$$= \frac{-(x^2 - a^2)}{a(x^2 + a^2 - 2axcos\gamma)^{\frac{3}{2}}}$$

(Eqn. 4-95)

Thus,

$$\Phi(\vec{x}) = \frac{1}{4\pi}\int \Phi(a,\theta',\emptyset')\frac{a(x^2 - a^2)}{(x^2 + a^2 - 2axcos\gamma)^{\frac{3}{2}}}d\Omega',$$

(Eqn. 4-96)

where, spherical coordinates, we have: $cos\gamma = cos\theta cos\theta' + sin\theta sin\theta' \cos(\varphi - \varphi')$.

Example 4.17. Let's apply this to the case where there is a conducting sphere with upper hemisphere at +V and lower at –V.

Solution

Recall the relation: $\int_0^\pi (sin\theta)d\theta' = \int_1^{-1}(-1)d(cos\theta')$:

$$\Phi(\vec{x}) = \frac{V}{4\pi}\int_0^{2\pi} d\varphi'\left\{\int_0^1 (cos\theta') - \int_{-1}^0 d(cos\theta')\right\}\frac{a(x^2 - a^2)}{(x^2 + a^2 - 2axcos\gamma)^{\frac{3}{2}}}d\Omega'$$

with general form for the potential:

$$\Phi(\vec{x}) = \frac{Va(x^2 - a^2)}{4\pi}\int_0^{2\pi} d\varphi'\int_0^1 d(cos\theta')\left[(a^2 + x^2 - 2axcos\gamma)^{-\frac{3}{2}} - (a^2 + x^2 - 2axcos\gamma)^{-\frac{3}{2}}\right]$$

Let's arrange our coordinates such that the hemispheres are aligned with the z-axis, then $cos\gamma = cos\theta'$ and we get:

$$\Phi(z) = V\left[1 - \frac{(z^2 - a^2)}{z\sqrt{z^2 + a^2}}\right].$$

Example 4.18. (Jackson 2.1) *A point charge q is brought to a position a distance d away from an infinite plane conductor held at zero potential. Using the method of images, find:*
(a) the surface charge density on the plane;
(b) the force on the charge;
(c) the total force on the plane (by integrating $2\pi\sigma^2$ over the plane);
(d) the work necessary to remove the charge q from its position to infinity;
(e) the potential energy between the charge and its image;

Solution:
(a) Have charge q at $r_1 = \sqrt{r^2 + d^2 - 2rdcos\theta}$ with mirror charge –q at$r_2 = \sqrt{r^2 + d^2 + 2rdcos\theta}$. The potential field is thus:

$$\Phi = \frac{q}{|r_1|} - \frac{q}{|r_2|}.$$

Let's move to the surface of the mirror with this potential field in the real (not virtual image) side, with electric field $\vec{E}_1 = \nabla\Phi$, while on the conductor side we simply have $\vec{E}_2 = 0$, and on the conductor surface $\vec{E}_1 \times \vec{n} = 0$. Using the "pillbox formula" $n_1 \cdot (\vec{E}_1 - \vec{E}_2) = 4\pi\sigma$, we get:

$$E_1 = 4\pi\sigma\ normal\ to\ surface.$$

Thus,

$$n \cdot (\nabla\Phi) = -4\pi\sigma$$

In polar coordinates:

$$\sigma = -\frac{1}{4\pi}\vec{n} \cdot \left(\hat{e}_r\frac{\partial\Phi}{\partial r} + \frac{\hat{e}_\theta}{r}\frac{\partial\Phi}{\partial\theta}\right)\Bigg|_{\theta = \frac{\pi}{2},\quad r = r},$$

and making use of $\vec{n} \cdot \vec{e}_r = \cos\theta$ and $\vec{n} \cdot \vec{e}_\theta = \sin\theta$:

$$\sigma = -\frac{1}{4\pi}(cos\theta)\left\{q\left[\frac{r + dcos\theta}{{r_1}^3} - \frac{r + dcos\theta}{{r_2}^3}\right]\right\}$$
$$-\frac{1}{4\pi}\frac{sin\theta q}{r}\left\{\frac{rdsin\theta}{r_1^3} + \frac{rdsin\theta}{r_2^3}\right\}$$

$$\sigma = -\frac{q}{4\pi}\left\{2\sin^2\theta\left(\frac{d}{r_1^3}\right)\right\}$$

Thus,

$$\sigma = -\frac{q}{2\pi}\frac{d}{(r^2 + d^2)^{\frac{3}{2}}}$$

(b) $F = -\frac{qq}{(2d)^2} = -\frac{q^2}{(2d)^2}\hat{n}$ (attractive).

(c) It is claimed that surface energy density is $W = \frac{1}{8\pi}|E|^2 = 2\pi\sigma^2$, let's start by showing this. Recall that

$$\frac{\partial W}{\partial r} = q\frac{\partial \Phi}{\partial r} = qE = F$$

Thus, using $W = q\Phi$, with the superposition of charge fields, with

$$W_i = q_i \sum \frac{q_i}{|x_i - x_j|},$$

individually, and

$$W = \frac{1}{2}\sum\sum \frac{q_i q_j}{|x_i - x_j|}$$

in total. Or, if charge distribution continuous rather than discrete, have:

$$W = \frac{1}{2}\int\int \frac{\rho\rho'}{R} d^3x d^3x' = \frac{1}{2}\int \rho\Phi d^3x = -\frac{1}{8\pi}\int \Phi\nabla^2\Phi d^3x$$
$$= \frac{1}{8\pi}\int |\nabla\Phi|^2 d^3x = \frac{1}{8\pi}|E|^2.$$

The surface energy density is thus:

$$W = \frac{1}{8\pi}|E|^2 = 2\pi\sigma^2$$

The integration across the mirror surface to get the total force acting on the plane is then straightforward:

$$F \cdot \vec{n} = \int 2\pi\sigma^2 dS = \int 2\pi\sigma^2\, 2\pi\rho d\rho = (2\pi)^2 q \int \sigma^2 \rho d\rho$$
$$= (2\pi)^2 q \int_0^\infty \left(-\frac{q}{2\pi}\right)^2 \rho d\rho$$
$$= q^2 \int_0^\infty \frac{d^2}{(\rho^2 + d^2)^3}\rho d\rho$$

Let's use $\rho = d\ sinhx$:

$$|F| = \frac{q^2}{d^2}\int_0^{\infty} \frac{sinhxcoshx}{(coshx)^6}dx = \frac{q^2}{(2d)^2}$$

(d) $W = -\int F dx = \int_d^{\infty} \frac{q^2}{4x^2} dx$ and we get:

$$W = \frac{q^2}{4d}$$

(e) $W = \frac{1}{2}\sum_i \sum_j \frac{q_i q_j}{|x_i - x_j|} = \frac{1}{2}\sum_i \Phi_i q_i$, and with $q_1 = 1, q_2 = -q, \Phi_1 = -\frac{q}{2d}, \Phi_2 = \frac{q}{2d}$ we get:

$$W = -\frac{q^2}{2d}$$

Example 4.19. *Consider a point charge q inside a hollow, grounded, conducting sphere of inner radius a.*
Find:
(a) the potential inside the sphere;
(b) te induced surface charge density;
(c) the magnitude and force acting on a charge q.

Solution
(a) Using the method of images we have the charge and its mirror appropriate to the spherical boundary condition giving a potential in the form:

$$\Phi(\vec{x}) = \frac{q}{|\vec{x} - \vec{y}|} + \frac{q'}{|\vec{x} - \vec{y}'|}.$$

Let's consider the vector in terms of unit normal directions since we know, by symmetry, that $\vec{y}$ and $\vec{y}'$ have the same unit normal, and using $\vec{x} = xn$, $\vec{y} = yn'$, and $\vec{y}' = y'n'$ →

$$\Phi(x) = \frac{q}{|x\vec{n} - y\vec{n}'|} + \frac{q'}{|x\vec{n} - \vec{y}'\vec{n}'|}$$

At $x = a$ we have $\Phi(a) = 0$: $\Phi(x = a) = \frac{q}{a\left|\vec{n} - \frac{y}{a}\vec{n}'\right|} + \frac{q'}{|y'|\left|\vec{n}\prime - \frac{a}{y'}\vec{n}\right|} = 0$.

This occurs when:

$$\frac{q}{a} = -\frac{q}{|y'|} \quad and \quad \frac{y}{a} = -\frac{a}{|y'|}.$$

The potential is, thus,

$$\Phi(x) = \frac{q}{|\vec{x} - \vec{y}|} - \frac{q\left(\frac{a}{y}\right)}{\left|\vec{x} - \left(\frac{a}{y}\right)^2 \vec{y}\right|}$$

Let's consider a shift to explicit polar coordinates, where our radial position remains inside the sphere, i.e. $r < a$. So, let $|\vec{x}| = r, |\vec{y}| = r', \vec{x} \cdot \vec{y} = rr' \cos\theta$ we then get for the potential inside the sphere:

$$\Phi(r,\theta) = q\left\{\frac{1}{\sqrt{r^2 + r'^2 - 2rr'\cos\theta}} - \frac{\left(\frac{q}{r'}\right)}{\sqrt{r^2 + (a^2/r')^2 - 2(a^2/r')r'\cos\theta}}\right\}.$$

(b) To get the induced surface charge density on the inner surface of the sphere we make use of the "pillbox relation":

$$\sigma = -\frac{1}{4\pi}\frac{\partial\Phi}{\partial n}\bigg|_{x = a^-}$$

We get:

$$\sigma = -\frac{q}{4\pi}\left\{\frac{+r - r'cos\theta}{(\;)^{\frac{3}{2}}} - \frac{(a/r')(r - (a^2/r')\cos\theta)}{(\;)^{\frac{3}{2}}}\right\}\bigg|_{r = 0}$$

$$\sigma = -\frac{q}{4\pi}\left\{\frac{1}{1 + \left(\frac{a}{r'}\right)^2 - 2\left(\frac{a}{r'}\right)(\cos\theta)^{\frac{3}{2}}}\right\}\left\{+\frac{1}{(r')^3}(r - r'cos\theta) \mp \frac{1}{r'a^2}\left(r - \left(\frac{a^2}{r'}\right)cos\theta\right)\right\}$$

$$\sigma = \frac{q}{4\pi}\frac{1}{(1 + (a/r')^2 - 2(a/r')\cos\theta)^{\frac{3}{2}}}\left\{\frac{a}{(r')^3}(1 - \left(\frac{r'}{a}\right)^2\right\}$$

(c) $|F| = \frac{|qq'|}{(r-r')^2} = \frac{ar'q^2}{\left(a^2-r'^2\right)^2}$ and is attractive.

Example 4.20. *Consider a potential problem in the half space defined by* $z \geq 0$, *with Dirichlet boundary conditions at* $z = 0$ *and at infinity.*
(a) What is the Green's Function?

(b) If the potential on the plane is V inside some radius a and zero outside (centered at the origin of cylindrical coordinates) find an integral expression for the potential at a point P.
(c) Show the potential along the axis of the circle;
(d) Show that at large distances

$$\Phi(\rho,\emptyset,z)=\frac{Va^2}{2}\frac{z}{(\rho^2+z^2)^{\frac{3}{2}}}\left\{1-\frac{3a^2}{4(\rho^2+z^2)}+\frac{5}{8}\frac{(3\rho^2a^2+a^4)}{(\rho^2+z^2)^2}+\cdots\right\}.$$

Solution

(a) The Green's Function is that of point source $\vec{x}'$ with mirror source $\vec{x}''$, due to b.c.'s, such that

$$G_D(\vec{x},\vec{x}')=\frac{1}{|\vec{x}-\vec{x}'|}+\frac{1}{|\vec{x}-\vec{x}''|}$$

and $G_D(\vec{x},\vec{x}')=0$ for $\vec{x}'$ at $z=0$ and at infinity. In cylindrical coordinates, $ds^2=d\rho^2+r^2d\theta^2+dz^2$,
this becomes:

$$G_D(\rho,\varphi,z)=\frac{1}{\left(\sqrt{\rho^2+\rho^2-2pp'cos\varphi+(z-z')^2}\right)^{\frac{3}{2}}}-\frac{1}{\left(\sqrt{\rho^2+\rho^2-2pp'cos\varphi+(z-z')^2}\right)^{\frac{3}{2}}}$$

Using the result that

$$\Phi(\vec{x})=\int_v\rho(\vec{x})\,G_D(\vec{x},\vec{x}')d^3x'-\frac{1}{4\pi}\oint_s\Phi(\vec{x}')\frac{\partial G_D}{\partial n'}da'$$

and

$$\left.\frac{\partial G_D}{\partial n'}\right|_{z'=0}=\left.\frac{\partial G}{\partial z'}\right|_{z'=0}=\frac{2z}{(\rho^2+\rho'^2-2\rho\rho'cos\varphi+z^2)^{\frac{3}{2}}}$$

Thus,

$$\Phi(x)=-\frac{1}{4\pi}\int_0^{2\pi}\int_0^a\frac{V2z}{(\rho^2+\rho'^2-2\rho\rho'cos\varphi+z^2)^{\frac{3}{2}}}\rho'd\rho'd\varphi$$

(c) Along the z-axis $\rho=0$ so:

$$\Phi(z) = -V \int_0^a \frac{z}{(\rho'^2 + z^2)^{\frac{3}{2}}} \rho' d\rho' = +Vz(\rho'^2 + z^2)^{-\frac{1}{2}} \Big|_0^a$$
$$= V\left[1 - \frac{z}{\sqrt{a^2 + z^2}}\right].$$

(d) At large distances $\rho^2 + z^2 \gg a^2$, we use

$$\frac{1}{(\rho^2 + \rho'^2 - 2\rho\rho' cos\varphi + z^2)^{\frac{3}{2}}} = \frac{\left\{1 - \frac{3}{2}\left(\frac{\rho'^2 - 2\rho\rho' cos\varphi}{\rho^2 + z^2}\right) + \frac{15}{8}\frac{(\rho'^2 - 2\rho\rho' cos\varphi)^2}{(\rho^2 + z^2)^2} + \cdots\right\}}{(\rho^2 + z^2)^{\frac{3}{2}}}$$

To get:

$$\Phi(\rho, \varphi, z) = \frac{Va^2}{2} \frac{z}{(\rho^2 + z^2)^{\frac{3}{2}}} \left\{1 - \frac{3a^2}{4(\rho^2 + z^2)} + \frac{5}{8} \frac{(3\rho^2 a^2 + a^4)}{(\rho^2 + z^2)^2} + \cdots\right\}.$$

Example 4.21. Consider a sphere of uniform charge, radius a, density ρ_0, with a cavity, spherical, centered at (d,0,0) with radius b, where $d < a$ and $b < a - d$ (for it to be a spherical cavity fully enclosed). What is the electric field along the x-axis?

Solution

For $x > a$

$$E = \left[\frac{Q_1}{4\pi\epsilon x^2} - \frac{Q_2}{4\pi\epsilon(x-d)^2}\right] x; \quad Q_1 = \frac{4}{3}\pi a^3 \rho_0, \quad Q_2 = \frac{4}{3}\pi b^3 \rho_0$$

Thus, $\varepsilon = \varepsilon_0$

$$E = \frac{\rho_0}{3\epsilon}\left(\frac{a^3}{x^2} - \frac{b^3}{(x-d)^2}\right) x$$

For $x < a$

$$E = \left[\frac{\rho_0}{3\epsilon}\left(\frac{a^3}{x^2} - \frac{b^3}{(x-d)^2}\right)(-\underline{x})\right]$$

Since the charge is attributed to a uniform density in the sphere it must not be able to flow to the boundaries as it would otherwise. The medium is nonconducting.

For inside the sphere, excluding the cavity, we have nearly the same equation as for $|x| > a$ but now x is inside the sphere. Superposition of a complete sphere with that of a negative sphere due to the cavity gives the desired results.

Thus, $E = \left[\frac{\rho_0}{3\epsilon}\left(|x| - \frac{b^3}{(x-d)^2}\right)\right] x$ in sphere for $x > d + b$.

The rest left to reader.

Example 4.22. Show that all of the equipotential surfaces for two line charges of opposite sign are cylinders whose traces in the perpendicular plane are circles.

Solution

We have

$$\Phi_- = \frac{-q_1}{2\pi\epsilon_0}\ln(r_-) + C_- \quad , \quad \Phi_+ = \frac{q_1}{2\pi\epsilon_0}\ln(r_+) + C_+$$

Thus

$$\Phi = \Phi_- + \Phi_+ = \frac{q_1}{2\pi\epsilon_0}\ln\left(\frac{r_-}{r_+}\right) + C$$

For equipotential surfaces $\Phi =$ constant, and

$$\frac{r_-^2}{r_+^2} = e^{4\pi\epsilon_0(\Phi+C)/q_1} \rightarrow \frac{(x+a)^2 + y^2}{(x-a)^2 + y^2} = e^{4\pi\epsilon_0(\Phi+C)/q_1}$$

Let $k = 2\pi\epsilon_0\,(\Phi + C)/q_1$, then

$$y^2 = -x^2 + 2ax\frac{\cosh k}{\sinh k} - a^2$$

which is

$$x^2 - 2ax\frac{\cosh k}{\sinh k} + y^2 = -a^2$$

the equation for a circle.

Example 4.23. Show that the potential in a charge free region can be determined, within an additive constant, by knowing the normal derivatives of the potential at all of the bounding surface.

Solution

Imagine there are two solutions Φ_1 and Φ_2. Since the normal derivatives of the potential are specified on the bounding surfaces (and is the same for both):

$$\nabla\Phi_1 \cdot dS = \nabla\Phi_2 \cdot dS.$$

Since Φ_1 and Φ_2 represent solutions: $\nabla^2\Phi_1 = 0, \nabla^2\Phi_2 = 0$ thus

$$\nabla^2(\Phi_1 - \Phi_2) = 0.$$

Recall the vector calculus relation:

$$\int_V \nabla \cdot [X\nabla Y] dV = \oint_S [X\nabla Y] \cdot dS$$

or

$$\int_V \nabla \cdot [(\Phi_1 - \Phi_2)\nabla(\Phi_1 - \Phi_2)] dV = \oint_S [(\Phi_1 - \Phi_2)\nabla(\Phi_1 - \Phi_2)] \cdot dS$$

which becomes:

$$\int_V (\Phi_1 - \Phi_2)\, \nabla^2(\Phi_1 - \Phi_2) dV + \int_v [\nabla(\Phi_1 - \Phi_2)]^2\, dV$$

$$= \oint_S (\Phi_1 - \Phi_2)\, [\nabla\Phi_1 \cdot \mathrm{dS} - \nabla\Phi_2 \cdot \mathrm{dS}].$$

Using the above relations we have

$$\int_v [\nabla(\Phi_1 - \Phi_2)]^2\, dV = 0 \rightarrow \quad (\Phi_1 - \Phi_2) = constant\ .$$

Since the potential is not specified on the surface we can be no more definite than this. Thus, the potential is uniquely determined except for an arbitrary additive constant.

4.5 Orthogonal Expansions and introduction to Laplace's equation (adopting notation/conventions of Jackson [32])

Before moving to more complex boundary conditions, with decomposition in terms of orthonormal functions, let's first review the properties of such functions. Consider a set of n functions of a variable ε in an interval (a, b): $u_n(\varepsilon)$. Those functions $u_n(\varepsilon)$ are orthogonal if they satisfy:

$$\int_a^b u_n^*(\varepsilon) u_m(\varepsilon) d\varepsilon = 0, \quad m \neq n.$$

(Eqn. 4-97)

If the functions also satisfy:

$$\int_a^b u_n^*(\varepsilon) u_m(\varepsilon) d\varepsilon = 1, \quad m = n,$$

(Eqn. 4-98)

then they are orthonormal. In other words, in Kronecker delta notation, we have for an orthonormal set of functions:

$$\int_a^b u_n^*(\varepsilon) u_m(\varepsilon) d\varepsilon = \delta_{nm}$$

(Eqn. 4-99)

From distribution theory [41], we know that we can express an arbitrary function $f(\varepsilon)$, that is square integrable in the interval (a, b), in terms of a series of orthonormal functions. Suppose this series representation can be made finite, with just N terms, for a specified allowable error, then we have for that mean square error:

$$M_N = \int_a^b \left| f(\varepsilon) - \sum_{n=1}^{N} a_n u_n(\varepsilon) \right|^2 d\varepsilon$$

(Eqn. 4-100)

which is minimized with choice of coefficients:

$$a_n = \int_a^b u_n^*(\varepsilon) f(\varepsilon) d\varepsilon.$$

(Eqn. 4-101)

Thus, we have

$$f(\varepsilon) = \int_a^b \left\{ \sum_{n=1}^{\infty} u_n^*(\varepsilon') u_n(\varepsilon) \right\} f(\varepsilon') d\varepsilon',$$

(Eqn. 4-102)

which means we have:

$$\sum_{n=1}^{\infty} u_n^*(\varepsilon') u_n(\varepsilon) = \delta(\varepsilon' - \varepsilon).$$

(Eqn. 4-103)

So far we've established the basics for a decomposition in terms of a series of orthonormal functions in terms of one variable. The Maxwell and Force relations are generally in 3D, of course, so lead to expressions with mixed coordinate system references. What is needed, for simplest advancement in complexity of the situations being considered is to now consider cases where the coordinate systems used imply sufficient symmetry, that 'separation of variables' is possible. This is where we can separate at least one degree of freedom, where a decomposition like described previously might then be employed. Let's first consider Laplace's equation in rectangular coordinates as an example of this.

Laplace's equation in rectangular coordinates

In rectangular coordinates Laplace's equation is:

$$\nabla^2 \Phi = 0 \Rightarrow \frac{\partial^2 \Phi}{\partial x^2} + \frac{\partial^2 \Phi}{\partial y^2} + \frac{\partial^2 \Phi}{\partial z^2} = 0.$$

(Eqn. 4-104)

We will find in Sec. 3.X, when we get to general geometries, that the general tensor form for Laplace's equation is:

$$\nabla^2 \Phi = \frac{1}{|g|^2} \partial_\mu (|g|^2 g^{\mu\nu} \partial_\nu \Phi)$$

(Eqn. 4-105)

and when $g_{\mu\nu} = \delta_{\mu\nu}$ (flat spacetime) we have

$$\nabla^2 \Phi = \partial^\mu \partial_\mu \Phi = \frac{\partial^2 \Phi}{\partial x^2} + \frac{\partial^2 \Phi}{\partial y^2} + \frac{\partial^2 \Phi}{\partial z^2}.$$

(Eqn. 4-106)

In this coordinate system we see trivial separation of terms in the $\{x, y, z\}$ variables, suggesting a potential function of the form: $\Phi(x, y, z) = X(x)Y(y)Z(z)$. If we substitute this into the $\nabla^2 \Phi = 0$ relation we get:

$$\frac{1}{X(x)} \frac{\partial^2 X}{\partial x} + \frac{1}{Y(y)} \frac{\partial^2 Y}{\partial y} + \frac{1}{Z(z)} \frac{\partial^2 Z}{\partial z} = 0$$

(Eqn. 4-107)

with one set of solutions given by

$$\Phi = e^{\pm i\alpha x} e^{\pm i\beta y} e^{\pm\sqrt{\alpha^2+\beta^2} z},$$

(Eqn. 4-108)

where we can regroup to have solutions in terms of standard trigonometry functions:

$$X = sin\alpha x, \quad Y = sin\beta y, \quad and \quad z = \sinh\left(\sqrt{\alpha^2 + \beta^2} z\right).$$

(Eqn. 4-109)

Laplace's equation in a box with side lengths $\{a, b, c\}$

Consider a box with side lengths $\{a, b, c\}$ oriented in the positive definite octant with one corner at the origin, will all sides at zero potential except for side $z = c$ which has potential $V(x, y)$. The set of solutions indicated above are precisely suited to this problem since we have zero potential at the $x = 0$, $y = 0$, and $z = 0$ faces. Let's now consider that at the $x = a$ face we also have zero potential, thus

$$\sin(\alpha a) = 0 \Rightarrow \alpha a = n\pi \Rightarrow \alpha_n = \frac{n\pi}{a}.$$

(Eqn. 4-110)

Similarly, we have

$$\sin(\beta b) = 0 \Rightarrow \beta b = n\pi \Rightarrow \beta_n = \frac{n\pi}{b}.$$

(Eqn. 4-111)

If we denote $\gamma_{nm} = \pi\sqrt{\frac{n^2}{a^2} + \frac{m^2}{b^2}}$, we can write:

$$\Phi(x,y,z) = \sum_{n,m=1}^{\infty} A_{nm}\ \Phi_{nm},$$
$$\text{where} \quad \Phi_{nm} = \sin(\alpha_n x)\sin(\beta_m y)\sinh(\gamma_{nm} z).$$
(Eqn. 4-112)

Now, at $z = c$ we have:

$$V(x,y) = \sum_{n,m=1}^{\infty} A_{nm} \sin(\alpha_n x)\, sin(\beta_m y) \sinh(\gamma_{nm} c).$$

(Eqn. 4-113)

We are given $V(x,y)$, however, so to arrive at the solution we need to invert the above relation, and this is where the completeness properties of orthonormal function introduced previously come into play:

$$A_{nm} = \frac{4}{ab\ sinh(\gamma_{nm}c)} \int_0^a dx \int_0^b dy\, V(x,y) \sin(\alpha_n x)\, sin(\beta_m y).$$

(Eqn. 4-114)

Laplace's equation in semi-infinite rectangular strip

Let's consider solutions to Laplace's equation in 2-D for the region given by the semi-infinite rectangular strip bounded by y>0 and 0<x<a, with potential zero at the x=0 and x=a boundaries, and V at the y=0 boundary (at the y=∞ boundary must have potential fall to zero). Since we have simple boundaries, in rectangular coordinates, we will proceed with rectangular coordinates with separation of variables with $= X(x)Y(y)$:

$$\nabla^2\Phi = \frac{\partial^2\Phi}{\partial x^2} + \frac{\partial^2\Phi}{\partial y^2} = 0 \Rightarrow \frac{1}{X}\frac{\partial^2 X}{\partial x} + \frac{1}{Y}\frac{\partial^2 Y}{\partial y} = 0.$$

(Eqn. 4-115)

We have zero boundaries in the x-direction, so expect sin(x) form in the set of general solutions. We have finite boundary value of V at y=0 falling to zero, so expect a $\exp(-y)$ form in the set too:

$$\sin(\alpha a) = 0 \Rightarrow \alpha a = n\pi \Rightarrow \alpha_n = \frac{n\pi}{a}.$$

(Eqn. 4-116)

Thus, the potential has the form:

$$\Phi(x,y) = \sum_{n=1}^{\infty} A_n \exp\left(-\frac{n\pi y}{a}\right) \sin\left(\frac{n\pi x}{a}\right),$$

(Eqn. 4-117)

where we have:

$$A_n = \frac{2}{a}\int_0^a \Phi(x,0) \sin\left(\frac{n\pi x}{a}\right) dx = \frac{4V}{\pi n}\begin{cases} 1\ for\ n\ odd \\ 0\ for\ n\ even \end{cases}$$

(Eqn. 4-118)

Thus,

$$\Phi(x,y) = \frac{4V}{\pi}\sum_{n=1}^{\infty}\frac{1}{n}\exp\left(-\frac{n\pi y}{a}\right)\sin\left(\frac{n\pi x}{a}\right)$$
$$= \frac{4V}{\pi} Im\left\{\sum_{n\ odd}\frac{1}{n}e^{\left(\frac{in\pi}{a}\right)(x+iy)}\right\}.$$

(Eqn. 4-119)

If we substitute with $z = e^{\left(\frac{in\pi}{a}\right)(x+iy)}$ we can simplify to:

$$\Phi(x,y) = \frac{2V}{\pi} Im\left[ln\left(\frac{1+z}{1-z}\right)\right].$$

(Eqn. 4-120)

Using the property that the imaginary part of a log is equal to the phase of its argument:

$$\Phi(x,y) = \frac{2V}{\pi}\tan^{-1}\left\{\frac{2Im(z)}{1-|z|^2}\right\} = \frac{2V}{\pi}\tan^{-1}\left(\frac{\sin\left(\frac{\pi x}{a}\right)}{\sinh\left(\frac{\pi y}{x}\right)}\right).$$

(Eqn. 4-121)

Laplace's equation for a corner:

Let's now consider another 2D example involving corners and edges of various types. For this it is best to work in polar coordinates: $ds^2 = d\rho^2 + \rho^2 d\varphi^2$, for which the metric is:

$$\|g_{uv}\| = \begin{pmatrix}1 & 0\\ 0 & \rho^2\end{pmatrix} \rightarrow \sqrt{|g|} = \rho.$$

(Eqn. 4-122)

Using the general form mentioned earlier,

$$\nabla^2\Phi = \frac{1}{|g|^2}\partial_m\left(|g|^{\frac{1}{2}}g^{\mu v}\partial_v\Phi\right) = 0$$

(Eqn. 4-123)

we then get:

$$\frac{1}{\rho}\partial_\rho(\rho\partial_\rho\Phi) + \frac{1}{\rho^2}\partial_\varphi^2\Phi = 0.$$

(Eqn. 4-124)

We can do separation of variables in ρ and φ, with $\Phi = R(\rho)\psi(\varphi)$, to get:

$$\frac{\rho}{R}\partial_\rho(\rho\partial_\rho R) + \frac{1}{\psi}\frac{\partial^2\psi}{\partial\varphi^2} = 0.$$

(Eqn. 4-125)

For which we have the solutions:

$$R(\rho) = a\rho^{v} + b\rho^{-v} \quad \text{and} \quad \psi(\varphi) = A\cos(v\varphi) + B\sin(v\varphi).$$

(Eqn. 4-126)

For $= 0$:

$$\rho\frac{\partial R}{\partial \rho} = const \Rightarrow R = b_0 ln\rho + a_0$$

(Eqn. 4-127)

$$\frac{\partial^2 \psi}{\partial \varphi^2} = 0 \Rightarrow \psi = A_0 + B_0\varphi.$$

(Eqn. 4-128)

If we restrict to single-valuedness on ψ we get $B_0 = 0$ and $v = n$ (integer valued), to get general solution:

$$\Phi(\rho, \varphi) = a_0 + b_0 ln\rho + \sum_{n=1}^{\infty} a_n\rho^n \sin(n\varphi + \alpha_n) + \sum_{n=1}^{\infty} b_n\rho^{-n} \sin(n\varphi + \beta_n).$$

(Eqn. 4-129)

Let's apply this to the case of a region in the first quadrant swept by angle φ above the x-axis. Let the boundaries have potential V. Let's evaluate the electric potential in the wedge region where $\nabla^2\Phi = 0$. From the boundary conditions, we can see that $b_n = 0, \quad A = 0, \quad b_0 = B_0 = 0$. Thus, we have:

$$\Phi(\rho, \varphi) = V + \sum_{m=1}^{\infty} a_m \rho^{\frac{m\pi}{\beta}} \sin\left(\frac{m\pi\varphi}{\beta}\right) \quad \simeq V + a_1\rho^{\frac{\pi}{\beta}} \sin\left(\frac{\pi\varphi}{\beta}\right) \; near\; \rho = 0.$$

(Eqn. 4-130)

The electric field components near $\rho = 0$ are then:

$$E_\rho(\rho, \varphi) = -\frac{\partial \Phi}{\partial \rho} = -\frac{\pi a_1}{\beta}\rho^{\frac{\pi}{\beta}-1} \sin\left(\frac{\pi\varphi}{\beta}\right)$$

(Eqn. 4-31)

and

$$E_\varphi(\rho, \varphi) = -\frac{1}{\rho}\frac{\partial \Phi}{\partial \varphi} = -\frac{\pi a_1}{\beta}\rho^{\frac{\pi}{\beta}-1} \cos\left(\frac{\pi\varphi}{\beta}\right).$$

(Eqn. 4-132)

The surface charge density near $\rho = 0$:

$$\sigma(\rho) = \frac{E_\varphi(\rho, 0)}{4\pi} \simeq -\frac{a_1}{4\beta}\rho^{\frac{\pi}{\beta}-1}.$$

(Eqn. 4-133)

Schwarz-Christoffel transform

Discussion – Schwarz-Christoffel transform makes all 2D trivially solvable by transform to separation of variables, solving, then transforming back. Thus, all 2D boundary problems with Laplace easily managed.

Example 4.24. *Consider the 2D potential problem that occurs when two straight parallel line charges are separated by a distance R with equal and opposite linear charge densities λ and $-\lambda$.*

(a) Show that the surface of constant potential V is a circular cylinder and find the coordinates of the axis of the cylinder and its radius in terms of R, λ, and V.

(b) Show that the capacitance per unit length C of two right-circular cylindrical conductors, with radii a and b, separated by d>a+b, is

$$C = \frac{1}{2\cosh^{-1}\left(\frac{d^2 - a^2 - b^2}{2ab}\right)}$$

(c) Verify agreement with (1.7), determine next non vanishing order.

(d) Repeat with the two cylinders one inside the other (verify concentric for result when d=0).

Solution

(a) Start with getting the potential from a single line charge:

$$\oint_S E \cdot \vec{n} d\vec{a} = 4\pi \int_v \rho d^3x = 4\pi Q \;\rightarrow\; E_r 2\pi r L = 4\pi Q$$

Thus,

$$E_r = \frac{2Q}{Lr} = \frac{2\lambda}{r} = -\nabla\Phi = -\frac{\partial\Phi}{\partial r} \;\rightarrow\; \Phi = -2\lambda\int\frac{dr}{r} = -2\lambda ln r + \Phi_0.$$

Using superposition for the potential from the two line charges:

$$\Phi = -2\lambda\{ln\rho' - ln\rho''\} + \Phi_0.$$

Let's now shift to Cartesian Coordinates to identify the circle in transverse dimensions easily:

$$\rho' = \sqrt{r^2 + \left(\frac{R}{2}\right)^2 + rRcos\varphi} = \sqrt{y^2 + \left(x - \frac{R}{2}\right)^2}$$

$$\rho'' = \sqrt{r^2 + \left(\frac{R}{2}\right)^2 + rRcos\varphi} = \sqrt{y^2 + \left(x + \frac{R}{2}\right)^2}$$

Thus

$$\Phi = \lambda ln\left(\frac{y^2 + \left(x + \frac{R}{2}\right)^2}{y^2 + \left(x - \frac{R}{2}\right)^2}\right)$$

Let $\Phi(Surface) = V$ then have:

$$e^{\frac{V}{\lambda}} = k = \frac{y^2+\left(x+\frac{R}{2}\right)^2}{y^2+\left(x-\frac{R}{2}\right)^2} \rightarrow \; y^2 + x^2 + xR\frac{1+k}{1-k} + \frac{R^2}{4} = 0,$$

and upon regrouping we get:

$$y^2 + \left(x + \frac{1+k}{1-k}\frac{R}{2}\right)^2 = \frac{R^2 k}{(1-k)^2}.$$

Thus we have:

$$Center = \frac{R}{2}\frac{e^{\frac{V}{\lambda}}+1}{e^{\frac{V}{\lambda}}-1} = \frac{R}{2}\coth\left(\frac{V}{2\lambda}\right) \quad \text{and} \quad Radius = R\frac{e^{\frac{V}{2\lambda}}}{\left|1-e^{\frac{V}{\lambda}}\right|} = \frac{R}{2\left|\sinh\left(\frac{V}{2\lambda}\right)\right|}.$$

(b) Here we approach the problem by reducing the two cylinder potential problem to a two parallel line charges problem. We approach it this way because we know from (a) that two parallel line charges generate a potential field with cylindrical symmetry. We then use the Radius and Center relations. At radius a on the a-cylinder we have potential V_a:

$$a = \frac{R}{2\left|\sinh\left(\frac{V_a}{2\lambda}\right)\right|}$$

Likewise for the b-cylinder:

$$b = \frac{R}{2\left|\sinh\left(\frac{V_b}{2\lambda}\right)\right|} = -\frac{R}{2\sinh\left(\frac{V_b}{2\lambda}\right)}.$$

If the cylinders have centers above, then the distance between the cylinder centers is

$$d = \frac{R}{2}\left\{\coth\left(\frac{V_a}{2\lambda}\right) - \coth\left(\frac{V_b}{2\lambda}\right)\right\}$$

It can be shown that

$$\frac{(d^2 - a^2 - b^2)}{2ab} = \cosh\left(\frac{V_a - V_b}{2\lambda}\right),$$

and we get from

$$C = \frac{\lambda}{\Delta V} = \frac{\lambda}{V_a - V_b}$$

that

$$C = \frac{1}{2\cosh^{-1}\frac{(d^2 - a^2 - b^2)}{2ab}}$$

as claimed.

(c) Now consider when $d^2 \gg a^2 + b^2$:Type equation here.

$$\cosh^{-1}\frac{(d^2-a^2-b^2)}{2ab} = \ln\left(\frac{d^2}{2ab}\left[1 - \frac{a^2+b^2}{d^2}\right]\left(2 - \frac{1}{2}\left[\frac{4a^2b^2}{d^4\left(1-\left(\frac{a^2+b^2}{d^2}\right)^2\right)} + ..\right]\right)\right)$$

$$= \ln\left(\frac{d^2}{ab}\right) - \left[\frac{a^2 + b^2}{d^2} + \frac{a^2b^2}{d^4} + \frac{1}{2}\left(\frac{a^2 + b^2}{d^2}\right)^2 \dots\right]$$

Thus,

$$C = \frac{1}{4\ln\left(\frac{d}{\sqrt{ab}}\right) - \frac{2(a^2 + b^2)}{d^2} - \frac{a^4 + 4a^2b^2 + b^4}{d^4} + \cdots}$$

(d) Now consider on cylinder inside another. The only change is $b \rightarrow -b$, thus

$$C = \frac{1}{2\cosh^{-1}\frac{(a^2 + b^2 - d^2)}{2ab}}$$

and for $d = 0$:

$$C = \frac{1}{2\cosh^{-1}\frac{(a^2 + b^2)}{2ab}} = \frac{1}{2\ln\left(\frac{b}{a}\right)}$$

Example 4.25. *Consider an insulated, conducting, spherical shell of radius 'a' in a uniform electric field* E_0.
If the sphere is cut into two hemispheres by a plane perpendicular to the field, find the force required to keep the hemispheres from separating when (a) the sphere is uncharged; (b) the sphere has charge Q.

Solution

(a) Let E_0 be in the $\hat{z}$ direction. We've seen that a uniform field near a given axis (here the z-axis) can be obtained with two charges, Q and $-Q$, placed distance $\pm R$ along that axis, relative to the current position (origin) of the uniform field considered. We then take the limit $R \to \infty$ and $Q \to \infty$ such that $E_0 = 2Q/R^2$ (shown next). For a conducting spherical boundary the field is modified by the presence of mirror charges inside the sphere. Using the relation on mirror charges for a sphere we then get for the electric potential:

$$\Phi = \frac{Q}{(r^2+R^2+2rRcos\theta)^{\frac{1}{2}}} - \frac{Q}{(r^2+R^2-2rRcos\theta)^{\frac{1}{2}}} \quad \text{(the charges outside the sphere)}$$

$$-\frac{aQ}{R\left(r^2+\frac{a^4}{R^2}+\frac{2a^2r}{R}cos\theta\right)^{\frac{1}{2}}} + \frac{aQ}{R\left(r^2+\frac{a^4}{R^2}-\frac{2a^2r}{R}cos\theta\right)^{\frac{1}{2}}} \quad \text{(the mirror charges inside the sphere).}$$

Thus, for $R \gg r$, and $R \to \infty$, only the lowest order term survives:

$$\Phi = \left[-\frac{2Q}{R^2}rcos\theta + \frac{2Q}{R^2}\frac{a^3}{r^2}cos\theta\right] + \cdots$$

Using $E_0 = 2Q/R^2$, we then have:

$$\Phi = -E_0\left(r - \frac{a^3}{r^2}\right)\cos\theta$$

The surface charge induced on the shell is given by the "pillbox formula":

$$\sigma = -\frac{1}{4\pi}\frac{\partial\Phi}{\partial r}\bigg|_{r=a} = \frac{3}{4\pi}E_0cos\theta.$$

The z-component of the force for given angle is $F_z = (2\pi\sigma^2)\cos\theta$. The total magnitude of the z-component of the force is given by:

$$|\vec{F}| = \int F_z da = \int_0^{\frac{\pi}{2}} 2\pi\sigma^2\cos\theta(2\pi r)rsin\theta d\theta$$

$$= (2\pi a)^2\left(\frac{3}{4\pi}\right)^2 E_0^2\int_0^{\frac{\pi}{2}}\cos^3\theta\, sin\theta d\theta$$

$$|\vec{F}| = \frac{36}{16}a^2E_0^2\left\{-\frac{\cos^4\theta}{4}\right\}\bigg|_0^{\frac{\pi}{2}} = \frac{9}{16}E_0^2a^2.$$

(b) now suppose the total charge on the shell is Q. In this case we can skip to the surface charge formula and modify it by the addition of $\frac{Q}{4\pi a^2}$ more surface charge: $\sigma_T = \frac{3}{4\pi} E_0 \cos\theta + \frac{Q}{4\pi a^2}$. Thus,

$$|\vec{F}| = \int_0^{\frac{\pi}{2}} 2\pi\left(\frac{3}{4\pi} E_0 cos\theta + \frac{Q}{4\pi a^2}\right)^2 cos\theta(2\pi r)rsin\theta d\theta$$

$$= \frac{q}{16} E_0^2 a^2 + \frac{Q^2}{8a^2} + \frac{1}{2} Q$$

$$\vec{F}_T = \vec{F}_{side\ 1} - \vec{F}_{side\ 2} = \frac{9}{8} E_0^2 a^2 + \frac{1}{4} \frac{Q^2}{a^2}.$$

Example 4.26. *Consider a parallel plate capacitor with separationD, where one place has a small hemispherical boss of radius a on its inner surface.*

The conductor with the boss is kept at zero potential, while the other is at a potential V such that in the far-field region the electric field between the plates appears to be E_0.

(a) Calculate the surface charge density on the plate with the boss (including angular dependence on θ where θ is the angle from the separator line connecting the center of the boss hemisphere to the nearest point on the other plate).

(b) Show the total charge on the boss has magnitude $(3/4)E_0 a^2$.

(c) Replace the conducting sheet at potential V with a charge Q directly above the boss on what was the plate separator line in (b). Let the charge be distance d from the center of the hemisphere. What is the charge induced on the boss?

Solution

(a) The scenario described allows the boundary with hemispherical boss to be replaced, via method of images, with a new parallel conducting plate, with potential $-V$, and with a spherical conductor ($V = 0$) that is the completion of the hemisphere. We, thus, have equivalence to parallel plates $\pm V$ at distance $\pm D$ on either side of a conducting sphere (at $V = 0$). For far-field we are told that the plates at $\pm V$ are equivalent to an electric field E_0 between the plates. We can thus relate the problem to that of a conducting sphere in a uniform electric field (in the $\hat{z}$ direction), where the electric potential is:

$$\Phi = -E_0\left(r - \frac{a^3}{r^2}\right)\cos\theta$$

We have the relations$4\pi\sigma = E$ and $\sigma = -\frac{1}{4\pi}\frac{\partial\Phi}{\partial r}$ to get:

$$\sigma_{plane} = \frac{1}{4\pi}|E_0|\left(1 + 2\frac{a^3}{r^3}\right)$$

For the surface charge density on the boss, at angle θ, we get:

$$\sigma_{boss}(\theta) = -\frac{1}{4\pi}\frac{\partial\Phi}{\partial r}\bigg|_{r=a} = \frac{1}{4\pi}E_0\left(1 + \frac{2a^3}{r^3}\right)cos\theta\bigg|_{r=a}$$
$$= \frac{3}{4\pi}E_0 cos\theta.$$

(b) Total charge on the boss is:

$$Q_{boss} = \int_S \sigma_{boss}\, asin\theta d\theta ad\varphi = \frac{3}{4\pi}E_0 2\pi(a^2)\int_0^{\frac{\pi}{2}} sin\theta cos\theta d\theta = \frac{3E_0 a^2}{4}$$

(c) We now repeat the problem with but instead of the parallel plate at potential V we have a charge Q placed as indicated directly above the outermost point of the boss. Again using the method of images, we eliminate the $V = 0$ conducting plate with a mirror charge, where the hemispherical conductor is completed to a spherical conductor as before. The two charges with the spherical conductor can then be replaced with the two charges and their mirror charges on a spherical conductor boundary as described previously. In fact, the expression for the potential is precisely that obtained previously (prior to taking the charges to infinity such that a uniform electric field is seen locally). Thus, for potential we already have:

$\Phi(\vec{x}) = \frac{Q}{(r^2+R^2+2rRcos\theta)^{\frac{1}{2}}} - \frac{Q}{(r^2+R^2-2rRcos\theta)^{\frac{1}{2}}}$ (original charge and first mirror charge off of plane mirror)

$+\frac{aQ}{R\left(r^2+\frac{a^4}{R^2}-\frac{2a^2r}{R}cos\theta\right)^{\frac{1}{2}}} - \frac{aQ}{R\left(r^2+\frac{a^4}{R^2}+\frac{2a^2r}{R}cos\theta\right)^{\frac{1}{2}}}$ (two mirror charges, respective to two above, off of sphere).

Recall that

$$4\pi\sigma_{boss} = -\nabla\Phi\bigg|_{r=a} = -\frac{\partial\Phi}{\partial r}\bigg|_{r=a}$$

Thus,

$$\sigma_{boss} = -\frac{1}{4\pi}\left\{\frac{-q(r + dcos\theta)(+1)}{(r^2 + d^2 + 2rdcos\theta)^{\frac{3}{2}}} + \frac{-q(r - dcos\theta(+1)}{(r^2 + d^2 - 2rdcos\theta)^{\frac{3}{2}}}\right\}\Bigg|_{r = a}$$

$$+\left\{\frac{aq}{d}\frac{\left(r - \frac{a^2}{d}\cos\theta\right)(-1)}{\left(r^2 + \frac{a^4}{d^2} - \frac{2a^2r}{d}cos\theta\right)^{\frac{3}{2}}} - \frac{aq}{d}\frac{\left(r + \frac{a^2}{d}\cos\theta\right)(-1)}{\left(r^2 + \frac{a^4}{d^2} + \frac{2a^2r}{d}cos\theta\right)^{\frac{3}{2}}}\right\}\Bigg|_{r = a}$$

and,

$$\sigma_{boss} = -\frac{1}{4\pi}\left\{\begin{array}{c}\dfrac{-q(a + dcos\theta)}{(a^2 + d^2 + 2adcos\theta)^{\frac{3}{2}}} + \dfrac{aq\left(1 + \frac{a}{d}\cos\theta\right)\left(\frac{d}{a}\right)^2}{(a^2 + d^2 + 2adcos\theta)^{\frac{3}{2}}} \\ + \dfrac{q(a - dcos\theta)}{(a^2 + d^2 - 2adcos\theta)^{\frac{3}{2}}} - \dfrac{aq\left(1 + \frac{a}{d}\cos\theta\right)\left(\frac{d}{a}\right)^2}{(a^2 + d^2 - 2adcos\theta)^{\frac{3}{2}}}\end{array}\right\}$$

which simplifies to:

$$\sigma_{boss} = \frac{1}{4\pi}\frac{q(a^2 - d^2)}{a}\left\{\frac{1}{(a^2 + d^2 + 2adcos\theta)^{\frac{3}{2}}} - \frac{1}{(a^2 + d^2 - 2adcos\theta)^{\frac{3}{2}}}\right\}$$

The total charge induced on the boss is then:

$$Q = \int_0^{\frac{\pi}{2}}\int_0^{2\pi} asin\theta d\theta ad\varphi\sigma = 2\pi a^2\int_0^{\frac{\pi}{2}}\sigma sin\theta d\theta$$

$$Q = 2\pi a^2\int_0^{\frac{\pi}{2}}\frac{1}{4\pi}\frac{q(a^2 - d^2)}{a}\left\{\frac{d(cos\theta)}{(a^2 + d^2 - 2adcos\theta)^{\frac{3}{2}}} - \frac{d(cos\theta)}{(a^2 + d^2 + 2adcos\theta)^{\frac{3}{2}}}\right\}$$

Thus,

$$Q = -q\left\{1 - \frac{d^2 - a^2}{d\sqrt{d^2 + a^2}}\right\}.$$

Example 4.27. *A line charge with a linear charge density λ' is placed parallel to, and a distance R away from, the axis of a conducting cylinder of radius b held at a fixed voltage such that the potential vanishes at infinity.*
Find:
(a) the magnitude and position of the image charge(s).;
(b) The potential at any point, expressed in polar coordinates, where the line from the center axis to the line charge is the positive x-axis direction;
(c) the induced surface charge density;
(d) the force on the charge.

Solution
To solve using the method of images we want to place an image linear charge density $\lambda'' = \alpha\lambda$ parallel to the λ' linear charge density at position inside the cylinder in the positive x-axis direct line formed by the perpendicular to the cylinder axis (origin) and to the (outside) linear charge axis.

We know the potential due to a linear charge density goes as radial distance from the line charge. In the relevant 2D polar coordinates $\{\theta, \rho\}$ for line charge at $\vec{\rho}'$ and observer at position $\vec{\rho}$ we have potential:

$$\Phi(\vec{\rho}) = -2\lambda'|\vec{\rho} - \vec{\rho}'| + C'.$$

And, for the mirror line charge will be at position $\vec{\rho}''$ with charge density $\lambda'' = \alpha\lambda$:

$$\Phi(\vec{\rho}) = -2\lambda''|\vec{\rho} - \vec{\rho}''| + C''.$$

The total potential in the cylinder exterior is thus:

$$\Phi(\vec{\rho}) = -2\lambda\, ln(|\vec{\rho} - \vec{\rho}'||\vec{\rho} - \vec{\rho}''|^{\alpha}) + \Phi_0.$$

For $\Phi(\vec{\rho})$ to remain finite at infinity (not to mention zero) we must have $\alpha = -1$:

$$\Phi(\vec{\rho}) = -2\lambda ln\left(\frac{|\vec{\rho} - \vec{\rho}'|}{|\vec{\rho} - \vec{\rho}''|}\right) + \Phi_0.$$

So, we've obtained the magnitude of the line charge, now let's determine the position of the line charge. This is done by satisfying the zero-potential at the conducting surface constraint: $\Phi(\rho = b) = 0$.
When $|\vec{\rho}| = b$ we can write $|\vec{\rho} - \vec{\rho}'| = \beta|\vec{\rho} - \vec{\rho}''|$. Writing in terms of the magnitudes (along the x-axis) we have two points where x-axis intercepts the cylinder, providing two relations:

$$(\rho' - b) = \beta(b - \rho'')\ and\ (\rho' + b) = \beta(\rho'' + b)$$

These can be grouped to give $\rho'' = \frac{b^2}{\rho'}$ for the position of the line charge and to solve for β:

$$\beta = \frac{\rho'}{b}.$$

(b) Thus, at the cylinder surface, we have:

$$\Phi(\rho = b) = \lambda ln\left\{\frac{|\vec{\rho} - \vec{\rho}''|^2}{|\vec{\rho} - \vec{\rho}'|^2}\right\} + \Phi_0 = \lambda ln\left\{\left(\frac{b}{\rho'}\right)^2\right\} + \Phi_0 = 0 \quad \rightarrow \quad \Phi_0$$
$$= -\lambda ln\left(\frac{b}{\rho'}\right)^2$$

The potential at any point is thus:

$$\Phi(\vec{\rho}) = \lambda ln\left[\frac{\rho^2 + \left(\frac{b^4}{\rho'^2}\right) - 2\rho\left(\frac{b^2}{\rho'}\right)cos\theta}{(\rho^2 + \rho'^2 - 2\rho\rho'\cos\theta)}\left(\frac{\rho'}{b}\right)^2\right]$$
$$= \lambda ln\left[\frac{\rho^2\rho'^2 - 2b^2\rho Rcos\theta + b^4}{b^2(\rho^2 - 2\rho R\cos\ \theta + R^2}\right]$$

Comparing with earlier derivations we have agreement:

$$G_D(\vec{\rho}, \vec{\rho'}) = \frac{\Phi(\vec{\rho})_{\lambda\ at\ \vec{\rho}'}}{\lambda} = ln\left[\frac{\rho^2\rho'^2 - 2b^2\rho\rho' cos\theta + b^4}{b^2(\rho^2 - 2\rho\rho'\cos\ \theta + \rho'^2}\right]$$

(c) For the induced surface charge density we have:

$$4\pi\sigma = -\nabla\Phi\Big|_{\rho = b} = -\frac{\partial}{\partial\rho}\left\{\lambda\ ln\left[\frac{\rho^2\rho'^2 - 2b^2\rho Rcos\theta + b^4}{b^2(\rho^2 - 2\rho R\cos\ \theta + R^2}\right]\right\}\Bigg|_{\rho = b}$$
$$= -\tau\left\{\frac{2\rho R^2 - 2b^2 R\cos\theta}{\rho^2 R^2 - 2b^2\rho R\cos\theta + b^4} - \frac{2\rho - 2R\cos\theta}{(\rho^2 - 2\rho R\cos\theta + R^2)}\right\}\Bigg|_{\rho=b}$$

$$\sigma = \frac{-\tau}{4\pi}\left\{\frac{2bR(R - b\cos\theta)}{b^2(R^2 + b^2 - 2bR\cos\theta)} - \frac{2(b - R\cos\theta)}{(R^2 + b^2 - 2bR\cos\theta)}\right\}$$

$$\sigma = \frac{-\tau}{2\pi b}\left[\frac{\left(1 - {b^2}/{R^2}\right)}{{b^2}/{R^2}\left(1 + {R^2}/{b^2} - 2(R/b)\cos\theta\right)}\right]$$

$R/b = 2 \qquad \sigma_2 = \frac{-\tau}{2\pi b}\left[\frac{3}{(5 - 4\cos\theta)}\right]$

$R/b = 4 \qquad \sigma_4 = \frac{-\tau}{2\pi b}\left[\frac{15}{(17 - 8\cos\theta)}\right]$

$\vec{x} = bR\frac{(b - R)^2}{(R^2 - b^2)^2}$

(d) $\vec{F} = \tau\vec{E}_{x=b} = \frac{+2\tau^2}{b}\left(\frac{\left(1 - {b^2}/{R^2}\right)}{\left(\frac{b^2}{R^2}\right)\left(1 - {R}/{b}\right)^2}\right) = \frac{2\tau^2}{b}\left(\frac{(R^2 - b^2)}{(b - R)^2}\right)$

Example 4.28. *Suppose you are told the potential on the surface of a cylinder and asked to do a series solution.*

Solution

By series solution:

$$\Phi(\rho, \phi) = a_0 + b_0 \ln\rho + \sum_{n=1}^{\infty} a_n \rho^n \sin(n\phi + a_n)$$

$$+ \sum_{n=1}^{\infty} b_n \rho^{-n} \sin(n\phi + \beta_n)$$

Since the origin is included in the region of consideration, all the b_n are zero.

$\Phi(\rho, \phi) = a_0 + \sum_{n=1}^{\infty} a_n \rho^n \sin(n\phi + \alpha_n) \qquad \leftarrow \qquad A_m =$
$-A_{-m} = \frac{1}{2i} a_m$
$= \sum_{m=-\infty}^{\infty} A_m \rho^{|m|} e^{im\phi}$

$A_m b^{|m|} = \frac{1}{2\pi}\int_0^{2\pi} \Phi\,(b, \phi')e^{im\phi'}d\phi'$ via orthogonality.

So,

$$\Phi(\rho,\phi) = \frac{1}{2\pi}\int_0^{2\pi} d\phi'\Phi\,(b,\phi') \sum_{m=-\infty}^{\infty} \left(\frac{\rho}{b}\right)^{|m|} e^{im(\phi-\phi')}$$

$$= \frac{1}{2\pi}\int_0^{2\pi} d\phi'\Phi\,(b,\phi')\left\{1 + \frac{(\rho/b)e^{i(\phi-\phi')}}{1-(\rho/b)e^{i(\phi-\phi')}} + \frac{(\rho/b)e^{i(\phi-\phi')}}{1-(\rho/b)e^{i(\phi-\phi')}}\right\}$$

Thus,

$$\Phi(\rho,\phi) = \frac{1}{2\pi}\int_0^{2\pi} d\phi'\Phi\,(b,\phi')\left\{\frac{b^2-\rho^2}{b^2+\rho^2-2b\rho\cos(\phi-\phi')}\right\}$$

(when $\rho > b$ the potential changes sign).

Now let's solve (2.8) via Green's functions:

$$\Phi = \Phi_1 + \Phi_2 = -2[\tau\ln\rho + \tau'\ln\rho'] + \Phi_0$$

with conditions: (i) $0 < R' < b$; (ii)$\Phi_0 = \text{const}$; (iii)$\tau = -\tau'$; (iv) $R'/_b = b/_R$; and (v) $\rho'/\rho = R/b$.

So, $\Phi_C = 2\tau\ln\left(\frac{R}{b}\right)$ ← grounded surface

Thus,

$\Phi = 2\tau\ln\left(\frac{\rho' R}{\rho b}\right)$ and $G(x,x')$ is then $G = 2\ln\left(\frac{\rho' R}{\rho b}\right)$. For the problem at hand let $\Phi_0' = 2\tau\ln k$ at the surface. Then,

$$\Phi - \tau G(x,x') + \Phi_0' = 2\tau\ln\left(\frac{\rho' R}{\rho b}k\right),$$

and

$$\frac{\rho' R}{\rho b} = \frac{R}{b}\,\frac{\sqrt{r^2 + b^4/R^2 - 2rb^2/R\cos\theta}}{\sqrt{r^2+R^2-2rR\cos\theta}}.$$

Since we have $\rho(x) = 0$ and Dirchlet boundary conditions:

$$\Phi(x) = -\frac{1}{4\pi}\oint \Phi\,(x)\frac{\partial G}{\partial n'}da', \qquad \frac{\partial G}{\partial n'} = \vec{n}'\cdot\nabla G = \left.\frac{\partial G}{\partial R}\right|_{R=b}.$$

Thus,

$$\left.\frac{\partial G}{\partial R}\right|_{R=b'} = \frac{-2(b^2-r^2)}{b(r^2+b^2-2rb\cos\theta)'}$$

and

$$\Phi(r,\theta) = \frac{1}{2\pi}\int_0^{2\pi} \Phi\,(b,\theta')\left\{\frac{(b^2-r^2)}{r^2+b^2-2rb\cos(\theta-\theta')}\right\}d\theta',$$

the results are the same.

Example 4.29. *A long hollow cylinder is split lengthwise by a gap. The halves are maintained at different potentials: V_1 and V_2, the cylinder radius is 'b'*
(a) show the potential inside;
(b) calculate the surface charge densities.

Solution
Using series solution or Poisson's integral equally messy but at the integrals are straightforward, so let's use that:

$$\Phi(\rho,\phi) = \frac{1}{2\pi}\int_0^{2\pi} \Phi\,(b,\phi')\left\{\frac{b^2-\rho^2}{b^2+\rho^2-2b\rho\cos(\phi'-\phi)}\right\}d\phi'.$$

So,

$$\Phi = \frac{V_1}{2\pi}\int_{-\pi/2}^{2/\pi}\frac{(b^2-r^2)d\theta'}{(r^2+b^2-2br\cos(\theta'-\theta)}$$

$$+\frac{V_2}{2\pi}\int_{\pi/2}^{-\pi/2}\frac{(b^2-r^2)d\theta'}{(r^2+b^2-2br\cos(\theta'-\theta))}$$

$$=\frac{V_1}{2\pi}\int_0^{\pi}\frac{(b^2-r^2)d\theta'}{(r^2+b^2)-2br\sin(\theta'-\theta)}$$

$$+\frac{V_2}{2\pi}\int_0^{\pi}\frac{(b^2-r^2)d\theta'}{(r^2+b^2)-(2br\sin(\theta'-\theta))}$$

Use $\int\frac{dx}{a+b\sin x}=\frac{2}{\sqrt{a^2+b^2}}\tan^{-1}\left\{\frac{a\tan^{x}/_{2}+b}{\sqrt{a^2+b^2}}\right\}$, $\tan^{-1}x+\tan^{-1}y=\tan^{-1}\left(\frac{x-y}{1-x+y}\right)$:

So, the 1st integral becomes:

$$\frac{V_1(b^2-r^2)}{2\pi}\int_0^{\pi}\frac{d\theta'}{(r^2+b^2)-2br\sin(\theta'-\theta)}$$

$$= \frac{V_1}{2\pi}(b^2 - r^2)\frac{2}{(b^2-r^2)}\left\{\tan^{-1}\left[\frac{(b^2-r^2)\tan(\pi/_2-\theta/_2)-2br}{b^2-r^2}\right] + \tan^{-1}\left[\frac{(b^2-r^2)\tan(-\theta/_2)+2br}{b^2-r^2}\right]\right\}$$

$$= \frac{V_1}{\pi}\tan^{-1}\{(b^2 - r^2)(b^2 + r^2)\left[\cot\left(\frac{\theta}{2}\right) + \tan\left(\frac{\theta}{2}\right)\right]$$

etc.

(b) $\quad \sigma(\phi) = -\frac{1}{4\pi}\frac{\partial\Phi}{\partial n}\Big|_{\rho=b}$ *and* $\vec{n} = -\hat{\rho} \rightarrow \sigma(\phi) = \frac{1}{4\pi}\frac{\partial\Phi}{\partial\rho}\Big|_{\rho=b}$.

Example 4.30. *A long hollow cylinder is split lengthwise into quarters by gaps. The quarters are maintained at alternating potentials: $+V$ and $-V$, with the cylinder radius equal to 'b'*

Solution

We have $\nabla^2\phi = 0$ inside the cylinder, e.g., Laplace's equation in cylindrical coordinates with cylindrical symmetry reducing to polar coordinates in 2D (with no charge inside). For polar coordinates:

$ds^2 = d\rho^2 + \rho^2 d\phi^2 \rightarrow |g| = \rho^2$ and $g_{\rho\rho} = 1, \rho_{\phi\phi} = \rho^2 \rightarrow g^{\phi\phi} = \frac{1}{\rho^2}$, and the rest zero. Thus:

$$\nabla^2\Phi = \frac{1}{|g|^{1/2}}\partial_\mu\left(|g|^{1/2}g^{\mu\nu}\partial_\nu\right) = 0,$$

with $\Phi(\rho,\phi) = R(\rho)\Psi(\phi)$ leads to:

$$\frac{1}{\rho}\partial_r(\rho\partial_r\Phi) + \frac{1}{\rho^2}\partial_\phi^2\Phi = 0$$

$$\frac{\rho}{R}\partial_r(\rho\partial_r R) + \frac{1}{\Psi}\partial_\phi^2\Psi = 0$$

Note that $\Psi = ae^{iv\phi}$ indicates that v is integer valued otherwise Ψ is not single valued, and $e^{iv(2\pi)} = 1$. Thus, have the solutions:

$$R = a\rho^v + b\rho^{-v}, \quad v \neq 0,$$

and, for $v = 0$ solution:

$$R = a_0 + b_0 \ln \rho \,, \Psi = A_0 + B_0\phi.$$

So,

$$\Phi(\rho,\phi) = a_0 + b_0 \ln \rho + \sum_{n=1}^{\infty} a_n\, \rho^n \sin(n\phi + \alpha_n)$$
$$+ \sum_{n=1}^{\infty} b_n\, \rho^{-n} \sin(n\phi + \beta_n) \quad v = \text{integer}$$

The region under consideration for the problem includes the origin, so solutions (finite) require $b's = 0$,

$$\Phi(\rho,\phi) = \sum_{\mathrm{m}=-\infty}^{\infty} A_m\, \rho^{|m|} e^{im\phi} \quad \rightarrow \quad \Phi(b,\phi) = \sum_{\mathrm{m}=-\infty}^{\infty} A_m\, \rho^{|m|} e^{im\phi}$$

↘

$A_m(b)^{|m|} = \frac{1}{2\pi}\int_0^{2\pi} \Phi(b,\phi')\, e^{-im\phi'} d\phi'$

$\frac{1}{2\pi}\int_0^{2\pi} \Phi(b,\phi')\, e^{-im\phi'} d\phi' = \sum_{\mathrm{m}=-\infty}^{\infty} A_m\, b^{|n|} \int_0^{2\pi} e^{i(n-m)}$

$= \sum_m A_m\, b^{|n|} \delta_{nm}$

So,

$= A_m b^{|n|}$ relable n →m

$$\Phi(\rho,\phi) = \sum_{\mathrm{m}=-\infty}^{\infty} \left(\frac{\rho}{b}\right)^{|\mathrm{m}|} \frac{1}{2\pi}\int_0^{2\pi} \Phi(b,\phi')\, e^{im(\phi-\phi')} d\phi'.$$

$$\Phi(\rho,\phi) = \frac{1}{2\pi}\int_0^{2\pi} d\phi' \Phi(b,\phi') \sum_{\mathrm{m}=-\infty}^{\infty} \left(\frac{\rho}{b}\right)^{|\mathrm{m}|} e^{im(\phi-\phi')}.$$

Now let's get the A_m:

$$A_m = \frac{1}{2\pi b^{|m|}}\int_0^{2\pi} \Phi(b,\phi')\, e^{-im\phi'} d\phi'$$

$= \frac{-V}{2\pi b^{|m|}}\left\{\int_{-\pi/2}^{0} e^{-im\phi'}\, d\phi' - \int_0^{\pi/2} e^{-im\phi'}\, d\phi' + \int_{\pi/2}^{\pi} e^{-im\phi'}\, d\phi' - \int_{\pi}^{3\pi/2} e^{-im\phi'}\, d\phi'\right\}$

$$= \frac{-V}{2\pi b^{|m|}} \left\{ e^{im\frac{\pi}{2}} \int_0^{\pi/2} e^{-im\phi'}\, d\phi' - \int_0^{\pi/2} e^{-im\phi'}\, d\phi' \right.$$

$$\left. + e^{-im\frac{\pi}{2}} \int_0^{\pi/2} e^{-im\phi'}\, d\phi' - e^{-im\pi} \int_0^{\pi/2} e^{-im\phi'}\, d\phi' \right\}$$

$$= \frac{-V}{2\pi b^{|m|}} \left[\cos\left(\frac{m\pi}{2}\right) - \left(1 + e^{-im\pi}\right)\right] \int_0^{\pi/2} e^{-im\phi'}\, d\phi'$$

$$= \frac{-V}{2\pi b^{|m|}} \left[e^{im\frac{\pi}{2}} + e^{-im\frac{\pi}{2}} - 1 - e^{-im\pi}\right] \frac{\left(e^{-im\frac{\pi}{2}-1}\right)}{(-im)}$$

$$= \frac{-V}{2\pi b^{|m|}} \left\{1 + e^{-im\pi} - e^{-im\frac{\pi}{2}} - e^{-im\frac{\pi}{2}} - e^{-im\frac{\pi}{2}} - e^{-im\frac{\pi}{2}} + e^{-im\pi}\right\}$$

$$= \frac{-2V}{2\pi b^{|m|}} \left\{1 + e^{-im\pi} - e^{-im\frac{\pi}{2}} - e^{-im\frac{\pi}{2}}\right\}$$

$\left(e^{-im\frac{\pi}{2}} + e^{im\frac{\pi}{2}}\right) \sim \cos\left(m\frac{\pi}{2}\right) \sim 0$ for m odd $\qquad m = 2(2n+1)$

± 1 for m even

$\left(e^{-im\frac{\pi}{2}} - 1\right) \sim 0$ for $m = 4n \to \underline{m = 2(2n+1)}$

$\searrow -2$ for $m = 2(2n+1)$

$$= \frac{2V}{2\pi b^{|m|}(im)} 2\cos\left(m\frac{\pi}{2}\right) e^{-im\frac{\pi}{4}} \left(e^{-im\frac{\pi}{4}} - e^{-im\frac{\pi}{4}}\right)$$

$$= \frac{2V}{2\pi b^{|m|}(m)} 2\cos\left(m\frac{\pi}{2}\right) e^{-im\frac{\pi}{4}} \left(-2\sin\left(m\frac{\pi}{4}\right)\right)$$

$\searrow m$ even $\qquad \searrow m = (2n+1)2$

$A_m = \frac{+4V}{im\pi b^{|m|}}$ *for* $m = 2(2n+1)$, $= 0$ otherwise.

Thus, we can answer with the following:

(a) $\Phi(\rho, \phi) = \sum_{m=-\infty}^{\infty} \left(\frac{4V}{im\pi b^{|m|}}\right) \rho^{|m|} e^{im\phi}$ for $m = 2(2n+1)$, e.g., $m = 2,6,10$.

$$= \left(\frac{4V}{\pi}\right)\left[\left(\frac{\rho}{b}\right)^2 \sin(2\phi) + \left(\frac{\rho}{b}\right)^6 \frac{\sin(6\phi)}{3} + \left(\frac{\rho}{b}\right)^{10} \frac{\sin(10\phi)}{5} + \cdots\right]$$

$$= \frac{4V}{\pi} \sum_{n=0}^{\infty} \left(\frac{\rho}{b}\right)^{4n+2} \frac{\sin[(4n+2)\phi]}{(2n+1)}$$

(b) Sum series:

$$\Phi(\rho,\phi) = \left(\frac{4V}{\pi}\right) \times \left\{ \sum_{n=0}^{\infty} \frac{\left(\frac{\rho}{b}\right)^{4n+2} e^{i(4n+2)\phi}}{2i(2n+1)} + C.C. \right\}$$

$$= \left(\frac{4V}{\pi}\right)\frac{1}{2i}\left\{\sum_{n=0}^{\infty} \frac{\left(\frac{\rho}{b}e^{i\phi}\right)^{4n+2}}{(2n+1)} - C.C.\right\} \quad let\ z = \left(\left(\frac{\rho}{b}\right)^2 e^{2i\phi}\right)$$

$$= \frac{1}{2i}\left(\frac{4V}{\pi}\right)\left[\sum_{n=0}^{\infty}\left(\frac{z^{2n+1}}{2n+1} - \frac{z^{*2n+1}}{2n+1}\right)\right]$$

$$= \frac{V}{i\pi}\left[\ln\left(\frac{1+z}{1-z}\right) - \ln\left(\frac{1+z^*}{1-z^*}\right)\right] = \frac{V}{i\pi}\ln\left\{\frac{1+(z-z^*)-|z|^2}{1-(z-z^*)-|z|^2}\right\}$$

$$= \frac{Vi}{\pi}\ln\left\{\frac{1-2i\left(\frac{\rho}{b}\right)^2 \sin 2\phi - \left(\frac{\rho}{b}\right)^4}{1+2i\left(\frac{\rho}{b}\right)^2 \sin 2\phi - \left(\frac{\rho}{b}\right)^4}\right\}$$

$$= \frac{2V}{\pi}\frac{i}{\pi}\ln\left[\frac{1-i\left(\frac{2b^2\rho^2 \sin 2\phi}{b^4-\rho^4}\right)}{1+i\left(\frac{2b^2\rho^2 \sin 2\phi}{b^4-\rho^4}\right)}\right]$$

$$= \frac{2V}{\pi}\tan^{-1}\left(\frac{2\rho^2 b^2 \sin 2\phi}{b^4-\rho^4}\right)$$

Example 4.31. *Obtain the two dimensional Dirichlet Green's function for the exterior of a cylinder of radius b.*

Solution

Standard 3-d Green's formula for Dirichlet cases:

$$\Phi(\vec{x}) = \int_V \rho(\vec{x}')\, G_D(\vec{x},\vec{x}')d^2x' - \frac{1}{4\pi}\oint_S \Phi(\vec{x}')\frac{\partial G_D(\vec{x},\vec{x}')}{\partial n'}da'$$

To calculate the capacitance let one of the cylinders have $\Phi = 0$ and put charge Q on the other. Thus, $\Phi = 0$ on $a_1 \Longrightarrow$ no "surface" term, assume unform line charge density on a slice: $\frac{\lambda}{2\pi a_2}$ .Then,

$$\Phi(\vec{x}) = \int \left(\frac{\lambda}{2\pi a_2}\right) G_D(\vec{x},\vec{x}')d\ell' = \frac{\lambda}{2\pi a_2}\int_0^{2\pi} G_D(\vec{\rho},\vec{\rho}')\, G_2 d\phi'$$

using Jackson's 2nd form $G_D = \ln\left[\frac{(\rho^2-b^2)(\rho'^2-b^2)+b^2a_2^2}{b^2a_2^2}\right]$, and setting $\rho = d \approx \rho', d \gg a_1, a_2$, simplifies to:

$$G_D \cong \ln\left(\frac{d4}{a_1^2 a_2^2}\right).$$

Thus,

$$\Phi(\text{center}) \approx \frac{\lambda}{2\pi}\ln\left(\frac{d^2}{a_1 a_2}\right)^2 \int_0^{2\pi} d\phi' = 4\lambda \ln\left(\frac{d}{a}\right) \qquad a = \sqrt{a_1 a_2}$$

Thus, $C = \frac{Q/L}{V} = \frac{\lambda}{V} \simeq \frac{1}{4\ln\left(\frac{d}{a}\right)}$ same as 1.7.

(c) If $\vec{\rho}$ is inside, $\vec{\rho}'$ and $\vec{\rho}''$ are swapped and τ and τ' swapped, this introduces two negative signs so the same formula works both interior and exterior.

Example 4.32. (a) Use **Example 4.31** and obtain the Poisson Integral form of the Dirichlet problem in a circle. (b) Obtain the solution using Cauchy's theorem instead.

(a) Exactly like (**Example 4.28**), (**Example 4.31**) radius b:

$$G(\rho,\phi,\rho',\phi') = \ln\left\{\frac{b^4 + \rho^2\rho'^2 - 2b^2\rho\rho'\cos(\phi-\phi')}{b^2\left[\rho^2 + \rho'^2 - 2\rho\rho'\cos(\phi-\phi')\right]}\right\}$$

$$\left.\frac{\partial G}{\partial \rho'}\right|_{\rho'=a} = \frac{2(b^2-\rho^2)}{b[b^2+\rho^2-2b\rho\cos(\phi-\phi')]}$$

Thus,

$$\Phi(\rho,\phi) = -\frac{1}{4\pi}\int_0^{2\pi} b\,d\phi'\,\Phi(b,\phi')\frac{\partial G(\rho,\phi,b,\phi')}{\partial n}$$

$$\Phi(\rho,\phi) = \frac{1}{2\pi}\int_0^{2\pi} d\phi'\,\Phi(b,\phi')\frac{(a^2-\rho^2)}{a^2+\rho^2-2a\rho\cos(\phi-\phi')}.$$

(b) Use Cauchy's theorem to derive the Poisson integral solution:
Cauchy's Theorem:

$$F(z) = \frac{1}{2\pi i}\oint_C \frac{F(z')dz'}{z'-z},$$

for z inside R (note, $F(z) = 0$ for z outside curve).

$$F(z) = \frac{1}{2\pi i}\oint_C F(z')\left\{\frac{1}{z'-z} - \frac{1}{z' - \left(\frac{b^2}{z^*}\right)}\right\}dz'$$

$$= \frac{1}{2\pi}\int_0^{2\pi} F(z')\left\{\frac{1}{be^{i\phi'} - \rho e^{i\phi}} - \frac{1}{be^{i\phi'} - \frac{b^2}{\rho}e^{i\phi}}\right\} be^{i\phi'}d\phi'$$

$$= \frac{1}{2\pi}(b^2 - \rho^2)\int_0^{2\pi} \frac{[U(x',y') + iV(x',y')]}{b^2 - \rho^2 - 2b\rho\cos(\phi - \phi')} d\phi'.$$

We then take real part to get Poisson's integral. Shift origin to take advantage of symmetry, now have:
$\Phi = 0$ for $x = \pm\frac{a}{2}, y = \pm\frac{a}{2}, \Phi = \mathrm{V}$ for $z = \pm\frac{a}{2}$:

(a) B.C.'s are now parity even → cos and cosh function

$$\Phi_{n,m} = \cos\left(\frac{n\pi x}{a}\right)\cos\left(\frac{m\pi y}{a}\right)\cosh\left(\frac{\pi}{a}\sqrt{n^2 + m^2}z\right)$$

Where bold *n, m* are odd numbers.

$$\Phi = \sum_{n,m \text{ odd}} A_{n,m} = \cos\left(\frac{n\pi x}{a}\right)\cos\left(\frac{m\pi y}{a}\right)\cosh\left(\frac{\pi}{a}\sqrt{n^2 + m^2}z\right)$$

We know $\Phi\left(x, y, \frac{a}{2}\right) = \mathrm{V} = \sum_{\substack{n,m \\ \text{odd}}} A_{n,m} =$
$\cosh\left(\frac{\pi}{2}\sqrt{n^2 + m^2}z\right)\cos\left(\frac{n\pi x}{a}\right)\cos\left(\frac{m\pi y}{a}\right)$

So, $A_{n'm'} = \frac{4V}{a^2\cosh\left(\frac{\pi}{2}\sqrt{n^2+m^2}\right)}\int_{-a/2}^{a/2}\cos\left(\frac{n'\pi x}{2}\right)dx\int_{-a/2}^{a/2}\cos\left(\frac{m'\pi y}{2}\right)dy$

$$= \frac{4V}{\pi^2 n'm'\cosh\left(\frac{\pi}{2}\sqrt{n^2+m^2}\right)}\left(2\sin\left(\frac{n'\pi}{2}\right)\right)\left(2\sin\left(\frac{m'\pi}{2}\right)\right)$$

$$= \frac{16V}{\pi^2 n'm'\cosh\left(\frac{\pi}{2}\sqrt{n'^2+m'^2}\right)}(-1)^{\frac{n'-1}{2}}(-1)^{(m'-1)/2}$$

$$\Phi(x, y, z) = \frac{16\mathrm{V}}{\pi^2}$$

$$= \sum_{\substack{n,m \\ \text{odd}}} \frac{(-1)^{(n+m-2)/2}}{nm\cosh\left(\frac{\pi}{2}\sqrt{n^2 + m^2}\right)}\cos\left(\frac{n\pi x}{a}\right)\cos\left(\frac{m\pi y}{a}\right)\cosh\left(\frac{\pi z}{a}\sqrt{n^2 + m^2}\right)$$

(b) $\Phi(0,0,0) = \frac{16\mathrm{V}}{\pi^2}\left\{\frac{1}{\cosh\left(\frac{\pi}{2}\sqrt{2}\right)} - \frac{2}{3\cosh\left(\frac{\pi}{2}\sqrt{10}\right)} + \frac{1}{9\cosh\left(\frac{\pi}{2}\sqrt{18}\right)} + \frac{2}{5\cosh\left(\frac{\pi}{2}\sqrt{26}\right)}\right\}$

$$= 0.33339V \simeq \frac{1}{3}V$$

(c) $\sigma\left(x, y, \frac{a}{2}\right) = -\frac{1}{4\pi}\frac{\partial \Phi}{\partial n}\bigg|_{z=\frac{a}{2}} = \frac{1}{4\pi}\frac{\partial \Phi}{\partial z}\bigg|_{z=\frac{a}{2}}$

4.6 Exercises

(Ex. 4.1) Re-derive Eqn.s 4-46.
(Ex. 4.2) Re-derive Eqn. 4-50.
(Ex. 4.3) Re-derive Eqn. 4-74.
(Ex. 4.4) Re-derive Eqn. 4-84.
(Ex. 4.5) Re-derive Eqn. 4-86.
(Ex. 4.6) Verify Eqn. 4-108.
(Ex. 4.7) Verify Eqn. 4-114.
(Ex. 4.8) Verify Eqn. 4-121.
(Ex. 4.9) Verify Eqn. 4-124.
(Ex. 4.10) Verify Eqn. 4-130.

Chapter 5. Fields – Laplace's Equation in 3D

In Chapters 4 thru 7 classic electromagnetism (EM) is covered, with extensive coverage on the Lorentz transform (EM is Lorentz invariant) in Ch. 8, which then leads to special relativity and general relativity and other topics in later chapters. This chapter (Ch. 5), is entirely focused on solving a class of ordinary differential equations (ODEs), known as the 3D Laplacian, that describes many of the electrostatic systems indicated in the preceding chapter. This will bring forth some of the main techniques for solving ODEs in general, thus the chapter is complemented by the material on advanced ODE's in App. D (and introductory ODEs in Book 1 [1] Appendix A). We will see the role of choice of coordinates and of symmetries; we will see the fully detailed solution of the classic Legendre and Associated Legendre ODEs; and we will see Green's function methods. For Ch. 5 there is, once again, use of notes and conventions from Jackson [32], however, there is also further details on ODEs from [43], especially in App. D, and from related coursework in applied Mathematics at Caltech (see problems and solutions in App. D). The physically motivated methods for solutions (method of mirrors, asymptotics) reveals that many of the methods to solve ODE's were developed by physicists. As mentioned previously (especially in Book 1 [1]), a physics solution often reduces to solving an ODE, and the set of such ODEs relevant to physics is limited (and most will be covered in this Series). Being able to solve the Laplacian class of ODE's is critical, and of these we only really need the 3D case. This is because we are generally working with 3 *spatial* dimensions, no more, and if symmetry allows us to reduce from 3D to 2D, then Sec. 2.2, App. A show that use of the Schwarz-Christoffel transform allows the most complex boundary conditions to be transformed into something trivial. So we need only focus on Laplacians in 3D to complete our understanding of this branch of ODE's relevant to physics.

Note on Examples and Exercises

The chapters on electromagnetism draw, foremost, from the excellent source, Classical Electrodynamics by J.D. Jackson [32], that was the required text for many years at Caltech. Many of the Jackson problems develop the physics theory further (e.g., Gauss's Theorem) or further develop the mathematical techniques (for ODEs, definition of delta function as limit, etc.). With my own notes from the courses at Caltech, and upon later preliminary exam review at the Univ. Wisconsin, the addition of the completed problems from various sources were acquired in a way that seemed to provide a 'mapping' of understanding. I wanted to convey that understanding here. Solutions to these key problems are carefully worked out in what follows, along with notes on the theory from Jackson as well as from [33-35]. Basic exercises are given to re-derive or verify the theory. The solutions to the exercises here, for the most part, are available via the aforementioned Jackson text (with the detailed derivations provided there). Many examples and related problems are done in order to cover material from basic undergraduate problems to advanced engineering (for antenna design, say), so the notes and problems also draw from undergraduate and graduate EM class notes/exams, and the other texts mentioned ([33-35]). The remarkable success and modern applications of EM will be made evident, but at the end of Ch. 7, when we carefully analyze the hydrogen atom, we will find that there is radiative collapse in the classical theory. This fundamental flaw in the theory, when dealing with even the simplest bound atomic system, will be corrected with the development of the quantum theory in Book 4 [15].

5.1 Spherical coordinates (notes/conventions from Jackson [32] Ch. 3)
Let's now consider Laplace's Equation in Spherical Coordinates:

$$\nabla^2\Phi = 0 \qquad ds^2 = dr^2 + r^2(d\theta^2 + \sin^2\theta\, d\phi^2)$$

(Eqn. 5-1)

$$\|g_{\mu\nu}\| = \begin{pmatrix} 1 & & 0 \\ & r^2 & \\ 0 & & r^2\sin^2\theta \end{pmatrix}, \qquad |g|^{1/2} = r^2\sin\theta$$

(Eqn. 5-2)

Thus,

$$\nabla^2\Phi = \frac{1}{|g|^{1/2}}\partial_\mu\left(|g|^{1/2}g^{\mu\nu}\partial_\nu\Phi\right)$$
$$= \frac{1}{r^2}\partial_r(r^2\partial_r\Phi) + \frac{1}{r^2\sin\theta}\partial_\theta(\sin\theta\,\partial_\theta\Phi) + \frac{1}{r^2\sin^2\theta}\partial_\phi^2\Phi = 0.$$

(Eqn. 5-3)

Since

$$\frac{1}{r^2}(2r(\partial_r\Phi) + r^2\partial_r^2\Phi) = \frac{2}{r}\partial_r\Phi + \partial_r^2\Phi = \frac{1}{r}\partial_r^2(r\Phi) = \frac{1}{r}\partial_r(\Phi + r\partial_r\Phi)$$

(Eqn. 5-4)

we have:

$$\nabla^2\Phi = \frac{1}{r}\frac{\partial^2(r\Phi)}{\partial r^2} + \frac{1}{r^2\sin\theta}\frac{\partial\left(\sin\theta\frac{\partial\Phi}{\partial\theta}\right)}{\partial\theta}\frac{1}{r^2\sin^2\theta}\frac{\partial^2\Phi}{\partial\phi^2}$$

(Eqn. 5-5)

Use $\Phi = \frac{U(r)}{r}P(\theta)Q(\phi)$ with $\nabla^2\Phi = 0$:

$$PQ\frac{\partial^2 U}{\partial r^2} + \frac{UQ}{r^2\sin\theta}\frac{d}{d\theta}\left(\sin\theta\frac{dP}{d\theta}\right) + \frac{UP}{r^2\sin^2\theta}\frac{d^2Q}{d\phi^2} = 0$$

$$r^2\sin^2\theta\left[\frac{1}{U}\frac{d^2U}{dr^2} + \frac{1}{r^2\sin\theta\, P}\frac{d}{d\theta}\left(\sin\theta\frac{dP}{d\theta}\right)\right] + \frac{1}{Q}\frac{d^2Q}{d\phi^2} = 0$$

(Eqn. 5-6)

Three separable equations result:

$$\frac{1}{Q}\frac{d^2Q}{d\phi^2} = 0 \quad \rightarrow \quad Q = e^{\pm im\phi}.$$

(Eqn. 5-7)

In order that Q be single valued, m must be an integer (if the full azimuthal range is allowed).

$$\frac{d^2U}{dr^2} - \frac{\ell(\ell+1)}{r^2}U = 0 \quad \rightarrow \quad U = Ar^{\ell+1} + Br^{-\ell}.$$

(Eqn. 5-8)

And,

$$\frac{1}{\sin\theta}\frac{d}{d\theta}\left(\sin\theta\frac{dP}{d\theta}\right) + \left[\ell(\ell+1) - \frac{m^2}{\sin^2\theta}\right]P = 0.$$

(Eqn. 5-9)

If we substitute $x = \cos\theta$ this becomes the Generalized Legendre Equation:

$$\frac{d}{dx}\left[(1-x^2)\frac{dP}{dx}\right] + \left[\ell(\ell+1) - \frac{m^2}{1-x^2}\right]P = 0$$

(Eqn. 5-10)

whose solutions called associated Legendre functions. The ordinary Legendre Equation is obtained when $m^2 = 0$:

$$\frac{d}{dx}\left[(1-x^2)\frac{dP}{dx}\right] + \ell(\ell+1)P = 0.$$

(Eqn. 5-11)

For the solution to the latter a power series is used: $P(x) = x^\alpha \sum_{j=0}^{\infty} a_j\, x^j$, which upon substitution becomes:

$$\sum_{j=0}^{\infty}\{(\alpha+j)(\alpha+j-1)a_j x^{\alpha+j-2} \\ - [(\alpha+j)(\alpha+j+1) - \ell(\ell+1)]a_j x^{\alpha+j}\} = 0.$$

(Eqn. 5-12)

Since the coefficient of each power of x must vanish separately, for $j = 0,1$ we find:

$$\text{if } a_0 \neq 0 \text{ then } \alpha(\alpha-1) = 0$$
$$\text{if } a_1 \neq 0 \text{ then } \alpha(\alpha+1) = 0$$

and for general j:

$$a_{j+2} = \left[\frac{(\alpha+j)(\alpha+j+1) - \ell(\ell+1)}{(\alpha+j+1)(\alpha+j+2)}\right] a_j$$

(Eqn. 5-13)

It is sufficient to choose either a_0 or a_1 different from zero, but not both. Let $a_0 \neq 0$ then $\alpha = 0$ or $\alpha = 1$. $(\alpha = 0) \Rightarrow$ even power series; and $(\alpha = 1) \Rightarrow$ odd power series.

Properties of series:

(a) converges for $x^2 < 1$ regardless of ℓ;
(b) diverges for $x = \pm 1$, unless it terminates.

Want a finite solution at $x = \pm 1$, so series must terminate. The recurrence relation will terminate only if ℓ is zero or a positive integer. And even then only the even or odd power series will terminate ($\alpha = 0$ or $\alpha = 1$).

Let's examine properties of Legendre polynomials, including:

- Completeness on $-1 \leq x \leq 1$
- Orthonormality
- Recurrence relations

Starting with the ordinary Legendre equation ($m^2 = 0$):

$$\frac{d}{dx}\left[(1-x^2)\frac{dP}{dx}\right] + \ell(\ell+1)P = 0$$

(Eqn. 5-14)

The solutions are Legendre polynomials, whose compact representation can be given via Rodriques' formula:

$$P_\ell(x) = \frac{1}{2^\ell \ell!}\frac{d^\ell}{dx^\ell}(x^2-1)^2.$$

(Eqn. 5-15)

Substituting we can show orthogonality:

$$\int_{-1}^{1} P_{\ell'}(x)\left\{\frac{d}{dx}\left[(1-x^2)\frac{dP_\ell}{dx}\right] + \ell(\ell+1)P_\ell(x)\right\} dx = 0$$

or

$$\int_{-1}^{1} \left[(x^2-1)\frac{dP_\ell}{dx}\frac{dP_{\ell'}}{dx} + \ell(\ell+1)P_{\ell'}(x)P_\ell(x)\right] dx = 0$$

Exchange indices and subtract:

$$[\ell(\ell+1) - \ell'(\ell'+1)] \int_{-1}^{1} P_{\ell'}(x)P_\ell(x)\, dx = 0$$

(Eqn. 5-16)

For $\ell \neq \ell' \Longrightarrow$ orthogonal.
For $\ell = \ell'$:

$$N_\ell \equiv \int_{-1}^{1} [P_\ell(x)]^2\, dx = \frac{1}{2^{2\ell}(\ell!)^2}\int_{-1}^{1} \frac{d^\ell}{dx^\ell}(x^2-1)\frac{d^\ell}{dx^\ell}(x^2-1)^\ell dx$$

$$= \frac{(-1)^\ell}{2^{2\ell}(\ell!)^2}\int_{-1}^{1} (x^2-1)^\ell \frac{d^{2\ell}}{dx^{2\ell}}(x^2-1)^\ell dx$$

$$= \frac{(2\ell)!}{2^{2\ell}(\ell!)^2}\int_{-1}^{1} (1-x^2)^\ell\, dx$$

(Eqn. 5-17)

Using $(1-x^2)^\ell = (1-x^2)(1-x^2)^{\ell-1} = (1-x^2)^{\ell-1} + \frac{x}{2\ell}\frac{d}{dx}(1-x^2)^\ell$

$$N_\ell = \left(\frac{2\ell-1}{2\ell}\right)N_{\ell-1} + \frac{(2\ell-1)!}{2^{2\ell}(\ell!)^2}\int_{-1}^{1} xd\left[(1-x^2)^\ell\right]$$

$$= \left(\frac{2\ell-1}{2\ell}\right)N_{\ell-1} - \frac{1}{2\ell}N_\ell$$

(Eqn. 5-18)

So, $(2\ell+1)N_\ell = (2\ell-1)N_{\ell-1}$, thus $(2\ell+1)N_\ell$ is independent of ℓ. Solving for $\ell = 0 \Rightarrow P_0(x) = 1 \Rightarrow N_0 = 2 \Rightarrow (2\ell+1)N_\ell = 2$, we get: $N_\ell = 2/(2\ell+1)$, thus:

$$\int_{-1}^{1} P_{\ell'}(x)P_\ell(x) = \frac{2}{2\ell+1}\delta_{\ell'\ell}$$

(Eqn. 5-19)

To make orthonormal, e.g.: $\int_a^b U_n^*(\varepsilon)U_m(\varepsilon)d\varepsilon = \delta_{nm}$, define as:

$$U_\ell(x) = \sqrt{\frac{2\ell+1}{2}}\,P_\ell(x).$$

(Eqn. 5-20)

For the Legendre series rep of any function $f(x)$ on the interval $-1 \le x \le 1$:

$$f(x) = \sum_{\ell=0}^{\infty} A_\ell P_\ell(x) \quad with \quad A_\ell = \frac{2\ell+1}{2}\int_{-1}^{1} f(x)\,P_\ell(x)dx.$$

(Eqn. 5-21)

The recurrence relations follow from the original Legendre equation and that generated by Rodrigues formula:

$$\frac{dP_{\ell+1}}{dx} - \frac{dP_{\ell-1}}{dx} - (2\ell+1)P_\ell = 0.$$

(Eqn. 5-22)

So,

$$(\ell+1)P_{\ell+1} - (2\ell+1)xP_\ell + \ell P_{\ell-1} = 0$$
$$\frac{dP_{\ell+1}}{dx} - \frac{dP_\ell}{dx} - (\ell+1)P_\ell = 0$$
$$(x^2-1)\frac{dP_\ell}{dx} - \ell_x P_\ell + \ell P_{\ell-1} = 0$$

(Eqn. 5-23)

For potential problems in conical holes or near a sharp point – lightning rods – Legendre functions of the first kind and order v: R. N. Hall, J. Appl. Phys. 20, 925 (1949) [117].

Additional Theorem for Spherical Harmonics.

$$P_\ell(\cos\theta) = \frac{4\pi}{2\ell+1}\sum_{m=-\ell}^{\ell} Y_{\ell m}^*(\theta',\phi')Y_{\ell m}(\theta,\phi)$$

(Eqn. 5-24)

Laplace Equation in Cylindrical coordinates gives rise to the classic Bessel Functions:

$$ds^2 = d\rho^2 + \rho^2 d\phi^2 + dz^2 \quad |g|^{1/2} = \rho$$

(Eqn. 5-25)

$$\nabla^2 \Phi = \frac{1}{|g|^{1/2}}\left(|g|^{1/2} g^{\mu\nu}\partial_\nu \Phi\right) = \frac{1}{\rho}\partial_\rho\left(\rho \partial_\rho \rho \Phi\right) + \frac{1}{\rho^2}\partial_\phi^2 \Phi + \partial_z^2 \Phi$$

$$= \frac{\partial^2 \Phi}{\partial \rho^2} + \frac{1}{\rho}\frac{\partial \Phi}{\partial \rho} + \frac{1}{\rho^2}\frac{\partial^2 \Phi}{\partial \phi^2} + \frac{\partial^2 \Phi}{\partial z^2}$$

(Eqn. 5-26)

BoundaryValue Problems with Azimuthal Symmetry (m=0):

When $m = 0$, the general solution is:

$$\Phi(r,\theta) = \sum_{\ell=0}^{\infty}\left[A_\ell r^\ell + B_\ell r^{-(\ell+1)}\right] P_\ell(\cos\theta).$$

(Eqn. 5-27)

Suppose we have $V(\theta)$ on the surface of a sphere, radius a, and we want the potential inside. We can write:

$$V(\theta) = \sum_{\ell=0}^{\infty} A_\ell a^\ell \, P_\ell(\cos\theta).$$

(Eqn. 5-28)

Consider a sphere centered at origin and cut by x-y plane, with upper hemisphere $+V$, and lowere hemisphere at $-V$ potential:

$$A_\ell = \frac{2\ell+1}{2}\frac{1}{a^\ell}\int_0^\pi V(\theta)\, P_\ell(\cos\theta)\sin\theta \, d\theta \qquad V(\theta)$$

$$= \begin{cases} +V & 0 \le \theta \le \frac{\pi}{2} \\ -V & \frac{\pi}{2} < \theta \le \pi \end{cases}$$

$$A_\ell = V\left(\frac{2\ell+1}{2a^\ell}\right)\left[\int_0^1 P_\ell(x)dx - \int_{-1}^0 P_\ell(x)dx\right]$$

$$= \frac{2\ell+1}{2a^\ell}(2V)\int_0^1 P_\ell(x)dx\,,$$

(Eqn. 5-29)

where only the ℓ odd terms are non-zero. To solve the integral, use $(2\ell+1)P_\ell(x) = \frac{dP_{\ell+1}}{dx} - \frac{dP_{\ell-1}}{dx}$:

$$(2\ell+1)\int P_\ell(x)dx = \int_0^1 \left(\frac{dP_{\ell+1}}{dx} - \frac{dP_{\ell-1}}{dx}\right)dx = [P_{\ell+1}(x) - P_{\ell-1}(x)]_0^1$$
$$= P_{\ell-1}(0) - P_{\ell+1}(0).$$

(Eqn. 5-30)

From Rodrigues' formula: $P_\ell(x) = \frac{1}{2^\ell \ell!}\frac{d^\ell}{dx^\ell}(x^2-1)^\ell$, we have

$$P_n(x) = \frac{1}{2^n}\sum_{m=0}^{n/2}(-1)^m\binom{n}{m}\binom{2n-2m}{n}x^{n-2m}.$$

(Eqn. 5-31)

Thus, $P_n(0) = a_0$ for even n, which is the $m = \frac{n}{2}$ term above:

$$P_n(0) = \frac{1}{2^n}(-1)^{n/2}\binom{n}{n/2}\binom{n}{n} = \frac{(-1)^{n/2}}{2^n\left(\frac{n}{2}\right)!}\frac{n!}{\left(\frac{n}{2}\right)!}.$$

(Eqn. 5-32)

Thus,

$$P_{\ell-1}(0) - P_{\ell+1}(0) = \frac{(-1)^{\frac{(\ell-1)}{2}}}{2^{(\ell-1)}}\frac{(\ell-1)!}{\left(\left(\frac{\ell-1}{2}\right)!\right)^2}\left\{1 + \frac{1}{4}\frac{\ell(\ell+1)}{\left(\frac{\ell+1}{2}\right)^2}\right\}$$
$$= \frac{(-1)^{(\ell-1)/2}}{2^{(\ell-1)}}\frac{(\ell-1)!}{\left(\frac{\ell-1}{2}\right)!^2}\left(\frac{2\ell+1}{\ell+1}\right)$$

and since: $2^{\ell-1}\left(\frac{\ell-1}{2}\right)! = 2^{\frac{(\ell-1)}{2}}(\ell-1)!!$, we can simply further:

$$P_{\ell-1}(0) - P_{\ell-1}(0) = \left(\frac{-1}{2}\right)^{\left(\frac{\ell-1}{2}\right)}\frac{(2\ell+1)(\ell-2)!!}{2\left(\frac{\ell+1}{2}\right)!}.$$

(Eqn. 5-33)

So,

$$A_\ell = \frac{V}{a^\ell}\left(\frac{-1}{2}\right)^{\left(\frac{\ell-1}{2}\right)}\frac{(2\ell+1)(\ell-2)!!}{2\left(\frac{\ell+1}{2}\right)!}.$$

(Eqn. 5-34)

Alternative Solution to problem using Poisson's integral for sphere:

$$\Phi(\vec{x}) = \frac{1}{4\pi}\int \Phi\,(a,\theta',\phi')\frac{a(x^2-a^2)}{(x^2+a^2-2ax\cos\gamma)^{3/2}}d\Omega',$$

(Eqn. 5-35)

which for our potential:

$$\Phi(x,\theta,\phi) = \frac{V}{4\pi}\int_0^{2\pi} d\phi'\left\{\int_0^1 d\,(\cos\phi') - \int_{-1}^0 d\,(\cos\phi')\right\}$$

(Eqn. 5-36)

Note that $\cos\gamma = \cos\theta\cos\theta' + \sin\theta\sin\theta'\cos(\phi-\phi')$. For potential along z-axis $\theta = 0$ and $\cos\gamma = \cos\theta'$, for which the integral is solvable.

Consider outside first:

$$\Phi(z) = \frac{V}{2}\left\{\frac{+a(x^2-a^2)}{-2ax}\right\}\left[\frac{1}{-\frac{1}{2}}\frac{1}{(a^2+x^2-2ax\cos\theta')^{1/2}}\Big|_0^1 - \frac{1}{-\frac{1}{2}}\frac{1}{(a^2+x^2-2ax\cos\theta')^{1/2}}\Big|_{-1}^0\right]\Bigg|_{x=z}$$

$$= \frac{V}{2}\left(\frac{+z^2-a^2}{z}\right)\left[\frac{1}{-(a-x)} - \frac{1}{\sqrt{(a^2+x^2)}} - \left(\frac{1}{\sqrt{(a^2+x^2)}}\right) + \frac{1}{(a+x)}\right]$$

$$\Phi(z) = V\left[\frac{a}{z} - \frac{(a^2-z^2)}{z\sqrt{(a^2+z^2)}}\right],\quad outside.$$

(Eqn. 5-37)

For the inside the signs change in the 1/x terms:

$$\Phi(z) = \frac{V}{2}\left\{\frac{(x^2-a^2)}{x}\right\}\left[\frac{1}{(x-a)} - \frac{2}{\sqrt{(a^2+\gamma^2)}} + \frac{1}{x+a}\right],$$

(Eqn. 5-38)

which simplifies to:

$$\Phi(z) = V\left[1 - \frac{(z^2-a^2)}{z\sqrt{z^2+a^2}}\right].$$

(Eqn. 5-39)

Note that they match at $z = a$. Since $\sqrt{a^2+z^2} \approx a\left(1+\frac{1}{2}\left(\frac{z}{a}\right)^2\right)$ as z→0, we can simplify to

$$\Phi(z) = \frac{3V}{2}\left(\frac{z}{a}\right)\quad \text{as } z \to 0,$$

(Eqn. 5-40)

which is a dipole potential! Note that the center of the sphere mirrors the potential seen asymptotically, where a dipole is also seen (asymptotically this makes more sense). Returning to issue at hand, we now have a closed form for the potential on axis:

$$\Phi(z) = V\left[1 - \frac{(z^2 - a^2)}{z\sqrt{(z^2 + a^2)}}\right],$$

(Eqn. 5-41)

and by matching this with $\Phi(r,\theta) = \sum_{\ell=0}^{\infty}\left[A_\ell r^\ell + B_\ell r^{-(\ell+1)}\right] P_\ell(\cos\theta)$ where $r = z, \cos\theta = 1$, we find an alternate way of obtaining the coefficients (with alternate expression):

$$\Phi(z = r) = \frac{V}{\sqrt{\pi}}\sum_{j=1}^{\infty}(-1)^{j-1}\frac{\left(2j - \frac{1}{2}\right)\Gamma\left(j - \frac{1}{2}\right)}{j!}\left(\frac{a}{r}\right)^{2j}$$

(Eqn. 5-42)

Thus

$$\Phi(r,\theta) = \frac{V}{\sqrt{\pi}}\sum_{j=1}^{\infty}(-1)^{j-1}\frac{\left(2j - \frac{1}{2}\right)\Gamma\left(j - \frac{1}{2}\right)}{j!}\left(\frac{a}{r}\right)^{2j} P_{2j-1}(\cos\theta)\,.$$

(Eqn. 5-43)

Now that we've obtained some general forms of solution, let's consider the very important example of a point charge. What is the expansion for a potential involving $\frac{1}{|\vec{x}-\vec{x}'|}$? Since there is azimuthal symmetry, the Laplace equation can be expanded as before, except at the point $\vec{x} - \vec{x}'$. Also, let $\vec{x}'$ be along the z-axis, let $r_<(r_>)$ be the smaller (larger) of $|\vec{x}|$ and $|\vec{x}'|$, and let γ be the angle between them, then:

$$\frac{1}{|\vec{x} - \vec{x}'|} \equiv \frac{1}{\left(r^2 + r'^2 - 2rr'\cos\gamma\right)^{\frac{1}{2}}} \quad or \quad \frac{1}{|\vec{x} - \vec{x}'|}$$

$$= \sum_{\ell=0}^{\infty}\left[A_\ell r^\ell + B_\ell r^{-(\ell+1)}\right] P_\ell(\cos\gamma)$$

(Eqn. 5-44)

Furthermore let $\vec{x}$ also lie on the z-axis ($\cos\gamma = 1$), now

$$\frac{1}{|\vec{x} - \vec{x}'|} = \frac{1}{|r - r'|} = \frac{1}{r_>}\sum_{\ell=0}^{\infty}\left(\frac{r_<}{r_>}\right)^\ell$$

$$if\ |\vec{x}| < |\vec{x}'|\ then\quad \frac{1}{|r-r'|} = \sum_{\ell=0}^{\infty} A_\ell r_<^\ell \rightarrow \text{determines } A_\ell, B_\ell = 0$$

$$If\ |\vec{x}| > |\vec{x}'|\ then\quad \frac{1}{|r-r'|} = \sum_{\ell=0}^{\infty} B_\ell r_>^{-(\ell+1)} \rightarrow \text{determines } B_\ell \text{ and} A_\ell$$
$$= 0$$

Either way we conclude:

$$\frac{1}{|\vec{x}-\vec{x}'|} = \sum_{\ell=0}^{\infty} \frac{r_<^\ell}{r_>^{\ell+1}} P_\ell(\cos\gamma)$$

(Eqn. 5-45)

Now let's consider the potential due to a ring charge of radius a, height z=b above x-y plane, such that the distance from origin to ring is c at a declination angle α:

$$c^2 = a^2 + b^2 \quad and \quad \alpha = \tan^{-1}\left(\frac{a}{b}\right).$$

The potential on the z-axis at distance r is then:

$$\Phi(z=r) = \frac{q}{(r^2+c^2-2cr\cos\alpha)^{1/2}} = \frac{q}{\left(r^2 - r'^2 - 2rr'\cos\alpha\right)^{1/2}}$$

For $r > c$:

$$\Phi(z=r) = q\sum_{\ell=0}^{\infty} \frac{c^\ell}{r^{\ell+1}} P_\ell(\cos\alpha)\,.$$

(Eqn. 5-46)

For $r < c$

$$\Phi(z=r) = q\sum_{\ell=0}^{\infty} \frac{r^\ell}{c^{\ell+1}} P_\ell(\cos\alpha)\,.$$

(Eqn. 5-47)

The potential at any point in space is then:

$$\Phi(r,\theta) = q\sum_{\ell=0}^{\infty} \frac{r_<^\ell}{r_>^{\ell+1}} P_\ell(\cos\alpha) P_\ell(\cos\theta)\,.$$

(Eqn. 5-48)

Example 5.1. *Two concentric spheres have radii a, b with b>a, and are each divided into two hemispheres by the same horizontal plane. The upper hemisphere of the inner sphere and the lower hemisphere of the outer sphere are maintained at potential V. The other hemispheres are at* *zero potential. Determine the potential in the region a<+ r<=b as a series in Legendre Polynomials. Include terms up to l=4. Check your solution against known results in the limiting cases b → ∞, and a → 0.*

With origin chosen at that of the concentric spheres, we clearly have radial dependence and dependence on the angle from the vertical axis, θ, but no dependence on the rotational angle (φ) about that vertical axis. Thus, we have azimuthal symmetry. The general potential solution expansion in terms of Legendre Polynomials is:

$$\Phi(r,\theta) = \sum_{l=0}^{\infty}\left[A_l r^l + B_l r^{-(l+1)}\right] P_l(\cos\theta).$$

Let's consider the inner sphere b.c.'s first:

$$\Phi(a,\theta) = \sum_{l=0}^{\infty}\left[A_l a^l + B_l a^{-(l+1)}\right] P_l(\cos\theta)$$

with standard inversion rule:

$$\left[A_l a^l + B_l a^{-(l+1)}\right] = \frac{(2l+1)}{2}\int_0^{\pi} \Phi(a,\theta) P_l(\cos\theta)\sin\theta d\theta$$

Substituting the indicated sphere potential:

$$\Phi(a,\theta) = \begin{cases} +V, & 0 \le \theta < \dfrac{\pi}{2} \\ 0, & \dfrac{\pi}{2} < \theta \le \pi \end{cases}$$

Thus, we have relations on unknown A_l and B_l from the r=a boundary:

$$\left[A_l a^l + B_l a^{-(l+1)}\right] = \left(\frac{2l+1}{2}\right) V \int_0^1 P_l(x)dx = V/2$$

Similarly for the r=b boundary:

$$\left[A_l b^l + B_l b^{-(l+1)}\right] = \left(\frac{2l+1}{2}\right) V \int_{-1}^{0} P_l(x)dx = -V/2$$

Before proceeding, however, let's simplify by choosing to shift the potential by adding a constant: $\Phi' = \Phi - V/2$. For such a choice the boundary conditions are odd functions of the radial coordinate, resulting in an expansion for Φ' that must entirely consist of only odd terms. Choosing new constants to conveniently center on the r=a boundary we now have:

$$\Phi'(r,\theta) = \sum_{\substack{l=1\\odd}} \left[\alpha_l \left(\frac{r}{a}\right)^l + \beta_l \left(\frac{a}{r}\right)^{l+1}\right] P_l(\cos\theta).$$

For r=a:

$$\Phi'(a,\theta) = \sum_{\substack{l=1\\odd}} [\alpha_l + \beta_l] P_l(\cos\theta)$$

Using $[\alpha_l + \beta_l] = \frac{2l+1}{2}\int_{-1}^{1} f(x)\, P_l(\cos\theta)$ together with the definition of the potential:

$$f(x) = \begin{cases} +\frac{V}{2} \; for \; x > 0 \\ -\frac{V}{2} \; for \; x < 0 \end{cases}$$

and we have:

$$[\alpha_l + \beta_l] = \frac{V}{2} R_l,$$

where the R_l constants are given by the Rodriguez formula ($R_1 = 3/2$; $R_3 = -7/8$; $R_5 = 11/16$). Let's now consider the r=b boundary:

$$\Phi'(b,\theta) = \sum_{\substack{l=1\\odd}} \left[\alpha_l \left(\frac{b}{a}\right)^l + \beta_l \left(\frac{a}{b}\right)^{l+1}\right] P_l(\cos\theta).$$

Furthermore, since $\Phi'(b,\theta) = -\,\Phi'(a,\theta)$, we must have:

$$\left[\alpha_l \left(\frac{b}{a}\right)^l + \beta_l \left(\frac{a}{b}\right)^{l+1}\right] = -\frac{V}{2} R_l = -[\alpha_l + \beta_l]$$

after regrouping:

$$\beta_l = -\alpha_l \frac{(1 + \left(\frac{b}{a}\right)^l}{1 + \left(\frac{a}{b}\right)^{l+1}}$$

Thus,

$$\frac{V}{2} R_l = \alpha_l \left[\frac{a^{2l+1} - b^{2l+1}}{a^l b^{l+1} + a^{2l+1}}\right]$$

The potential is then:

$$\Phi'(r,\theta) = \sum_{\substack{l=1 \\ odd}} \frac{V}{2}\left[\frac{a^l b^{l+1} + a^{2l+1}}{a^{2l+1} - b^{2l+1}}\right] R_l \left\{\left(\frac{r}{a}\right)^l - \left(\frac{a^l b^{l+1} + b^{2l+1}}{a^l b^{l+1} - a^{2l+1}}\right)\left(\frac{a}{r}\right)^{l+1}\right\} P_l(\cos\theta)$$

where the first three R_l constants are given above.

Example 5.2. *A spherical surface of radius R has charge uniformly distributed over its surface with a density of $Q/4\pi R^2$, except for a spherical cap at the north pole, defined by the cone $\theta = \alpha$. (a) what is the potential inside? outside? (b) what is the magnitude and direction of the electric field at the origin? (c) Discuss the limiting forms of the potential and electric field as the spherical cap becomes very small or very large.*

Using the axis of the cone as the vertical z-axis, we clearly have azimuthal symmetry. We also have azimuthal symmetry for a ring charge. For a ring of charge at the boundary of the cone, for example, we have

$$\Phi(r,\theta) = q\sum_{l=0}^{\infty} \frac{r_<^l}{r_>^{l+1}} P_l(\cos\alpha) P_l(\cos\theta)$$

where $\{r_<, r_>\}$ are the values {c,r} according to which is smaller, where r is the radial coordinate of the potential and c is the radial coordinate of the charge ring. Let's consider rings of charge of width $Rd\theta'$ at angle θ' such that the ring circumference is $2\pi R\sin\theta'$, we then get for the charge ring:

$$\Phi(r,\theta) = \frac{Q}{2}\sin\theta'\,d\theta' \sum_{l=0}^{\infty} \frac{r^l}{R^{l+1}} P_l(\cos\theta') P_l(\cos\theta).$$

Let's integrate the charge rings from cone edge at angle $\theta = \alpha$ to the maximum at $\theta = \pi$:

$$\frac{Q}{2}\int_{\alpha}^{\pi} P_l(x)(\cos\theta')\sin\theta'\,d\theta' == \frac{Q}{2}\int_{-1}^{\cos\alpha} P_l(x)\,dx$$

Using the relation

$$P_l(x) = \frac{1}{2l+1}\left(\frac{dP_{l+1}}{dx} - \frac{dP_{l-1}}{dx}\right)$$

we get:

$$\int_{-1}^{\cos\alpha} P_l(x)dx = \frac{1}{2l+1}[P_{l+1}(x) - P_{l-1}(x)]_{-1}^{\cos\alpha}$$

$$= \frac{1}{2l+1}[P_{l+1}(\cos\alpha) - P_{l-1}(\cos\alpha)] \; for \; l \neq 0$$

For $l = 0$: $\int_{-1}^{\cos\alpha} P_l(x)dx = P_1(\cos\alpha) + 1$. Thus:

$$\Phi(r,\theta) = \frac{Q}{2}\sum_{l=0}^{\infty}\frac{1}{2l+1}(P_{l+1}(\cos\alpha)$$
$$- P_{l-1}(\cos\alpha))\frac{r^l}{R^{l+1}}P_l(\cos\theta), \quad for \; r < R$$

For $r > R$, simply switch $\frac{r^l}{R^{l+1}} \Rightarrow \frac{R^l}{r^{l+1}}$.

(b) Let's now consider the electric field. Recall that $E = -\nabla\Phi$ implies that

$$E(r=0) = -\nabla\phi|_{r=0} = -\left(\frac{\partial\phi}{\partial r}\hat{r} + \frac{1}{r}\frac{\partial\phi}{\partial\theta}\hat{\theta}\right)\Big|_{r=0}$$

Since the latter expression isn't well-defined at precisely the place of interest (r=0) we must approach the value r=0 with care. Let's consider an expansion in small r, the main contribution to Φ at first order in r, is then:

$$\Phi_1 = \frac{Q}{2}\left(\frac{1}{3}\right)[P_2(\cos\alpha) - P_0(\cos\alpha)]\frac{r}{R^2}P_1(\cos\theta)$$
$$= \frac{Q}{6}\cdot\frac{1}{2}(3\cos^2\alpha + 3)\frac{r}{R^2}\cos\theta.$$

Evaluating the derivatives we get:

$$\frac{\partial\Phi}{\partial r} = \frac{Q}{12}(3\cos^2\alpha - 3)\frac{\cos\theta}{R^2} \text{ and } \frac{\partial\Phi}{\partial\theta} = \frac{Q}{12}(3\cos^2\alpha + 1)\frac{(-\sin\theta)}{R^2}r.$$

The E field is thus:

$$E(r=0) = \frac{Q}{12}\frac{(3\cos^2\alpha - 3)}{R^2}(-\cos\theta\,\hat{r} + \sin\theta\hat{\theta})$$
$$= -\hat{z}\left\{\frac{Q(3\cos^2\alpha - 3)}{12\,R^2}\right\}.$$

(c) Consider $\alpha \to 0$: redoing the expression from (a), separating the l=0 and l>0 terms, first consider for l=0:

$$P_1\left(1-\frac{1}{2}\alpha^2\right) = 1-\frac{1}{2}\alpha^2, \quad P_{-1} = \left(1-\frac{1}{2}\alpha^2\right) = -1.$$

Thus,

$$\Phi(r,\theta) = \frac{Q}{2}\left(2-\frac{1}{2}\alpha^2\right)\frac{1}{R} - \frac{Q\alpha^2}{4R}\sum_{l=0}^{\infty}\left(\frac{r}{R}\right)^l P_l(\cos\theta)$$

$$= \frac{Q}{R} - \frac{Q\alpha^2}{4R}\sum_{l=1}^{\infty}\left(\frac{r}{R}\right)^l P_l(\cos\theta).$$

Notice that the result is the potential of a uniform spherical shell with total charge Q minus that of a point charge located at z=+R.

For the E-field at r=0 as $\alpha \to 0$:

$$E(r=0)\Bigg|_{\alpha \to 0} = \hat{z}\left(\frac{Q\alpha^2}{4R^2}\right).$$

Example 5.3. *A thin, flat, conducting, circular disc of radius R is located in the x-y plane with its center at the origin, and it is maintained at a fixed potential V. The charge density on the disc is proportional to* $(R^2-\rho^2)^{-1/2}$*, where* ρ *is the radial distance on the disc:*
(a) find the potential for r>R.
(b) find the potential for r<R.
(c) find the capacitance of the disc.

Solution
Once again, we make use of the result for the potential due to a ring of charge:

$$\Phi_{ring}(r,\theta) = (dq)\sum_{l=0}^{\infty}\frac{r_<^l}{r_>^{l+1}}P_l(\cos\alpha)\,P_l(\cos\theta).$$

As before, the ring radius and the observer radial distance are first ordered and the smaller is $r_<$ and the larger is $r_>$. Let's first get the total charge (eventually needed for Q=CV anyway):

Suppose the surface charge density is

$$\sigma(\rho) = \frac{A}{\sqrt{R^2-\rho^2}}$$

with total charge Q then determining the value of the constant A:

$$\int_0^R \frac{A}{\sqrt{R^2-\rho^2}} 2\pi\rho d\rho = 2\pi A = Q$$

So, we have: $dq = \frac{Q/(2\pi)}{\sqrt{R^2-\rho^2}}(2\pi\rho)d\rho$ and $\alpha = \frac{\pi}{2} \Rightarrow \cos\alpha = 0, P_l(\cos\alpha) = 0$ for all odd l's. We also have that $P_{2n}(0) = \frac{(-1)^n}{2^{2n}}\frac{(2n)!}{n!n!}$. Let's start with the case where $r_< = \rho, r_> = r$:

$$\Phi(r,\theta)_{r>R} = Q\sum_{l=0}^{\infty} \frac{(-1)^l}{2^{2l}}\frac{(2l)!}{l!\,l!}\frac{1}{r^{(2l+1)}}\left[\int_0^R \frac{\rho^{2l+1}}{\sqrt{R^2-\rho^2}} d\rho\right] P_{2l}(\cos\theta)$$

$$= Q\sum_{l=0}^{\infty} \frac{(-1)^l}{2^{2l}}\frac{(2l)!}{(l!)(l!)}\frac{(2l)!!}{(2l+1)!!}\left(\frac{R}{r}\right)^{2l+1} P_{2l}(\cos\theta)$$

$$= Q\sum_{l=0}^{\infty} \frac{(-1)^l}{2^{2l}}\frac{(2l)!}{(l!)(l!)}\frac{(2l)!!}{(2l+1)!!}\left(\frac{R}{r}\right)^{2l+1} P_{2l}(\cos\theta)$$

$$= Q\sum_{l=0}^{\infty} \frac{(-1)^l}{2l+1}\left(\frac{R}{r}\right)^{2l+1} P_{2l}(\cos\theta)$$

We know that $\lim_{r\to R}\Phi(r,\theta=\pi/2) = V$, and for $\theta = \pi/2, \cos\theta = 0$, we have the relation:

$$P_{2l}(0) = \frac{(-1)^l}{2^{2l}}\frac{(2l)!}{(l!)(l!)}$$

Thus,

$$\Phi\left(r,\theta=\frac{\pi}{2}\right)_{r>R} = Q\sum_{l=0}^{\infty}\frac{(-1)^l}{2l+1}\frac{(-1)^l}{2^{2l}}\frac{(2l)!}{l!\,l!}\left(\frac{R}{r}\right)^{2l+1}$$

$$= Q\sum_{l=0}^{\infty}\frac{(2l-1)!!}{(2l)!!\,(2l+1)}\left(\frac{R}{r}\right)^{2l+1}$$

$$= Q\left[\left(\frac{R}{r}\right) + \frac{1}{2\cdot 3}\left(\frac{R}{r}\right)^3 + \frac{1\cdot 3}{2\cdot 4\cdot 5}\left(\frac{R}{r}\right)^5 + \frac{1\cdot 3\cdot 5}{2\cdot 4\cdot 6\cdot 7}\left(\frac{R}{r}\right)^7 + \cdots\right]$$

$$= Q\sin^{-1}\left(\frac{R}{r}\right)\ for\ r > R$$

For $r = R$, $\sin^{-1}(1) = \frac{\pi}{2}$, and we have:

$$V = \pi Q/2.$$

So, the solution for $r \geq R$ is:

$$\Phi(r,\theta)_{r\geq R} = \frac{2V}{\pi}\sum_{l=0}^{\infty}\frac{(-1)^l}{(2l+1)}\left(\frac{R}{r}\right)^{2l+1} P_{2l}(\cos\theta).$$

(b) To find the potential for $r < R$, let's start from the solution in (a) for $r > R$. Let's consider the behavior when crossing from $\theta = 0$ with $r > R$ to $r < R$. Start by summing series for $\theta = 0 \Rightarrow P_{2l}(\cos\theta) = 1$:

$$\Phi(r,\theta)_{r>R} = \frac{2V}{\pi}\left[\left(\frac{R}{r}\right) - \frac{1}{3}\left(\frac{R}{r}\right)^3 + \frac{1}{5}\left(\frac{R}{r}\right)^5 + \frac{1}{7}\left(\frac{R}{r}\right)^7 + \cdots\right]$$
$$= \frac{2V}{\pi}\tan^{-1}\left(\frac{R}{r}\right).$$

Notice that the sum form of the Legendre Polynomial diverges for $r < R$ while the inverse tangent does not. This allows us to uniquely continue the functional dependence according to $\tan^{-1}\left(\frac{R}{r}\right)$ for $r < R$ along the $\theta = 0$ z-axis. Let's now do an expansion of the inverse tangent form of the potential for $r < R$, $\theta = 0$, that is valid in that domain, particularly when $r = 0$:

$$\Phi(r,\theta=0)_{r<R} = \frac{2V}{\pi}\left[\left(\frac{\pi}{2}\right) - \frac{r}{R} + \frac{1}{3}\left(\frac{r}{R}\right)^3 - \frac{1}{5}\left(\frac{r}{R}\right)^5 + \cdots\right],$$

which, when rewritten in terms of matching Legendre expansion, uniquely arrives at the solution for $0 \leq \theta \leq \pi/2$ (recall we crossed at $\theta = 0$, so derivation tied to region above x-y plane):

$$\Phi(r,\theta)_{r<R,0\leq\theta\leq\pi/2} = V - \frac{2V}{\pi}\sum_{l=0}^{\infty}\frac{(-1)^l}{(2l+1)}\left(\frac{r}{R}\right)^{2l+1} P_{2l+1}(\cos\theta).$$

Now, symmetry above and below the disc implies that Φ is even, but since P_{2l+1} is odd, we must have for the potential below the x-y axis:

$$\Phi(r,\theta)_{r<R,\pi/2\leq\theta\leq\pi} = V + \frac{2V}{\pi}\sum_{l=0}^{\infty}\frac{(-1)^l}{(2l+1)}\left(\frac{r}{R}\right)^{2l+1} P_{2l+1}(\cos\theta).$$

(c) For capacitance we have the definition $Q = CV$ and recall $V = \pi Q/2$, thus: $C = 2/\pi$.

Example 5.4. *The surface of a hollow conducting sphere of inner radius a is divided into and event number longitudinal "orange slices", where* *the segments alternate in potential according to $\pm V$. (a) For 2n segments, what is the potential inside? (b) For case n=1 (two hemispheres), get terms up to l=3 (show agreement with prior results).*

Solution

(a) We no longer have azimuthal symmetry, so general spherical harmonics are needed:

$$\Phi(r,\theta,\varphi) = \sum_{l=0}^{\infty}\sum_{m=l}^{l}\left[A_{lm}r^{l} + B_{lm}r^{-(l+1)}\right]Y_{lm}(\theta,\varphi).$$

The potential can't be singular at the origin (without a source there, mean value theorem), thus $B_{lm} = 0$ and we have:

$$\Phi(r,\theta,\varphi) = \sum_{l,m} A_{lm}\left(\frac{r}{a}\right)^{l} Y_{lm}(\theta,\varphi),$$

where we can determine the coefficients from:

$$A_{lm} = \int d\Omega\, Y_{lm}^{*}(\theta,\varphi)\Phi(a,\theta,\varphi).$$

Again asserting the mean value theorem: $\Phi(r=0) = 0 = A_{00}$. If n (the number of dividing planes) is odd, Φ is odd, Y_{lm} is odd, thus only odd l is allowed. Similarly, if n is even only even l are allowed. Also note that $\Phi(r,\theta,\varphi) = \Phi(r,\pi-\theta,\varphi)$, which indicates that Φ is an even function of $\cos\theta$. $Y_{lm}(\theta,\varphi)$ is an even function of $\cos\theta$ only if $P_l^m(x)$ is an even function of x. The parity of $P_l^m(x)$ is $(-1)^{l+m}$, so $l+m = even$. So, if n is even, so are l and m, and if n is odd, so are l and m.

Also note that since Φ is a periodic function in φ with period $(2\pi/n)$ we have a relation imposed on m such that $m = 0,\ \pm n,\ \pm 2n,\ \pm 3n, \ldots$, and there is also the relation m/n is an integer.

Consider one of the dividing planes to be the x-z plane, such that

$$\Phi(r,\theta,\varphi) = +V \quad for \quad 0 < \varphi < \frac{\pi}{n}$$

We then alternate:

$$\Phi(r,\theta,\varphi) = -V \quad for \quad \frac{\pi}{n} < \varphi < \frac{2\pi}{n}$$

then:

$$\Phi(r,\theta,\varphi) = +V \quad for \quad \frac{2\pi}{n} < \varphi < \frac{3\pi}{n}$$

etc. Recall the integral with spherical harmonics can be rewritten in terms of the modified Legendre Polynomials:

$$\int d\Omega\, Y_{lm}^*(\theta,\varphi)\phi(\theta,\varphi)$$
$$= \int_0^{\pi} \sin\theta d\theta \sqrt{\frac{2l+1}{4\pi}\frac{(l-m)!}{(l+m)!}} P_l^m(\cos\theta) \int_0^{2\pi} e^{-im\varphi}\Phi(\varphi)d\varphi.$$

Let's focus on the φ integral first:

$$\int_0^{2\pi} e^{-im\varphi}\Phi(\varphi)d\varphi = n\left[\int_0^{2\pi} e^{-im\varphi}\Phi(\varphi)d\varphi\right]$$
$$= n\left[\int_0^{\frac{\pi}{n}} e^{-im\varphi}d\theta - V\int_{\frac{\pi}{n}}^{\frac{2\pi}{n}} e^{-im\varphi}d\varphi\right]$$
$$= \frac{inV}{m}\left\{2e^{-i\pi\frac{m}{n}} - e^{-2\pi i\frac{m}{n}} - 1\right\} = \frac{inV}{m}2\left[e^{-i\pi\frac{m}{n}} - 1\right]$$
$$= \begin{cases} -\dfrac{4inV}{m} & if\ \dfrac{m}{n}\ is\ odd \\ 0 & if\ \dfrac{m}{n}\ is\ even \end{cases}$$

Thus,

$$A_{lm} = +\frac{4inV}{m}\int_0^{\pi} d(\cos\theta) \sqrt{\frac{2l+1}{4\pi}\frac{(l-m)!}{(l+m)!}} P_l^m(\cos\theta)$$

$(for\, \frac{m}{n} odd,\ \ zero\, for\ \frac{m}{n}\ even)$

(b) Consider $n = 1$, then l is odd and $m = \pm 1, \pm 3, \pm 5$:

$$\Phi(r,\theta,\varphi) = A_{11}\left(\frac{r}{a}\right)Y_{11} + A_{1,-1}\left(\frac{r}{a}\right)Y_{1,-1} + A_{3,3}\left(\frac{r}{a}\right)^3 Y_{3,3}$$
$$+ A_{3,1}\left(\frac{r}{a}\right)^3 Y_{3,1} + A_{3,-1}\left(\frac{r}{a}\right)^3 Y_{3,-1} + A_{3,-3}\left(\frac{r}{a}\right)^3 Y_{3,-3}$$
$$+ \cdots$$

Φ is real, thus $Y_{l_1-m} = (-1)^m Y_{l_1-m}^* \Rightarrow A_{l_1-m} = (-1)^m A_{l-m}^*$. And we have:

$$A_{1,1} = -\frac{4inv}{m}\int_0^\pi \sin\theta\, \mathrm{d}\,\theta \left(-\sqrt{\frac{3}{8\pi}}\sin\theta\right) = 4iV\sqrt{\frac{3}{8\pi}}\int_0^\pi \sin^2\theta\, d\,\theta$$

$$= iV\sqrt{\frac{3\pi}{2}}$$

$$A_{1,-1} = A_{1,1}^{*} = iV\sqrt{\frac{3\pi}{2}}$$

and

$$A_{3,1} = -4iV\int_0^\pi \sin\theta\, \mathrm{d}\,\theta\left(-\frac{1}{4}\sqrt{\frac{21}{4\pi}}\sin\theta\,[5\cos^2\theta - 1]\right) = \frac{\sqrt{21\pi}}{16}iV$$

$$A_{3,-1} = -A_{3,1}^{*} = \frac{\sqrt{21\pi}}{16}iV$$

and

$$A_{3,3} = \frac{-4iV}{3}\int_0^\pi\left(-\frac{1}{4}\right)\sqrt{\frac{35}{4\pi}}\sin^4\theta\, d\theta = \frac{\pi}{8}\sqrt{\frac{35}{4\pi}}iV = A_{3,-3}$$

So, to order $l = 3$ we have:

$$\Phi(r,\theta,\emptyset) = \left(\frac{r}{a}\right)\left[iV\sqrt{\frac{3\pi}{2}}\left\{-\sqrt{\frac{3}{8\pi}}\sin\theta\, e^{i\varphi} + \sqrt{\frac{3}{8\pi}}\sin\theta\, e^{-i\varphi}\right\}\right]$$

$$+\left(\frac{r}{a}\right)^3\left[iV\frac{\sqrt{21\pi}}{16}\left\{-\frac{1}{4}\sqrt{\frac{21}{4\pi}}\sin\theta\,(5\cos^2\theta - 1)e^{i\varphi}\right.\right.$$

$$\left.\left.+\frac{1}{4}\sqrt{\frac{21}{4\pi}}\sin\theta\,(5\cos^2\theta - 1)e^{-i\varphi}\right\}\right]$$

$$+iV\frac{\sqrt{35\pi}}{16}\left\{-\frac{1}{4}\sqrt{\frac{35}{4\pi}}\sin^3\theta\, e^{3i\varphi} + \frac{1}{4}\sqrt{\frac{35}{4\pi}}\sin^3\theta\, e^{-3i\varphi}\right\}$$

Example 5.5. *A hollow sphere has potential at its surface given by* $V(\theta,\varphi)$. *Find the potential inside the sphere.*

Solution

For the potential inside the sphere we have:

$$\Phi(x) = \sum_{l=0}^{\infty}\sum_{m=-l}^{l} A_{lm}\left(\frac{r}{a}\right)^l Y_{lm}(\theta,\varphi)$$

$$= \sum_{l=0}^{\infty}\sum_{m=-l}^{l} \int d\Omega' Y_{lm}^{*}(\theta',\varphi')V(\theta',\varphi')\left(\frac{r}{a}\right)^l Y_{lm}(\theta,\varphi).$$

$$= \int d\Omega' V(\theta',\varphi)\sum_{l=0}^{\infty}\sum_{m=-l}^{l}\left(\frac{r}{a}\right)^l Y_{lm}^{*}(\theta',\varphi')Y_{lm}(\theta,\varphi)$$

$$= \int d\Omega' V(\theta',\varphi)\sum_{l=0}^{\infty}\sum_{m=-l}^{l}\left(\frac{r}{a}\right)^l P_l(\cos\gamma)\left(\frac{2l+1}{4\pi}\right)$$

where $\cos\gamma = \cos\theta\cos\theta' + \sin\theta\sin\theta'\cos(\varphi-\varphi')$. Using a relation on Legendre Polynomials like before:

$$\Phi(x) = \frac{1}{4\pi}\int d\Omega' V(\theta',\varphi')\left\{\left[\sum_{l=0}^{\infty}\left(\frac{r}{a}\right)^l\left(\frac{dP_{l+1}(\cos\gamma)}{d(\cos\gamma)} - \frac{dP_{l-1}(\cos\gamma)}{d(\cos\gamma)}\right)\right] + 1\right\}$$

Note:

$$\left[\sum_{l=0}^{\infty}\left(\frac{r}{a}\right)^l\left(\frac{dP_{l+1}(\cos\gamma)}{d(\cos\gamma)} - \frac{dP_{l-1}(\cos\gamma)}{d(\cos\gamma)}\right)\right]$$

$$= \frac{d}{d(\cos\gamma)}\left[\frac{a^2}{r}\sum_{l=0}^{\infty}\frac{r^{l+1}}{a^{l+2}}P_{l+1}(\cos\gamma) - r\sum_{l=1}^{\infty}\frac{r^{l-1}}{a^l}P_{l-1}(cos\gamma)\right]$$

$$= \frac{d}{d(\cos\gamma)}\left[\frac{a^2}{r}\frac{1}{|\vec{a}-\vec{r}|} - r\frac{1}{|\vec{a}-\vec{r}|} - \frac{a^2}{r}\left(\frac{r}{a^2}cos\gamma + \frac{1}{a}\right)\right]$$

So,

$$\Phi(x) = \frac{1}{4\pi}\int d\Omega' V(\theta',\varphi')\left\{\frac{d}{d(\cos\gamma)}\left[\frac{a^2-r^2}{r}\frac{1}{|\vec{a}-\vec{r}|}\right]\right\}$$

$$= \frac{1}{4\pi}\int d\Omega' V(\theta',\varphi')\left(\frac{a^2-r^2}{r}\right)\frac{d}{d(\cos\gamma)}\left(\frac{a}{\sqrt{r^2+a^2-2\,ar\,\cos\gamma}}\right)$$

$$= \frac{1}{4\pi}\int d\Omega' V(\theta',\varphi')\left(\frac{a^2-r^2}{r}\right)\left(\frac{ar}{(r^2+a^2-2\,ar\,\cos\gamma)^{\frac{3}{2}}}\right)$$

$$= \frac{a(a^2-r^2)}{4\pi}\int d\Omega' V(\theta',\varphi')(r^2+a^2-2arcos\gamma)^{-\frac{3}{2}}.$$

So, we have a second way to express the potential:

$$\Phi(x) = \frac{a(a^2 - r^2)}{4\pi} \int d\Omega' \, V(\theta', \varphi')(r^2 + a^2 - 2ar cos\gamma)^{-\frac{3}{2}}$$

Example 5.6. *Two point charges, $\pm q$ are placed at positions $\pm a$ on the z-axis. (a) Find the potential as an expansion in spherical harmonics; (b) keep $qa = p/2$ (a finite constant) as $a \to 0$, and find the potential for everywhere except $r = 0$. (This is a dipole at the origin directed along the z-axis.) (c) consider the dipole in part (b) with a surrounding, grounded, spherical shell at radius b. Find the potential inside by linear superposition.*

Solution

(a) The potential of the two charges is ($r > a$):

$$\Phi = q\left(\frac{1}{r_1} - \frac{1}{r_2}\right) = q\left\{\frac{1}{[r^2 + a^2 - 2ar\cos\theta]^{\frac{1}{2}}} - \frac{1}{[r^2 + a^2 - 2ar\cos\theta]^{\frac{1}{2}}}\right\}$$

$$= \frac{q}{r}\left\{\left[1 - 2\left(\frac{a}{r}\right)cos\theta + \left(\frac{a}{r}\right)^2\right]^{-\frac{1}{2}} - \left[1 + 2\left(\frac{a}{r}\right)cos\theta + \left(\frac{a}{r}\right)^2\right]^{-\frac{1}{2}}\right\}$$

$$= \frac{q}{r}\left[\sum_{l=0}^{\infty}\left(\frac{a}{r}\right)^l P_l(cos\theta) - \sum_{l=0}^{\infty}(-1)^l\left(\frac{a}{r}\right)^l P_l(cos\theta)\right]$$

$$= \frac{2q}{r}\left[P_1(\cos\theta)\left(\frac{a}{r}\right) + P_3(\cos\theta)\left(\frac{a}{r}\right)^3 + \cdots\right]$$

$$\Phi = 2q\sum_{l=0}^{\infty}\frac{a^{2l+1}}{r^{2l+2}}P_{2l+1}(cos\theta) \quad for \quad r > a.$$

The potential for r < a then is of the form regular at the origin with matching coefficients with the above, such that we simply have:

$$\Phi = 2q\sum_{l=0}^{\infty}\frac{r^{2l+1}}{a^{2l+2}}P_{2l+1}(cos\theta) \quad for\, r < a.$$

(b) Let's take $a \to 0$ while keeping $qa = p/2$ constant in the above form valid for $r > a$:

$$\lim_{a\to 0}\Phi = \frac{2(qa)}{r^2}P_1(\cos\theta) = \frac{p}{r^2}P_1(\cos\theta).$$

Thus, in the limit of the definition for a dipole, we find the dipole potential to be:

$$\Phi = \frac{p}{r^2} P_1(\cos\theta).$$

(c) Inside a grounded shell radius b we get the added charge induced on the shell such that it is at zero potential. Let's describe the potential inside the sphere for the induced shell charge alone, i.e. with the previous inner form but new coefficients (to be determined):

$$\Phi_{shell\ charge} = q \sum_{l=0}^{\infty} A^{2l+1}\, r^{2l+1} P_{2l+1}(cos\theta).$$

At $r = b$ the grounded shell must have a superposition of contributions such that the shell has zero potential (at all orders of r), i.e., we must have agreement on coefficients such that:

$$2q\left(\frac{a^{2l+1}}{b^{2l+2}}\right) = -q\, A_{2l+1} b^{2l+1} \Rightarrow A_{2l+1} = -2\left(\frac{a}{b}\right)^{2l+1} \frac{1}{b^{2l+2}}\,.$$

Thus,

$$\Phi_{total} = 2q \sum_{l=0}^{\infty} P_{2l+1}(\cos\theta) \left\{\frac{a^{2l+1}}{r^{2l+2}} - \left(\frac{a}{b}\right)^{2l+1} \frac{r^{2l+1}}{b^{2l+2}}\right\} \quad for\, r > a$$

and for $r < a$we get:

$$\Phi_{total} = 2q \sum_{l=0}^{\infty} P_{2l+1}(\cos\theta) \left\{\frac{r^{2l+1}}{a^{2l+2}} - \left(\frac{a}{b}\right)^{2l+1} \frac{r^{2l+1}}{b^{2l+2}}\right\} \quad for\, r < a.$$

Example 5.7. *Consider three charges on the z-axis: a -2q charge at the origin and two q charges at* $\pm a$*.Have this be inside a grounded conducting sphere of radius b (b>a). (a) First consider the description for the potential when there is no grounded sphere and show the limiting form as* $a \to 0$ *while keeping* $qa^2 = Q$ *(use spherical coordinates). (b) Repeat for description of potential inside the sphere when it is added (for* $r > a$ *and* $r < a$ *cases, and as* $a \to 0$ *while keeping* $a^2 = Q$ *).*

Solution

(a) The potential is:

$$\Phi = \frac{q}{|\vec{r} - a\hat{z}|} - \frac{2q}{r} + \frac{q}{|\vec{r} + a\hat{z}|}$$

Shifting to spherical coordinates we have:

$\frac{1}{|\vec{r}-a\hat{z}|} = \sum_{l=0}^{\infty} \frac{r_<^l}{r_>^{l+1}} P_l(cos\gamma)$, $\gamma = \theta$ and $\frac{1}{|\vec{r}+a\hat{z}|} = \sum_{l=0}^{\infty} \frac{r_<^l}{r_>^{l+1}} P_l(cos\gamma')$, $\gamma' = \pi - \theta$.

Since $\gamma' = \pi - \gamma$, we have $P_l(\cos\gamma') = (-1)^l P_l(cos\theta)$, allowing grouping to get:

$$\Phi(\vec{r}) = q \sum_{l=0}^{\infty} (1 + (-1)^l) \frac{r_<^l}{r_>^{l+1}} P_l(cos\theta) - \frac{2q}{r}$$

$$= 2q \sum_{\substack{l=0 \\ Even}}^{\infty} \frac{r_<^l}{r_>^{l+1}} P_l(cos\theta) - \frac{2q}{r}$$

First consider $r_< = a$:

$$\Phi(\vec{r}) = 2q \sum_{\substack{l=0 \\ Even}}^{\infty} \frac{a^l}{r^{l+1}} P_l(cos\theta) - \frac{2q}{r}.$$

Now let $a \to 0$ while $qa^2 = Q$:

$$\lim_{a\to 0,\ qa^2=Q} \Phi(\vec{r}) = \frac{2Q}{r^3} P_2(cos\theta).$$

(b) The potential due to the charge induced on the sphere alone has the form:

$$\Phi_{induced\ sphere} = \sum_{l=0}^{\infty} a_l \left(\frac{r}{b}\right) P_l(cos\theta).$$

For $a < r < b$ we then have by superposition:

$$\Phi_{total} = 2q \sum_{\substack{l=2 \\ even}}^{\infty} \frac{a^l}{r^{l+1}} P_l(cos\theta) + \sum_{l=0}^{\infty} a_l \left(\frac{r}{b}\right) P_l(cos\theta).$$

Notice that the $l = 0$ terms are cancelled in the first sum. Let's take $r \to b$, where the potential should be zero at the grounded sphere: $\Phi_{tot}(r = b) = 0$, in other words:

$$\sum_{\substack{l=2 \\ even}}^{\infty} \left[2q \left(\frac{a}{b}\right)^l \frac{1}{b} P_l(cos\theta) + a_l P_l(\cos\theta)\right] + a_0 = 0$$

examining by order: $a_0 = 0$ and $a_{odd} = 0$, while for the others:

$$\left(\frac{a}{b}\right)^l \frac{1}{b}(2q) + a_l = 0 \to \quad a_l = -2q\frac{1}{b}\left(\frac{a}{b}\right)^l.$$

Thus, we have for the potential:

$$\Phi_{total} = 2q \sum_{\substack{l=2 \\ even}}^{\infty} \left(\frac{a^l}{r^{l+1}} - \frac{a^l r^l}{b^{2l+1}} \right) P_l(cos\theta) \ for\ a < r < b$$

and

$$\Phi_{total} = 2q \sum_{\substack{l=2 \\ even}}^{\infty} \left(\frac{r^l}{a^{l+1}} - \frac{a^l r^l}{b^{2l+1}} \right) P_l(cos\theta) + 2q\left(\frac{1}{a} - \frac{1}{r}\right) \ for\ r < a.$$

Notice in the latter, the $l = 0$ terms don't cancel like before. Returning to the $a < r < b$ form for evaluation in limit $a \to 0$ while $qa^2 = Q$ we get:

$$\Phi(\vec{r}) = \frac{2Q}{r^3} P_2(\cos\theta) \left\{1 - \frac{r^5}{b^5}\right\}.$$

Example 5.8. *Consider a Cylinder, radius b, with axis on the z-axis, with bottom at z=0 and top at z=L.*
Suppose the potential on the ends is zero while on the sides it has potential $V(\phi, z)$.

Solution

As before we have $\Phi = R(\rho)Q(\phi)Z(z)$ with:

$$\frac{d^2Q}{d\phi^2} + m^2 Q = 0 \Rightarrow Q = e^{\pm im\phi},$$

and m integer-valued for Q to be single-valued. Similarly we have

$$\frac{d^2Z}{dz^2} + k^2 Z = 0 \Rightarrow Z = e^{\pm ikz},$$

and the boundary conditions $z(0) = 0 = Z(L) \to Z = \sin\left(\frac{n\pi z}{L}\right)$.

We, thus, get the radial differential equation in the form of the modified Bessel equation:

$$\frac{d}{d\rho}\left(\rho\frac{dR}{d\rho}\right) - \left(k^2\rho + \frac{m^2}{\rho}\right) R = 0$$

The solutions consist of the modified Bessel functions: $I_m(k\rho), k_m(k\rho)$. Since $\rho = 0$ is in the region of interest we only need $I_m(k\rho)$:

$$\Phi(\rho, \phi, z) = \sum_{m,n} I_m\left(\frac{n\pi\rho}{L}\right) \sin\left(\frac{n\pi z}{L}\right) [A_{mn}\cos(m\phi) + B_{mn}\sin(m\phi)].$$

To find A_{mn}, B_{mn} expand $V(\phi, z) = \Phi(b, \phi, z)$ in a Fourier-Bessel series:

$$\int_0^{2\pi} d\phi \int_0^L dz \sin\left(\frac{n\pi z}{L}\right) \begin{Bmatrix} \cos(m\phi) \\ \sin(m\phi) \end{Bmatrix} V(\phi, z)$$

$$= \int_0^{2\pi} d\phi \int_0^L dz \sin\left(\frac{n\pi z}{L}\right) \begin{Bmatrix} \cos(m\phi) \\ \sin(m\phi) \end{Bmatrix} \Phi(b, \phi, z)$$

$$= I_m\left(\frac{n\pi b}{L}\right) \underbrace{\int_0^L \sin^2\left(\frac{n\pi\delta}{L}\right) dx}_{=L/2} \underbrace{\begin{Bmatrix} A_{mn} \\ B_{mn} \end{Bmatrix} \int_0^{2\pi} \begin{matrix} \cos^2(m\phi) \\ \sin^2(m\phi) \end{matrix} d\phi}_{\substack{A_{0,n}=\pi \ if \ m\neq 0 \\ A_{0,n}=0 \ if \ m=0 \ and \ sin \\ A_{0,n}=2\pi \ if \ m=0 \ and \ cos.}}$$

Thus,

$$\int_0^{2\pi} d\phi \int_0^L dz \sin\left(\frac{n\pi z}{L}\right) \begin{Bmatrix} \cos(m\phi) \\ \sin(m\phi) \end{Bmatrix} V(\phi, z) = \frac{\pi L}{2} I_m\left(\frac{n\pi b}{L}\right) \begin{Bmatrix} A_{mn} \\ B_{mn} \end{Bmatrix}_{m\neq 0},$$

$$B_{0,n} = 0,$$

We, thus, have:

$$\begin{matrix} A_{m,n} \\ B_{m,n} \end{matrix}\Bigg\} = \frac{2}{\pi L I_m\left(\frac{n\pi b}{L}\right)} \int_0^{2\pi} d\phi \int_0^L dz \sin\left(\frac{n\pi z}{L}\right) \begin{Bmatrix} \cos(m\phi) \\ \sin(m\phi) \end{Bmatrix} V(\phi, z)$$

with special cases:

$$A_{0,n} = \frac{2}{\pi L I_0\left(\frac{n\pi b}{L}\right)} \int_0^{2\pi} d\phi \int_0^L dz \sin\left(\frac{n\pi z}{L}\right) V(\phi, z)$$

and

$B_{0,n} = 0.$

5.2 Spherical Harmonics

Let's now consider the behavior of Fields in a Conical Hole or near a Sharp Point. For this problem we explore boundary value issues, not source issues but again with azimuthal symmetry. We not only have azimuthal symmetry, we have only a limited range of θ. We now seek solutions finite and *single valued on the range* of $x = \cos\theta$ of $\cos\beta \leq x \leq 1$. Since we demand regularity at $x = 1$ (no charge there) it is convenient to make a series expansion around $x = 1$ instead of $x = 0 \rightarrow$ this is done by the change of variable:

$$\varepsilon = \frac{1}{2}(1 - x) \quad d\varepsilon = -\frac{1}{2}dx$$

The Legendre Eqn. becomes:

$$\frac{d}{dx}\left[(1 - x^2)\frac{dP}{dx}\right] + \ell(\ell + 1)P = 0.$$

(Eqn. 5-49)

Recall that we obtained this form, with ℓ integer valued, due to boundary conditions on the full range allowed. The boundary conditions are no longer at full range, thus ℓ need no longer be an integer. The notation will be changed to reflect this with $\ell \rightarrow V$:

$$\frac{d}{-2d\varepsilon}\left[4(-\varepsilon)(1 - \varepsilon)\frac{dP}{-2d\varepsilon}\right] + V(V + 1)P = 0$$

$$\rightarrow \quad \frac{d}{d\varepsilon}\left[\varepsilon(1 - \varepsilon)\frac{dP}{d\varepsilon}\right] + V(V + 1)P = 0.$$

(Eqn. 5-50)

Let's substitute a power series solution $P(\varepsilon) = \varepsilon^\alpha \sum_{j=0}^{\infty} a_j\, \varepsilon^j$:

$$\frac{d}{d\varepsilon}\left[(\alpha + j)\left(\sum_{j=0}^{\infty} a_j\, \varepsilon^{j+\alpha-1} - \sum_{j=0}^{\infty} a_j\, \varepsilon^{j+\alpha+1}\right)\right] + V(V + 1)\sum_{j=0}^{\infty} a_j\, \varepsilon^{\alpha+j} = 0$$

$$\sum_{j=0}^{\infty} a_j\left[(\alpha + j)^2\varepsilon^{j+\alpha-1} - (\alpha + j)(\alpha + j + 1)\varepsilon^{j+\alpha}\right] + V(V + 1)\sum_{j=0}^{\infty} a_j\, \varepsilon^{\alpha+j} = 0$$

(Eqn. 5-51)

Lowest coefficient of ε is $\varepsilon^{(\alpha-1)} \cdot \alpha^2$for $j = 0 \Rightarrow \alpha = 0$. So,

$$\sum_{j=0} a_j\left[j^2\varepsilon^{j-1} - j(j + 1)\varepsilon^j + V(V + 1)\varepsilon^j\right] = 0$$

becomes

$$a_{j+1}(j + 1)^2 - a_j[j(j + 1) - V(V + 1)] = 0\,.$$

The relation on coefficients is thus:

$$\frac{a_{j+1}}{a_j} = \frac{(j - V)(j + V - 1)}{(j + 1)^2}.$$

(Eqn. 5-52)

Let $a_0 = 1$ to normalize solution to units at $\varepsilon = 0(\cos\theta = 1)$, thus

$$P_V(\varepsilon) = 1 + \frac{(-V)(V+1)}{1!\,1!}\varepsilon + \frac{(-V)(V+1)(-V+1)(V+2)}{2!\,2!}\varepsilon^2 + \cdots$$

(Eqn. 5-53)

If V is zero, or a positive integer, the series terminates. For $V = \ell = 0,1,2$ ... we simply regain the Legendre polynomials of that order. For V non-integer get Legendre function of the first kind and order V. The series above is an example of a hypergeometric function ${}_2F_1(a, b; c; z)$ whose series expansion is:

$${}_2F_1(a, b; c; z) = 1 + \frac{ab}{c}\frac{8}{1!} + \frac{a(a+1)b(b+1)}{c(c+1)}\frac{z^2}{2!} + \cdots$$

(Eqn. 5-54)

Comparison shows:

$$P_V(x) = {}_2F_1\left(-V, V+1; 1; \frac{1-x}{2}\right).$$

(Eqn. 5-55)

The basic solution to the Laplace boundary value problem for the conical conducting surface will involve the $Ar^V P_V(\cos\theta)$ terms (since $V > 0$ at origin). Since the potential must vanish at $\theta = \beta$ for all r: $P_V(\cos\beta) = 0$, which has an infinite number of solutions V_k. Thus

$$\Phi(r, \theta) = \sum_{k=1}^{\infty} A_k\, r^{V_k} P_{V_k}(\cos\theta)$$

(Eqn. 5-56)

For the potential near the origin the potential is dominated by the smallest root (call it V):

$$\Phi(r, \theta) \simeq Ar^V P_V(\cos\theta).$$

(Eqn. 5-57)

Subsequently:

$$\left.\begin{aligned} E_r &= -\frac{\partial\Phi}{\partial r} = -vAr^{V-1}P_V(\cos\theta) \\ E_\theta &= -\frac{1}{8}\frac{\partial\Phi}{\partial\theta} = Ar^{V-1}\sin\theta\, P_V'(\cos\theta) \\ \sigma(r) &= -\frac{1}{4\pi}E_\theta|_{\theta=\beta} \simeq -\frac{A}{4\pi}r^{V-1}\sin\beta\, P_V'(\cos\beta) \\ &\to 0 \end{aligned}\right\}\text{vary as } r^{V-1}\text{as } r$$

(Eqn. 5-58)

And,

$$V \simeq \frac{2.405}{\beta} - \frac{1}{2}\, for\ \beta \ll 1.$$

(Eqn. 5-59)

Let's now consider the fully general solution $m \neq 0$, for this we have the Associated Legendre Functions and need to introduce the Spherical Harmonics $Y_{\ell m}(\theta, \phi)$.

Associated Legendre Functions and Spherical Harmonics $Y_{\ell m}(\theta, \phi)$

When $m \neq 0$ and the full range in θ allowed then ℓ must be zero or a positive integer and m must be integer with values $-\ell, (\ell - 1) \dots 0, \dots., (\ell - 1), \ell$. The Associated Legendre Functions can be written for positive m as:

$$P_\ell^m(x) = (-1)^m (1 - x^2)^{m/2} \frac{d^m}{dx^m} P_\ell(x)\,,$$

(Eqn. 5-60)

where there are differing phase conventions and this follows Condon-Shortley Theory of atomic spectra. A form valid for any sign of m is then:

$$P_\ell^m(x) = \frac{(-1)^m}{2^\ell \ell!} (1 - x^2)^{m/2} \frac{d^{\ell+m}}{dx^{\ell+m}} (x^2 - 1)^\ell\,.$$

(Eqn. 5-61)

Note that $P_\ell^{-m}(x)$ and $P_\ell^m(x)$ are proportional since the equation depends only on m^2, there relation can be determined and is:

$$P_\ell^{-m}(x) = (-1)^m \frac{(\ell - m)!}{(\ell + m)!} P_\ell^m(x)\,.$$

(Eqn. 5-62)

As before, with regular Legendre functions, we have the orthogonality relations:

$$\int_{-1}^{1} P_{\ell'}^m(x)\, P_\ell^m(x) = \frac{2}{2\ell + 1} \frac{(\ell + m)!}{(\ell - m)!} \delta_{\ell'\ell}\,.$$

(Eqn. 5-63)

It is convenient to combine the angular factors and construct orthonormal functions over the unit sphere. These functions are called spherical harmonics, so by definition:

$$Y_{\ell m}(\theta, \phi) = \sqrt{\frac{2\ell + 1}{4\pi} \frac{(\ell - m)!}{(\ell + m)!}}\, P_\ell^m(\cos\theta) e^{im\phi}$$

(Eqn. 5-64)

and we have a m and –m relation as before:

$$Y_{\ell,-m}(\theta,\phi) = (-1)^m Y^*_{\ell m}(\theta,\phi)$$

(Eqn. 5-65)

We then have normalization and orthogonality:

$$\int_0^{2\pi} d\phi \int_0^{\pi} \sin\theta \, d\theta Y^*_{\ell' m'}(\theta,\phi) Y_{\ell m}(\theta,\phi) = \delta_{\ell'\ell}\delta_{m'm}$$

(Eqn. 5-66)

as well as completeness:

$$\sum_{\ell=0}^{\infty}\sum_{m=-\ell}^{\ell} Y^*_{\ell m}(\theta',\phi')\, Y_{\ell m}(\theta,\phi) = \delta(\phi - \phi)\delta(\cos\theta - \cos\theta')\,.$$

(Eqn. 5-67)

For the special case $m = 0$:

$$Y_{\ell 0}(\theta,\phi) = \sqrt{\frac{2\ell+1}{4\pi}} P_\ell(\cos\theta)\,.$$

(Eqn. 5-68)

Thus, an arbitrary function $g(\theta,\phi)$ can be expanded in spherical harmonics:

$$g(\theta,\phi) = \sum_{\ell=0}^{\infty}\sum_{m=-\ell}^{\ell} A_{\ell m}\, Y_{\ell m}(\theta,\phi), \;\; where \quad A_{\ell m} = \int d\Omega\, Y^*_{\ell m}(\theta,\phi) g(\theta,\phi).$$

(Eqn. 5-69)

Note that $[g(\theta,\phi)]_{\theta=0} = \sum_{\ell=0}^{\infty}\sqrt{\frac{2\ell+1}{4\pi}} A_{\ell 0}$ since all terms with $m \neq 0$ vanish at $\theta = 0$, and

$$A_{\ell 0} = \sqrt{\frac{2\ell+1}{4\pi}} \int d\Omega\, P_\ell(\cos\theta) g(\theta,\phi)\,.$$

(Eqn. 5-70)

The general solution with spherical harmonics:

$$\Phi(r,\theta,\phi) = \sum_{\ell=0}^{\infty}\sum_{m=-\ell}^{\ell} \left[A_{\ell m} + B_{\ell m} r^{-(\ell+0)}\right] Y_{\ell m}(\theta,\phi)$$

(Eqn. 5-71)

Addition theorem for Spherical Harmonics

We need to understand $P_\ell(\gamma)$ in terms of Spherical Harmonics. We will also confirm the relation

$$\cos\gamma = \cos\theta\cos\theta' + \sin\theta\sin\theta'\cos(\phi - \phi').$$

Two coordinate vectors $\vec{x}$ and $\vec{x}'$ wiith spherical coordinates (r, θ, ϕ) and (r', θ', ϕ') have an angle γ between them. The addition theorem expresses a Legendre polynomial of order ℓ in the angle γ in terms of products of the spherical harmonics of the angles θ, ϕ and θ', ϕ':

$$P_\ell(\cos\gamma) = \sum_{\ell'=0}^{\infty}\sum_{m=-\ell'}^{\ell'} A_{\ell' m}(\theta', \phi') Y_{\ell' m}(\theta, \phi)$$

(Eqn. 5-72)

Choose $\vec{x}'$ on z-axis, $\gamma \to \theta$ polar angle and

$$\nabla'^2 P_\ell(\cos\gamma) + \frac{\ell(\ell+1)}{r^2} P_\ell(\cos\gamma) = 0$$

(Eqn. 5-73)

If axes are rotated $\nabla'^2 = \nabla^2$ and r is uncharged, consequently $P_\ell(\cos\gamma)$ still satisfies an equation of the above form, a spherical harmonic of order ℓ (only $\ell' = \ell$ terms). So,

$$P_\ell(\cos\gamma) = \sum_{m=-\ell}^{\ell} A_m(\theta', \phi') Y_{\ell m}(\theta, \phi)$$

(Eqn. 5-74)

and

$$A_m(\theta', \phi') = \int Y_{\ell m}^*(\theta, \phi) P_\ell(\cos\gamma) d\Omega\,.$$

(Eqn. 5-75)

Viewed as $m = 0$ coefficient in expansion of $\sqrt{\frac{4\pi}{2\ell+1}} Y_{\ell m}^*(\theta, \phi) = g(\theta, \phi)$ from (previous eqn)

$$A_{\ell 0} = \sqrt{\frac{2\ell+1}{4\pi}} \int d\Omega\, P_\ell(\cos\theta) g(\theta, \phi)$$

(Eqn. 5-76)

and using,

$$g(\theta, \phi) = \sqrt{\frac{4\pi}{2\ell+1}} Y_{\ell m}^*(\theta, \phi)$$

$$[g(\theta,\phi)]_{\theta=0} = \sum_{\ell=0}^{\infty} \sqrt{\frac{2\ell+1}{4\pi}} A_{\ell 0}$$

with $A_{\ell 0}$ identified with $A_m, \ell = \ell'$:

$$A_m = \sqrt{\frac{4\pi}{2\ell+1}} \left\{ \sqrt{\frac{4\pi}{2\ell+1}} Y_{\ell m}^*[\theta(\gamma,\beta), \phi(\gamma,\beta)] \right\}_{\gamma=0}$$

(Eqn. 5-77)

So, as $\gamma \to 0 \quad (\theta,\phi) \to (\theta',\phi')$ so $A_m = \frac{4\pi}{2\ell+1} Y_{\ell m}^*(\theta',\phi')$, and

$$P_\ell(\cos\gamma) = \frac{4\pi}{2\ell+1} \sum_{m=-\ell}^{\ell} Y_{\ell m}^*(\theta',\phi')\, Y_{\ell m}(\theta,\phi)$$

(Eqn. 5-78)

We also have

$$\sum_{m=-\ell}^{\ell} [Y_{\ell m}(\theta,\phi)]^2 = \frac{2\ell+1}{4\pi}$$

(Eqn. 5-79)

and:

$$\frac{1}{|\vec{x}-\vec{x}'|} = 4\pi \sum_{\ell=0}^{\infty} \sum_{m=-\ell}^{\ell} \frac{1}{2\ell+1} \frac{r_<^\ell}{r_>^{\ell+1}} Y_{\ell m}^*(\theta',\phi') Y_{\ell m}(\theta,\phi)$$

(Eqn. 5-80)

That completes the descriptions in spherical coordinates, let's now turn to systems more easily described using cylindrical coordinates.

5.3 Spherical Green's function

Let's start with an expansion of Green Function in Spherical Coordinates. In the case of no boundary surfaces:

$$\frac{1}{|\vec{x}-\vec{x}'|} = \sum_{\ell=0}^{\infty} \sum_{m=-\ell}^{\ell} \frac{4\pi}{2\ell+1} \frac{r_<^\ell}{r_>^{\ell+1}} Y_{\ell m}^*(\theta',\phi') Y_{\ell m}(\theta,\phi).$$

(Eqn. 5-81)

The Green's function for the "exterior" problem with a spherical boundary at $r = a$ appears to simply be:

$$G(\vec{x}-\vec{x}') = \frac{1}{|\vec{x}-\vec{x}'|} - \frac{a}{x'\left(\vec{x}-\frac{a^2}{x'^2}\vec{x}'\right)} = \sum_{\ell=0}^{\infty} \sum_{m=-\ell}^{\ell} \frac{4\pi}{2\ell+1} \left[\frac{r_<^\ell}{r_>^{\ell+1}} - \frac{a}{r'} \frac{\left(\frac{a^2}{r'}\right)^\ell}{r^{\ell+1}} \right] Y^* Y$$

$$= \sum_{\ell=0}^{\infty} \sum_{m=-\ell}^{\ell} \frac{4\pi}{2\ell+1} \left[\frac{r_<^\ell}{r_>^{\ell+1}} - \frac{1}{a} \left(\frac{a^2}{rr'} \right)^{\ell+1} \right] Y_{\ell m}^*(\theta', \phi') Y_{\ell m}(\theta, \phi)$$

(Eqn. 5-82)

Let's show this to be correct via a systematic construction of $G(\vec{x} - \vec{x}')$, where:

$$\nabla_x^2 G(\vec{x} - \vec{x}') = -4\pi\delta(\vec{x} - \vec{x}'),$$

(Eqn. 5-83)

and in spherical coordinates:

$$\delta(\vec{x} - \vec{x}') = \frac{1}{r^2} \delta(r - r') \delta(\phi, \phi') \delta(\cos\theta - \cos\theta')$$

(Eqn. 5-84)

or

$$\delta(\vec{x} - \vec{x}') = \frac{1}{r^2} \delta(r - r') \sum_{\ell=0}^{\infty} \sum_{m=-\ell}^{\ell} Y_{\ell m}^* (\theta', \phi') Y_{\ell m}(\theta, \phi).$$

(Eqn. 5-85)

So,

$$G(\vec{x} - \vec{x}') = \sum_{\ell=0}^{\infty} \sum_{m=-\ell}^{\ell} A_{\ell m} (r|r', \theta', \phi') Y_{\ell m}(\theta, \phi)$$

(Eqn. 5-86)

where $A_{\ell m}(r|r', \theta', \phi') = g_\ell(r, r') Y_{\ell m}^*(\theta', \phi')$. Thus

$$\frac{1}{r} \frac{d^2}{dr^2} (r g_\ell) - \frac{\ell(\ell+1)}{r^2} g_\ell(r, r') = -\frac{4\pi}{r^2} \delta(r - r').$$

(Eqn. 5-87)

For $r \neq r'$ the solution is simply:

$$g_\ell(r, r') = \begin{cases} A r^\ell + B r^{-(\ell+1)} & r < r' \\ A' r^\ell + B' r^{-(\ell+1)} & r > r' \end{cases}$$

(Eqn. 5-88)

Consider, for a general case, boundary surfaces to be concentric spheres at $r = a, r = b$. Then have

$$g_\ell(r, r') = \begin{cases} A \left(r^\ell - \dfrac{a^{2\ell+1}}{r^{\ell+1}} \right) & r < r' \\ B \left(\dfrac{1}{r^{\ell+1}} - \dfrac{r^\ell}{b^{2\ell+1}} \right) & r > r' \end{cases}$$

(Eqn. 5-89)

For which a combined solution would be:

$$g_\ell(r, r') = C \left(r_<^\ell - \frac{a^{2\ell+1}}{r_<^{\ell+1}} \right) \left(\frac{1}{r_>^{\ell+1}} - \frac{r_>^\ell}{b^{2\ell+1}} \right)$$

(Eqn. 5-90)

but need to determine C:

$$\left\{\frac{d}{dr}[rg_\ell(r,r')]\right\}_{r'+\epsilon} - \left\{\frac{d}{dr}[rg_\ell(r,r')]\right\}_{r'-\epsilon} = -\frac{4\pi}{r'}$$

$$\left\{\frac{d}{dr}[rg_\ell(r,r')]\right\}_{r'+\epsilon} = \frac{-C}{r'}\left[1-\left(\frac{a}{r'}\right)^{2\ell+1}\right]\left[\ell+(\ell+1)\left(\frac{r'}{b}\right)^{2\ell+1}\right]$$

and

$$\left\{\frac{d}{dr}[rg_\ell(r,r')]\right\}_{r'-\theta} = \frac{-C}{r'}\left[\ell+1+\ell\left(\frac{a}{r'}\right)^{2\ell+1}\right]\left[1-\left(\frac{r'}{b}\right)^{2\ell+1}\right]$$

Thus,

$$C = \frac{4\pi}{(2\ell+1)\left[1-\left(\frac{a}{b}\right)^{2\ell+1}\right]}$$

(Eqn. 5-91)

and

$$G(\vec{x}-\vec{x}') = 4\pi\sum_{\ell=0}^{\infty}\sum_{m=-\ell}^{\ell}\frac{Y_{\ell m}^*(\theta',\phi')Y_{\ell m}(\theta,\phi)}{(2\ell+1)\left[1-\left(\frac{a}{b}\right)^{2\ell+1}\right]}\left(r_<^\ell-\frac{a^{2\ell+1}}{r_<^{\ell+1}}\right)\left(\frac{1}{r_>^{\ell+1}}-\frac{r_>^\ell}{b^{2\ell+1}}\right),$$

(Eqn. 5-92)

which covers the previous examples as special cases. Let's now examine some problems with these results in hand.

Solutions with the Spherical Green Function Expansion

Using $\Phi(\vec{x}) = \int_v \rho(\vec{x}')G(\vec{x}-\vec{x}')d^3x - \frac{1}{4\pi}\oint_S \Phi(\vec{x}')\frac{\partial G}{\partial n'}da'$, consider the potential inside a sphere, radius b:

$$G(\vec{x}-\vec{x}') = 4\pi\sum_{\ell=0}^{\infty}\sum_{m=-\ell}^{\ell}\frac{Y_{\ell m}^*(\theta',\phi')Y_{\ell m}(\theta,\phi)}{(2\ell+1)\left[1-\left(\frac{a}{b}\right)^{2\ell+1}\right]}\left(r_<^\ell-\frac{a^{2\ell+1}}{r_<^{\ell+1}}\right)\left(\frac{1}{r_>^{\ell+1}}-\frac{r_>^\ell}{b^{2\ell+1}}\right)$$

If $a \to 0$

$$G(\vec{x}-\vec{x}') = 4\pi\sum_{\ell=0}^{\infty}\sum_{m=-\ell}^{\ell}\frac{Y_{\ell m}^*(\theta',\phi')Y_{\ell m}(\theta,\phi)}{(2\ell+1)}r_<^\ell\left(\frac{1}{r_>^{\ell+1}}-\frac{r_>^\ell}{b^{2\ell+1}}\right)$$

Let's consider a ring of radius "a", charge Q, inside a grounded, conducting sphere of radius b. We have azimuthal symmetry ($m = 0$), so with

$$\Phi(\vec{x}) = \int_v \rho\,(\vec{x}')\left\{4\pi \sum_{\ell=0}^{\infty} \frac{1}{(2\ell+1)} Y_{\ell m}^*(\theta',\phi')Y_{\ell m}(\theta,\phi)r_<^\ell\left(\frac{1}{r_>^{\ell+1}} - \frac{r_>^\ell}{b^{2\ell+1}}\right)\right\} d^3x$$

and $Y_{\ell 0} = \sqrt{\frac{2\ell+1}{4\pi}}P_\ell(\cos\theta)$, and for a ring charge:

$$\rho(\vec{x}') = \frac{Q}{2\pi a}\frac{1}{a}\delta(r'-a)\delta(\cos\theta')$$

So,

$$\Phi(\vec{x}) = Q\sum_{\ell=0}^{\infty} P_\ell(0)\, r_<^\ell\left(\frac{1}{r_>^{\ell+1}} - \frac{r_>^\ell}{b^{2\ell+1}}\right)P_\ell(\cos\theta).$$

Using the relations

$$P_{2n+1}(0) = 0 \quad and \quad P_{2n}(0) = \frac{(-1)^n(2n-1)!!}{2^n n!},$$

get:

$$\Phi(\vec{x}) = Q\sum_{n=0}^{\infty}\frac{(-1)^n(2n-1)!!}{2^n n!} r_<^{2n}\left(\frac{1}{r_>^{2n+1}} - \frac{r_>^{2n}}{b^{4n+1}}\right)P_{2n}(\cos\theta).$$

Similarly, consider a uniform line charge, length 2b, total charge Q, inside a grounded, conducting sphere of radius b:
Same as before but now have:

$$\int \rho(\vec{x}')\, r'\sin\theta'\, d\phi r' d\theta' dr' = Q$$

thus

$$\rho(\vec{x}') = \frac{Q}{2b}\frac{1}{2\pi r'^2}[\delta(\cos\theta'-1) + \delta(\cos\theta'+1)]$$

And

$$\Phi(\vec{x}) = \frac{Q}{2b}\sum_{\ell=0}^{\infty}[P_\ell(1) + P_\ell(-1)]\,P_\ell(\cos\theta)\int_0^b r_<^\ell\left(\frac{1}{r_>^{\ell+1}} - \frac{r_>^\ell}{b^{2\ell+1}}\right)dr'.$$

To solve further, must break up integral into $0 \le r' < r$ and $r \le r' = b$:

$$\int_0^b [\ldots]dr' = \frac{(2\ell+1)}{\ell(\ell+1)}\left[1-\left(\frac{r}{b}\right)^{\ell}\right] \quad \begin{array}{l}\ell=0 \\ \text{indeterminant} \Rightarrow \text{L'Hop}\end{array}$$

$$\Rightarrow \int_0^b = \ln\left(\frac{b}{r}\right)$$

with $P_\ell(-1) = (-1)^\ell$ then gives:

$$\Phi(\vec{x}) = \frac{Q}{b}\left\{\ln\left(\frac{b}{r}\right) + \sum_{j=1} \frac{(4j+1)}{2j(2j+1)}\left[1-\left(\frac{r}{b}\right)^{2j}\right]P_{2j}(\cos\theta)\right\}.$$

and surface charge density is $\sigma(\theta) = \frac{1}{4\pi}\frac{\partial\Phi}{\partial r}\Big|_{r=b}$.

Example 5.9. *Consider two concentric spheres, radii a,b (b>a), each split (with gaps) into hemispheres in the x-y plane. The upper inner hemisphere and the lower outer sphere are at V potential, the rest at zero. Solve for the potential using Green's functions.*

Solution

$$\Phi(\vec{x}) = \int_v \rho(\vec{x}')G_D(\vec{x},\vec{x}')d^3x' + \frac{1}{4\pi}\oint_S \left\{\frac{\partial\Phi}{\partial n'}G_D(\vec{x},\vec{x}') - \Phi(\vec{x}')\frac{\partial G_D(\vec{x},\vec{x}')}{\partial n'}\right\}da'$$

Here $\rho(\vec{x}') = 0$ and $G_D(\vec{x},\vec{x}') = 0$ for $|\vec{x}'| = a$ or b (definition). We know:

$$G_D(\vec{x},\vec{x}') = \sum_{\ell,m} \frac{4\pi}{2\ell+1}\frac{Y^*_{\ell m}(\theta',\phi')Y_{\ell m}(\theta,\phi)}{\left[1-\left(\frac{a}{b}\right)^{2\ell+1}\right]}\left(r_<^\ell - \frac{a^{2\ell+1}}{r_<^{\ell+1}}\right)\left(\frac{1}{r_>^{\ell+1}} - \frac{r_>^\ell}{b^{2\ell+1}}\right)$$

At $|\vec{x}'| = a, \vec{n}' = -\hat{r}'$ ($\hat{n}'$ is a unit normal away from the region of interest.)

$|\vec{x}'| = b, \vec{n}' = -\hat{r}'$

Azimuthal symmetry $\Rightarrow m = 0, Y_{\ell 0}(\theta,\phi) = \sqrt{\frac{2\ell+1}{4\pi}}P_\ell(\cos\theta)$

At $|\vec{x}'| = b, r = r_<, r' = r_> = b$, thus:

$$\left.\frac{\partial G'(\vec{x},\vec{x}')}{\partial n'}\right|_{r'=b} = \left.\frac{\partial G'}{\partial r'}\right|_{r'=b}$$

$$= \sum_{\ell=0}^{\infty} \frac{P_\ell(\cos\theta)P_\ell(\cos\theta')}{1-\left(\frac{a}{b}\right)^{2\ell+1}} \left(r^\ell - \frac{a^{2\ell+1}}{r^{\ell+1}}\right)\left(\frac{-(2\ell+1)}{b^{\ell+2}}\right)$$

At $|\vec{x}'| = a, r = r_>, r' = r_< = a$, thus:

$$\left.\frac{\partial G'(\vec{x},\vec{x}')}{\partial n'}\right|_{r'=a} = \left.\frac{\partial G'}{\partial r'}\right|_{r'=a}$$

$$= \sum_{\ell=0}^{\infty} \frac{P_\ell(\cos\theta)P_\ell(\cos\theta')}{1-\left(\frac{a}{b}\right)^{2\ell+1}} \left(-(2\ell+1)a^{\ell+1}\right)\left(\frac{1}{r^{\ell+1}} - \frac{r^\ell}{b^{\ell+2}}\right)$$

Meanwhile:

$$\Phi(\vec{x}) = \frac{1}{4\pi}\oint_S \Phi(\vec{x}')\frac{\partial G(\vec{x},\vec{x}')}{\partial n'}da',$$

which becomes:

$$\Phi(\vec{x}) = \frac{V}{4\pi}\sum_{\ell=0}^{\infty}\frac{P_\ell(\cos\theta)}{1-\left(\frac{a}{b}\right)^{2\ell+1}}\left\{\begin{array}{c}(2\ell+1)a^{\ell+1}\left(\frac{1}{r^{\ell+1}} - \frac{r^\ell}{b^{\ell+2}}\right)2\pi a^2\int_0^1 P_\ell(x')dx' + \\ \frac{(2\ell+1)}{b^{\ell+2}}\left(r^\ell - \frac{a^{2\ell+1}}{r^{\ell+1}}\right)2\pi b^2\int_{-1}^0 P_\ell(x')dx'\end{array}\right\}$$

$$\Phi(\vec{x}) = \frac{V}{2}\sum_{\ell=0}^{\infty}\frac{(2\ell+1)P_\ell(\cos\theta)}{1-\left(\frac{a}{b}\right)^{2\ell+1}}\left\{\left(\frac{a^{\ell+1}}{r^{\ell+1}} - \frac{a^{\ell+1}r^\ell}{b^{2\ell+1}}\right)\int_0^1 P_\ell(x')dx' + \left(\frac{r^\ell}{b^\ell} - \frac{a^{2\ell+1}}{b^\ell r^{\ell+1}}\right)\int_{-1}^0 P_\ell(x')dx'\right\}$$

For $\ell = 0$, integrals = 1, and get $\frac{V}{2}\frac{(1)(1)}{\left(1-\frac{a}{b}\right)}\left\{\frac{a}{r} - \frac{a}{b} + 1 - \frac{a}{r}\right\} = \frac{V}{2}$.

For $\ell \neq 0 \int_{-1}^{1} P_\ell(x)dx = 0 \rightarrow$ for ℓ even, $\int_{-1}^{1} P_\ell(x)dx = 2\int_0^1 P_\ell\, dx = 2\int_{-1}^{0} P_\ell\, dx = 0$

So for ℓ even, $\int_0^1 P_\ell\, dx = 0 = \int_{-1}^{0} P_\ell\, dx$

For ℓ odd $\int_{-1}^{0} P_\ell(x)dx = -\int_0^1 P_\ell(x)dx$, thus

$$\Phi(\vec{x}) = \frac{V}{2} + \frac{V}{2}\sum_{\ell\ \text{odd}}\left[(2\ell + 1)\int_0^1 P_\ell(x')dx'\right]\left\{\frac{\frac{a^{\ell+1}}{r^{\ell+1}} - \frac{a^{\ell+1}r^\ell}{b^{2\ell+1}} - \frac{r^\ell}{b^\ell} + \frac{a^{2\ell+1}}{b^\ell r^{\ell+1}}}{1-\left(\frac{a}{b}\right)^{2\ell+1}}\right\} \times P_\ell(\cos\theta).$$

Example 5.10. *A line charge of length 2d with a total charge of Q and a linear charge density varying as* $(d^2 - z^2)$, *with z distance from midpoint. A grounded conducting spherical shell is centered on the midpoint of the line charge, where the radius b of the shell is greater than the length z.*
(a) Find the potential inside the shell.
(b) Calculate the surface charge density.
(c) Describe the limit $d \ll b$.

Solution
(a) The linear charge density goes as $(d^2 - z^2)$ for $z < d$, so have delta functions in θ and step functions in r:
$\rho(\vec{x}) = k\frac{(d^2-z^2)}{r^2}[\delta(\cos\theta - 1) + \delta(\cos\theta + 1)]\Theta(d - r)$

$$Q = \int \rho(\vec{x})\, d^3x$$

$$= k\int \frac{(d^2 - r^2\cos^2\theta)}{r^2}[\delta(\cos\theta - 1) + \delta(\cos\theta + 1)]\theta(d - r)r^2 dr d(\cos\theta)d\phi$$

$= k(2\pi)2\int_0^d (d^2 - r^2)\, dr = 4\pi k\left(\frac{2}{3}d^3\right) \Rightarrow k = \frac{3Q}{8\pi d^3}$

So,

$$\rho(\vec{x}) = \frac{3Q}{8\pi d^3}\frac{d^2 - z^2}{r^2}[\delta(\cos\theta - 1) + \delta(\cos\theta + 1)]\Theta(d - r)$$

$$\Phi(\vec{x}) = \int_v \rho(x')\, G_D(\vec{x},\vec{x}')d^3x'$$

So,

$$\Phi(\vec{x}) = \sum_{\ell m}\frac{4\pi}{2\ell+1}Y_{\ell m}(\theta,\phi)\left(\frac{3Q}{8\pi d^3}\right)\int\left(\frac{d^2 - r'^2}{r'^2}\right)[\delta^- + \delta^+]\,\Theta(d - r)Y'^*_{\ell m}r_<^\ell\left(\frac{1}{r_>^{\ell+1}} - \frac{r_>^\ell}{b^{2\ell+1}}\right)$$

$$Y^{*'}_{\ell 0} = \sqrt{\frac{2\ell+1}{4\pi}}P_\ell(\cos\theta'),\ \ Y_{\ell 0}(\theta,\phi)$$
$$= \sqrt{\frac{2\ell+1}{4\pi}}P_\ell(\cos\theta)\, r'^2\, dr'd(\cos\theta')d\phi$$

So,

$$\Phi(\vec{x}) = \sum_{\ell=0}^{\infty} P_\ell(\cos\theta)\left(\frac{3Q}{4d^3}\right)\int_0^d\left(d^2 - r'^2\right)r_<^\ell\left(\frac{1}{r_>^{\ell+1}} - \frac{r_>^\ell}{b^{2\ell+1}}\right)dr'\big(P_\ell(1) + P_\ell(-1)\big)$$

and since

$$P_\ell(1) + P_\ell(-1) = 1 + (-1)^\ell = \begin{cases}2 \text{ if } \ell \text{ even}\\ 0 \text{ if } \ell \text{ odd}\end{cases}$$

We have:

$$\Phi(\vec{x}) = \frac{3}{2}\frac{Q}{d^3}\sum_{\ell=0\ even}^{\infty} P_\ell(\cos\theta)\int_0^d\left(d^2 - r'^2\right)r_<^\ell\left(\frac{1}{r_>^{\ell+1}} - \frac{r_>^\ell}{b^{2\ell+1}}\right)dr'$$

There are two cases: $r \geq d$ *and* $r < d$:

Case 1: $r \geq d \rightarrow r_< = r', r_> = r$

$$\int_0^d\left(d^2 - r'^2\right)r'^\ell dr' = \left[d^2\frac{r'^{\ell+1}}{\ell+1} - \frac{r'^{\ell+3}}{\ell+3}\right]_{r'=0}^d = d^{\ell+3}\left[\frac{1}{\ell+1} - \frac{1}{\ell+3}\right]$$

$$\Rightarrow \Phi(r,\theta) = 3Q \sum_{\ell=0\ even}^{\infty} P_\ell(\cos\theta) \frac{d^\ell}{(\ell+1)(\ell+3)} \left\{\frac{1}{r^{\ell+1}} - \frac{r^\ell}{b^{2\ell+1}}\right\} \quad \text{for} \quad d \le r \le b$$

(b) $\sigma(\theta,\phi) = -\frac{1}{4\pi}\frac{\partial\Phi}{\partial n}$ where $\hat{n} = -\hat{r}$

$$\sigma(\theta,\phi) = -\frac{1}{4\pi}\frac{\partial\Phi}{\partial r}\bigg|_{r=b}$$

$$= \frac{3Q}{4\pi} \sum_{\ell=0\ even}^{\infty} \frac{P_\ell(\cos\theta)}{(\ell+1)(\ell+3)} \left[-(\ell+1)\frac{d^\ell}{b^{\ell+2}} - \ell\frac{d^\ell b^{\ell-1}}{b^{2\ell+1}}\right]$$

$$= \frac{3Q}{4\pi}\sum_{\ell=0\ even}^{\infty} \frac{P_\ell(\cos\theta)}{(\ell+1)(\ell+3)} \left[\frac{-(2\ell+1)d^\ell}{b^{\ell+2}}\right]$$

$$\sigma(\theta,\phi) = -\frac{3Q}{4\pi b^2} \sum_{\substack{\ell=0 \\ even}}^{\infty} \frac{(2\ell+1)}{(\ell+1)(\ell+3)} \left(\frac{d}{b}\right)^\ell P_\ell(\cos\theta)$$

$$Qind = \oint \sigma(\theta,\phi)\, da = b^2 \oint \sigma\, d\Omega \quad \text{but} \quad \oint P_\ell(\cos\theta)\, d\Omega = 4\pi\delta_{l0}$$

$$Qind = \frac{3Q}{4\pi b^2}\frac{1}{(1)(3)}(4\pi b^2) = -Q$$

Case 2: $r \le d$

$\int_0^d dr' \rightarrow \int_0^r dr' + \int_r^d dr'$, since roles of $r_>$ and $r_<$ switch at $r' = r$

$r' = r_<, \mathrm{r} = r_>$ as before: $\int_0^r \left(d^2 - r'^2\right) r'^\ell dr' = d^2\left(\frac{r^{\ell+1}}{\ell+1}\right) - \frac{r^{\ell+3}}{\ell+3}$.

5.4 Cylindrical Coordinates

Laplace Equation in Cylindrical Coordinates and Bessel Functions

The distance measure in cylindrical coordinates is:

$$ds^2 = d\rho^2 + \rho^2 d\phi^2 + dz^2\,, \tag{Eqn. 5-93}$$

thus,

$$g_{\mu\nu} = \begin{pmatrix} 1 & & 0 \\ & \rho^2 & \\ 0 & & 1 \end{pmatrix} \quad and \quad |g|^{1/2} = \rho\,. \tag{Eqn. 5-94}$$

We can thus write:

$$\nabla^2\Phi = \frac{1}{|g|^{1/2}}\partial_\mu\left(|g|^{1/2}g^{\mu\nu}\partial_\nu\Phi\right) = \frac{1}{\rho}\frac{\partial}{\partial\rho}\left(\rho\frac{\partial}{\partial\rho}\Phi\right) + \frac{1}{\rho^2}\frac{\partial^2\Phi}{\partial\phi^2} + \frac{\partial^2\Phi}{\partial z^2}$$

(Eqn. 5-95)

and

$$\nabla^2\Phi = \frac{\partial^2\Phi}{\partial\rho^2} + \frac{1}{\rho}\frac{\partial\Phi}{\partial\rho} + \frac{1}{\rho^2}\frac{\partial^2\Phi}{\partial\phi^2} + \frac{\partial^2\Phi}{\partial z^2}.$$

(Eqn. 5-96)

As usual, for separation of variables, we seek a form $\Phi = R(\rho)\phi(\phi)Z(z)$, with resulting separable equations:

$$\frac{1}{Z(z)}\frac{\partial^2 Z(z)}{\partial z^2} = K^2 \rightarrow Z(z) = e^{\pm Kz},$$

(Eqn. 5-97)

$$\frac{1}{Q}\frac{\partial^2 Q}{\partial\phi^2} = -v^2 \rightarrow Q(\phi) = e^{\pm iv\phi}$$

$$\rightarrow v\ must\ be\ integer\ if full\ azimuth,$$

(Eqn. 5-98)

and

$$\frac{\partial^2 R}{\partial\rho^2} + \frac{1}{\rho}\frac{\partial R}{\partial\rho} + \left(K^2 - \frac{v^2}{\rho^2}\right)R = 0\,.$$

(Eqn. 5-99)

Let $x = K\rho$, then we have:

$$\frac{d^2R}{dx^2} + \frac{1}{x}\frac{dR}{dx} + \left(1 - \frac{v^2}{x^2}\right)R = 0,$$

(Eqn. 5-100)

which is known as the Bessel equation, with solutions known as Bessel functions of order v. Let's solve via power series: $R(x) = x^\alpha \sum_{j=0}^{\infty} a_j\, x^j$:

$$(\alpha+j)(\alpha+j-1)\sum_{j=0}^{\infty} a_j\, x^{\alpha+j-2} + (\alpha+j)\sum_{j=0}^{\infty} a_j\, x^{\alpha+j-2} + \sum_{j=0}^{\infty} a_j\, x^{\alpha+j} - v^2\sum_{j=0}^{\infty} a_j\, x^{\alpha+j-2} = 0$$

Thus,

$$a_j[(\alpha+j)(\alpha+j-1) + (\alpha+j) - v^2] = 0 \quad \text{for} \quad j = 0{,}1.$$

For:

$j = 0: \alpha(\alpha+1) + \alpha - V^2 = \alpha^2 - V^2 = 0 \Longrightarrow \alpha = \pm v$

$j = 1: a_1(\alpha(\alpha + 1) + (\alpha + 1) - V^2) = (2\alpha + 1)a_1 = 0 \Rightarrow a_1 = 0$
$\Rightarrow$ all odd powers = zero

$$[(\alpha + j)^2 - v^2]a_j = -a_{j-2} \Rightarrow \quad a_{2j} = \frac{-a_{2j-2}}{4j(j + \alpha)}$$

Iteration yields

$$a_{2j} = \frac{(-1)^j \Gamma(\alpha + 1)}{2^{2j} j!\, \Gamma(j + \alpha + 1)} a_0.$$

(Eqn. 5-101)

Conventional to choose $a_0 = [2^\alpha \Gamma(\alpha + 1)^{-1}]$. Then: two solutions:

$$J_v(x) = \left(\frac{x}{2}\right)^v \sum_{j=0}^{\infty} \frac{(-1)^j}{j!\, \Gamma(j + v + 1)} \left(\frac{x}{2}\right)^{2j}$$

and:

$$J_{-v}(x) = \left(\frac{x}{2}\right)^{-v} \sum_{j=0}^{\infty} \frac{(-1)^j}{j!\, \Gamma(j - v + 1)} \left(\frac{x}{2}\right)^{2j}$$

For v NOT an integer the above equations are linearly independent. If v is integer valued they are dependent, in fact:

$$J_{-m}(x) = (-1)^m J_m(x)$$

(Eqn. 5-102)

for m integer valued. Another linearly independent solution is needed for m integer. It is customary to replace J_{-v} by $N_v(x)$, the Newmann function (or Bessel function of the 2nd kind); even if v is not an integer:

$$N_v(x) = \frac{J_v(x) \cos v\pi - J_{-v}(x)}{\sin v\pi}.$$

(Eqn. 5-103)

Another convenient set of solutions is given in terms of the Bessel functions of the 3rd kind, also called the first and second Hankel functions:

$$H_V^{(1)}(x) = J_V(x) + iN_V(x)$$

(Eqn. 5-104)

$$H_V^{(2)}(x) = J_V(x) - iN_V(x)$$

(Eqn. 5-105)

Recursion relations for all of the $J_v, N_v, H_v^{(1)}, H_v^{(2)}$, written with the generic form "Ω_v":

$$\Omega_{v-1}(x) + \Omega_{v+1}(x) = \frac{2v}{x}\Omega_v(x),$$

(Eqn. 5-106)

$$\Omega_{v-1}(x) - \Omega_{v+1}(x) = 2\frac{d\Omega_v(x)}{dx}.$$

(Eqn. 5-107)

(For $K^2 \rightarrow -K^2$ we get the modified Bessel functions.)

The general solution, in separable form, is then:

$$\Phi = RQZ,$$

(Eqn. 5-108)

with

$$Q(\phi) = A \sin m\,\phi + B \cos m\,\phi$$
$$Z(z) = ae^{KZ} + be^{-KZ}$$
$$R(\rho) = (J_m(K\rho) + DN_m(K\rho)$$

(Eqn. 5-109)

Assuming that the set of Bessel functions is complete, we can expand an arbitrary function of ρ on the interval $0 \leq \rho \leq a$ in a Fourier-Bessel series:

$$f(\rho) = \sum_{n=1}^{\infty} A_{vn} J_v\left(x_{vn}\frac{\rho}{a}\right),$$

(Eqn. 5-110)

where $x_{vn}$is the nth root of $J_v(x)$, and

$$A_{vn} = \frac{2}{a^2 J_{v+1}^2(x_{vn})}\int_0^a \rho f(\rho) J_v\left(\frac{x_{vn}\rho}{a}\right) d\rho.$$

(Eqn. 5-111)

Let's now consider a few boundary–value problems in cylindrical coordinates.

Boundary – Value Problems in Cylindrical Coordinates

Consider the potential on the side and bottom of the cylinder to be zero while the top has $\Phi = V(\rho, \phi)$:

The general solution has $\Phi = RQZ$:

$$Q(\phi) = A \sin(m\phi) + B \cos(m\phi)$$
$$Z(z) = \sinh kz$$
$$R(\rho) = CJ_m(k\rho) + DN_n(k\rho)$$

(Eqn. 5-112)

Since we want the potential inside the cylinder, which includes origin, we have $D = 0$. Since the potential also vanishes at $\rho = a \rightarrow k_{mn} = \frac{x_{mn}}{a}$ (where $x_{mn}$is the nth root of $J_m(x)$). Thus,

$$\Phi = \sum_{m=0}^{\infty}\sum_{n=1}^{\infty} J_m(k_{mn}\rho)\, sinh(k_{mn}z)(A_{mn}\sin m\,\phi + B_{mn}\cos m\,\phi).$$

(Eqn. 5-113)

At $z = L$:

$$V(\rho,\phi) = \sum_{m,n} sinh(k_{mn}L)J_m(k_{mn}\rho)(A_{mn}\sin m\,\phi + B_{mn}\cos m\,\phi)$$

$$A_{mn} = \frac{2\,cosech(k_{mn}L)}{\pi a^2 J_{m+1}^2(k_{mn}a)}\int_0^{2\pi} d\phi \int_0^a d\rho\, \rho V(\rho,\phi)J_m(k_{mn}\rho)\sin m\phi$$

$$B_{mn} = \frac{2\,cosech(k_{mn}L)}{\pi a^2 J_{m+1}^2(k_{mn}a)}\int_0^{2\pi} d\phi \int_0^a d\rho\, \rho V(\rho,\phi)J_m(k_{mn}\rho)\cos m\phi$$

(Eqn. 5-114)

(For $m = 0$ use $\frac{1}{2}B_{0n}$ in the series (normalization factor different is all.)

Example 5.11. *Suppose in 5.10 we have the cylinder split vertically with opposite potentials:*

$$V(\phi,z) = \begin{cases} V & \text{for } -\frac{\pi}{2} < \phi < \frac{\pi}{2} \\ -V & \text{for } \frac{\pi}{2} < \phi < \frac{3\pi}{2} \end{cases}$$

(a) Find the potential inside the cylinder.
(b) Consider $L \gg b$ and $z = L/2$, and the result as a function of ρ and ϕ. Compare with (2.9).

Solution

(a) We are starting with:

$$\Phi(\rho,\phi,z) = \sum_{m,n} I_m\left(\frac{n\pi\rho}{L}\right)\sin\left(\frac{n\pi z}{L}\right)[A_{mn}\cos(m\phi) + B_{mn}\sin(m\phi)],$$

$$\left.\begin{matrix} A_{m,n} \\ B_{m,n} \end{matrix}\right\} = \frac{2}{\pi L I_m\left(\frac{n\pi b}{L}\right)}\int_0^{2\pi} d\phi \int_0^L dz \sin\left(\frac{n\pi z}{L}\right)\begin{Bmatrix} \cos(m\phi) \\ \sin(m\phi) \end{Bmatrix} V(\phi,z),$$

$$A_{0,n} = \frac{2}{\pi L I_0\left(\frac{n\pi b}{L}\right)} \int_0^{2\pi} d\phi \int_0^L dz \sin\left(\frac{n\pi z}{L}\right) V(\phi, z),$$

and,

$B_{0,n} = 0.$

Since $V(-\phi, z) = V(\phi, z) \Rightarrow \Phi$ even in $\phi \Rightarrow B_{mn} = 0$. Let's simplfy further by doing the z integral (since no z dependence):

$$V(\phi, z) = V(\phi) \Rightarrow \int_0^L dz \sin\left(\frac{n\pi z}{L}\right) = \left[-\frac{\cos\left(\frac{n\pi z}{L}\right)}{\frac{n\pi}{L}}\right]_0^L$$

$$= \frac{L}{n\pi}(1-(-1)^n) = \begin{cases} \frac{2L}{n\pi} & n \text{ odd} \\ 0 & n \text{ even} \end{cases}$$

Thus,

$$A_{mn} = \frac{2}{\pi L I_m\left(\frac{n\pi b}{L}\right)}\left(\frac{2L}{n\pi}\right) \int_0^{2\pi} d\phi\, V(\phi) \cos(m\phi) \leftarrow n \text{ odd}$$

$= \frac{16V(-1)^{(m-1)/2}}{\pi^2 mn I_m\left(\frac{n\pi b}{L}\right)}$ for m and n both odd, = 0 otherwise.

The potential is thus:

$$\Phi(\rho, \phi, z) = \frac{16V}{\pi^2} \sum_{\substack{m,n \\ odd}} \frac{(-1)^{(m-1)/2}}{mn} \frac{I_m\left(\frac{n\pi\rho}{L}\right)}{I_m\left(\frac{n\pi b}{L}\right)} \cos(m\phi) \sin\left(\frac{n\pi z}{L}\right).$$

(b) Assuming $L \gg b, z = L/2$

For $I_v(x), x \ll 1$: $I_v(x) \to \frac{1}{\Gamma(v+1)}\left(\frac{x}{2}\right)^v$, so $I_m\left(\frac{n\pi b}{L}\right) \to \frac{1}{\Gamma(m+1)}\left(\frac{\frac{n\pi b}{L}}{2}\right)^m$, or:

$$I_m\left(\frac{n\pi\rho}{L}\right) \to \frac{1}{\Gamma(m+1)}\left(\frac{\frac{n\pi\rho}{L}}{2}\right)^m \qquad \text{since } \rho < b.$$

$$\Phi\left(\rho, \phi, z = {}^1/_2\right)\Big|_{L\gg b}$$

$$= \frac{16V}{\pi^2} \sum_{\substack{m,n \\ odd}} \frac{(-1)^{(m-1)/2}}{mn} \left(\frac{\rho}{b}\right)^m \cos(m\phi)\, (-1)^{(n-1)/2}$$

$$= \frac{4V}{\pi}\left[\left(\frac{\rho}{b}\right)\cos\phi - \frac{1}{3}\left(\frac{\rho}{b}\right)^3\cos(3\phi) + \frac{1}{5}\left(\frac{\rho}{b}\right)^5\cos(5\phi) - \cdots\right]$$

Results are in agreement with (2.9) for the infinite cylinder.

Example 5.12. *Consider an infinite thin plane sheet of conducting material with a hole in it (circular, radius a). Filling the hole, but with (circular) gap is a disc of the same material. The disc is at potential V while the rest of the plane is at zero potential.*

(a) Find the potential at any point above the plane.
(b) Find the potential z above the center of the disc.
(c) Find the potential z above the edge of the disc.

Solution

Starting with

$Q(\phi) = A\sin m\phi + B\cos m\phi$

$Z(z) = e^{-kz}$

$R(\rho) = CJ_m(k\rho) + DN_n(k\rho)$

We have:

$$\Phi(\rho,\phi,z) = \sum_{m=0}^{\infty}\int_0^{\infty} dk\, e^{-kz}J_m(k\rho)[A_m(k)\sin(m\phi) + B_m(k)\cos(m\phi)]$$

where

$$V(\rho,\phi) = \sum_{m=0}^{\infty}\int_0^{\infty} dk\, J_m(k\rho)[A_m(k)\sin(m\phi) + B_m(k)\cos(m\phi)]$$

leads to:

$$\left.\begin{matrix} A_m(k) \\ B_m(k) \end{matrix}\right\} = \frac{k}{\pi}\int_0^{\infty} d\rho\rho\int_0^{2\pi} d\phi\, V(\rho,\phi)J_m(k\rho)\begin{cases}\sin m\phi \\ \cos m\phi \quad \text{for } m = 0, B_m(k) = \frac{1}{2}B_0\end{cases}$$

Since there is Azimuthal symmetry, $m = 0, A_0 = 0$:

$$\Phi(\rho,\phi,z) = \int_0^{\infty} dk\, e^{-kz}J_0(k\rho)B_0$$

We can solve for B_0:

$$\int_0^\infty \rho J_0(k'\rho)\Phi(\rho,\phi,z)d\rho = \int_0^\infty dk\, B_0 \underbrace{\int_0^\infty \rho d\rho\, J_0(k\rho)J_0(k'\rho)}_{\frac{1}{k}\delta(k'-k)} = \frac{1}{k}B_0$$

Thus,

$$B_0 = \int_0^a (k\rho)\, V J_0(k\rho)d\rho.$$

Since $xJ_0 = \frac{d}{dx}[xJ_1] \rightarrow\ (k\rho)J_0(k\rho)kd\rho$, we have:

$$B_0 = V(\rho)J_1(k\rho)|_0^a = V_aJ_1(ka).$$

and

$$\Phi(\rho,\phi,z) = \int_0^\infty dk\, e^{-kz}V_aJ_1(ka)J_0(t\rho).$$

(b) $\Phi(\rho,\phi,z) = \int_0^\infty dk\, e^{-kz}V_aJ_1(ka)\underbrace{J_0(0)}_{1}$

$= V_a\int_0^\infty dk\, e^{-kz}J_1(ka)$ use $J_0' = -J_1$

$= V_a\int_0^\infty dk\, e^{-kz}(-J_0'(ka))$

$= -V_a\int_0^\infty dk\,(-z)e^{-kz}\left(-\frac{J_0(ka)}{a}\right) + V_ae^{-kz}\left(-\frac{J_0(ka)}{a}\right)\Big|_{k=0}^\infty$

$= V\{1 - z\int_0^\infty dk\, e^{-kz}J_0(ka)\}$

Thus,

$$\Phi(\rho,\phi,z) = V\left\{1 - \frac{z}{\sqrt{a^2+z^2}}\right\}.$$

(c) $\Phi(a,\phi,z) = V_a\int_0^\infty dk\, e^{-kz}J_0(ka)J_1(ka) =$
$-V_a\int_0^\infty dk\, e^{-kz}J_0(ka)J_0'(ka)$

$= -V_a\left\{e^{-kz}J_0(ka)\left(\frac{J_0(ka)}{a}\right)\Big|_0^\infty - \int_0^\infty dk\left\{\frac{-z}{a}e^{-kz}J_0^2(ka) + \underbrace{e^{-kz}J_0(ka)J_0'(ka)}\right\}\right\}$

$\nearrow -\Phi(a,\phi,z)$

$2\Phi(a,\phi,z) = V - Vz\int_0^\infty e^{-kz}J_0^2(ka)dk,$ *and* $\int_0^\infty e^{-kz}J_0^2(ka)dk = \frac{2}{\pi\sqrt{z^2+4a^2}}K\left(\frac{2a}{\sqrt{z^2+4a^2}}\right),$

where K is the complete elliptic integral. Thus,

$$\Phi(a,\phi,z) = \frac{V}{2}\left[1 - \frac{2z}{\pi\sqrt{z^2+4a^2}}K\left(\frac{2a}{\sqrt{z^2+4a^2}}\right)\right].$$

5.5 Cylindrical Green's function

Expansion of Green Functions in Cylindrical Coordinates:

Starting point:

$$\nabla_x^2 G(\vec{x}-\vec{x}') = \frac{4\pi}{\rho}\delta(\rho-\rho')\delta(\phi-\phi')\delta(z-z')$$

(Eqn. 5-115)

In cylindrical coordinates the delta function can be expressed:

$$\delta(z-z') = \frac{1}{2\pi}\int_{-\infty}^{\infty} dk\, e^{ik(z-z')} = \frac{1}{\pi}\int_{0}^{\infty} dk\cos[k(z-z')]$$

and

$$\delta(\phi-\phi') = \frac{1}{2\pi}\sum_{m=-\infty}^{\infty} e^{im(\phi-\phi')}.$$

Thus we can write:

$$G(\vec{x}-\vec{x}') = \frac{1}{2\pi^2}\sum_{m=-\infty}^{\infty}\int_0^{\infty} dk\, e^{im(\phi-\phi')}\cos[k(z-z')]\, g_m(k,\rho,\rho')$$

(Eqn. 5-116)

The difficult radial differential equation is then:

$$\frac{1}{\rho}\frac{d}{d\rho}\left(\rho\frac{dg_m}{d\rho}\right) - \left(k^2+\frac{m^2}{\rho^2}\right)_{g_m} = -\frac{4\pi}{\rho}\delta(\rho-\rho').$$

(Eqn. 5-117)

For $\rho \neq \rho'$ the equation is the modified Bessel equation with solution consisting of modified Bessel functions $I_m(k\rho)$ and $K_m(k\rho)$ where $I_\upsilon(x) = i^{-\nu}J_\upsilon(ix)$ and $K_\nu(x) = (\pi/2)i^{\nu+1}H_\upsilon{}^{(1)}(ix)$. Suppose $\Psi_1(k\rho)$ is some linear comb. of I_m and K_m which is a solution for $\rho < \rho'$ and $\Psi_2(k\rho)$ is for $\rho > \rho'$. Then symmetry of the Green function in ρ and ρ' requires that:

$$g_m(k,\rho,\rho') = \Psi_1(k\rho_<)\Psi_2(k\rho_>)$$

(Eqn. 5-118)

where Normalization of $\Psi_1 \Psi_2$ is determined by discontinuity:

$$\left.\frac{dg_m}{d\rho}\right|_+ - \left.\frac{dg_m}{d\rho}\right|_- = -\frac{4\pi}{\rho'} = k[\Psi_1 \Psi_2^1 - \Psi_2 \Psi_1^1] = kW[\Psi_1, \Psi_2]$$

(Eqn. 5-119)

This is a Sturm – Louisville type equation: $\frac{d}{dx}\left[p(x)\frac{dy}{dx}\right] + g(x)y = 0 \Rightarrow W \simeq \frac{1}{p(x)}$, so want $W[\Psi_1(x), \Psi_2(x)] = -\frac{4\pi}{x}$. For no boundary surfaces: want $g_m(k, \rho, \rho')$ finite at $\rho = 0$ and vanishing as $\rho \to \infty$ consequently $\Psi_1(k\rho) = \mathrm{AI}_m(k\rho)$ and $\Psi_2(k\rho) = \mathrm{K}_m(k\rho)$. Since Wronskian is proportional to $(1/x)$ for all x, it does not matter where we evaluate it. Thus, use the limiting forms to obtain: $W[I_m(x), k_m(x)] = -\frac{1}{x}$ $so\ A = 4\pi$, then $G(\vec{x} - \vec{x}') = \frac{1}{|\vec{x} - \vec{x}'|}$ in unbound space:

$$\frac{1}{|\vec{x} - \vec{x}'|} = \frac{2}{\pi} \sum_{m=-\infty}^{\infty} \int_0^{\infty} dk\, e^{im(\phi - \phi')} \cos[k(z - z')]\, I_m(k\rho_<) k_m(k\rho_>)$$

$$= \frac{4}{\pi}\int_0^{\infty} dk \cos[k(z - z')] \left\{\frac{1}{2} I_0(k\rho_<) K_0(k\rho_>) + \sum_{m=1}^{\infty} \cos[m(\phi - \phi')]\, I_m(k\rho_<) K_m(k\rho_>)\right\}$$

(Eqn. 5-120)

Let $\vec{x} = 0 \Rightarrow m = 0 \Rightarrow \frac{1}{\sqrt{\rho^2 + z^2}} = \frac{2}{\pi}\int_0^{\infty} \cos kz\, K_0(k\rho) dA.$

$\downarrow$

Replace ρ^2 by $R^2 = \rho^2 + \rho'^2 - 2\rho\rho' \cos(\phi - \phi')$ which is $\frac{1}{|\vec{x} - \vec{x}'|}$ in the earlier eqn with $z' = 0$, comparig the right hand sides to both equations we get:

$$\frac{2}{\pi}\int_0^{\infty} dk \cos kz \left\{ I_0(k\rho_<) K_0(k\rho_>) + 2\sum_{m=1}^{\infty} \cos[m(\phi - \phi')]\, I_m(k\rho_<) K_m(k\rho_>) \right\}$$

$$= \frac{2}{\pi}\int_0^{\infty} dk \cos kz\, k_0\left(k\sqrt{\rho^2 + \rho'^2 - 2\rho\rho' \cos(\phi - \phi')}\right)$$

Thus,

$$K_0\left(k\sqrt{\rho^2+\rho'^2-2\rho\rho'\cos(\phi-\phi')}\right)$$
$$= I_0(k\rho_<)K_0(k\rho_>)$$
$$+2\sum_{m=1}^{\infty}\cos[m(\phi-\phi')]\,I_m(k\rho_<)K(k\rho_>)$$

If we take k→0 we obtain the expansion for the Green functions for polar coordinates:

$$\ln\left(\frac{1}{\sqrt{\rho^2+\rho'^2-2\rho\rho'\cos(\phi-\phi')}}\right)$$
$$=\ln\left(\frac{1}{\rho_>}\right)+\sum_{m=1}^{\infty}\frac{1}{m}\left(\frac{\rho_<}{\rho_>}\right)^m\cos[m(\phi-\phi')].$$

(Eqn. 5-121)

Mixed Boundary Conditions, Conducting Plane with a Circular Hole: Consider an infinitely thin, grounded, conducting plane with a circular hole radius "a", and with E-field far from the hole being normal to the plane, constant in magnitude, and having different values on either side: Let $E_z = -E_0\ for\ z > 0$ and $E_z = -E_1\ for\ z < 0$. Thus, have

$$\Phi = \begin{cases} E_0 z + \Phi^{(1)} & (z > 0) \\ E_1 z + \Phi^{(1)} & (z < 0) \end{cases}.$$

(Eqn. 5-122)

If $\Phi^{(1)}$were zero, then with $\sigma = -\frac{1}{4\pi}\frac{\partial\Phi}{\partial n}$ we see that $\sigma_{top} = -E_0/4\pi$ and $\sigma_{bottom} = E_1/4\pi$. $\Phi^{(1)}$ can be thought of as resulting from a rearrangement of surface charge in the neighborhood of the hole. Since the charge density is located at $z = 0$ we have:

$$\Phi^{(1)}(x,y,z) = \int \frac{\sigma^{(1)}(x',y')dx'dy'}{\sqrt{(x-x')^2+(y-y')^2+z^2}}$$

(Eqn. 5-123)

Thus,

$$\Phi^{(1)}\text{ even in} \Rightarrow \begin{cases} E_x^{(1)}, E_y^{(1)} \text{ even in } z \longrightarrow \text{total } x \text{ and } y \text{ comp. of field} \\ E_z^{(1)} \text{ odd in } z \longrightarrow \text{not total z comp.} \end{cases}$$

Even though $E_z^{(1)}$ is odd in z it does not vanish at $z = 0$ – rather it is discontinuous there so that the total z component of the E-field might be continuous across $z = 0$:

$$-E_0 + E_z^{(1)}\Big|_{z=0^+} = -E_1 + E_z^{(1)}\Big|_{z=0^-}$$

Since $E_z^{(1)}\Big|_{z=0^+} = E_z^{(1)}\Big|_{z=0^-}$ we have

$$E_z^{(1)}\Big|_{z=0^+} = \frac{1}{2}(E_0 - E_1) \qquad (\text{for } 0 \le \rho < a).$$

$$\left.\begin{array}{l}\text{Potential not known in opening, E} - \text{field known} \\ \text{Potential known as surface (0), E} - \text{field not known on surface}\end{array}\right\} \to \text{mixed b.c.'s.}$$

Have

$$\frac{\partial \Phi^{(1)}}{\partial z}\Bigg|_{z=0^+} = -\frac{1}{2}(E_0 - E_1) \quad for \quad 0 \le \rho < a$$

(Eqn. 5-124)

and

$$\Phi^{(1)}\Big|_{z=0} = 0 \quad for \quad a \le \rho < \infty.$$

(Eqn. 5-125)

Example 5.13. *Consider parallel conducting planes at z=0 and z=L that are held at zero. Use Dirichlet Green's functions in cylindrical coordinates to solve for the potential between the plates.*

(a) Show $G_D(\vec{x},\vec{x}') = \frac{4}{L}\sum_{n=1}^{\infty}\sum_{m=-\infty}^{\infty} e^{im\phi}\sin\left(\frac{n\pi z}{L}\right)\sin\left(\frac{n\pi z'}{L}\right)I_m\left(\frac{n\pi}{L}\rho_<\right)k_m\left(\frac{n\pi}{L}\rho_>\right)$,

(b) Show $G_D(\vec{x},\vec{x}') = 2\sum_{m=-\infty}^{\infty}\int_0^{\infty} dk\, e^{im(\phi-\phi')}J_m(k\rho)J_m(k\rho')\frac{\sinh(kz_<)\sinh[k(L-z_>)]}{\sinh(kL)}$.

Solution

(a) Want $\nabla^2 G_D(\vec{x},\vec{x}') = -4\pi\delta(\vec{x},\vec{x}') \quad if \quad 0 \le z \le L$

With $G_D(\vec{x},\vec{x}') = 0$ for $z\ or\ z' = 0$ or L or $\rho\ or\ \rho' \to \infty$. So have form:

$$G_D(\vec{x},\vec{x}') = \sum_{m=-\infty}^{\infty} e^{im\phi}\sum_{n=1}^{\infty}\sin\left(\frac{n\pi z}{L}\right)A_{mn}(\rho;\rho'\phi'z')$$

We also have:

$$\nabla^2 G_D = -4\pi\delta = -4\pi\delta(\phi-\phi')\delta(z-z')\frac{\delta(\rho-\rho')}{\rho},$$

and since:

$$\frac{1}{2\pi}\sum_{m=-\infty}^{\infty} e^{im(\phi-\phi')} = \delta(\phi-\phi')$$

and

$$\frac{2}{L}\sum_{n=1}^{\infty}\sin\left(\frac{n\pi z}{L}\right)\sin\left(\frac{n\pi z'}{L}\right)=\delta(z-z')$$

We can write:

$$\nabla^2 G_D=\frac{1}{\rho}\left(\frac{\partial}{\partial\rho}\rho\frac{\partial G_D}{\partial\rho}\right)+\frac{1}{\rho^2}\frac{\partial^2 G_D}{\partial\phi^2}+\frac{\partial^2 G_D}{\partial z^2}$$

$$=\sum_{m=-\infty}^{\infty}\sum_{n=1}^{\infty}e^{im\phi}\sin\left(\frac{n\pi z}{L}\right)\left\{\frac{1}{\rho}\frac{\partial}{\partial\rho}\rho\frac{\partial A_{m,n}}{\partial\rho}-\frac{m^2}{\rho^2}A_{m,n}-\frac{n^2\pi^2}{L^2}A_{m,n}\right\}$$

$$=-\frac{4}{L}\sum_{m=-\infty}^{\infty}\sum_{n=1}^{\infty}e^{im\phi}\sin\left(\frac{n\pi z}{L}\right)\sin\left(\frac{n\pi z'}{L}\right)\frac{\delta(\rho-\rho')}{\rho}$$

Let's multiply by $e^{im\phi}\sin\left(\frac{n\pi z}{L}\right)$, and integrate by $\int_0^{2\pi}d\phi\int_0^L dz$, where:

$$A_{m,n}(\rho;\rho'\phi'z')=-\frac{4}{L}e^{im\phi}\sin\left(\frac{n\pi z'}{L}\right)A_{mn}(\rho).$$

And we arrive at:

$$\frac{1}{\rho}\frac{\partial}{\partial\rho}\left(\rho\frac{\partial A_{m,n}}{\partial\rho}\right)-\left(\frac{n^2\pi^2}{L^2}+\frac{m^2}{\rho^2}\right)A_{m,n}=\frac{\delta(\rho-\rho')}{\rho}$$

For $\rho\neq\rho'$ this is the modified Bessel eqn. with $I_m\left(\frac{n\pi\rho}{L}\right)$ and $K_m\left(\frac{n\pi\rho}{L}\right)$:

For $\rho<\rho'$: $A_{m,n}(\rho;\rho')=\alpha_{m,n}(\rho')I_m\left(\frac{n\pi\rho}{L}\right)+\beta_{m,n}(\rho')K_m\left(\frac{n\pi\rho}{L}\right)$

For $\rho>\rho'$: $A_{m,n}(\rho;\rho')=\gamma_{m,n}(\rho')I_m\left(\frac{n\pi\rho}{L}\right)+\delta_{m,n}(\rho')k_m\left(\frac{n\pi\rho}{L}\right)$

Symmetry in G requires $A_{m,n}(\rho;\rho')=\alpha'_{m,n}I_m\left(\frac{n\pi\rho_<}{L}\right)k_m\left(\frac{n\pi\rho_>}{L}\right)$, thus:

$$G_D(\vec{x},\vec{x}')=\frac{4}{L}\sum_{n=1}^{\infty}\sum_{m=-\infty}^{\infty}e^{im\phi}\sin\left(\frac{n\pi z}{L}\right)\sin\left(\frac{n\pi z'}{L}\right)I_m\left(\frac{n\pi}{L}\rho_<\right)k_m\left(\frac{n\pi}{L}\rho_>\right).$$

(b) Let's use a Hankel transform in ρ, instead of a Fourier series in z. Expand G_D as a Fourier series in ϕ and a Hankel transform in ρ:

$$G_D(\vec{x},\vec{x}')=\sum_{m=-\infty}^{\infty}e^{im\phi}\int_0^{\infty}kdk\,J_m(k\rho)H_m(k,z;\rho'\phi'z')$$

$$\nabla^2 G_D = \sum_{m=-\infty}^{\infty} \int_0^{\infty} kdk\, e^{im\phi} J_m(k\rho) \left\{ \left[\left(-k^2 + \frac{m^2}{\rho^2} \right) - \frac{m^2}{\rho^2} \right] H_m + \frac{\partial^2 H_m}{\partial z^2} \right\}$$

$$\left(\text{since } \frac{1}{\rho} \frac{d}{d\rho} \left[\rho \frac{dJ_m(k\rho)}{d\rho} \right] = \left(\frac{m^2}{\rho^2} - k^2 \right) J_m(k\rho) \right)$$

$$\nabla^2 G_D = \sum_{m=-\infty}^{\infty} \int_0^{\infty} kdk\, e^{im\phi} J_m(k\rho) \left\{ \frac{\partial^2 H_m}{\partial z^2} - k^2 H_m \right\}$$
$$= -4\pi\delta(\vec{x},\vec{x}') = -4\pi\delta(\phi - \phi')\frac{\delta(\rho-\rho')}{\rho}\delta(z - z')$$
$$= -4\pi \left[\frac{1}{2\pi} \sum_{m=-\infty}^{\infty} e^{im(\phi-\phi')} \right] \left[\int_0^{\infty} kdk\, J_m(k\rho) J_m(k\rho') \right] \delta(z - z')$$

Multiply by $e^{-im'\phi} \rho J_m(k'\rho)$ then integrate $\int_0^{2\pi} d\phi \int_0^{\infty} d\rho$. Since $\int_0^{\infty} \rho\, J_m(k\rho) J_m(k') d\rho = \frac{\delta(k-1)}{k}$ we get:

$$\frac{\partial^2 H_{m'}}{\partial z^2}(k'z; \rho'\phi'z') - k'^2 H_{m'}(k'z; \rho'\phi'z')$$
$$= -2e^{-im'\phi'} J_{n'}(k'\rho')\delta(z - z')$$

Drop primes: $H_m(kz; \rho'\phi'z') = -2e^{-im'^m\phi'} J_{n'^m}(k\rho') H(kz; z')$

With $\frac{\partial^2 H(kz,z')}{\partial z^2} - k^2 H(kz; z') = \delta(z - z')$

Then, for $z < z' : H(kz; z') = \alpha_k(z')e^{kz} + \beta_k(z')e^{-kz}$
$z > z' : H(kz; z') = \gamma_k(z')e^{kz} + \delta_k(z')e^{-kz}$

$G_D = 0$ when $z = 0$ or $L \Longrightarrow H(kz; z') = 0$ when $z = 0$ or L
$z < z' : H = \alpha_k(z') \sinh(kz)$
$z > z' : H = \beta_k(z') \sinh(k(L - z))$

From symmetry $\Longrightarrow H(kz; z') = \alpha_k \sinh(kz_<) \sinh(k(L - z_>))$.

Now, the junction condition gives:

$$\left(\frac{\partial^2}{\partial z^2} - k^2 \right) H = \delta(z - z') \text{ and taking } \int_{z'-\varepsilon}^{z'+\varepsilon} dz:$$

$$\alpha_k\{\sinh(kz')\cosh[k(L-z')](-k)-\cosh(kz')(k)\sinh(k(L-z'))\} = 1$$

So,

$$\alpha_k = \frac{-1}{k\sinh(kL)}$$

Thus,

$$G_D = \sum_{m=-\infty}^{\infty}\int_0^{\infty} kdk\, e^{im\phi}J_m(k\rho)[2e^{-im\phi'}J_m(k\rho')]\left(\frac{-1}{k\sinh(kL)}\right)\sinh(kz_<)\sinh[k(L - z_>)]$$

$$G_D(\vec{x},\vec{x}') = 2\sum_{m=-\infty}^{\infty}\int_0^{\infty} dk\, e^{im(\phi-\phi')}J_m(k\rho)J_m(k\rho')\frac{\sinh(kz_<)\sinh[k(L-z_>)]}{\sinh(kL)}$$

5.6 Exercises

(Ex. 5.1) Re-derive Eqn. 5-10.
(Ex. 5.2) Re-derive Eqn. 5-13.
(Ex. 5.3) Re-derive (verify) Eqn. 5-48.
(Ex. 5.4) Re-derive (verify) Eqn. 5-49.
(Ex. 5.5) Re-derive (verify) Eqn.s 5-58 and 5-59.
(Ex. 5.6) Re-derive Eqn. 5-80.
(Ex. 5.7) Re-derive Eqn. 5-92.
(Ex. 5.8) Re-derive Eqn. 5-96.
(Ex. 5.9) Re-derive Eqn. 5-121.

Chapter 6. Fields – Multipoles, Dielectrics, and Magnetostatics

In Chapters 4 thru 7 classic electromagnetism (EM) is covered, with extensive coverage on the Lorentz transform (EM is Lorentz invariant) in Ch. 8, which then leads to special relativity and general relativity and other topics in later chapters. This chapter (Ch. 6), generalizes the formalism to macroscopic matter, starting with multipole expansions. Once we have macroscopic matter we can also describe magnetostatics.

Note on Examples and Exercises
The chapters on electromagnetism draw, foremost, from the excellent source, Classical Electrodynamics by J.D. Jackson [32], that was the required text for many years at Caltech. Many of the Jackson problems develop the physics theory further (e.g., Gauss's Theorem) or further develop the mathematical techniques (for ODEs, definition of delta function as limit, etc.). With my own notes from the courses at Caltech, and upon later preliminary exam review at the Univ. Wisconsin, the addition of the completed problems from various sources were acquired in a way that seemed to provide a 'mapping' of understanding. I wanted to convey that understanding here. Solutions to these key problems are carefully worked out in what follows, along with notes on the theory from Jackson as well as from [33-35]. Basic exercises are given to re-derive or verify the theory. The solutions to the exercises here, for the most part, are available via the aforementioned Jackson text (with the detailed derivations provided there). Many examples and related problems are done in order to cover material from basic undergraduate problems to advanced engineering (for antenna design, say), so the notes and problems also draw from undergraduate and graduate EM class notes/exams, and the other texts mentioned ([33-35]). The remarkable success and modern applications of EM will be made evident, but at the end of Ch. 7, when we carefully analyze the hydrogen atom, we will find that there is radiative collapse in the classical theory. This fundamental flaw in the theory, when dealing with even the simplest bound atomic system, will be corrected with the development of the quantum theory in Book 4 [15].

6.1 Macroscopic matter: Multipoles and Dielectric Materials

(Continuing to use notes/conventions of Jackson [32].)

Linear superposition and non-vacuum conditions

As we add sources and their fields, up to and including macroscopic aggregates of matter, we must consider the limits of linear superposition that permit such simple additivity. The classical theory is linear, thus linear superposition, so we are asking for a peek ahead, into the quantum electrodynamic theory, to see when there is a nonlinear effect. Such non-linear effect can arise from quantum theory because a photon can 'scatter' off another photon (Delbruck scattering [44], where two photons transform to an electron-positron pair and then back to two photons). Thus, even the electrodynamic vacuum condition will have some nonlinearity. Here is the expression for this nonlinear feature due to quantum electrodynamics (ca. 1935) [45], that is valid for slowly varying fields, where we also see our first example of electric and magnetic permeability tensors (here precisely, theoretically, known for the quantum vacuum):

$$D_i = \sum_k \epsilon_{ik} E_k,$$

(Eqn. 6-1)

$$B_i = \sum_k \mu_{ik} H_k \quad (traditional\ form)$$

(Eqn. 6-2)

$$H_\alpha = \sum_\beta \mu'_{\alpha\beta} B_\rho \quad (natural\ form)$$

(Eqn. 6-3)

$$where\ \epsilon_{ik} = \delta_{ik} + \frac{e^4\hbar}{45\pi m^4 c^7}\left[2(E^2 - B^2)\delta_{ik} + 7B_i B_k\right] + \cdots,$$

(Eqn. 6-4)

$$\mu_{ik} = \delta_{ik} + \frac{e^4\hbar}{45\pi m^4 c^7}\left[2(B^2 - E^2)\delta_{ik} + 7E_i E_k\right] + \cdots,$$

(Eqn. 6-5)

from which it can be seen that the non-linear effect is incredibly small for anything at the scale we will be considering. For this effect to be relevant we would probably be considering something at atomic scale and it would likely not be 'slowly-varying'. So, for what follows it will be safe to assume the strong validity of linear superposition.

Let's now consider fields due to macroscopic aggregates of matter. What is relevant from Maxwell's equations now is the average of fields for the relevant average of their sources. This is can expressed in terms of new 'macroscopic' fields ($\{\vec{D},\vec{H}\}$) that result from elementary sources in a background medium that is not vacuum, and has electric and magnetic permeability tensors relating macroscopic fields to vacuum fields:

$$\nabla \cdot \vec{D} = 4\pi\rho$$

(Eqn. 6-6)

$$\nabla \times \vec{H} - \frac{1}{c}\frac{\partial D}{\partial t} = \frac{4\pi}{c}\vec{J}$$

(Eqn. 6-7)

$$\nabla \times \vec{E} + \frac{1}{c}\frac{\partial \vec{B}}{\partial t} = 0$$

(Eqn. 6-8)

$$\nabla \cdot \vec{B} = 0$$

(Eqn. 6-9)

typically with:

$$D_\alpha = \sum_\beta \varepsilon_{\alpha\beta} E_\beta$$

(Eqn. 6-10)

and

$$H_\alpha = \sum_\beta \mu'_{\alpha\beta} B_\beta$$

(Eqn. 6-11)

for materials other than ferroelectrics or ferromagnets, and for weak enough fields. Note that the constitutive relations, $\vec{D} = \vec{D}[\vec{E},\vec{B}]$ and $\vec{H} = \vec{H}[\vec{E},\vec{B}]$, are necessary to move from a solution of the homogenous equations to the inhomogenous equations.

As noted in Jackson's text, to be generally correct the constitutive relations should be understood as holding for the Fourier Transform (FT) in space and time on the field quantities (because the linear connection, although linear, can be non-local):

$$D_\alpha(\vec{x},t) = \sum_\beta \int d^3x' \int dt' \epsilon_{\alpha\beta}(\vec{x}',t') E_\beta(\vec{x}-\vec{x}',t-t')$$

(Eqn. 6-12)

where $\epsilon_{\alpha\beta}(\vec{x}',t')$ can be localized around $\vec{x}' = 0$ and $t' = 0$, but is non-vanishing for some neighborhood around the origin. Using the relation that FT on convolution gives product on FT's we get the relation on FT's $D_\alpha(\vec{k},\omega)$, $E_\beta(\vec{k},\omega)$, and $\epsilon_{\alpha\beta}(\vec{k},\omega)$ using the standard FT definition

$$f(\vec{k},\omega) = \int d^3x \int dt f(\vec{x},t) e^{-i\vec{k}\cdot\vec{x}+i\omega t}$$

(Eqn. 6-13)

to get

$$D(\vec{k},\omega) = \sum_\beta \epsilon_{\alpha\beta}(\vec{k},\omega) E_\beta(\vec{k},\omega).$$

(Eqn. 6-14)

So, we are back to a 'strong form' of linearity even in the context of macroscopic media. This will hold in almost every situation considered. The dielectric constants, $\epsilon_{\alpha\beta}(\vec{k},\omega)$, will generally have a significant frequency dependence, however, according to the atomic structure underlying macroscopic media. Consider sufficiently low frequencies that all charges (regardless of inertia) are responsive to applied fields (less than $100KHz$ [1]). For solids we generally see a diagonal dielectric tensor $\epsilon = \epsilon_{\alpha\beta} \approx 2 - 20$. For distilled water (which consists of molecular dipoles) we have $\epsilon = 88$ at 0°C and $\epsilon = 56$ at 100°C. At optical frequencies, on the other hand $(400 - 790THz)$, only the electrons are responsive, and for solids we generally see dielectric coefficient $\epsilon \approx 1.7 - 10$, and water has $\epsilon \approx 1.77 - 1.80$ over the visible range, almost independent of temperature between 0°C and 100°C.

In quantum physics, Book 4 [15], we find that the elementary particles have an internal angular momentum, or spin, that is non-zero. We will obtain precise mathematical expressions for how spin factors into the interactions and scattering in Book 4 [15]. The theoretical origins of spin won't be explained until Book 5 [12], however, and for spin statistics, not until Book 5 part II. At the classical field level in Books 2 & 3 [116,5], however, we need only be concerned with macroscopic material responses to the applied magnetic fields (as with the electric fields), with the breakdown of materials into the categories diamagnetic, paramagnetic, and ferromagnetic.

Atoms and molecules, and the substances they comprise, are diamagnetic if they have an internal angular momentum that is zero. For these substances, the response to applied field follows from Lenz's law – e.g. a bulk magnetization opposing the applied field: ${\mu'}_{\alpha\alpha} > 1$ and with the strongest diamagnetic effect for bismuth, which has: ${\mu'}_{\alpha\alpha} = 1 +$

1.8×10^{-4}. If, on the other hand, the atom or molecular basis consists of elements with a net angular momentum (chemically, they have unpaired electrons), then they are paramagnetic. Now the alignment will favor the applied field, which then leads to $\mu'_{\alpha\alpha} < 1$, with the smallest value of $\mu'_{\alpha\alpha} \approx 0.99$. Clearly, $\mu'_{\alpha\alpha} \approx 1$ throughout, so the diamagnetic coefficient will generally be assumed to be 1 and not discussed further until Ch. 7. Note that paramagnetic atoms also have the property that they don't strongly interact among themselves (so they each react to the external field in an isolated way favoring alignment). Once there is significant interactions between the non-zero spin elements we get into the domain of ferromagnetic substances, including statistical mechanical effects like spontaneous magnetization. So, further discussion on these effects is not given until Ch. 7

Charge distributions seen as multipole expansions
From the expansions studied previously, let's now consider, specifically, systems with more than one charge, e.g. charge distributions, we then have multipole expansions. We have encountered point charge and dipole moment at lowest order thus far. Let's consider the multipole in spherical coordinates (to retain a point 'particle' perspective, with 'internal' multipole implied, an issue that comes back in Ch. 4 when discussing particle spin). We, therefore, want to examine:

Multipole Expansion

$$\Phi(\vec{x}) = \int \frac{\rho(\vec{x}')}{|\vec{x} - \vec{x}'|} d^3x'$$
$$= 4\pi \sum_{\ell,m} \frac{1}{2\ell + 1} \left[\int Y^*_{\ell m}(\theta', \phi') r'^{\ell} \rho(\vec{x}') d^3x' \right] \frac{Y_{\ell m}(\theta, \phi)}{r^{\ell+1}}$$

(Eqn. 6-15)

Interested in potential outside distribution, so $r_< = r', r_> = r$, thus:

$$\Phi(\vec{x}) = \sum_{\ell,m} \frac{4\pi}{2\ell + 1} q_{\ell m} \frac{Y_{\ell m}(\theta, \phi)}{r^{\ell+1}} \text{ where } q_{\ell m}$$
$$= \int Y^*_{\ell m}(\theta', \phi') r'^{\ell} \rho(\vec{x}') d^3x'.$$

(Eqn. 6-16)

where $q_{\ell m}$ are the multipole moments and $q_{\ell,-m} = (-1)^m q^*_{\ell m}$. If we expand $\Phi(\vec{x})$ in rectangular coordinates this becomes:

$$\Phi(\vec{x}) = \frac{q}{r} + \frac{\vec{p} \cdot \vec{x}}{r^3} + \frac{1}{2}\sum_{i,j} Q_{ij} \frac{x_i x_j}{r^5} + \cdots.$$

(Eqn. 6-17)

which follows from direct Taylor series expansion of $^1/_{|\vec{x} - \vec{x}'|}$. The E-field is thus:

$$\vec{E}(\vec{x}) = \frac{3\vec{n}(\vec{p} \cdot \vec{n}) - \vec{p}}{|\vec{x} - \vec{x}_0|^3}, \qquad \textit{where } \vec{n} \textit{ is the unit vector from } \vec{x} \text{ to } \vec{x}_0$$

(Eqn. 6-18)

The values of $q_{\ell m}$ for the lowest nonvanishing multipole moment of any charge distribution are independent of the choice or origin of the coordinates, but all higher multipole moments do in general depend on the location of the origin.

In terms of a Rectangular Coordinate Expansion, we have the following:

$$\text{electric dipole moment:} \, \vec{p} = \int \vec{x}' \rho(\vec{x}')d^3x'$$

$$\left(\text{monopole moment is total charge } q = \int \rho \, d^3x'\right)$$

$$\text{electric quadrupole moment! } Q_{ij} = \int \left(3x_i'x_j' - r'^2 \delta_{ij}\right) \rho(\vec{x}')d^3x'$$

(Eqn. 6-19)

Example 6.1. *Calculate the multipole moments* $q_{\ell m}$ *for when four charges with magnitude* $|q|$ *are placed in (x,y,z) coordinates such that there are positive charges at (a,0,0) and (0,a,0) and negative charges at (−a,0,0) and (0,−a,0).*

Solution

We have by definition:

$$q_{\ell m} = \int r^\ell Y^*_{\ell m}(\theta, \phi)\rho(r, \theta, \phi)d^3x.$$

The charge distribution is odd about the origin (can invert sign and have same type of solutions) so $q_{\ell m}$ will be nonzero only for ℓ odd, also, $\theta = \frac{\pi}{2} \rightarrow \cos\theta = 0$, so $q_{\ell m}$ will only be non-zero when P_ℓ^m is an even function of $\cos\theta$. Parity of P_ℓ^m is $(-1)^{\ell+m}$, and we know ℓ must be odd so m must be odd.

Since $Y_{\ell,-m}(\theta,\phi) = (-1)^m Y^*_{\ell m}(\theta,\phi) = -Y^*_{\ell m}(\theta,\phi)\ for\ m\ odd \Rightarrow$
$q_{\ell,-m} = -q^*_{\ell m}$

$$q_{\ell m} = qa^\ell \left[Y^*_{\ell m}\left(\frac{\pi}{2},0\right) + Y^*_{\ell m}\left(\frac{\pi}{2},\frac{\pi}{2}\right) - Y^*_{\ell m}\left(\frac{\pi}{2},\pi\right) - Y^*_{\ell m}\left(\frac{\pi}{2},\frac{3\pi}{2}\right)\right]$$
$$= 2qa^\ell \left[Y^*_{\ell m}\left(\frac{\pi}{2},0\right) + Y^*_{\ell m}\left(\frac{\pi}{2},\frac{\pi}{2}\right)\right] \text{ for } \ell \text{ odd}$$

Now $Y^*_{\ell m}\left(\frac{\pi}{2},0\right) = \sqrt{\frac{2\ell+1}{4\pi}\frac{(\ell-m)!}{(\ell+1)!}}P_\ell^m(0)$
$Y^*_{\ell m}\left(\frac{\pi}{2},\frac{\pi}{2}\right) = Y^*_{\ell m}\left(\frac{\pi}{2},0\right)e^{-im\pi/2}$
So,

$$q_{\ell m} = 2qa^\ell \sqrt{\frac{2\ell+1}{4\pi}\frac{(\ell-m)!}{(\ell+1)!}}P_\ell^m(0)\left[1+\left(e^{-i\frac{\pi}{2}}\right)^m\right] \qquad \ell, m \text{ odd}$$

Thus, at lowest order:

$$q_{11} = \sqrt{\frac{3}{8\pi}}2aq(1+i); \quad q_{31} = \sqrt{\frac{21}{4\pi}}a^2q(1+i); \quad q_{33}$$
$$= \sqrt{\frac{35}{4\pi}}a^3q(1+i)\ etc.$$

Multipole Expansion of the Energy of a Charge Distribution in an External Field

Consider a localized charge distribution $\rho(\vec{x})$ in an external potential $\Phi(\vec{x})$. Its electrostatic energy is given by:

$$W = \int \rho(\vec{x})\,\Phi(\vec{x})d^3x.$$

(Eqn. 6-20)

If Φ is slowly varying in the region of where Φ is non-negligable, it can be expanded in a Taylor series: (around a suitably chosen origin):

$$\Phi(\vec{x}) = \Phi(0) + \vec{x}\cdot\nabla\Phi(0) + \frac{1}{2}\sum_i\sum_j x_i - x_j\frac{\partial^2\Phi(0)}{\partial x_i \partial x_j} + \cdots.$$

(Eqn. 6-21)

Since $\vec{E} = -\vec{\nabla}\Phi$:

$$\Phi(\vec{x}) = \Phi(0) - \vec{x} \cdot \vec{E}(0) + \frac{1}{2} \sum_i \sum_j x_i x_j \frac{\partial E_j(0)}{\partial x_i} + \cdots.$$

(Eqn. 6-22)

Since $\nabla \cdot \vec{E} = 0$ we can make x^2 form traceless, subtract $\frac{1}{6} r^2 \vec{\nabla} \cdot E$:

$$\Phi(\vec{x}) = \Phi(0) - \vec{x} \cdot \vec{E}(0) + \frac{1}{6} \sum_{i,j} (3x_i x_j - r^2 \delta_{ij}) \frac{\partial E_j(0)}{\partial x_i} + \cdots.$$

(Eqn. 6-23)

So,

$$W = q\Phi(0) - \vec{p} \cdot \vec{E}(0) + \frac{1}{6} \sum_{i,j} Q_{ij} \frac{\partial E_j(0)}{\partial x_i} + \cdots.$$

(Eqn. 6-24)

From the above result we can see the interactions of various multipoles with the external field that contribute to the potential energy. Thus, there is potential energy from:
(i) a charge with the external potential
(ii) a dipole moment with the electric field (potential gradient)
(iii) a quadrupole moment with electric field gradient
Etc.

Interaction energy between two dipoles $\vec{p}_1$ and $\vec{p}_2$ can now be understood from the $-\vec{p} \cdot \vec{E}(0)$ term with the $\vec{E}(\vec{x})$ taken to be due to a dipole (with equation above). Thus:

$$W_{12} = \frac{\vec{p}_1 \cdot \vec{p}_2 - 3(\vec{n} \cdot \vec{p}_1)(\vec{n} \cdot \vec{p}_2)}{|\vec{x}_1 - \vec{x}_2|^3},$$

(Eqn. 6-25)

where $\vec{n}$ is the unit vector in the direction $(\vec{x}_1 - \vec{x}_2)$, it is assumed $\vec{x}_1 \neq \vec{x}_2$.

Electrostatics with Ponderable Media

Ponderable Media consists sources and materials, such as charges, conductors, dielectrics, etc.

Averaging of $\vec{\nabla} \times \vec{E}_{micro} = 0$ still yields $\vec{\nabla} \times \vec{E} = 0$. Thus, the electric field is still derivable from a potential $\Phi(\vec{x})$ in electrostatics. The dominant molecular multipole interaction with an applied field is the dipole. Therefore, let's focus on the production in a medium of an electric polarization $\vec{p}$ (denoting a dipole moment per unit volume) given by:

$$\vec{p}(\vec{x}) = \sum_i N_i \langle p_i \rangle,$$

(Eqn. 6-26)

where the index is over the various types of molecule. Similarly, the charge density at the macroscopic level is:

$$\rho(\vec{x}) = \sum_i N_i \langle e_i \rangle + \rho_{excess}.$$

(Eqn. 6-27)

If we consider no higher macroscopic multipole moment densities (than dipole), and consider a small volume element ΔV centered at $\vec{x}'$, then we have for the contribution to potential from that volume element:

$$\Delta\Phi(\vec{x},\vec{x}') = \frac{\rho(\vec{x}')}{|\vec{x}-\vec{x}'|}\Delta V + \frac{\vec{p}(\vec{x}')\cdot(\vec{x}-\vec{x}')}{|\vec{x}-\vec{x}'|^3}\Delta V$$

(Eqn. 6-28)

Integrating:

$$\Phi(\vec{x}) = \int d^3x' \left[\frac{\rho(\vec{x}')}{|\vec{x}-\vec{x}'|} + \vec{p}(\vec{x}')\cdot\vec{\nabla}'\left(\frac{1}{|\vec{x}-\vec{x}'|}\right)\right]$$
$$= \int d^3x' \frac{1}{|\vec{x}-\vec{x}'|}\left[\rho(\vec{x}') - \vec{\nabla}'\cdot\vec{p}(\vec{x}')\right].$$

(Eqn. 6-29)

(using integration by parts and surface terms go to zero). Thus, we have:

$$\nabla\cdot\vec{E} = 4\pi[\rho - \nabla\cdot\vec{p}].$$

(Eqn. 6-30)

More natural for the ponderable media description is to consider a new effective field $\vec{D} = \vec{E} + 4\pi\vec{p}$, since we can then write:

$$\nabla\cdot\vec{D} = 4\pi\rho,$$

(Eqn. 6-31)

and effectively have the same mathematical formalism as for bare charge in vacuum.

A constitutive relation connecting $\vec{D}$ and $\vec{E}$ is necessary before a solution for the electrostatic potential or fields can be obtained. Assume a linear response of medium (excludes ferroelectrics), which is reasonable for field strengths not too large. Also, assume medium is isotropic, so:

$$\vec{p} = \chi_e\vec{E},$$

(Eqn. 6-32)

where χ_e is a constant, known as the electric susceptibility. Thus,

$$\vec{D} = \epsilon\vec{E}$$

(Eqn. 6-33)

where $\vec{D} = \vec{E} + 4\pi x_e \vec{E} = (1 + 4\pi x_e)\vec{E}$. The constant ϵ $(= 1 + 4\pi x_e)$ is called the dielectric constant or relative electric permittivity.

Now the divergence equations can be written:

$$\nabla \cdot E = 4\pi \left(\frac{\rho}{\epsilon}\right)$$

(Eqn. 6-34)

All problems in the electrically responsive medium are reduced to those of the previous analyses, but with the electric fields produced by charges reduced by a factor of $1/\epsilon$. Since $C = Q/V' = Q/(V/\epsilon) = \epsilon\, Q/V$, one immediate consequence is that the capacitance of a capacitor is increased by a factor of ϵ if the empty space between the electrodes is filled with a dielectric with dielectric constant ϵ.

If there are different media juxtaposed we must consider the question of boundary conditions on $\vec{D}$ and $\vec{E}$ at the interfaces between media:

$$(\vec{D}_2 - \vec{D}_1) \cdot \vec{n}_{21} = 4\pi\sigma,$$

(Eqn. 6-35)

$$(\vec{E}_2 - \vec{E}_1) \times \vec{n}_{21} = 0,$$

(Eqn. 6-36)

where

$\vec{n}_{21}$ – the unit normal directed from region 1 to .2,

and

σ is the macroscopic surface charge density on the boundary surface (<u>not</u> including the polarization charge).

Let's now consider some boundary value problems with dielectrics.

Boundary – Value Problems with Dielectrics

Consider the problem of a point charge in a semi-infinite dielectric a distance d (z-direction) from a different semi-infinite dielectric. To solve, consider solutions to equations:

$$\epsilon_1 \vec{\nabla} \cdot \vec{E} = 4\pi\rho \quad z > 0$$
$$\epsilon_2 \vec{\nabla} \cdot \vec{E} = 0 \quad z < 0$$
$$\vec{\nabla} \times \vec{E} = 0 \quad \text{everywhere}$$

(Eqn.s 6-37)

Subject to the b.c.'s (boundary conditions) at $z = 0$:

$$\begin{aligned}(\vec{D}_2 - \vec{D}_1)\cdot \vec{n}_{21} &= 4\pi\sigma \\ (\vec{E}_2 - \vec{E}_1)\times \vec{n}_{21} &= 0\end{aligned} \Rightarrow \lim_{z\to 0^+}\begin{Bmatrix}\epsilon_1 E_z\\ E_x\\ E_y\end{Bmatrix} = \lim_{z\to 0^-}\begin{Bmatrix}\epsilon_2 E_z\\ E_x\\ E_y\end{Bmatrix}$$

(Eqn.s 6-38)

Since $\nabla\times E = 0$ everywhere, E is derivable in the usual way from a potential Φ. Using image method; start by guessing q' at A', such that:

$$For\ z > 0 \quad \Phi_R = \frac{1}{\epsilon_1}\left(\frac{q}{R_1} + \frac{q'}{R_2}\right),$$

where $R_1 = \sqrt{\rho^2 + (d-z)^2}$ and $R_2 = \sqrt{\rho^2 + (d+z)^2}$.

For $z < 0$ simplest assumption is that $\Phi_L = \frac{1}{\epsilon_2}\frac{q''}{R_1}$ i.e., q'' at A.

Normal: $\epsilon_1\left(\frac{-\partial\Phi_R}{\partial z}\right)\Big|_{z=0} = \epsilon_2\left(\frac{-\partial\Phi_L}{\partial z}\right)\Big|_{z=0} \Rightarrow \epsilon_1\left\{\frac{1}{\epsilon_1}\left(\frac{d(q)}{(\rho^2+(d-z)^2)^{3/2}} - \frac{d(q')}{(\rho^2+d^2)^{3/2}}\right)\right\} = \epsilon_2\frac{1}{\epsilon_2}\frac{d(q'')}{(\rho^2+d^2)^{3/2}}$

So, $q - q' = q''$.

Tangent: $\frac{\partial\Phi_R}{\partial\rho}\Big|_{z=0} = \frac{\partial\Phi_L}{\partial\rho}\Big|_{z=0} \Rightarrow \frac{1}{\epsilon_1}\left(\frac{-q\rho}{(\rho^2+d^2)^{3/2}} - \frac{-q'\rho}{(\rho^2+d^2)^{3/2}}\right) = \frac{1}{\epsilon_2}\frac{-q''\rho}{(\rho^2+d^2)^{3/2}}$

So, $\frac{1}{\epsilon_1}(q + q') = \frac{1}{\epsilon_2}q''$. Using $q - q' = q''$ and $q + q' = \frac{\epsilon_1}{\epsilon_2}q''$, we then get

$$q' = q - q'' = \frac{\epsilon_1 + \epsilon_2 - 2\epsilon_2}{\epsilon_1 + \epsilon_2}q = \left(\frac{\epsilon_1 - \epsilon_2}{\epsilon_1 + \epsilon_2}\right)q.$$

Now, polarization-charge density is given by $-\vec{\nabla}\cdot\vec{p}$. Inside either dielectric, $\vec{p} = \chi_e\vec{E}$, so that $-\vec{\nabla}\cdot\vec{p} = -\chi_e\vec{\nabla}\cdot\vec{E} = 0$, except at q. At the surface x_e takes a discontinuous jump: $\Delta\chi_e = \frac{1}{4\pi}(\epsilon_1 - \epsilon_2)$ at $z = 0$. This implies that there is a polarization surface – charge density on the plane $z = 0$:

$$\sigma_{pol} = -(p_2 - p_1)\cdot\vec{n}_{21}$$

Since

$$4\pi\vec{p}_i + \vec{E} = \epsilon_i\vec{E} \Rightarrow p_i = \left(\frac{\epsilon_i - 1}{4\pi}\right)\vec{E} = -\left(\frac{\epsilon_i - 1}{4\pi}\right)\nabla\Phi,$$

we can write:

$$p_1 \cdot \vec{n}_{21} = -\left(\frac{\epsilon_1 - 1}{4\pi}\right) \frac{\partial \left\{ \frac{1}{\epsilon_1}\left(\frac{q}{R_1} + \frac{q'}{R_2}\right)\right\}}{\partial z}\Bigg|_{z=0}$$

$$= -\left(\frac{\epsilon_1 - 1}{4\pi}\right)\frac{q}{\epsilon_1}\left(\frac{d}{(\rho^2 + d^2)^{3/2}}\right)\left[\frac{2\epsilon_2}{\epsilon_1 + \epsilon_2}\right]$$

Thus,

$$p_2 \cdot \vec{n}_{21} = -\left(\frac{\epsilon_2 - 1}{4\pi}\right)\frac{1}{\epsilon_2}\left(\frac{2\epsilon_2}{\epsilon_1 + \epsilon_2}\right) q \frac{\partial \frac{1}{R_1}}{\partial z}$$

$$= -\left(\frac{\epsilon_2 - 1}{4\pi}\right)\frac{q}{\epsilon_2}\left(\frac{2\epsilon_2}{\epsilon_1 + \epsilon_2}\right)\left\{\frac{-d}{(\rho^2 + d^2)^{3/2}}\right\}$$

and

$$\sigma_{pol} = -\frac{q}{4\pi}\frac{(\epsilon_2 - \epsilon_1)}{\epsilon_1(\epsilon_2 + \epsilon_1)}\frac{d}{(\rho^2 + d^2)^{3/2}}.$$

(Eqn. 6-39)

In the limit $\epsilon_2 \gg \epsilon_1$ the dielectric ϵ_2 behaves much like a conductor in that the electric field inside it becomes very small and the surface – charge density approaches the value appropriate to a conducting surface:

$$\sigma_{pol}\Big|_{\epsilon_2 \gg \epsilon_1} = \frac{-q/\epsilon_1}{2\pi}\frac{d}{(\rho^2 + d^2)^{3/2}}.$$

(Eqn. 6-40)

Consider the problem of a dielectric sphere of radius a in a uniform electric field (along z-axis)

We want to know potential everywhere, problem is one of solving the Laplace equation with the proper boundary conditions at $r = a$. We clearly have azimuthal symmetry, so:

Inside: $\Phi_{in} = \sum_{\ell=0}^{\infty} A_\ell r^\ell P_\ell(\cos\theta)$

Outside: $\Phi_{out} = \sum_{\ell=0}^{\infty}\left[B_\ell r^\ell + C_\ell r^{-(\ell+1)}\right] P_\ell(\cos\theta)$

The boundary condition at infinity is $\Phi \to -E_0 z = -E_0 r\cos\theta$, thus the only non-vanishing B_ℓ is $B_1 = -E_0$ the other coefficients are determined from the b.c. at $r = a$:

Tangential E: $-\frac{1}{a}\frac{\partial\Phi_{in}}{\partial\theta}\Big|_{r=a} = -\frac{1}{a}\frac{\partial\Phi_{out}}{\partial\theta}\Big|_{r=a}$

(Eqn. 6-41)

Normal D: $\quad -\epsilon \frac{\partial \Phi_{in}}{\partial r}\Big|_{r=a} = -\frac{\partial \Phi_{out}}{\partial r}\Big|_{r=a}$

(Eqn. 6-42)

Let's substitute the series:

$$-\frac{1}{a}\frac{\partial\left[\sum_{\ell=0}^{\infty} A_\ell\, r^\ell P_\ell(\cos\theta)\right]}{\partial\theta}\Bigg|_{r=a} = -\frac{1}{a}\frac{\partial\left[-E_0 r\cos\theta + \sum_{\ell=0}^{\infty} C_\ell r^{-(\ell+1)}\, P_\ell(\cos\theta)\right]}{\partial\theta}\Bigg|_{r=a}$$

$$\sum_{\ell=0}^{\infty} A_\ell\, a^\ell \frac{\partial P_\ell(\cos\theta)}{\partial\theta} = -E_0 a\frac{\partial P_\ell(\cos\theta)}{\partial\theta} + \sum_{\ell=0}^{\infty} C_\ell a^{-(\ell+1)}\frac{\partial P_\ell(\cos\theta)}{\partial\theta}$$

From which we get: $A_0 = C_0\frac{1}{a},\ A_1 a = -E_0 a + Cp^{-2},\ A_\ell a^\ell = C_\ell a^{-(\ell+1)}$ for $\ell > 1$.
Thus, $A_1 = -E_0 + \frac{C_1}{a^3}$ and $A_\ell = \frac{C_\ell}{a^{2\ell+1}}$ for $\ell \neq 1$ (relations i and ii)

For the 2nd relation:

$$-\epsilon\sum_{\ell=0}^{\infty} A_\ell\, \ell a^{\ell-1} P_\ell(\cos\theta) = -\sum_{\ell=0}^{\infty}\left[B_\ell \ell a^{\ell-1} + C_\ell(-[\ell+1])a^{-(\ell+2)}\right]P_\ell(\cos\theta)$$

again, have only $B_1 = -E_0$. Let's get relations at various ℓ values:

$\ell = 0$: $\quad C_0(-1)a^{-2} = 0 \Rightarrow C_0 = 0$

$\ell = 1$: $\quad -\epsilon A_1 = -[-E_0 + C_1(-2)a^{-3}] \rightarrow\ \epsilon A_1 = -E_0 - 2\frac{C_1}{a^3}$ (relation iii).

$\ell \neq 1$: $\quad -\epsilon A_\ell \ell a^{\ell-1} = +C_\ell(\ell+1)a^{-(\ell+2)} \ \rightarrow\ \ \epsilon\ell A_\ell = -(\ell+1)\frac{C_\ell}{a^{2\ell+1}}$ (relation iv).

(i) + (iii) $\Rightarrow 2A_1 + \epsilon A_1 = -3E_0 \Rightarrow A_1 = -\left(\frac{3}{2+\epsilon}\right)E_0$

(ii) + (iv) $\Rightarrow [\epsilon\ell + (\ell+1)]\left(\frac{C_\ell}{a^{2\ell+1}}\right) = 0 \quad \text{for} \quad \ell \neq$
1 $\Rightarrow C_\ell = 0 \quad \text{for } \ell \neq 1$
$\Rightarrow A_\ell = 0 \quad \text{for } \ell \neq 1$

$$C_1 = (A_1 + E_0)a^3 = \left\{1 - \frac{3}{2+\epsilon}\right\} E_0 a^3 = \left(\frac{\epsilon - 1}{\epsilon + 2}\right) E_0 a^3 \Rightarrow C_1$$
$$= \left(\frac{\epsilon - 1}{\epsilon + 2}\right) E_0 a^3$$

Thus,

$$\Phi_{in} = -\left(\frac{3}{2+\epsilon}\right) E_0 r \cos\theta$$

(Eqn. 6-43)

and

$$\Phi_{out} = -E_0 r \cos\theta + \left(\frac{\epsilon - 1}{\epsilon + 2}\right) E_0 \frac{a^3}{r^2} \cos\theta.$$

(Eqn. 6-44)

Note that $E_{in} = \frac{3}{\epsilon+2} E_0 < E_0$ if $\epsilon > 1 \Rightarrow$ potential less inside, thus E-field is partially screened inside the sphere. Outside the sphere the potential is that of the applied field and a dipole moment:

$$p = \left(\frac{\epsilon - 1}{\epsilon + 2}\right) a^3 E_0,$$

(Eqn. 6-45)

where the dipole is oriented to oppose the field. The dipole moment can be interpreted as the volume integral of the polarization $\vec{p}$; where the polarization is

$$\vec{p} = \left(\frac{\epsilon - 1}{4\pi}\right) \vec{E} = \frac{3}{4\pi}\left(\frac{\epsilon - 1}{\epsilon + 2}\right) \vec{E}_0$$

(Eqn. 6-46)

and $\int \vec{p}\, dv = p$ as expected. For polarization surface-charge density:

$$\sigma_{pol} = -(\vec{p}_2 - \vec{p}_1) \cdot \vec{n}_{21} \quad and \quad \vec{p}_2 = 0 \quad (\epsilon = 1) \quad and \quad \vec{n}_{21} = \hat{r}.$$

(Eqn. 6-47)

Thus,

$$\sigma_{pol} = \frac{(\vec{p} \cdot \vec{r})}{r} = \frac{3}{4\pi}\left(\frac{\epsilon - 1}{\epsilon + 2}\right) |E_0| \cos\theta.$$

(Eqn. 6-48)

Example 6.2. *Consider two concentric conducting spheres, radii a, b, with b > a, with inner sphere having charge Q and outer sphere charge – Q. The hemispherical shell above the x-y plane is filled with dielectric* ϵ.
(a) Obtain the E-field everywhere between the spheres.
(b) Calculate he surface charge distribution on the inner sphere.
(c) Calculate the polarization-charge density induced on the surface of the dielectric at r=a.

Solution
(a) Want E field everywhere between spheres:

For $(a \le r \le b)$ and $(0 \le \theta \le \pi_2)$

$$\Phi_1(r,\theta) = \sum_{\ell=0}^{\infty}\left[A_{1\ell}r^{\ell} + B_{1\ell}r^{-(\ell+1)}\right] P_\ell(\cos\theta)$$

For $(a \le r \le b)$ and $(\pi \le \theta \le 2\pi)$

$$\Phi_2(r,\theta) = \sum_{\ell=0}^{\infty}\left[A_{2\ell}r^{\ell} + B_{2\ell}r^{-(\ell+1)}\right] P_\ell(\cos\theta)$$

Since spheres are conductors E_θ must vanish at $r = a, r = b$:

$$E_\theta(r,\theta) = \frac{1}{r}\frac{\partial\Phi}{\partial\theta}$$

$$\left.\begin{aligned} E_{1\theta}(a,\theta) &= \sum_{\ell=0}^{\infty}\left(A_{1\ell}a^{\ell-1} + B_{1\ell}a^{-\ell+2}\right)\frac{\partial P_\ell}{\partial\theta} = 0 \\ E_{1\theta}(b,\theta) &= \sum_{\underset{\ne 0}{\ell=0}}^{\infty}\left(A_{1\ell}b^{\ell-1} + B_{1\ell}b^{-\ell+2}\right)\frac{\partial P_\ell}{\partial\theta} = 0 \end{aligned}\right\} \Rightarrow \begin{cases} A_{1\ell} = 0 \\ B_{1\ell} = 0 \end{cases} \textit{for } \ell$$

Same result for $\begin{cases} A_{2\ell} \\ B_{2\ell} \end{cases}$.

So,

$$\Phi_1(\vec{r}) = A_1 + \frac{B_1}{r} \qquad \Phi_2(\vec{r}) = A_2 + \frac{B_2}{r}$$

$$\vec{E}_1(\vec{r}) = -\frac{B_1}{r^2}\hat{r} \qquad \vec{E}_2(\vec{r}) = -\frac{B_2}{r^2}\hat{r}$$

Assume charge densities σ_1 and σ_2 on inner surface: Gauss's Law:

$D_1(2\pi r^2) = 4\pi(2\pi a^2\sigma_1) \Rightarrow D_1 = (4\pi\sigma_1 a^2)/r^2 \Rightarrow E_1 = (4\pi\sigma_1 a^2/\theta r^2)$. Similarly, $E_2 = (4\pi\sigma_2 a^2)/r^2$.

Tangential E is continuous: $(4\pi\sigma_1 a^2)/(\epsilon r^2) = (4\pi\sigma_2 a^2)/r^2 \Rightarrow \sigma_1/\epsilon = \sigma_2$.

So, for the surface charge distribution on the inner sphere: $2\pi a^2(\epsilon + 1)\sigma_2 = Q \Rightarrow \sigma_2 = Q/[2\pi a^2(1+\epsilon)]$, similarly, $\sigma_1 = \epsilon\, Q/[2\pi a^2(1+\epsilon)]$.

(c) For the E-field:

$$E_1 = \frac{2Q\hat{r}}{(1+\epsilon)r^2}; \quad E_2 = \frac{2Q}{(1+\epsilon)}\frac{\hat{r}}{r^2} \quad \text{(same)}.$$

Let's now consider Molecular Polarizability and Electric Susceptibility.

Molecular Polarizability and Electric Susceptibility

A full analysis of the relation between molecular polarizability and electric susceptibility requires quantum mechanical models. Fortunately, simple dielectric properties can be obtained with the existing classical model. The electric field, being additive, can be decomposed into parts. So, let's split the E-field contribution felt at a position x (internal field $\vec{E}_i$) to that of nearby molecules ($\vec{E}_{near}$) and that of an average background (viewed as macroscopic field), where the background is decomposed into a macroscopic $\vec{E}$ part and a (macroscopic) material polarization part $\vec{E}_p$ (due to polarization $\vec{P}$). The potential felt at position x is thus decomposed into

$$\vec{E}_{Total} = \vec{E} + \vec{E}_i, \quad \text{(Eqn. 6-49)}$$

where the internal field is related to the "near" field by:

$$\vec{E}_i = \vec{E}_{near} - \vec{E}_p. \quad \text{(Eqn. 6-50)}$$

Note, we are effectively saying use the improved approximation where the locally variable polarization, giving $\vec{E}_{near}$, is swapped in for the average background polarization.

Consider a spherical volume $V = \frac{4\pi R^3}{3}$, total dipole moment in V is:

$$\vec{p} = \frac{4\pi R^3}{3}\vec{P}.$$

(Eqn. 6-51)

The average electric field inside the sphere due to this polarization (thus $\vec{E}_p$) is:

$$\vec{E}_p = \frac{1}{\frac{4}{3}\pi R^3}\int_{r<R} E d^3x = \frac{-4\pi}{3}\vec{P}$$

(Eqn. 6-52)

Note the sign change due to definitions in reference to

$$W = q\Phi(0) - \vec{p}\cdot\vec{E}(0) + \frac{1}{6}\sum_{i,j} Q_{ij}\frac{\partial E_j(0)}{\partial x_i} + \cdots.$$

(Eqn. 6-53)

discussed previously, where the latter has alternating signs in the expansion.

So,

$$\vec{E}_i = \frac{4\pi}{3}\vec{P} + \vec{E}_{near}.$$

(Eqn. 6-54)

In most materials $\vec{E}_{near} \simeq 0$. In a cubic lattice or a completely random (homogeneous and isotropic) there is the exact result $\vec{E}_{near} = 0$. So the focus is mainly on $\vec{P}$. Now $\vec{P} = N\langle\vec{p}_{mol}\rangle$, where we now reference the average dipole moment of surrounding molecules (not total like $\vec{p}$). If we define molecular polarizability γ_{mol} (a macroscopic quantity), we can write $\langle\vec{p}_{mol}\rangle = \gamma_{mol}(\vec{E} + \vec{E}_i)$, which then gives:

$$\vec{P} = N\gamma_{mol}\left(\vec{E} + \frac{4\pi}{3}\vec{P}\right),$$

(Eqn. 6-55)

using $\vec{P} = \chi_e\vec{E}$ and $\epsilon = 1 + 4\pi\chi_e$ this becomes:

$$\chi_e = \frac{N\gamma_{mol}}{1 - \frac{4\pi}{3}N\gamma_{mol}},$$

(Eqn. 6-56)

or

$$N\gamma_{mol} = \frac{3}{4\pi N}\left(\frac{\epsilon - 1}{\epsilon + 1}\right),$$

(Eqn. 6-57)

which is known as the Clausius – Mossotti equation (at optical frequencies $\epsilon = n^2$ and this is then known as the Lorentz-Lorenz equation).

Electrostatic Energy in Dielectric Media

For a system of charges in free space: $W = \frac{1}{2}\int \rho(\vec{x})\,\Phi(\vec{x})d^3x$. For dielectrics, however, work is also done to produce a certain polarization, etc. So, for a system of charges in a dielectric we need to restart the derivation in terms of variational pieces of energy:

$$\delta W = \int \delta\rho(\vec{x})\,\Phi(\vec{x})d^3x,$$

(Eqn. 6-58)

where we take the potential to be that due to $\rho(\vec{x})$ already present (a similar analysis is done for a dust-shell in-fall analysis later). We also want to relate the charge density to a "D" field since a dielectric is present, so use $\nabla \cdot D = 4\pi\rho \Rightarrow \delta\rho = \frac{1}{4\pi}\nabla \cdot (\delta\vec{D})$. The variation of the Work is thus

$$\delta W = \frac{1}{4\pi}\int E \cdot \delta D d^3x,$$

(Eqn. 6-59)

which gives rise to the integral form:

$$W = \frac{1}{4\pi}\int d^3x \int_0^D E \cdot \delta D.$$

(Eqn. 6-60)

If medium is linear $\vec{E} \cdot \delta\vec{D} = \frac{1}{2}\delta(\vec{E} \cdot \vec{D})$, a solution is readily obtained:

$$W = \frac{1}{8\pi}\int \vec{E} \cdot \vec{D}\, d^3x,$$

(Eqn. 6-61)

where $E = \nabla\Phi$ and $\nabla \cdot D = 4\pi\rho$. Thus, our original equation is valid microscopically only in the behavior of the medium is linear. Otherwise the energy may have hysteresis effects, etc.

What is the change in energy when a dielectric object, with a linear response, is placed in an electric field whose sources are fixed?

Consider an initial electric field $\vec{E}_0$ due to $\rho_0(\vec{x})$ were dielectric = $\epsilon_0(x)$. For initial energy we have:

$$W_0 = \frac{1}{8\pi}\int E_0 \cdot D_0\, d^3x \qquad D_0 = \epsilon E_0.$$

Introduce dielectric object with volume V_1: $E_0 \to E, \epsilon = \epsilon_1$ inside V_1, ϵ_2 outside V_1 and $\vec{D} = \epsilon\vec{E}$, the new energy is:

$$W_1 = \frac{1}{8\pi}\int E \cdot D\, d^3x.$$

The Energy difference:

$$W = \frac{1}{8\pi}\int \left(\vec{E} \cdot \vec{D} - \vec{E}_0 \cdot \vec{D}_0\right) d^3x$$

$$= \frac{1}{8\pi}\int \left(\vec{E} \cdot \vec{D}_0 - \vec{D} \cdot \vec{E}_0\right) d^3x$$

$$+ \frac{1}{8\pi}\int \left(\vec{E} + \vec{E}_0\right) \cdot \left(\vec{D} - \vec{D}_0\right) d^3x$$

Focusing on the last term, let's call it I. Since $\nabla \times (E + E_0) = 0 \to E + E_0 = -\nabla\Phi'$, thus:

$$I = -\frac{1}{8\pi}\int \nabla\Phi \cdot (D - D_0) d^3x.$$

Thus, $I = \frac{1}{8\pi}\int \Phi\nabla \cdot (D - D_0) d^3x = 0$ since $\nabla \cdot (D - D_0) = 0$ since $\rho_0(\vec{x})$ is assumed unaltered by the insertion of the dielectric. So,

$$W = \frac{1}{8\pi}\int (E \cdot D_0 - D \cdot E_0)\, d^3x = \frac{-1}{8\pi}\int_{V_1} (\epsilon_1 - \epsilon_0)\, E \cdot E_0 d^3x.$$

For $\epsilon_0 = 1$, recall $\vec{P}_1 = \left(\frac{\epsilon_1 - 1}{4\pi}\right)\vec{E}_1$, so

$$W = \frac{1}{2}\int_{V_1} \vec{P} \cdot \vec{E}_0\, d^3x \Rightarrow w = -\frac{1}{2}\vec{P} \cdot \vec{E}_0,$$

where w is the energy density.

A dielectric will tend to move towards regions of increasing field $\vec{E}_0$. An example of this in the LIGO project is that dust is sucked into a laser beam, charged, then fired into mirrors as a degradation process. The above holds when charges are held fixed. The opposite occurs when potential held fixed.

Motion of dielectrics in E-fields produced by fixed potentials (energy is supplied by an external source, a battery for example)

We now start with

$$\delta W = \frac{1}{2}\int (\rho\partial\Phi + \Phi\delta\rho)\, d^3x.$$

(Eqn. 6-62)

Note that if dielectric properties unchanged with the variation then the two terms on right are equal.
The process of altering the dielectric properties in some way in the presence of electrodes at fixed potentials can be viewed as taking place in 2 steps:
(1) Electrodes disconnected from batteries $\Rightarrow$ charges held fixed $\Rightarrow$ $\delta\rho = 0$:

$$\delta W_1 = \frac{1}{2}\int \rho\delta\Phi_1\, d^3x = -\frac{1}{2}\int_{V_1} \vec{P}\cdot\vec{E}_0\, d^3x$$

(2) Batteries connected to restore to original potentials $\Rightarrow$ flow of charge $\delta\rho_2$ accompanied by change in potential $\delta\Phi_2 = -\delta\Phi_1$:

$$\delta W_2 = \frac{1}{2}\int (\rho\delta\Phi_2 + \Phi\delta\rho_2)\, d^3x = -2\delta W_1$$

Consequently:

$$\delta W = -\frac{1}{2}\int \rho\partial\Phi_1\, d^3x\,.$$

Example 6.3. *Two coaxial cylindrical surfaces are lowered into a liquid dielectric (*$\epsilon = 1 + 4\pi\chi_e$*), with radii a, b, with* $b > a$. *When a potential V is applied, the liquid rises h between the electrodes. Express the susceptibility,* χ_e*, of the liquid in terms of V, h, and the geometry of the electrodes.*

Let's start by finding the Potential between electrodes. We have:

$$\oint_S \vec{E}\cdot d\vec{A} = \int q\, dV$$

Thus, $(2\pi\rho)LE_r = q_T$ where $E_r = Q/2\pi\rho$ and $Q = q_T/L$ charge per unit length. Thus:

$$\Phi = -\int_{\rho}^{ref} \vec{E} = \frac{Q}{2\pi}\ln\rho + const \rightarrow \Phi = \Phi_b - \Phi_a = \frac{Q}{2\pi}\ln\left(\frac{b}{a}\right) = V$$

Thus,

$$V = \frac{Q}{2\pi}\ln\left(\frac{b}{a}\right) \rightarrow Q^2 = \frac{4\pi^2V^2}{\left[\ln\left[\frac{b}{a}\right]\right]^2}\,.$$

For the energy change with the rise of the liquid dielectric:

$$\Delta W = \frac{1}{8\pi}\int_V \left(\vec{E}\cdot\vec{D}_0 - \vec{D}\cdot\vec{E}_0\right) d^3x \qquad \vec{D} = \epsilon\vec{E};\;\; \vec{D}_0 = \epsilon_0\vec{E}_0;\;\; \epsilon_0 = 1$$

Thus,

$$\Delta W = -\frac{1}{8\pi}\int_V (\epsilon - 1)\,\vec{E}\cdot\vec{E}_0 d^3x,$$

and since $|\vec{E}| = |\vec{E}_0| = Q/2\pi\rho$:

$$\Delta W = -\frac{(\epsilon - 1)}{8\pi}\frac{Q^2}{4\pi^2}\int_0^{2\pi} d\phi \int_0^h dz \int_a^b \frac{1}{\rho}d\rho = -\frac{(\epsilon - 1)Q^2 h}{16\pi^2}\ln\left(\frac{b}{a}\right)$$

Thus,

$$\Delta W = -\frac{(\epsilon - 1)}{4}\frac{hV^2}{\ln\left(\frac{b}{a}\right)} = -\frac{\pi x_e hV^2}{\ln\left(\frac{b}{a}\right)}.$$

The minus sign indicates that the total energy will be lower by sucking in the dielectric, where equilibrium is obtained against increase in gravitational P.E. Thus,

$$\Delta W + mgh = 0 \qquad m = \rho h\pi(b^2 - a^2)$$

So, $\frac{\pi\chi_e hV^2}{\ln\left(\frac{b}{a}\right)} = \rho h\pi(b^2 - a^2)gh$, and we get:

$$\chi_e = \frac{(b^2 - a^2)\rho gh\ln\left(\frac{b}{a}\right)}{V^2}.$$

Example 6.4. Find the capacitance for a parallel-plate capacitor which is partly filled with dielectric, where there is d separation of plates, where the volume is $d(A_1 + A_0)$ and the dielectric is in the area A_1 region with $A_1 = xy$, and x is length along the side with variable dielectric. So, $A_0 = (D - x)y$, where D is the overall length of the x-side.

Solution

It is left to the reader to show that $\frac{1}{C} = \frac{d}{\varepsilon_1 A_1 + \varepsilon_0 A_0}$, together with $U = \frac{Q^2}{2C}$ we then have:

$$U = \frac{Q^2 d}{2\varepsilon_1 A_1 + 2\varepsilon_0 A_0} = \frac{Q^2 d}{\varepsilon_1 yx + \varepsilon_0 y(D - x)}$$

Thus,

$$\frac{\partial U}{\partial x} = -\frac{Q^2 d}{2}\cdot\frac{\varepsilon_1 y - \varepsilon_0 y}{\left(\varepsilon_1 yx + \varepsilon_0 y(D - x)\right)^2}$$

and

$$F_x = -\frac{\partial U}{\partial x} = \frac{Q^2 d}{2} \cdot \frac{(\varepsilon_1 - \varepsilon_0)}{(\varepsilon_1 A_1 + \varepsilon_0 A_0)^2} = \frac{Q^2 dy}{2} \cdot \frac{(\varepsilon_1 - \varepsilon_0)}{C^2 d^2}$$
$$= \frac{y(\varepsilon_1 - \varepsilon_0)V^2}{2d}.$$

Since $\varepsilon_1 > \varepsilon_0$ the force is positive and the dielectric is "sucked in."

6.2 Magnetostatics (using notes/convention from Jackson [32] Ch. 5) Thus far we have worked with electrostatics as an outgrowth of Coulomb, which was much like the gravitational field. There were sources, and if charges were opposite, giving attraction, much of the mathematics worked out the same as with gravitation. While staying with the statics, to not get into unfamiliar territory too quickly, we are already confronted with something very different when examining magnetism. There is no magnetic charge. For electric charge recall the conservation argument that gave the charge conservation equation:

$$\frac{\partial \rho}{\partial t} + \nabla \cdot \vec{J} = 0.$$

(Eqn. 6-63)

In magnetostatics there is no equivalent of magnetic charge, thus there is only the non-divergence of some kind of 'flow': $\nabla \cdot \vec{J} = 0$. For this reason we generally don't speak of a magnetic field like with electric field, here we speak of a magnetic flux density or a magnetic induction. Let's begin by showing this odd property (of no magnetic monopoles), like Maxwell did, from the experimentally established Biot-Savart and Ampere's Laws.

The magnitude of magnetic flux density can be defined by the mechanical torque measured on a magnetic dipole:

$$\vec{N} = \vec{\mu} \times \vec{B}$$

(Eqn. 6-64)

where $\vec{\mu}$ is the magnetic moment of the dipole. Based on observations Ampere establishes the experimental law relating $\vec{\mathrm{B}}$ (magnetic-flux density) to current (called Biot-Savart law for special case that follows). We have the relation between current in a straight wire and $\vec{\mathrm{B}}$:

$$dB = KI\frac{(d\vec{\ell} \times \vec{x})}{|x|^3},$$

(Eqn. 6-65)

where $d\vec{\ell}$ is in direction of current flow. Biot-Savart law is a special form of this: $\vec{B} = kq\frac{\vec{V}\times\vec{x}}{|x|^3}$.

For current measured in esu, flux in emu, $K = \frac{1}{c}$, thus the Gaussian system of units are more convenient for problems involving magnetism, so Gaussian units are used in Ch. 5.

Let's integrate $d\vec{B}$ for a straight line:

$$|B| = \frac{IR}{c}\int_{-\infty}^{\infty}\frac{d\ell}{(R^2+\ell^2)^{3/2}} = \frac{2I}{cR},$$

(Eqn. 6-66)

thus, $|B| = 2I/cR$, a form of the Biot-Savart law.

For the force on a current element due to a magnetic-flux density $\vec{B}$:

$$d\vec{F} = \frac{I_1}{c}\left(d\vec{\ell}_1 \times \vec{B}\right)$$

(Eqn. 6-67)

where the force is experienced by current element $I_1 d\vec{\ell}_1$. The total force is thus:

$$\vec{F}_{12} = \frac{I_1 I_2}{c^2}\oiint\frac{d\vec{\ell}_1\times\left(d\vec{\ell}_2\times\vec{x}_{12}\right)}{|\vec{x}_{12}|^3} = -\frac{I_1 I_2}{c^2}\oiint\frac{\left(d\vec{\ell}_1\cdot d\vec{\ell}_2\right)}{|\vec{x}_{12}|^3}\vec{x}_{12}\,.$$

(Eqn. 6-68)

For two parallel wires $\frac{dF}{d\ell} = \frac{2I_1 I_2}{c^2 d}$.

If a current density $\vec{J}(\vec{x})$ is in an external magnetic-flux density $\vec{B}(\vec{x})$, then the total force on current distribution is:

$$\vec{F} = \frac{1}{c}\int \vec{J}(\vec{x})\times\vec{B}(\vec{x})d^3x,$$

(Eqn. 6-69)

and the total torque is:

$$\vec{N} = \frac{1}{c}\int \vec{x}\times\left(\vec{J}\times\vec{B}\right)d^3x.$$

(Eqn. 6-70)

Obtaining Differential Equations of Magnetostatics and Ampere's Law

Let's start with the expression from Biot-Savart when there is a current density $\vec{J}(\vec{x})$:

$$B(\vec{x}) = \frac{1}{c}\int \vec{J}(\vec{x}') \times \frac{(\vec{x}-\vec{x}')}{|\vec{x}-\vec{x}'|^3} d^3x' = \frac{1}{c}\nabla \times \int \frac{J(\vec{x}')}{|\vec{x}-\vec{x}'|} d^3x'$$

$$\Rightarrow \underline{\nabla \cdot B = 0}$$

(Eqn. 6-71)

From $\nabla \times \vec{B} = \frac{1}{c}\nabla \times \nabla \times \int \frac{J(x')}{|x-x'|} d^3x'$ and $\nabla \times \nabla \times A = \nabla(\nabla \cdot A) - \nabla^2 A$, we have:

$$\nabla \times \vec{B} = \frac{1}{c}\vec{\nabla}\int \vec{J}(\vec{x}') \cdot \nabla\left(\frac{1}{|\vec{x}-\vec{x}'|}\right) d^3x' - \frac{1}{c}\int J(\vec{x}')\,\nabla^2\left(\frac{1}{|\vec{x}-\vec{x}'|}\right) d^3x'$$

$$= \frac{1}{c}\vec{\nabla}\int J(\vec{x}') \cdot \nabla'\left(\frac{1}{|\vec{x}-\vec{x}'|}\right) d^3x' + \frac{4\pi}{c}\vec{J}(\vec{x}) = \frac{1}{c}\vec{\nabla}\int \frac{\vec{\nabla}' \cdot J(\vec{x}')}{|\vec{x}-\vec{x}'|} d^3x' + \frac{4\pi}{c}\vec{J}(\vec{x}) = \frac{4\pi}{c}\vec{J}(\vec{x}),$$

since $\vec{\nabla} \cdot \mathrm{J} = 0$. So,

$$\vec{\nabla} \times \vec{\mathrm{B}} = \frac{4\pi}{\mathrm{c}}\vec{\mathrm{J}}.$$

(Eqn. 6-72)

Ampere's Law is an integral form of the above:

$$\int_S \nabla \times B \cdot \vec{n}da = \frac{4\pi}{c}\int_S \vec{J} \cdot \vec{n}da$$

$$\oint_C B \cdot d\vec{\ell} = \frac{4\pi}{c}\int_S J \cdot nda = \frac{4\pi}{c}\mathrm{I}$$

Thus,

$$\oint_C \vec{B} \cdot d\vec{\ell} = \frac{4\pi}{c}\mathrm{I}$$

(Eqn. 6-73)

Ampere's Law can be employed advantageously, like Gauss's law, in highly symmetric situations.

Vector Potential

Whenever you have $\nabla \cdot \vec{B} = 0$ it is possible to make use of a vector potential $\vec{B} = \vec{\nabla} \times A$. Starting with the differential laws of magnetostatics:

$$\nabla \times \vec{B} = \frac{4\pi}{\mathrm{C}}\vec{\mathrm{J}} \quad and \quad \nabla \cdot \vec{B} = 0,$$

(Eqn. 6-74)

we can see that if $\vec{J} = 0$ we can have $\vec{B} = -\nabla\Phi_m$ where Φ_m is a magnetic scalar potential and previous techniques from electrostatics come to bear. Let's now proceed with $\vec{J} \neq 0$ and $\vec{B} = \vec{\nabla} \times A$. From $B(x) = \frac{1}{c}\nabla \times \int \frac{J(\vec{x}')}{|\vec{x}-\vec{x}'|} d^3x'$ we get:

$$A(x) = \frac{1}{c}\int \frac{J(\vec{x}')}{|\vec{x}-\vec{x}'|} d^3x' + \nabla\Psi(\vec{x})$$

(Eqn. 6-75)

where $\vec{\nabla} \times \nabla\Psi = 0$. Thus, $\vec{A} \rightarrow \vec{A} + \nabla\Psi$ makes no change in the magnetic flux density and is our first example of what is known is a gauge transformation. Given this freedom, it is possible to choose the gauge such that $\nabla \cdot A = 0$ and using $\vec{B} = \nabla \times \vec{A}$:

$$\nabla \times \vec{B} = \nabla \times \nabla \times A = \frac{4\pi}{C}J \quad \Rightarrow \quad \nabla(\nabla \cdot A) - \nabla^2 A = \frac{4\pi}{C}J \quad \Rightarrow \quad \nabla^2 A = -\frac{4\pi}{C}J$$

(Eqn. 6-76)

where each rectangular component of the vector potential satisfies the Poisson equation.

There is a surprising difference between cylindrically symmetric magnetostatic problems and corresponding cylindrically symmetric electrostatic problems: Associated Legendre polynomials appear together with ordinary Legendre polynomials. This can be traced to the vector character of the current and the vector potential, as opposed to the scalar properties of charge and electrostatic potential. Put another way, the electrostatic had "completeness" in terms of associated Legendre polynomial when the electric charge (monopole) was allowed. When there was no monopole in the magnetostatic case, completeness requiring a mixing of associated and ordinary Legendre polynomials isn't a surprise. Furthermore, the natural extension, via analyticity between the two Legendre polynomial solution families (associated and ordinary) to encompass both suggest an underlying construct consisting of a mixture of E and B 3D vector fields as some other construct (to be elaborated when seen together comprising an antisymmetric 4D tensor in the special relativity formulation of EM that follows).

As with the moment expansion for electrostatics, the moment expansion for magnetostatics is given by the multipole expansion. The magnetic dipole moment being the lowest order term (there being no monopole) is then found to be:

$$\vec{m} = \frac{I}{2c}\oint \vec{x} \times d\vec{\ell}$$

(Eqn. 6-77)

For a planar current loop the magnitude of the magnetic moment is:

$$|\vec{m}| = \frac{I}{C} \times (Area).$$

(Eqn. 6-78)

Force and Torque on magnetic moment (leading order)

The force given by $F = \nabla(m \cdot B)$ is valid even for time varying fields. For steady state fields $\vec{\nabla} \times \vec{B} = 0$ and this expression becomes:

$$\vec{F} = \vec{m} \cdot \vec{\nabla}\vec{B}.$$

(Eqn. 6-79)

The leading order term in the torque is:

$$\vec{N} = \vec{m} \times \vec{B}(0).$$

(Eqn. 6-80)

The potential energy of a magnetic moment in a uniform field is:

$$U = -m \cdot B,$$

(Eqn. 6-81)

which follows from $F = \nabla(m \cdot B)$.

Macroscopic Equations and B.C.'s on B and H

Consider the macroscopic magnetization:

$$\vec{M}(x) = \sum_i N_i \langle m_i \rangle.$$

(Eqn. 6-82)

This implies an effective current density of:

$$\vec{J}_m = c\vec{\nabla} \times \vec{M}.$$

(Eqn. 6-83)

Non-vacuum material satisfies:

$$\vec{D} = \epsilon\vec{E}$$

(Eqn. 6-84)

$$\vec{H} = \vec{B} - 4\pi\vec{M}.$$

(Eqn. 6-85)

Note how E-field with leading order monopole contribution has a multiplicative factor (a scalar, geometric, dilation factor) while B-field with leading order dipole contribution has an additive factor (a vector contribution). Also note the sign conventions.

Maxwell's macroscopic equations become:

$$\left.\begin{aligned}\nabla \times \vec{H} &= \frac{4\pi}{c}\vec{J} \\ \nabla \cdot \vec{B} &= 0\end{aligned}\right\} \quad where\ \vec{B} = \mu\vec{H}$$

(Eqn.s 6-86)

The Junction conditions become:

$$(\vec{B}_2 - \vec{B}_1) \cdot \vec{n} = 0$$

$$\vec{n} \times (\vec{H}_2 - \vec{H}_1) = \frac{4\pi}{C}\vec{k}$$

(Eqn.s 6-87)

Methods of Solving Boundary-Value Problems in Magnetostatics

Starting with the basic equations: $\nabla \cdot B = 0, \nabla \times H = \frac{4\pi}{c}\vec{J}$, etc., the general method for solving problems in magnetostatics is to make use of a vector potential or a scalar potential, and in so doing simplify to familiar electrostatic–type solutions already known. The most general method is to use a vector potential since $\nabla \cdot B = 0 \Longrightarrow substitute\ B = \nabla \times A$, and with linear media $B = \mu H$, we get:

$$\nabla \times \left(\frac{1}{\mu}\nabla \times A\right) = \frac{4\pi}{C}\vec{J}.$$

Thus,

$$\nabla(\nabla \cdot A) - \nabla^2 A = \frac{4\pi\mu}{C}J.$$

In Coulomb gauge ($\nabla \cdot A = 0$):

$$\nabla^2 A = -\frac{4\pi\mu}{c}J.$$

If $\vec{J} = 0$, however, a magnetic scalar potential can be introduced instead:

$$\nabla \times H = 0 \Longrightarrow introduce\ H = -\nabla\Phi_m.$$

For linear response medium we get:

$$\nabla \cdot (\mu\nabla\Phi_m) = 0,$$

or

$$\nabla^2\Phi_m = 0$$

in regions where μ is piecewise constant.

In Hard Ferromagnets ($\vec{M}$ given and $\vec{J} = 0$) we can use the scalar potential approach and have an effective magnetic charge density:

$$\nabla \cdot B = \nabla \cdot (H + 4\pi\vec{M}) = 0 \Longrightarrow \nabla^2\Phi_m = -4\pi\rho_m,$$

where $\rho_m = -\nabla \cdot \vec{M}$ is the effective magnetic charge density. Thus,

$$\Phi_m(x) = -\int \frac{\nabla' \cdot M(\vec{x}')}{|x - x'|} d^3x = -\nabla \cdot \int \frac{M(\vec{x}')}{|\vec{x} - \vec{x}'|} d^3x$$

$$\simeq -\nabla\left(\frac{1}{\mathrm{r}}\right) \cdot \int M(\vec{x}')\, d^3x' = \frac{\vec{m} \cdot \vec{x}}{r^3} \qquad \vec{m} = \int \vec{M}\, d^3x$$

$$\sigma_m = \vec{n} \cdot \vec{M}$$

The general vector potential approach with $\nabla^2 A = -\frac{4\pi}{c} J_m$ has:

$$A(x) = \int \frac{\nabla' \times M(\vec{x}')}{|\vec{x} - \vec{x}'|} d^3x'.$$

(Eqn. 6-88)

Example 6.5. *Consider the magnetic induction at the point P with coordinate $\vec{x}$ produced by an increment of current $Id\vec{\ell}'$ at $\vec{x}'$ given by $dB = \frac{I}{c} d\vec{\ell}' \times \frac{(\vec{x}-\vec{x}')}{|\vec{x}-\vec{x}'|^3}$ and determine the total magnetic induction when in the presence of a closed loop carrying current I.*

Solution

From $dB = \frac{I}{c} d\vec{\ell}' \times \frac{(\vec{x}-\vec{x}')}{|\vec{x}-\vec{x}'|^3}$ for the magnetic induction at the point P with coordinate $\vec{x}$ produced by an increment of current $Id\vec{\ell}'$ at $\vec{x}'$ we can show that

$$\vec{B} = \frac{I}{C} \nabla \Omega$$

This corresponds to a magnetic scalar potential $\Phi_m = -I\,\Omega/C$.

The line integral for the closed loop is written

$$\vec{B}(\vec{r}) = \frac{I}{C} \oint_C d\vec{\ell}' \times \frac{(\vec{r} - \vec{r}')}{|\vec{r} - \vec{r}'|^3},$$

for which we can use the vector relation: $\oint_C d\vec{\ell} \times \vec{b} = \int_S (\hat{n} \times \vec{\nabla}) \times \vec{b}\, dS$ to get:

$$\vec{B}(\vec{r}) = -\frac{I}{c} \vec{\nabla} \int_S \frac{(\vec{r} - \vec{r}')}{|\vec{r} - \vec{r}'|^3} \cdot \hat{n}\, dS'.$$

Let

$$\Omega = \int_S \frac{(\vec{r} - \vec{r}')}{|\vec{r} - \vec{r}'|^3} \cdot \hat{n}\, ds,$$

i.e., Ω is the solid angle subtended by the loop at P. Thus:

$$\vec{B}(\vec{r}) = \frac{I}{c}\vec{\nabla}\Omega,$$

where the loop need not be planar.

Example 6.6. Consider a square-shaped loop (wire) of circulation in the x-y plane such that the circulation (right-hand rule) is in the positive z-axis direction. The side lengths of the square are '2a'. What is the magnetic field on the z-axis (e.g., at a distance z from the x-y plane).

Solution

On the z-axis, the non z-axis directed contributions from the loop cancel and we are left with a purely axial component (invoking symmetry). The magnitude of the magnetic field due to a conducting segment of length "2a" (centered vis-à-vis nearest point) was determined previously to be:

$$H_{segment} = \frac{I_0}{2\pi r}\frac{1}{[(r/a)^2 + 1]^{1/2}}.$$

Here we have the superposition of four such components, and then there projections in the *z* direction.

$$H = \frac{2I_0}{\pi r}\frac{1}{[(r/a)^2 + 1]^{1/2}}$$

and recall that $r = (a^2 + z^2)^{1/2}$, thus

$$H_z = (\cos\theta)H = \frac{2I_0}{\pi}\cdot\frac{a^2}{(a^2 + z^2)(2a^2 + z^2)^{1/2}}.$$

Example 6.7. Consider the classic Helmholtz Coil configuration consisting of two parallel coaxial loops where we seek to maximize the region between them with uniform-field. What is the relation between the loop radii a and the loop axial separation of d to maximize this region?

Solution

Recall the magnitude of the axial magnetic field due to a circular loop (coaxial), where loop is centered at the origin and in the x-y plane, and the current circulation (right hand rule) is in the positive z-axis direction, satisfies:

$$H_z = \frac{Ia^2}{2}\frac{1}{(a^2 + z^2)^{3/2}},$$

where z is the distance from the center of the loop on the (coaxial) z-axis. We have two such loops, one placed like above, and another, parallel, with center at $z = d$, so the total axial magnetic field is:

$$H_z = \frac{Ia^2}{2}\left[\frac{1}{(a^2+z^2)^{3/2}} + \frac{1}{(a^2+(z-d)^2)^{3/2}}\right].$$

Also recall the classic Taylor expansion:

$$f(x) = f(a) + f'(a)(x-a) + \frac{f''(a)(x-a)^2}{2!} \ldots,$$

and to have this be constant in a large region we need $f'(a) = 0$ and $f''(a) = 0$. Thus, we start with the first derivative:

$$\frac{\partial H_z}{\partial z} = \frac{Ia^2}{2}\left[\frac{-3z}{(a^2+z^2)^{-5/2}} + \frac{-3(z-d)}{(a^2+(z-d)^2)^{-5/2}}\right]$$

This is zero when $z = d/2$. Let's now consider the second derivative when $z = d/2$:

$$\frac{\partial^2 H_z\left(\frac{d}{2}\right)}{\partial z^2} = \frac{Ia^2}{2}\cdot\frac{1}{\left(a^2+\frac{d^2}{4}\right)^{-5/2}}\left[\frac{15d^2}{2}\cdot\frac{1}{\left(a^2+\frac{d^2}{4}\right)} - 6\right]$$

If we substitute $a = d$, the square bracket expression becomes zero. Thus, to achieve the largest region of uniform field separate the coils by a distance equal to their radii magnitude.

Example 6.8. Let's crush a thin metal tube by passing a current through it. Let's assume crush-failure occurs when the magnetic field strength reaches $9Wb/m^2 = 9T$, and that the tube radius is 2cm. What is the magnitude of the force per unit area? What is the direction of the force?

Solution

At what current will we obtain a magnetic field strength of $B = 9T$ (where $B = \mu_0 H$)? Recall

$$\oint H\cdot d\ell = 2\pi r H_\phi = I$$

So crush failure when

$$I = 2\pi(.02m)\cdot\frac{9}{4\pi\times10^{-7}}A/m = 9\times10^5 A$$

Since we have $df = Id\ell\times B$:

$$F_{(on\ the\ area)} = \int I\,d\ell\times B = IB\,(1m)$$

and

$$P_{(per\ unit\ area)} = \frac{F_{(on\ the\ area)}}{(1m) \times (2\pi r)} = \frac{IB}{2\pi r}$$

Thus, the force per unit area is

$$\frac{(9 \times 10^5 A) \cdot (9T)}{2\pi(.02m)} = 6.45 \times 10^7\ N/m^2.$$

We recover direction of the force (inward) by Lenz's rule, where stability requires reduction in flux area coupling, thus 'crushing'.

Example 6.9. Consider a coaxial transmission line with inner conductor having radius a and out having radius b. There is a permeability of $\mu_1\ for\ a < r < d$ and 'air' for $b < r < d$ (with μ_0). For $r < a$ denote the permeability by μ. What is the inductance per unit length.

Solution

We have:

$$H_\phi(r) = \frac{Ir}{2\pi a^2} \qquad r < a \qquad \mu$$

$$H_\phi(r) = \frac{I}{2\pi r} \qquad a < r < d \qquad \mu_1$$

$$H_\phi(r) = \frac{I}{2\pi r} \qquad b < r < d \qquad \mu_0$$

Thus, the energy can be obtained in three parts:

$$(i) \quad \int_0^a \frac{\mu}{2}\left(\frac{Ir}{2\pi a^2}\right)^2 2\pi r dr = \frac{\mu I^2}{4\pi a^4} \cdot \frac{a^4}{4} = \frac{\mu I^2}{16\pi}$$

$$(ii) \quad \int_a^d \frac{\mu_1}{2} \cdot \left(\frac{Ir}{2\pi a^2}\right)^2 2\pi r dr = \int_a^d \frac{\mu_1}{2}, \frac{I^2}{2\pi} \frac{dr}{r} = \frac{\mu_1 I^2}{4\pi} \ln\left(\frac{d}{a}\right)$$

$$(iii) \quad \int_d^b \frac{\mu_0}{2}\left(\frac{I}{2\pi r}\right)^2 2\pi r dr = \frac{\mu_0 I^2}{4\pi} \ln\left(\frac{b}{d}\right)$$

Thus,

$$\frac{1}{2} LI^2 = I^2 \left(\mu_0 \left(\frac{1}{16\pi} + \frac{1}{4\pi} \ln\left(\frac{b}{d}\right)\right) + \mu_1 \left(\frac{1}{4\pi} \ln\left(\frac{d}{a}\right)\right)\right)$$

and the inductance per unit length is thus:

$$L/unit\ length = \mu\left(\frac{1}{8\pi}\right) + \mu_0\left(\frac{1}{2\pi}\ln\left(\frac{b}{d}\right)\right) + \mu_1\left(\frac{1}{2\pi}\ln\left(\frac{d}{a}\right)\right).$$

Example 6.10. Consider a line current running parallel to a superconducting plane (superconducting in the sense that a mirror problem is established vis-à-vis magnetic field lines). Not how this "mirrors" a similar problem with electric field of a line charge parallel to a regular conducting mirror. Thus, $B_n$at the surface is zero, as is the tangential H_x at the surface. What is the surface current density?

Solution

We have an effective parallel-wire transmission line problem. The vector potential due to a wire of current is:

$$A(r) = \int \frac{\mu I'(r')d\ell}{4\pi R}.$$

Let's orient the coordinates and wires such that the superconducting plane is the y-z plane. The wires are a distance $\pm d$ in the x direction. The wire at $x = +d$ runs parallel to the y-axis with current I, the wire at $x = -d$ has current – I. We have for finite wires:

$$A_y(x,z) = \int_{-L}^{L} \frac{\mu I dy'}{4\pi\sqrt{(x-d)^2 + z^2 + y'^2}} - \int_{-L}^{L} \frac{\mu I dy''}{4\pi\sqrt{(x+d)^2 + z^2 + y''^2}}$$

and, as $L \to \infty$,

$$A_y(x,z) = \frac{I\mu}{2\pi}\ln\left[\frac{(x+d)^2 + z^2}{(x-d)^2 + z^2}\right].$$

Using $B(r) = \nabla \times A(r)$, and $H(r) = \frac{1}{\mu}B(r)$:

$$H_x = \frac{1}{\mu}\frac{\partial A_y}{\partial z} = \frac{I}{2\pi}\left[\frac{z}{(x+d)^2 + z^2} - \frac{z}{(x-d)^2 + z^2}\right]$$

and

$$H_z = -\frac{1}{\mu}\frac{\partial A_y}{\partial x} = \frac{I}{2\pi}\left[\frac{(x-d)}{(x-d)^2 + z^2} - \frac{(x+d)}{(x+d)^2 + z^2}\right]$$

Thus,

$$H_x(x = 0^+) = 0$$

as expected, and

$$H_z(x = 0^+) = -\frac{I}{\pi}\left(\frac{d}{d^2 + z^2}\right).$$

Since $J_s = \hat{n} \times H$ on all surfaces, we have the surface current to be:

$$J_s = -\frac{I}{\pi}\left(\frac{d}{d^2 + z^2}\right)\hat{x}\,.$$

Example 6.11. Consider the parallel plate transmission line running in the z-direction with plate thickness h and separation d (in the x-direction) and current $\pm I$ (in the z-direction), with width in the y-direction w. What is the internal inductance per unit length?

Solution

Using $I = \oint H \cdot d\ell$ and considering the edge effect negligible, we want to determine let's make our line integral run in a rectangle enclosing a length w in the z-direction of the upper plate and a variable extent(x) into the plate (with its thickness d). Thus, for $H \;\; 0 < x < d$:

$$enclosed\ current = Hw$$

Thus,

$$H = \frac{I}{w} \cdot \frac{x}{d}$$

Since potential energy $U_H = \int_V \frac{\mu}{2} H^2 dV$ and considering internal inductance per unit length.

$$U_H = \int_0^d \frac{\mu}{2} H^2 w dx = \frac{\mu}{2} w \frac{I^2}{w^2} \frac{1}{d^2} \int_0^d x^2\, dx = \frac{1}{6}\mu \frac{I^2 d}{w}$$

Twice this is the total potential energy due to both conductors and since $U_H = LI^2/2$ we have:

$$L = \frac{2}{3}\mu \frac{d}{w}.$$

Example 6.12. Consider a wire of radius c with a uniform current density. Inside the wire is a parallel cylindrical void with radius b (that doesn't include the origin). The enclosing wire is directed in the z-direction, with x-y cross-section, and relative to that coordinate system the void region is centered at x=a (y=0). We have $b < a < c$ and $b < c - a$ for this to be consistent. What is $H(x)$ for $0 < x < \infty$?

Solution

Use superposition to find *H* field. Without the void and with the uniform current density J, we have

$$I = (\pi c^2)J \, .$$

Using a concentric circle with radius $r > c$:

$$\oint H \cdot d\ell = 2\pi r H_\phi = I,$$

and we have for H field outside the wire:

$$H_\phi = \frac{I}{2\pi r}$$

Once inside the conductor I becomes a function of radius: $I(r) = \left(\frac{r}{c}\right)^2 I$, so this is changed to

$$H(r) = \frac{I(r)}{2\pi r} = \frac{Ir}{2\pi c^2} \, .$$

Or, if considering only the positive x axis this becomes $H_\phi(x) = \frac{Ix}{2\pi c^2}$.

The void represents a superposition of a negative current, uniform density $-J$, upon the original conductor, with analysis much like before. Considering only the positive x axis, we have $r = (x - a)$ for the void which is displaced from the origin. Now to combine, or superpose:

For $x \geq c$:

$$H(x) = \frac{J}{2}\left(\frac{c^2}{x} - \frac{b^2}{(x-a)}\right)\hat{y}$$

For $a + b \leq x \leq c$:

$$H(x) = \frac{J}{2}\left(x - \frac{b^2}{(x-a)}\right)\hat{y}$$

For $a - b \leq x \leq a + b$:

$$H(x) = \frac{aJ}{2}\hat{y}$$

For $0 \leq x \leq a - b$:

$$H(x) = \frac{J}{2}\left(x - \frac{b^2}{(x-a)}\right)\hat{y} \, .$$

Note:

> Magnetostatics vs electrostatics in cylindrical coordinates: electrostatics has only associated Legendre polynomials while magnetostatics requires both Assoc. and Ordinary Legendre polynomials (can reach one from the other by analytic continuation).One of many signs of finding the larger theory by analytic continuation.

6.3 Exercises

(Ex. 6.1) Compute $\frac{e^4\hbar}{45\pi m^4 c^7}$, discuss strength of nonlinear EM interactions.

(Ex. 6.2) Re-derive Eqn. 6-14.

(Ex. 6.3) Verify Eqn. 6-18.

(Ex. 6.4) Verify Eqn. 6-25.

(Ex. 6.5) Re-derive Eqn.s 6-35 and 6-36.

(Ex. 6.6) Re-derive Eqn.s 6-37 and 6-38.

(Ex. 6.7) Re-derive Eqn. 6-43.

(Ex. 6.8) Re-derive Eqn. 6-44.

(Ex. 6.9) Re-derive Eqn. 6-48.

(Ex. 6.10) Re-derive Eqn. 6-57 (the Clausius-Mossotti Eqn.).

(Ex. 6.11) Re-derive Eqn. 6-62.

(Ex. 6.12) Re-derive Eqn. 6-88.

Chapter 7. Fields – Electrodynamics

In Chapters 4 thru 7 classic electromagnetism (EM) is covered, with extensive coverage on the Lorentz transform (EM is Lorentz invariant) in Ch. 8, which then leads to special relativity and general relativity and other topics in later chapters. This chapter (Ch. 7), is the last of the main EM sequence and is on dynamical electromagnetic systems in general, with careful analysis of frequency dispersion effects as well as broadcast/radiation and surface/interface problems (transmission lines, microwave waveguide). The material draws from Jackson [32] Ch.s 6-9; Ramo [34] Ch.s 5-12.

Note on Examples and Exercises
The chapters on electromagnetism draw, foremost, from the excellent source, Classical Electrodynamics by J.D. Jackson [32], that was the required text for many years at Caltech. Many of the Jackson problems develop the physics theory further (e.g., Gauss's Theorem) or further develop the mathematical techniques (for ODEs, definition of delta function as limit, etc.). With my own notes from the courses at Caltech, and upon later preliminary exam review at the Univ. Wisconsin, the addition of the completed problems from various sources were acquired in a way that seemed to provide a 'mapping' of understanding. I wanted to convey that understanding here. Solutions to these key problems are carefully worked out in what follows, along with notes on the theory from Jackson as well as from [33-35]. Basic exercises are given to re-derive or verify the theory. The solutions to the exercises here, for the most part, are available via the aforementioned Jackson text (with the detailed derivations provided there). Many examples and related problems are done in order to cover material from basic undergraduate problems to advanced engineering (for antenna design, say), so the notes and problems also draw from undergraduate and graduate EM class notes/exams, and the other texts mentioned ([33-35]). The remarkable success and modern applications of EM will be made evident, but at the end of Ch. 7, when we carefully analyze the hydrogen atom, we will find that there is radiative collapse in the classical theory. This fundamental flaw in the theory, when dealing with even the simplest bound atomic

system, will be corrected with the development of the quantum theory in Book 4 [15].

7.1 Time-varying EM Fields

Faradays observations of time-varying electromagnetic fields, in mathematical terms, will be done now but with direct mathematical expression according to Maxwell.

Given that the magnetic induction in the neighborhood of the circuit is $\vec{B}$. Define the magnetic flux "linking" the circuit by:

$$F = \int_S \vec{B} \cdot \vec{n}\, da.$$

(Eqn. 7-1)

The electromagnetic force around the circuit is

$$\mathcal{E} = \oint_C \vec{E}' \cdot d\vec{\ell}$$

(Eqn. 7-2)

where $\vec{E}'$ is the electric field at $d\vec{\ell}$ in the coordinate system where $d\vec{\ell}$ is at rest.

Faraday observed that:

$$\mathcal{E} = -K\frac{dF}{dt}$$

(Eqn. 7-3)

The sign is specified by Lenz's law, which states that the induced current (and accompanying magnetic flux) is in such a direction as to oppose the change of flux through the circuit. Determination of the constant K by using Galilian invariance$(\vec{x}' = \vec{x} + \vec{v}t;\ t' = t)$ in the appropriate nonrelativistic limit:

$$\oint_C \vec{E}' \cdot d\vec{\ell} = -K\frac{d}{dt'}\int_S \vec{B} \cdot n\, da$$

(Eqn. 7-4)

Since, $\frac{d\vec{B}}{dt'} = \frac{\partial \vec{B}}{\partial t'} + (V \cdot \vec{\nabla})\vec{B} = \frac{\partial \vec{B}}{\partial t'} + \nabla \times (B \times \vec{V}') + \vec{\nabla}(\vec{\nabla} \cdot \vec{B})$, with $\vec{\nabla} \cdot \vec{B} = 0$ this becomes:

$$\oint_C \vec{E}' \cdot d\vec{\ell} = -K\left\{\int_S \frac{\partial B \cdot \vec{n}}{\partial t'} da + \int_S \nabla \times (B \times \vec{V}') \cdot \vec{n} da\right\}$$

or

$$\oint_C [\vec{E}' - K(\vec{V}' \times \vec{B})] \cdot d\vec{\ell} = -K \int_s \frac{\partial B}{\partial t'} \cdot \vec{n} da.$$

In the lab frame (where $\vec{E}$ is the field in the lab frame):

$$\oint_C \vec{E}' \cdot d\vec{\ell} = -K \int_s \frac{\partial \vec{B}}{\partial t} \cdot \vec{n} da$$

(Eqn. 7-5)

Galilean invariance yields $\frac{\partial}{\partial t} = \frac{\partial}{\partial t'}$, etc., due to assumption of absolute time, this implies in turn that the R.H.S. of the two equations are equal, thus the L.H.S.'s, so:

$$\vec{E}' = \vec{E} + K(V + \vec{B}).$$

(Eqn. 7-6)

In order to determine *K* merely consider the significance of $\vec{E}'$: In prime frame force experienced on charge is $q\vec{E}'$. In lab frame charge represents a current $\vec{J} = q\vec{V}\delta(\vec{x} - x_0)$ and so from the magnetic force law: $d\vec{F} = \frac{I_1}{c}(d\vec{\ell}_1 \times \vec{B})$ or $\vec{F} = \frac{1}{c}\int \vec{J} \times Bd^3x$, so we see that the charge (current) experiences an additional force (aside from $q\vec{E}$) in agreement with Ampere's law if $k = 1/c$ in Gaussian units.

In Gaussian units, Faraday's law therefore reads:

$$\oint_C \vec{E}' \cdot d\vec{\ell} = -\frac{1}{c}\frac{d}{dt}\int_s \vec{B} \cdot \vec{n} da$$

(Eqn. 7-7)

As a by-product we have found that the electric field $\vec{E}'$ in a coordinate frame moving with a velocity $\vec{V}$ relative to the laboratory is:

$$\vec{E}' = \vec{E} + \frac{1}{c}(\vec{V} \times \vec{B}).$$

(Eqn. 7-8)

Interestingly, even though derived using Galilean invariance ($V \ll C$), the result is valid in the relativistic formulation as well, and is known as Lorentz's law:

$$F = q\left[\vec{E} + \frac{1}{c}(\vec{V} + \vec{B})\right].$$

(Eqn. 7-9)

Holding the circuit fixed in a chosen reference frame – in order to have $\vec{E}$ and $\vec{B}$ defined in the same frame – Faraday's law can be put in differential form by use of Stoke's theorem to get the differential form of Faraday's law:

$$\int_S \left(\nabla \times \vec{E} + \frac{1}{C}\frac{\partial B}{\partial t}\right) \cdot \vec{n} da = 0 \Rightarrow \nabla \times \vec{E} + \frac{1}{c}\frac{\partial \vec{B}}{\partial t} = 0$$

(Eqn. 7-10)

In the integral form, and more obviously in the differential form, we have a far-reaching generalization of Faraday's law. The "circuit" can be thought of as any closed geometrical path in space, not necessarily coincident with an electric circuit. (In the differential form, of course, there isn't even a reference to a circuit.)

Energy in the Magnetic Field:

Determining energy in magnetic fields is not as simple at the energy in an electric field – since we can't analogously "fix the charges" and make a simple build-up. The creation of a steady-state configuration of currents and associated fields involves an initial transient period during which the currents and fields are brought from zero to the final values. For such time-varying fields there are induced electromotive forces which cause the sources of current to do work, these contributions must be considered when describing the energy of the field since the field is by definition the total work done to establish it.

Consider a single circuit, current *I*. If the flux through the circuit changes there is an $\mathcal{E}$ induced on the circuit. If the current is kept at *I* work is done $\frac{dW}{dt} = -I\mathcal{E} = \frac{1}{C} I \frac{dF}{dt}$ (work done by sources). Thus,

$$\delta W = \frac{1}{C} I \delta F.$$

(Eqn. 7-11)

Break up current distribution into a network of elementary current loops, elemental tubes of cross-sectional area $\Delta\sigma$ then:

$$\Delta(\delta W) = \frac{J\Delta\sigma}{C}\int_S \vec{n} \cdot \delta B da = \frac{J\Delta\sigma}{C}\int_S \left(\nabla \times \delta\vec{A}\right) \cdot \vec{n} da = \frac{J\Delta\sigma}{C}\oint_C \delta\vec{A} \cdot d\vec{\ell}$$

(Eqn. 7-12)

Since $J\Delta\sigma d\vec{\ell} = \vec{J} d^3x$, by definition:

$$\delta W = \frac{1}{C}\int \delta\vec{A} \cdot \vec{J} d^3x,$$

(Eqn. 7-13)

and use $\vec{\nabla} \times \vec{H} = \frac{4\pi}{C}\vec{J}$:

$$\delta W = \frac{1}{4\pi}\int \delta\vec{A}\cdot(\vec{\nabla}\times\vec{H})d^3x$$

$$= \frac{1}{4\pi}\int \left[\vec{H}\cdot(\vec{\nabla}\times\delta\vec{A}) + \vec{\nabla}\cdot(\vec{H}\times\delta\vec{A})\right] d^3x.$$

(Eqn. 7-14)

Since we have a localized field distribution the last term is zero, thus:

$$\delta W = \frac{1}{4\pi}\int \vec{H}\cdot\delta\vec{B}d^3x,$$

(Eqn. 7-15)

where, in this form the relation is also applicable to all magnetic media. If medium is para- or diamagnetic, we have a linear relation between $\vec{H}$ and $\vec{B}$, so

$$\vec{H}\cdot\delta\vec{B} = \frac{1}{2}\delta(\vec{H}\cdot\vec{B}).$$

(Eqn. 7-16)

For which

$$W = \frac{1}{8\pi}\int \vec{H}\cdot\vec{B}\,d^3x = \frac{1}{2C}\int \vec{J}\cdot\vec{A}\,d^3x.$$

(Eqn. 7-17)

Magnetic energy change in energy when an object of permeability μ_1 is placed in a magnetic field whose current sources are fixed:

$$W = \frac{1}{8\pi}\int_{V_1}\left(\vec{B}\cdot\vec{H}_0 - \vec{H}\cdot\vec{B}_0\right)d^3x = \frac{1}{8\pi}\int_{V_1}(\mu_1-\mu_0)\,H\cdot H_0 d^3x.$$

(Eqn. 7-18)

Thus,

$$W = \frac{1}{8\pi}\int_{V_1}\left(\frac{1}{\mu_0}-\frac{1}{\mu_1}\right)B\cdot B_0 d^3x = \frac{1}{2}\int_{V_1} M\cdot B_0 d^3x.$$

(Eqn. 7-19)

Example 7.1. Consider a system of current carrying elements in empty space.

(a) Show that for current density $\vec{J}(\vec{x})$, the total energy in the magnetic field is:

$$W = \frac{1}{2c^2}\int d^3x\int d^3x'\frac{\vec{J}(\vec{x})\cdot\vec{J}(\vec{x}')}{|\vec{x}-\vec{x}'|},$$

(b) For n circuits, each with their own current I_n, show that energy can be expressed as

$$W = \frac{1}{2}\sum_{i} L_i I_i^2 + \sum_{i}\sum_{j>i} M_{ij} I_i I_j.$$

Solution

(a) System of current carrying elements in empty space, the total energy in the magnetic field is:

$$W = \frac{1}{2c}\int \vec{J}\cdot\vec{A}\,d^3x \quad , \quad \vec{A}(\vec{x}) = \frac{1}{c}\int \frac{\vec{J}(\vec{x}')}{|\vec{x}-\vec{x}'|}d^3\vec{x}',$$

substitution then gives:

$$W = \frac{1}{2c^2}\int d^3x \int d^3x' \frac{\vec{J}(\vec{x})\cdot\vec{J}(\vec{x}')}{|\vec{x}-\vec{x}'|}.$$

(b) If the current configuration consists of *n* circuits carrying currents $I, \dots I_n$ then

$$W = \frac{1}{2c^2}\sum_{ij} I_i I_j \oiint \frac{d\vec{s}_i\cdot d\vec{s}_j}{|\vec{x}_i-\vec{x}_j|} = \frac{1}{2}\sum_{i} I_i^2 L_i + \frac{1}{2}\sum_{i}\sum_{j\neq i} M_{ij} I_i I_j$$

$$= \frac{1}{2}\sum_{i} L_i I_i^2 + \frac{1}{2}\sum_{i}\sum_{j>i}(M_{ij}+M_{ji}) I_i I_j = \frac{1}{2}\sum_{i} L_i I_i^2 + \sum_{i}\sum_{j>i} M_{ij} I_i I_j$$

So

$$L_{ij} = \frac{1}{c^2}\oiint \frac{d\vec{s}_i\cdot d\vec{s}_j}{|\vec{x}_i-\vec{x}_j|} \begin{array}{ll} = M_{ij} & (i\neq j) \\ = L_i & (i=j) \end{array}$$

Maxwell's displacement Current, the completion of Maxwell Equations

Recall Ampere's law (magnetostatics), when $\nabla\cdot\vec{J} = 0$ is $\nabla\times\vec{H} = \frac{4\pi}{C}\vec{J}$. Using the continuity equation and Coulomb's law we can generalize Ampere's law to the dynamic case (Maxwell's contribution).

From

$$\nabla\cdot J + \frac{\partial\rho}{\partial t} = 0 \quad , \quad \nabla\cdot D = 4\pi\rho$$

Substitution gives:

$$\nabla\cdot J + \frac{\partial}{\partial t}\left(\frac{1}{4\pi}\nabla\cdot D\right) = 0 \Rightarrow \nabla\cdot\left(J + \frac{1}{4\pi}\frac{\partial\vec{D}}{\partial t}\right) = 0$$

Thus, for a time-varying form, Maxwell replaced $J \to J + \frac{1}{4\pi}\frac{\partial\vec{D}}{\partial t}$ in Ampere's law to get:

$$\nabla\times\vec{H} = \frac{4\pi}{C}J + \frac{1}{C}\frac{\partial\vec{D}}{\partial t},$$

(Eqn. 7-20)

where Maxwell called the added term the displacement current. Maxwell's equations, given the presence of ponderable media, and the related constitutive relations are thus:

$$\nabla \cdot D = 4\pi\rho \qquad \nabla \times H = \frac{4\pi}{c}J + \frac{1}{c}\frac{\partial \vec{D}}{\partial t}$$

$$\nabla \times E = -\frac{1}{c}\frac{\partial B}{\partial t} \qquad \nabla \cdot B = 0$$

Vector and Scalar Potentials

Since $\vec{\nabla} \cdot B = 0$ still holds we substitute $\vec{B} = \vec{\nabla} \times \vec{A}$:

$$\nabla \times \left(E + \frac{1}{c}\frac{\partial A}{\partial t}\right) = 0 \Rightarrow E + \frac{1}{c}\frac{\partial \vec{A}}{\partial t} = -\nabla\Phi \Rightarrow \vec{E} = -\nabla\Phi - \frac{1}{c}\frac{\partial \vec{A}}{\partial t}$$

(Eqn. 7-21)

Dynamic behavior of A and Φ then follows from the inhomogeneous equations. Let's consider (restrict to vacuum Maxwell for simplicity):

$$\nabla^2\Phi + \frac{1}{c}\frac{\partial}{\partial t}(\vec{\nabla} \cdot \vec{A}) = -4\pi\rho$$

(Eqn. 7-22)

$$\nabla^2 A - \frac{1}{c^2}\frac{\partial^2 \vec{A}}{\partial t^2} - \vec{\nabla}\left(\vec{\nabla} \cdot \vec{A} + \frac{1}{c}\frac{\partial \Phi}{\partial t}\right) = -\frac{4\pi}{c}\vec{J}$$

(Eqn. 7-23)

This reduces the set of four Maxwell equations to two equations. Let's return to considering Gauge transformation: ($\vec{B}$ and $\vec{E}$ are left unchanged by these)

$$\vec{A} \to \vec{A}' = \vec{A} + \vec{\nabla}\Omega \qquad \Phi \to \Phi' = \Phi - \frac{1}{c}\frac{\partial \Omega}{\partial t}$$

(Eqn. 7-24)

The freedom allowed by the gauge transformation allows us to choose (A, Φ) such that:

$$\nabla \cdot \vec{A} + \frac{1}{c}\frac{\partial \Phi}{\partial t} = 0$$

(Eqn. 7-25)

(known as the Lorentz condition). Then the pair of 2nd order equations uncouple:

$$\nabla^2\Phi - \frac{1}{c^2}\frac{\partial^2 \Phi}{\partial t^2} = -4\pi\rho$$

(Eqn. 7-26)

$$\nabla^2\vec{A} - \frac{1}{c^2}\frac{\partial^2 \vec{A}}{\partial t^2} = -\frac{4\pi}{c}\vec{J}$$

(Eqn. 7-27)

Even for equations which satisfy the Lorentz condition there is arbitrariness, if we consider

$$\vec{A} \rightarrow \vec{A} + \vec{\nabla}\Omega$$

(Eqn. 7-28)

$$\Phi \rightarrow \Phi - \frac{1}{c}\frac{\partial \Omega}{\partial t}$$

(Eqn. 7-29)

we see that

$$\nabla^2\Omega - \frac{1}{c^2}\frac{\partial^2\Omega}{\partial t^2} = 0,$$

(Eqn. 7-30)

indicating a wave phenomenon traveling at the speed 'c'.

Poynting's Theorem (1884)

The rate of electric fields doing work on charge q moving at $\vec{v}$ is $q\vec{v}\cdot\vec{E}$. The magnetic field does no work since the magnetic force is perpendicular to the velocity. For a continuous distribution of charge, the total rate of doing work is:

$$\int_V \vec{J}\cdot\vec{E}d^3x = \frac{1}{4\pi}\int_V \left[c\vec{E}\cdot\left(\vec{\nabla}\times\vec{H}\right) - \vec{E}\cdot\frac{\partial\vec{D}}{\partial t}\right]d^3x$$

Using $\nabla\cdot(E\times H) = H\cdot\left(\vec{\nabla}\times\vec{E}\right) - \vec{E}\cdot\left(\nabla\times\vec{H}\right)$:

$$\int_V \vec{J}\cdot\vec{E}d^3x = -\frac{1}{4\pi}\int_V \left[c\vec{\nabla}\cdot\left(\vec{E}\times\vec{H}\right) + \vec{E}\cdot\frac{\partial\vec{D}}{\partial t} + \vec{H}\cdot\frac{\partial\vec{B}}{\partial t}\right]d^3x$$

Assume the medium is linear in its electrical and magnetic properties (disallows hysteresis, etc.).

Assume that the static EM energy equations based on the fields are valid for the time-varying fields.

Then we have EM energy density:

$$u = \frac{1}{8\pi}(E\cdot D + B\cdot H) \rightarrow \quad -\int_V \vec{J}\cdot\vec{E}d^3x$$

$$= \int_V \left[\frac{\partial u}{\partial t} + \frac{c}{4\pi}\vec{\nabla}\cdot\left(\vec{E}\times\vec{H}\right)\right]d^3x$$

(Eqn. 7-31)

From this we have a continuity equation for energy density and energy flow:

$$\frac{\partial u}{\partial t} + \vec{\nabla}\cdot\vec{S} = -\vec{J}\cdot\vec{E}, \qquad \vec{S} = \frac{c}{4\pi}\left(\vec{E}\times\vec{H}\right),$$

(Eqn. 7-32)

where $\vec{S}$ is the "Poynting" Vector representing energy flow. For EM momentum density we have:

$$\vec{g} = \frac{1}{4\pi c}(\vec{E} \times \vec{B}) = \frac{\vec{S}}{c^2} \quad and \quad \vec{L}_{em} = \int \vec{x} \times \vec{g} d^3x.$$

(Eqn. 7-33)

Gauge Transformations, Lorentz Gauge, Coulomb Gauge

Coulomb, radiation, or transverse gauge have vector potential with $\nabla \cdot A = 0$, which gives:

$$\nabla^2\Phi - \frac{1}{c}\frac{\partial\Phi}{\partial t}(\nabla \cdot A) = -4\pi\rho \quad \rightarrow \quad \nabla^2\Phi = -4\pi\rho,$$

(Eqn. 7-34)

and

$$\nabla^2\vec{A} - \frac{1}{c}\frac{\partial^2\vec{A}}{\partial t^2} - \nabla\left(\nabla \cdot \vec{A} + \frac{1}{2}\frac{\partial\Phi}{\partial t}\right) = -\frac{4\pi}{c}J \rightarrow \nabla^2\vec{A} - \frac{1}{c^2}\frac{\partial^2\vec{A}}{\partial t^2}$$
$$= -\frac{4\pi}{c}J + \frac{1}{c}\nabla\frac{\partial\Phi}{\partial t}.$$

(Eqn. 7-35)

From $\nabla^2\Phi = -4\pi\rho \Rightarrow \Phi(\vec{x},\mathrm{t}) = \int \frac{\rho(\vec{x}',t)}{|\vec{x}-\vec{x}'|}d^3x'$ we have an instantaneous Coulomb potential due to $\rho(\vec{x},t)$. Also, the term $1/c\;\nabla\;\partial\Phi/\partial t$, if irrotational in the sense of arranging $\nabla \times \{\ldots\} = 0$, then this suggests a decomposition: $J = J_\ell + J_t$, where J_ℓ is longitudinal or irrotational and J_t is transverse or solenoidal. To make further progress on solving for A, let's rewrite the identity $\nabla \times (\nabla \times A) = \nabla(\nabla \cdot A) - \nabla^2 A$ with the substitution $A = \int \frac{J(\vec{x}')}{|\vec{x}-\vec{x}'|}d^3x'$:

$$\nabla \times \nabla \times \int \frac{\vec{J}d^3x'}{|\vec{x}-\vec{x}'|} - \nabla\left(\nabla \cdot \int \frac{\vec{J}}{|\vec{x}-\vec{x}'|}d^3x'\right) = -\nabla^2\int \frac{J(\vec{x}')}{|\vec{x}-\vec{x}'|}d^3x'$$

$$\nabla \times \nabla \times \int \frac{\vec{J}d^3x'}{|\vec{x}-\vec{x}'|} - \nabla\left(\int J(\vec{x}') \cdot \nabla\left(\frac{1}{|\vec{x}-\vec{x}'|}\right)d^3x'\right)$$
$$= -\int J(\vec{x}')\nabla^2\left(\frac{1}{|\vec{x}-\vec{x}'|}\right)d^3x'$$

using

$$\nabla' \cdot \left(\frac{\vec{J}(\vec{x}')}{|\vec{x}-\vec{x}'|}\right) = \frac{\nabla' \cdot \vec{J}(\vec{x}')}{|\vec{x}-\vec{x}'|} + \vec{J}(\vec{x}') \cdot \nabla\left(\frac{1}{|\vec{x}-\vec{x}'|}\right)$$

where localized current analysis allows dropping the surface term, thus:

$$\nabla \times \nabla \times \int \frac{\vec{J}(\vec{x}')}{|\vec{x}-\vec{x}'|} d^3x' - \nabla \int \frac{\nabla' \cdot \vec{J}(\vec{x}')}{|\vec{x}-\vec{x}'|} d^3x' = 4\pi J(x)$$

and we do indeed have

$$\vec{J}(\vec{x}) = \frac{1}{4\pi} \nabla \times \nabla \times \int \frac{\vec{J}(\vec{x}')}{|\vec{x}-\vec{x}'|} d^3x' - \frac{1}{4\pi} \nabla \int \frac{\nabla' \cdot \vec{J}(\vec{x}')}{|\vec{x}-\vec{x}'|} d^3x'$$
$$= \vec{J}_t(\vec{x}) + \vec{J}_\ell(\vec{x}),$$

where

$$\vec{J}_\ell(\vec{x}) = -\frac{1}{4\pi} \nabla \int \frac{\nabla' \cdot \vec{J}(\vec{x}')}{|\vec{x}-\vec{x}'|} d^3x' \quad and \quad \vec{J}_t(\vec{x})$$
$$= \frac{1}{4\pi} \nabla \times \nabla \times \int \frac{\vec{J}(\vec{x}')}{|\vec{x}-\vec{x}'|} d^3x'.$$

So, $\vec{J}(\vec{x}) = \vec{J}_t(\vec{x}) + \vec{J}_\ell(\vec{x})$

Since $\Phi(\vec{x},t) = \int \frac{\rho(\vec{x}',t)}{|\vec{x}-\vec{x}'|} d^3x'$, and from continuity on charge $\frac{\partial \rho}{\partial t} + \nabla \cdot \vec{J} = 0$, we get:

$$\nabla \frac{\partial \Phi}{\partial t} = \nabla \int \frac{\frac{\partial \rho}{\partial t}(\vec{x}',t)}{|\vec{x}-\vec{x}'|} d^3x' = \nabla \int \frac{-\nabla' \cdot \vec{J}}{|\vec{x}-\vec{x}'|} d^3x' = 4\pi \vec{J}_\ell \,.$$

(Eqn. 7-36)

This allows us to simplify to

$$\nabla^2 \vec{A} - \frac{1}{c^2} \frac{\partial^2 \vec{A}}{\partial t^2} = -\frac{4\pi}{c} \vec{J}_t$$

(Eqn. 7-37)

Since the transverse radiation fields are given by the vector potential alone, this is known as "radiation gauge", with the instantaneous Coulomb potential contributing only to the near fields. In the above form we have a wave equation (inhomogeneous), so let's review the Green's function for the wave equation before proceeding.

Green Functions for the wave equation

Consider

$$\nabla^2 \Psi - \frac{1}{c^2} \frac{\partial^2 \Psi}{\partial t^2} = -4\pi f(\vec{x},t).$$

(Eqn. 7-38)

We immediately shift to Fourier transforms (since the dispersion relation is a trivial complication in the FT space). Define:

$$\Psi(\vec{x},t) = \frac{1}{2\pi}\int_{-\infty}^{\infty} \Psi(\vec{x},\omega)e^{-i\omega t}d\omega$$

(Eqn. 7-39)

with inverse:

$$\Psi(\vec{x},\omega) = \int_{-\infty}^{\infty} \Psi(\vec{x},t)e^{-i\omega t}dt$$

(Eqn. 7-40)

and

$$f(\vec{x},t) = \frac{1}{2\pi}\int_{-\infty}^{\infty} f(\vec{x},\omega)e^{-i\omega t}d\omega$$

(Eqn. 7-41)

with inverse:

$$f(\vec{x},\omega) = \int_{-\infty}^{\infty} f(\vec{x},t)e^{-i\omega t}dt.$$

(Eqn. 7-42)

We then get

$$(\nabla^2 + K^2)\Psi(\vec{x},\omega) = -4\pi f(\vec{x},\omega),$$

(Eqn. 7-43)

which is the inhomogeneous Helmholtz wave equation. The appropriate Green's function is:

$$(\nabla^2 + K^2)G_K(\vec{x},\vec{x}') = -4\pi\delta(\vec{x}-\vec{x}').$$

(Eqn. 7-44)

For no boundaries, $G \propto \vec{R} = \vec{x} - \vec{x}'$, with spherical symmetry $G \propto |R|$, using Laplacian in spherical coordinates and examining the radial equation:

$$\frac{1}{R}\frac{d^2}{dR^2}(RG_k) + k^2G_k = 4\pi\delta(\vec{R}).$$

(Eqn. 7-45)

Everywhere except $R = 0$ we have:

$$\frac{d^2}{dR^2}(RG_k) + k^2(RG_k) = 0 \Rightarrow RG_k(R) = Ae^{ikR} + Be^{-ikR}$$

(Eqn. 7-46)

as R → 0, and from electrostatics $\lim_{kR\to 0} G_k = \frac{1}{R}$, we get:

$$G_k = AG_k^+ + BG_k^- \quad \text{where} \quad G_k^{(\pm)} = \frac{e^{\pm ikR}}{R} \quad \text{with} \quad A + B = 1.$$

(Eqn. 7-47)

The choice of A and B depends on the boundary conditions in time. For the time dependent Green function:

$$\left(\nabla_x^2 - \frac{1}{c^2}\frac{\partial^2}{\partial t^2}\right) G^{(\pm)}(\vec{x},t;\vec{x}',t') = -4\pi\delta(\vec{x}-\vec{x}')\delta(t-t'),$$

(Eqn. 7-48)

and using the inverse FT's, the source term becomes $-4\pi\delta(\vec{x}-\vec{x}')e^{i\omega t'}$. So,

$$G^{(\pm)}(R,\tau) = \frac{1}{2\pi}\int_{-\infty}^{\infty} \frac{e^{\pm ikR}}{R} e^{-i\omega\tau} d\omega \quad , \quad \tau = t - t'$$

(Eqn. 7-49)

For a nondispersive medium where $k = \omega/c$, the integral is a delta function:

$$G^{(\pm)}(R,\tau) = \frac{1}{R}\delta\left(\tau \mp \frac{R}{c}\right).$$

(Eqn. 7-50)

Thus,

$$G^{(\pm)}(\vec{x},t;\vec{x}',t') = \frac{\delta\left(t' - \left[t \mp \frac{|\vec{x}-\vec{x}'|}{c}\right]\right)}{|\vec{x}-\vec{x}'|},$$

(Eqn. 7-51)

known as the retarded Green's function.

So,

$$\Psi^{(\pm)}(\vec{x},t) = \iint G^{(\pm)}(\vec{x},t;\vec{x}',t') f(\vec{x}',t') d^3x'dt'$$

(Eqn. 7-52)

where the general solution is:

$$\Psi(\vec{x},t) = \Psi_{\text{in}}(\vec{x},t) + \iint G^{(+)} f d^3x'dt',$$

(Eqn. 7-53)

where $\Psi_{\text{in}}(\vec{x},t)$ satisfies the homogeneous equation and use of $G^{(+)}$ guarantees that at early time there is no contribution from the integral part.

Poynting's Theorem and Conservation of Energy and Momentum

Recall that the rate of doing work by EM field:

$$-\vec{J}\cdot\vec{E} = \frac{\partial u}{\partial t} + \vec{\nabla}\cdot\vec{S},$$

(Eqn. 7-54)

where $\vec{S} = (c/4\pi)\vec{E} \times \vec{H}$ is the Poynting Vector. Let's now consider a system involving EM fields and charged particles. Denote the total energy of the particles within the volume V as E_{mech} and assume that no particles move out of the volume:

$$\frac{dE_{mech}}{dt} = \int_v \vec{J} \cdot \vec{E}\, d^3x.$$

(Eqn. 7-55)

Then, the energy flow is given by

$$\frac{dE}{dt} = \frac{d}{dt}\left(E_{mech} + E_{field}\right) = \int_v \vec{J} \cdot \vec{E}\, d^3x + \int_v \frac{\partial u}{\partial t} d^3x = \int_v -\vec{\nabla} \cdot \vec{S}\, d^3x$$
$$= \oint_s \vec{n} \cdot \vec{S}\, da.$$

(Eqn. 7-56)

Thus,

$$\frac{dE}{dt} = \oint_s \vec{n} \cdot \vec{S}\, da.$$

(Eqn. 7-57)

Since we are dealing directly with the particles (not ponderable media) we can use microscopic fields E and B, thus

$$E_{field} = \int_v u\, d^3x = \frac{1}{8\pi} \int_v (E^2 + B^2)\, d^3x.$$

(Eqn. 7-58)

Consider next conservation of linear momentum, starting with the Lorentz relation:

$$\vec{F} = \vec{q}\left(E + \frac{\vec{v}}{c} \times \vec{B}\right),$$

which can be rewritten:

$$\frac{d\vec{P}_{mech}}{dt} = \int_v \left(\rho\vec{E} + \frac{1}{c}\vec{J} \times \vec{B}\right) d^3x.$$

Using the relations

$$\rho = \frac{1}{4\pi} \nabla \cdot E \quad and \quad \vec{J} = \frac{c}{4\pi}\left(\nabla \times B - \frac{1}{c}\frac{\partial E}{\partial t}\right),$$

we get:

$$\frac{d\vec{P}_{mech}}{dt} + \frac{d}{dt}\int_v \frac{1}{4\pi c}(\vec{E}\times\vec{B})d^3x$$
$$= \frac{1}{4\pi}[E(\nabla\cdot E) - E\times(\nabla\times E) + B(\nabla\cdot B)$$
$$- B\times(\nabla\times B)]d^3x$$

The LHS (Left hand side) tells us $\vec{P}_{field} = \frac{1}{4\pi c}(\vec{E}\times\vec{B})d^3x$ and the EM momentum density is $\vec{g} = \frac{1}{4\pi c}(\vec{E}\times\vec{B})$. The RHS gives us the EM stress tensor:

$$T_{\alpha\beta} = \frac{1}{4\pi}\left[E_\alpha E_\beta + B_\alpha B_\beta - \frac{1}{2}(E\cdot E + B\cdot B)\delta_{\alpha\beta}\right].$$

So,

$$\frac{d}{dt}\left(\vec{P}_{mech} + \vec{P}_{field}\right)_\alpha = \sum_\beta \int_v \frac{\partial}{\partial x_\beta} T_{\alpha\beta} d^3x = \oint_s \sum_\beta T_{\alpha\beta}\, n_\beta da,$$

where $\sum_\beta T_{\alpha\beta}\, n_\beta$ corresponds to the α component of force per unit area across S. Thus,

$$\frac{d}{dt}\left(\vec{P}_{mech} + \vec{P}_{field}\right)_\alpha = \oint_s \sum_\beta T_{\alpha\beta}\, n_\beta da.$$

(Eqn. 7-59)

The above formula can be used to calculate the forces acting on material objects in EM fields by enclosing the object with a boundary surface S and adding up the total electromagnetic force. Note that the stress tensor is not symmetric for anisotropic media.

Example 7.2. *Two nonpermeable parallel wires, separated by a distance d, form a transmission line with one wire having radius a, the other radius b, connected at their termination end such that current flows down one wire and back up the other. Assume a uniform distribution for the current looping through the wires, what is the self inductance per unit length?*

Solution:

The current density in the radius a wire is: $\vec{J}_a = I_z/\pi a^2\, \hat{e}_z$. Using this we have near wire a:

$$\nabla\times\vec{H} = \frac{4\pi}{c}\vec{J} = \frac{4I_z}{ca^2}\hat{e}_z.$$

Switching to integral form:

$$\int \nabla \times \vec{H} \cdot d\vec{a} = \oint \vec{H} \cdot d\vec{\ell} = 2\pi r \vec{H} = \frac{4\pi}{c} I_z.$$

Thus,

$$\vec{H}_{outside} = \frac{2I_z}{cr} \hat{e}_\phi \quad (r > a),$$

and

$$\vec{H}_{inside} = \frac{2rI_z}{ca^2} \hat{e}_\phi \quad (r < a).$$

For nonpermeable medium: $|\vec{B}| = |\vec{H}|$.

Let's get the vector potential, use of standard formulas then apply. Since $\vec{B} = \vec{\nabla} \times \vec{A}$, we have $\vec{A}(x) = \frac{1}{c} \int \frac{J(\vec{x}')d^3x'}{|\vec{x} - \vec{x}'|}$, thus $\vec{A}$ is parallel to the current. Thus, the only comp. of $\vec{A}$ is A_z. So,

$$\vec{\nabla} \times \vec{A} = \frac{1}{r} \frac{\partial A_z}{\partial \phi} \hat{e}_r - \frac{\partial A_z}{\partial r} \hat{e}_\phi$$

The only comp. of $\vec{B}$ is along $\hat{e}_\phi$, so

$$B\hat{e}_\phi = -\frac{\partial A_z}{\partial r} \hat{e}_\phi,$$

and for region inside or outside the wire we have:

$$A_z = \frac{2I_z}{c} \int_a^r \frac{dR}{R} = -\frac{2I_z}{c} \ln\left(\frac{r}{a}\right) \quad (r > a)$$

and

$$A_z = \frac{2I_z}{c} \int_a^r \frac{R}{a^2} dR = -\frac{I_z}{c} \left(\frac{r}{a}\right)^2 \quad (r < a).$$

The vector potential arbitrary up to an additive gradient of a scalar function, continuity of A_z at surface of wire resolves arbitrariness:

$$A_{z(in)}\Big|_{r=a} = A_{z(out)}\Big|_{r=a}.$$

Thus,

$$\frac{I_z}{c} \left(\frac{r}{a}\right)^2 \Big|_{r=a} + const = -\frac{2I_z}{c} \ln\left(\frac{r}{a}\right)\Big|_{r=a} + cont'.$$

The vector potential is thus:

$$A_{z(in)} = -\frac{I_z}{c} \left(\frac{r}{a}\right)^2 \; (r < a)$$

$$A_{z(out)} = -\frac{2I_z}{c} \left[\ln\left(\frac{r}{a}\right) + \frac{1}{2}\right]$$

Now,

$$W = \frac{1}{2c}\int \vec{J}(\vec{x}) \cdot \vec{A}(\vec{x}) d^3x$$

and the self-inductance is identified from the terms that result (left as exercise.)

Example 7.3. *Two nonpermeable coaxial conductors, separated by a distance (b-a), form a transmission line with the inner conductor being a wire having radius a, while the outer conductor is a cylindrical shell with radius b, connected at their termination end such that current flows down one conductor and back up the other. Assume a uniform distribution for the current looping through the wire or shell, what is the self inductance per unit length?*

Solution

Focusing on the inner wire conductor, we know that system energy can be expressed as

$$W = \frac{1}{2} I^2 L_1,$$

where L_1 is the self-inductance desired. Label the wire region 1, the region between the wire and the shell as region 2, the shell as region 3 and the region outside the shell as region 4. We also have system energy in terms of the fields:

$$W = \frac{1}{8\pi}\int \vec{H} \cdot \vec{B}\, d^3x = \frac{\mu^{-1}}{8\pi}\int B^2\, d^3x,$$

where $D = \epsilon E$ and $H = \frac{1}{\mu} B$.

Region I: $|\vec{B}| = \frac{2\mu^{-1} I\rho}{a^2 c}$

$$\oint \frac{\vec{B}}{\mu} \cdot d\vec{\ell} = \int_S \frac{4\pi}{c}\vec{J}$$

$$2\pi\rho|B| = \frac{4\pi}{c}\frac{\pi\rho^2}{\pi a^2} I \times \mu$$

$$|B| = \frac{2I\rho}{ca^2}\mu$$

$$W = \frac{\mu^{-1}}{8\pi}\left(\frac{2\mu I}{a^2 c}\right)^2 \int_0^{2\pi}\int_0^a \rho^3\, d\rho d\phi dz$$

Energy per unit length in 1:

$$W_1 = \frac{\mu^{-1}}{8\pi}\left(\frac{2\mu I}{a^2 c}\right)^2 2\pi\frac{a^4}{4} = \frac{\mu I^2}{4c^2}$$

Region 2: $|\vec{B}| = ?$

$$\oint \frac{\vec{B}}{\mu_0} \cdot d\vec{\ell} = \int_S \frac{4\pi}{c}\vec{J}$$

$$2\pi\rho\frac{1}{\mu_0}|B| = \frac{4\pi}{c} I$$

$W_2 = \frac{\mu^{-1}}{8\pi} \int_0^{2\pi} \int_0^b \left(\frac{2I\mu}{\rho c}\right)^2 \rho d\rho d\phi$ $\qquad$ $|B| = \frac{2I\mu_0}{\rho c}$

$= \frac{\mu^{-1}}{8\pi} 2\pi \left(\frac{2I\mu}{c}\right)^2 \int_a^b \frac{d\rho}{\rho} = \frac{I^2\mu}{c^2} \ln(b/a)$

Region 3: $\qquad |\vec{B}| = ?$ $\qquad$ $\oint \frac{\vec{B}}{\mu} \cdot d\vec{\ell} = \int_S \frac{4\pi}{c} \vec{J}$

$2\pi\rho \frac{1}{\mu} |B| = \frac{4\pi}{c} \left(I - I\frac{(\rho^2-b^2)}{(c^2-b^2)}\right)$

$W_3 = \frac{\mu^{-1}}{8\pi} \int_0^{2\pi} \int_b^c \left(\frac{2\mu I}{c}\right)^2$ $\qquad$ $|B| = \frac{2\mu}{c\rho} I \left(1 - \frac{(\rho^2-b^2)}{(c^2-b^2)}\right)$

$= \frac{\mu I^2}{c^2} \left[\frac{c^4}{(c^2-b^2)^2} \ln\left(\frac{c}{b}\right) - \frac{3c^2-b^2}{4(c^2-b^2)}\right]$

Region 4:
$W_4 = 0 \quad (B_4 = 0)$

Thus (assuming $\mu = \mu_0$),

$$L_1 = 2\,(W_1 + W_2 + W_3)/I^2$$

$$= \frac{2\mu_0}{c}\left\{\frac{1}{4} + \ln\left(\frac{b}{a}\right) + \frac{c^4}{c^2 - b^2} \ln\left(\frac{c}{b}\right) - \frac{3c^2 - b^2}{4(c^2 - b^2)}\right\}.$$

Example 7.4. A betatron has a toroidal region with constant E-field and the axial B-field is $B_z(r,t) = Ctr^n$ for $t \geq 0$. Find the induced E-field. Find the velocity v of a particle at time t where that particle is assumed to stay at radius a. What is the magnetic force on the charge? For what values of n will there be a path of constant radius as assumed?

Solution
Recall

$$\oint E \cdot d\ell = -\frac{\partial}{\partial t} \int_S B \cdot dS.$$

Take the line integral to be a circle of radius *a* in the betatron, centered on the *z* axis:

$$2\pi aE = -\frac{\partial}{\partial t}\int B_z \cdot dA = -\frac{\partial}{\partial t}\int_0^a\int_0^{2\pi} Ctr^n \, rdrd\theta$$

$$= -\frac{\partial}{\partial t}Ct2\pi\int_0^a r^{n+1}\,dr = -2\pi C\frac{a^{n+2}}{n+2}$$

So

$$|E| = C\frac{a^{n+1}}{n+2}$$

and E is tangential to the circle circulating clockwise when viewed from above.

Since$\vec{F} = q\vec{E}$ and $\vec{F} = m\vec{a}$ we have

$$\vec{a} = \frac{q}{m}\vec{E}\,.$$

Thus

$$|v| = \int_0^t \frac{q}{m}Edt = \frac{q}{m}C\frac{a^{n+1}}{n+2}t$$

and is directed tangentially to the circle at radius a. Since

$$E = v \times B,$$

the magnetic force on a charged particle is:

$$F = qv \times B,$$

and if $v \perp B$ (like here) we have

$$F = qVB = \frac{q^2C^2}{m}\frac{a^{2n+1}}{n+2}t^2$$

Centrifugal Force $= mv^2/a$, so for this to balance the magnetic force:

$$qvB = \frac{mv^2}{a} \quad\rightarrow\quad aqB = mv \quad\rightarrow\quad aqCta^n = m\left(\frac{q}{m}C\frac{a^{n+1}}{n+2}t\right),$$

which requires $n + 2 = 1$, thus $n = -1$.

Example 7.5. Consider a parallel plate transmission line as described in a previous example with plate separation a, width w, and ignoring internal inductance. (i) What is the characteristic impedance? Consider a computer circuit with such lines having $w = 5\mu m$ and $a = 1\mu m$. The relative permittivity is 2.5 and the relative permeability is ~1.0 (typical). (ii) What is the characteristic impedance and wave velocity? (iii) Halve the dielectric thickness, what happens?

Solution

(i) Along a unit length we have $C = \frac{\varepsilon w}{a} F$ (F is the units Farads) and $L = \frac{\mu a}{w} H/m$ (with Hbeing the unit of Henrys). Together with $Z = Lv$ and $v = (LC)^{-1/2}$, we get:

$$Z = \sqrt{\frac{L}{C}}\Omega$$

(again, showing units explicitly, here Ohms Ω). Thus,

$$Z = \frac{a}{w}\sqrt{\frac{\mu}{\epsilon}} \quad and \quad v = \frac{1}{\sqrt{\varepsilon\mu}} \ (in\ units\ of\ c).$$

(ii) $Z_0 = \frac{1}{5}\sqrt{\frac{1}{2.5}}(377\Omega) = 47.8\Omega$ and $v = \frac{1}{\sqrt{\varepsilon\mu}} = .6325c = 1.9 \times 10^8\ m/s$.

(iii) If $a = .5\mu m$ we see that the impedance is simply halved, while the wave velocity is unchanged.

7.2 Plane wave propagation

Plane waves in Nonconducting Medium:

Start with Maxwell equations:

$$\nabla \cdot B = 0 \qquad \nabla \times \vec{E} = -\frac{1}{c}\frac{\partial B}{\partial t}$$

$$\nabla \times H = +\frac{1}{c}\frac{\partial \vec{D}}{\partial t} + \frac{4\pi}{c}\vec{J} \qquad \nabla \cdot \vec{D} = 4\pi\rho$$

For no sources in an infinite medium they become:

$$\nabla \cdot E = 0 \quad \nabla \times E = -\frac{1}{c}\frac{\partial \vec{B}}{\partial t}$$

$$\nabla \cdot B = 0 \quad \nabla \times B = -\frac{\mu E}{c}\frac{\partial \vec{E}}{\partial t}$$

Combining we arrive at the wave equations:

$$\nabla^2 E = -\frac{\mu\epsilon}{c^2}\frac{\partial^2 E}{\partial t^2} = 0 \quad and \quad \nabla^2 B - \frac{\mu\epsilon}{c^2}\frac{\partial^2 B}{\partial t^2} = 0,$$

where the phase velocity is $v = \frac{c}{\sqrt{\mu\epsilon}}$, so both equations satisfy:

$$\nabla^2 u - \frac{1}{v^2}\frac{\partial^2 u}{\partial t^2} = 0,$$

(Eqn. 7-60)

which has simple $e^{-\vec{k}\cdot\vec{x}=i\omega t}$ plane wave solutions, where $k = \frac{\omega}{v} = \sqrt{\mu\epsilon}\frac{\omega}{c}$. For waves propagating in the x-direction the solution is

$$u_k(x,t) = Ae^{ikx-i\omega t} + Be^{-ikx-i\omega t} = Ae^{ik(x-vt)} + Be^{-ik(x+vt)},$$

By completeness of Fourier Transforms we can construct a general solution from $u_k(x,t)$. For a dispersive medium, i.e., $\mu\epsilon$ a function of frequency, we return to the original differential equation Fourier transform analysis, where we arrived at the Helmholtz wave equation:

$$\nabla^2 u + \mu\epsilon\frac{\omega^2}{c^2}u = 0$$

(Eqn. 7-61)

When we reconstitute the wave as a function of $\vec{x}$ and t, the dispersion produces modifications. (The wave changes shape as it propagates.)

Considering the vector nature of the EM fields and the requirement of satisfying the Maxwell eqn's, let's assume solution:

$$\vec{E}(\vec{x},t) = \vec{\mathcal{E}}e^{ik\vec{n}\cdot x-i\omega t}$$

$$\vec{B}(\vec{x},t) = \vec{\beta}e^{ik\vec{n}\cdot x-i\omega t}$$

(eventually take real part). Each component satisfies the wave eqn. provided $k^2\vec{n}\cdot\vec{n} = \mu\epsilon\frac{\omega^2}{c^2}$ so it is necessary that $\vec{n}\cdot\vec{n} = 1$ (then $k^2 = \mu\epsilon\frac{\omega^2}{c^2}$ and $\omega = \frac{ck}{\sqrt{\mu\epsilon}}$). Then the relations $\nabla\cdot E = 0$ and $\nabla\cdot B = 0$ require that

$$n\cdot\vec{\mathcal{E}} = 0 \quad and \quad n\cdot\vec{\beta} = 0.$$

So, both $\vec{E}$ and $\vec{B}$ are transverse, and we have:

$\nabla\times E + \frac{1}{c}\frac{\partial B}{\partial t} = 0 \Rightarrow \nabla\times E + \frac{-i\omega}{c}B = 0 \Rightarrow n\times\vec{\mathcal{E}} = \frac{1}{\sqrt{\mu\epsilon}}\vec{B} = 0 \Rightarrow$
$\vec{B} = \sqrt{\mu\epsilon}n\times\vec{\mathcal{E}}$.

Similarly,

$$\vec{\mathcal{E}} = -\frac{1}{\sqrt{\mu\epsilon}}\vec{n}\times\vec{B}.$$

(Eqn. 7-62)

Consider $\vec{\mathcal{E}} = \vec{\epsilon}_1 E_0$, and $\vec{B} = \vec{\epsilon}_2\sqrt{\mu\epsilon}E_0$, in the basis $(\vec{\epsilon}_1, \vec{\epsilon}_2, \vec{n})$, let's compute the Poynting vector:

$$\vec{S} = \frac{1}{2}\frac{c}{4\pi}(\vec{E}\times\vec{H}^*),$$

(Eqn. 7-63)

where there is a ½ factor because of time averaging of the harmonic terms. Thus,

$$\vec{S} = \frac{c}{8\pi}\sqrt{\frac{\epsilon}{\mu}}|E_0|^2\vec{n}$$

(Eqn. 7-64)

And

$$u = \frac{1}{16\pi}\left(\epsilon\vec{E}\cdot\vec{E}^* + \frac{1}{\mu}\vec{B}\cdot\vec{B}^*\right) = \frac{\epsilon}{8\pi}|E_0|^2.$$

(Eqn. 7-65)

Notice $\frac{|\vec{S}|}{u} = \frac{c}{\sqrt{\mu\epsilon}} = v$ is speed of energy flow.

Suppose $\vec{k} = \vec{n}K$ is now complex, where the formation is then modified by $\vec{n} = \vec{n}_R + i\vec{n}_I$. The wave now possesses exponential growth or decay in some directions. It is called an inhomogeneous plane wave.
Requirement that $\vec{n}\cdot\vec{n} = 1$ now implies:

$$n_R^2 - n_I^2 = 1 \text{ real part}$$
$$\vec{n}_R\cdot\vec{n}_I = 0 \text{ Im. Part}$$

So $\vec{n}_R$ and $\vec{n}_I$ must be orthogonal and we can write in basis $\vec{e}_1, \vec{e}_2$:

$$\vec{n} = \vec{e}_1\cosh\theta + i\vec{e}_2\sinh\theta,$$

and with the restriction $n\cdot\vec{\mathcal{E}} = 0$ we have:

$$\vec{\mathcal{E}} = (i\vec{e}_1\sinh\theta - \vec{e}_2\cosh\theta)A + \vec{e}_3A'.$$

(Eqn. 7-66)

Linear and Circular Polarization
Generally polarized wave:

$$\vec{E}(\vec{x},t) = (\epsilon_1E_1 + \epsilon_2E_2)e^{i\vec{k}\cdot\vec{x}-i\omega t}.$$

(Eqn. 7-67)

E_1 and E_2 are complex to allow for a phase difference between waves of different polarization. If E_1 and E_2 have same phase then linear polarization. If E_1 and E_2 have different phase: elliptical polarization.

Reflection and Refraction of Electromagnetic Waves at a Plane Interface between Dielectrics
Two classes of reflection, refraction properties:
(1) Kinematic properties
  (a) Angle of reflection equals angle of incidence
  (b) Snell's law: $n\sin i = n'\sin r$
(2) Dynamic properties

(a) Intensities of reflected and refracted radiation
(b) Phase changes and polarization

Consider the following system (since $v = \frac{c}{\sqrt{\epsilon\mu}}$, $n = \sqrt{\epsilon\mu}$):

Incident $\begin{cases} \vec{E} = \vec{E}_0 e^{i\vec{k}\cdot\vec{x}-i\omega t} \\ \vec{B} = \sqrt{\mu\epsilon}\,\frac{\vec{K}\times\vec{E}}{k} \end{cases}$

Refracted $\begin{cases} \vec{E}' = \vec{E}_0' e^{i\vec{k}'\cdot\vec{x}-i\omega t} \\ \vec{B}' = \sqrt{\mu'\epsilon'}\,\frac{\vec{K}'\times\vec{E}'}{k'} \end{cases}$

Reflected $\begin{cases} \vec{E}'' = \vec{E}_0'' e^{i\vec{k}''\cdot\vec{x}-i\omega t} \\ \vec{B}'' = \sqrt{\mu\epsilon}\,\frac{\vec{K}''\times\vec{E}''}{k} \end{cases}$

$$|k| = |k''| = k = \frac{\omega}{c}\sqrt{\mu\epsilon}$$

$$|k'| = k' = \frac{\omega}{c}\sqrt{\mu'\epsilon'}$$

The mere existence of boundary conditions at $z = 0$ which must be satistifed at all points on the plane at all times, implies that the spatial (and time) variation all fields must be the same at $z = 0$, independent of the nature of the boundary conditions. Consequently the phase factors at $z = 0$ must all be equal:

$$(k \cdot \vec{x})_{z=0} = \left(\vec{k}' \cdot \vec{x}\right)_{z=0} = \left(\vec{k}'' \cdot \vec{x}\right)_{z=0}$$

The above relation demonstrates that all three wave vectors must lie in a plane, (defined to be plane of incidence). All the kinematic relations follow from this relation:

$$k \sin i = k' \sin r = k'' \sin r'$$

Since $k = k'' \Rightarrow i = r'$ part (a) and $\frac{\sin i}{\sin r} = \frac{k'}{k}$ part (b).

The dynamic properties are contained on the b.c.'s, where Normal $\vec{D}$ and $\vec{B}$ are continuous and Tangential $\vec{E}$ and $\vec{H}$ are continuous:

$$\left.\begin{array}{c} \left[\epsilon(\vec{E}_0 + \vec{E}_0'') - \epsilon'\vec{E}_0'\right]\cdot\vec{n} = 0 \\ \left[\vec{K}\times\vec{E}_0 + \vec{K}''\times\vec{E}_0'' - \vec{K}'\times\vec{E}_0'\right]\cdot\vec{n} = 0 \\ \left(\vec{E}_0 + \vec{E}_0'' - \vec{E}_0'\right)\times\cdot\,\vec{n} = 0 \\ \left[\frac{1}{\mu}\left(\vec{K}\times\vec{E}_0 + \vec{K}''\times\vec{E}_0''\right) - \frac{1}{\mu'}\left(\vec{K}'\times\vec{E}_0'\right)\right]\times\cdot\,\vec{n} = 0 \end{array}\right\} \begin{array}{c}(i)\\(ii)\\(iii)\\(iv)\end{array}$$

(Eqn.s 7-68)

Consider two cases for the above relations:

(1) $\vec{E}$ wave is linearly polarized perpendicular to the plane of incidence. So, all $\vec{E}$ fields are all parallel to the surface. Eqn. (i) yields nothing. (ii) duplicates (iii), and equations (iii) and (iv) give:

$$E_0 + E_0'' - E_0' = 0$$

$$\sqrt{\frac{\epsilon}{\mu}}(E_0 - E_0'')\cos i - \sqrt{\frac{\epsilon'}{\mu'}}E_0'\cos r = 0$$

where

$$k'' = k = \frac{\omega}{c}\sqrt{\mu\epsilon} \quad , and \quad k' = \frac{\omega}{c}\sqrt{\mu'\epsilon'}.$$

So, for $\vec{E}\perp$, plane of incidence:

$$\left.\begin{array}{l} \dfrac{E_0'}{E_0} = \dfrac{2n\cos i}{n\cos i + \frac{\mu}{\mu'}\sqrt{n'^2 - n^2\sin^2 i}} \\ \dfrac{E_0''}{E_0} = \dfrac{E_0'}{E_0} - 1 - \dfrac{n\cos i - \frac{\mu}{\mu'}\sqrt{n'^2 - n^2\sin^2 i}}{n\cos i + \frac{\mu}{\mu'}\sqrt{n'^2 - n^2\sin^2 i}} \end{array}\right\}$$

(Eqn.s 7-69)

For optical freq. it is usually permitted to put $\mu/\mu' = 1$. The results for E_0'/E_0, etc., are most often employed in optical contexts with real n and n', but they are also valid for complex dielectric constants.

If $\vec{\mathbf{E}}$ is parallel to the plane of incidence:

$$\cos i\,(E_0 - E_0'') - E_0'\cos r = 0$$

$$\sqrt{\frac{\epsilon}{\mu}}(E_0 + E_0'') - \sqrt{\frac{\epsilon'}{\mu'}}E_0' = 0$$

So, for $\vec{E}\|$ to plane of incidence:

$$\frac{E_0'}{E_0} = \frac{2nn'\cos i}{\frac{\mu}{\mu'}n'^2\cos i + n\sqrt{n'^2 - n^2\sin^2 i}}$$

$$\frac{E_0''}{E_0} = \frac{E_0'}{E_0} - 1 - \frac{\frac{\mu}{\mu'}n'^2\cos i - n\sqrt{n'^2 - n^2\sin^2 i}}{\frac{\mu}{\mu'}n'^2\cos i + n\sqrt{n'^2 - n^2\sin^2 i}}$$

(Eqn.s 7-70)

For normal incidence both cases yield:

$$\frac{E_0'}{E_0} = \frac{2}{\sqrt{\frac{\mu\epsilon'}{\mu'\epsilon}} + 1} \underset{\mu' = \mu}{=} \frac{2n}{n' + n}$$

$$\frac{E_0''}{E_0} = \frac{\sqrt{\frac{\mu\epsilon'}{\mu'\epsilon}} - 1}{\sqrt{\frac{\mu\epsilon'}{\mu'\epsilon}} + 1} = \frac{n' - n}{n' + n}$$

(Eqn. 7-71)

If $n' > n$ there is a phase reversal for the relected wave.

Polarization by Reflection, and Total Internal Reflection

Let $\mu' = \mu$ for convenience, for $\vec{E} \perp$ plane of incidence the only way $E_0'' = 0$ is if $n = n'$, i.e., no boundary. But, for $\vec{E} \|$ to plane of incidence $E_0'' = 0$ when

$$n'^2\cos i_B - n\sqrt{n'^2 - n^2\sin^2 i_B} = 0 \rightarrow \quad i_B = \tan^{-1}\left(\frac{n'}{n}\right)$$

Brewster's angle, reflected radiation is completely plane polarized with polarization vector perpendicular to the plane of incidence. The phenomena of total internal reflection follows directly from Snell's law:

$$n\sin i = n'\sin r.$$

If $r = \pi/2$ then $\sin i = \frac{n'}{n}$ ($n > n'$ required, so in denser medium for internal reflection). Thus

$$i_0 = \sin^{-1}\left(\frac{n'}{n}\right).$$

For waves incident at $i = i_0$, the refracted wave is propagated parallel to the surface. For $i > i_0, \sin r > i(n > n')r$ is a complex angle with a purely imaginary cosine: $\cos r = i\sqrt{\left(\frac{\sin i}{\sin i_0}\right)^2 - 1}$. Since

$$e^{i\vec{k}'\cdot\vec{x}} = e^{ik'(x\sin r + z\cos r)} = e^{-k'[(\sin i/\sin i_0)^2-1]_z^{1/2}} = e^{ik'(\sin i/\sin i_0)}$$

for $i > i_0$ the refracted wave is propagated only parallel to the surface and is attenuated exponentially beyond the interface.

Lack of energy flow across the surface for $i \geq i_0$ can be verified by calculating the time-avg. normal component:

$$S \cdot n = \frac{c}{8\pi} Re\left[n \cdot (\vec{E}' \times \vec{H}'^*)\right] \qquad \vec{H}' = \frac{c}{\mu'\omega} n(\vec{K}' \times \vec{E}') \quad K = \frac{\omega}{c}\sqrt{\mu'\epsilon'}$$

$$S \cdot n = \frac{c^2}{8\pi\omega\mu'} Re\left[(\vec{n} \cdot \vec{K}')|E_0'|^2\right]$$

But $\vec{n} \cdot \vec{K}' = K' \cos r$ is purely imaginary, so $S \cdot n = 0$.

Frequency Dispersion Characteristics of Dielectrics, Conductors, and Plasmas

Model for $\epsilon(\omega)$ (dispersion)

Classical model of electron as a space oscillator, assume applied electric field ≈ local field (low density approx. in reality), and let the electric field be harmonic → giving a harmonic force:

$$m[\ddot{\vec{x}} + \gamma\dot{\vec{x}} + \omega_0^2\vec{x}] = -e\vec{E}(\vec{x}, t)$$

(Eqn. 7-72)

For $\vec{E} \propto e^{i\omega t} \Rightarrow m[-\omega^2\vec{x} - \gamma i\omega\vec{x} + \omega_0^2\vec{x}] = -e\vec{E}$. Thus,

$$\vec{p} = -e\vec{x} = \frac{e^2}{m}(\omega_0^2 - \omega^2 - i\omega\gamma)^{-1}\vec{E}$$

(Eqn. 7-73)

Recall:

$$\vec{D} = \epsilon\vec{E} = \vec{E} + 4\pi\vec{p} \quad and \quad \vec{p} = \chi_e E$$

(Eqn. 7-74)

Suppose there are N molecules per unit volume then we can write:

$$\vec{P} = N\vec{p}.$$

(Eqn. 7-75)

Suppose there are (j) different families of electrons in each molecule, with total electrons per moleculeZ and number of electrons in molecule per family is given by f_j (thus $\sum f_j = Z$ electrons per molecule).We find that:

$$\epsilon = 1 + 4\pi\chi_e = 1 + \frac{4\pi Ne^2}{m}\sum_j f_j\left(\omega_j^2 - \omega^2 - i\omega\gamma_j\right).$$

(Eqn. 7-76)

Let's now interpret this result for various scenarios.

Anomalous dispersion and Resonant Absorption

For $\omega <$ all $\omega_j \Rightarrow \chi_e > 0 \Rightarrow \epsilon > 1$. For smallest $\omega_j < \omega <$ largest $\omega_j \Rightarrow \chi_e$: goes from positive to zero to negative $\Rightarrow \epsilon$ goes from > 1 to 1 to < 1. In the neighborhood of any ω_j there is violent behavior, χ_e becomes large and purely imaginary for $\omega = \omega_j$. Normal dispersion is associated with an increase in $Re\ \epsilon(\omega)$ with ω. Anomalous dispersion with the reverse. Only where there is anomalous dispersion is $Im\ \epsilon(\omega)$ appreciable. Positive $Im\ \epsilon(\omega)$ represents dissipation of energy from the electromagnetic wave into the medium, regions where $Im\ \epsilon(\omega)$ are large are called regions of resonant absorption.

Low frequency Behavior, Electric Conductivity

If smallest $\omega_j = 0 \Rightarrow$ then a fraction f_0 of the electrons are "free", and the dielectric constant is singular at $\omega = 0$. Exhibiting the free electron contribution separately we have:

$$\epsilon(\omega) = \epsilon_0 + i\frac{4\pi Ne^2 f_0}{m\omega(\gamma_0 - i\omega)}.$$

(Eqn. 7-77)

To explain the singular nature of

$$\epsilon(\omega) = \epsilon_0 + i\frac{4\pi Ne^2 f_0}{m\omega(\gamma_0 - i\omega)},$$

(Eqn. 7-78)

return to the Maxwell-Ampere relation:

$$\nabla \times \vec{H} = \frac{4\pi}{c}\vec{J} + \frac{1}{c}\frac{d\vec{D}}{dt}$$

(Eqn. 7-79)

Assuming the medium obeys Ohm's law $\vec{J} = \sigma\vec{E}$ and has a "normal" dielectric constant ϵ_0:

$$\nabla \times \vec{H} = -i\frac{\omega}{c}\left(\epsilon_0 + i\frac{4\pi\sigma}{\omega}\right)\vec{E}$$

(Eqn. 7-80)

Comparing the equations:

$$\sigma = \frac{f_0 N e^2}{m(\gamma_0 - i\omega)}$$

(Eqn. 7-81)

The distinction between dielectrics and conductors is an artificial one, at least away from $\omega = 0$. If the medium posseses free electrons it is a conductor at low frequencies, otherwise an insulator. At nonzero frequencies the "conductivity" contribution to $\epsilon(\omega)$ merely appears as a resonant amplitude like the rest. The dispersive properties of the medium can be attributed as well to a complex dielectric constant as to a frequency – dependent conductivity and a dielectric constant.

Waves in a Conducting or Dissipative Medium

Using $k = \sqrt{\mu\epsilon}\frac{\omega}{c}$ and $\epsilon(\omega) = \epsilon_0 + i\frac{4\pi\sigma}{\omega}$ (and dropping the subscript):

$$k^2 = \mu\epsilon\frac{\omega^2}{c^2}\left(1 + i\frac{4\pi\sigma}{\omega\epsilon}\right).$$

(Eqn. 7-82)

Let $k = \beta + 1\frac{\alpha}{2}$:

$$\left.\begin{matrix}\beta \\ \alpha/2\end{matrix}\right\} = \sqrt{\mu\epsilon}\frac{\omega}{c}\left[\frac{\sqrt{1 + \left(\frac{4\pi\sigma}{\omega\epsilon}\right)^2} \pm 1}{2}\right]^{1/2}.$$

(Eqn. 7-83)

Let's consider some cases.

<u>Poor conductor</u>: $\frac{4\pi\sigma}{\omega\epsilon} \ll 1.$

$k = \beta + i\frac{\alpha}{2} \simeq \sqrt{\mu\epsilon}\frac{\omega}{c} + i\frac{2\pi}{c}\sqrt{\frac{\mu}{\epsilon}}\sigma$ attenuation independant of ω, $Rek \gg Imk$

(Eqn. 7-84)

<u>Good conductor</u>: $\left(\frac{4\pi\sigma}{\omega\epsilon}\right) \gg 1$

$k = \beta + i\frac{\alpha}{2} \simeq (1+i)\frac{\sqrt{2\pi\omega\mu\sigma}}{c} \rightarrow \delta = \frac{2}{\alpha} \simeq \frac{c}{\sqrt{2\pi\omega\mu\sigma}},$

where δ is known as "skin depth."

Waves propagating as $\exp(i\vec{k}\cdot\vec{x} - i\omega t)$ are damped, transverse waves: Start with

$$\vec{E} = \vec{E}_0 e^{-\frac{\alpha}{2}\vec{n}\cdot\vec{x}} e^{i\beta\vec{n}\cdot\vec{x} - i\omega t} \quad and \quad H = H_0 e^{-\frac{\alpha}{2}\vec{n}\cdot\vec{x}} e^{i\beta\vec{n}\cdot\vec{x} - i\omega t}.$$

Recall Faraday's law:

$\nabla \times E -= -\frac{1}{c}\frac{\partial B}{\partial t} \Rightarrow \frac{i\omega}{c} H_0 = \vec{n} \times \vec{E}_0 \left(i\beta - \frac{\alpha}{2}\right),\ k = \sqrt{\mu\epsilon}\frac{\omega}{c}$, thus:

$$H_0 = \frac{c}{\mu\omega}\left(\beta + i\frac{\alpha}{2}\right)\vec{n} \times \vec{E}_0$$

Note how $\vec{H}$ and $\vec{E}$ are out of phase in a conductor:

$$|k| = \sqrt{\mu\epsilon}\frac{\omega}{c}\left[1 + \left(\frac{4\pi\sigma}{\omega\epsilon}\right)^2\right]^{1/4} \qquad \phi = \tan^{-1}\left(\frac{\alpha}{2\beta}\right) = \frac{1}{2}\tan^{-1}\left(\frac{4\pi\sigma}{\omega\epsilon}\right)$$

and

$$H_0 = \sqrt{\frac{\epsilon}{\mu}\left[1 + \left(\frac{4\pi\sigma}{\omega\epsilon}\right)^2\right]^{1/4}} e^{i\phi}\vec{n} \times \vec{E}_0$$

So, field energy is almost entirely magnetic in nature in a good conductor. For a good conductor the waves also show an exponential damping with distance (the skin depth):

$$\delta = \frac{2}{\alpha} \simeq \frac{c}{\sqrt{2\pi\omega\mu\sigma}}$$

(Eqn. 7-85)

where this expression is for Gaussian units (for copper $\delta \simeq .80cm$ at $60Hz$ and $\delta \simeq .71 \times 10^{-3}cm$ at $100 \times 10^6 Hz$). In MKSA units $\delta = \sqrt{\frac{2}{\mu\omega\sigma}}$.

Example 7.6. Consider transverse electromagnetic (TEM) waves in a dielectric where $H_z = 0$ and $E_z = 0$. (i) Show propagation is at te speed of light. (ii) Show both E and H satisfy the Laplace equation in x and y.

Solution

(i) Start with Maxwell's Equations:

$$\nabla \times E = -\mu \frac{\partial H}{\partial t} \quad \rightarrow \quad \begin{cases} -\dfrac{\partial E_y}{\partial z} = -\mu \dfrac{\partial H_x}{\partial t} \\ \dfrac{\partial E_x}{\partial z} = -\mu \dfrac{\partial H_y}{\partial t} \\ 0 - \mu\, \partial H_z / \partial t \end{cases}$$

and

$$\nabla \times H = \epsilon \frac{\partial E}{\partial t} \quad \rightarrow \quad \begin{cases} -\dfrac{\partial H_y}{\partial z} = \epsilon \dfrac{\partial E_x}{\partial t} \\ \dfrac{\partial H_x}{\partial z} = \epsilon \dfrac{\partial E_y}{\partial t} \\ 0 = \epsilon\, \partial E_z / \partial t \end{cases}$$

Combining equations:

$$\frac{\partial^2 E_x}{\partial z^2} = -\mu \frac{\partial^2 H_y}{\partial z \partial t} \quad \text{and} \quad \frac{-\partial^2 H_y}{\partial t \partial z} = \varepsilon \frac{\partial^2 E_x}{\partial t^2}$$

Giving

$$\frac{\partial^2 E_x}{\partial z^2} = \mu\epsilon \frac{\partial^2 E_x}{\partial t^2}$$

This is the classical one-dimensional wave equation for which the velocity (in the z direction) is

$$v = \frac{1}{\sqrt{\mu\epsilon}}$$

Since we have $\mu = \mu_0$ and $\epsilon = \epsilon_0$ we have

$$v = \frac{1}{\sqrt{\mu_0 \epsilon_0}} = c\,.$$

Similarly for E_y.

(ii) For a source free dielectric $\nabla \cdot D = 0$ and if ϵ is not a function of space coordinates, $\nabla \cdot E = 0$ also.
Then

$$\nabla^2 E = \mu\epsilon \frac{\partial^2 E}{\partial t^2}$$

Similarly

$$\nabla^2 H = \epsilon\mu\, \partial^2 H / \partial t^2$$

and the rest is left to the reader.

7.3 Surface Fields

Consider a perfect conductor

$$\vec{n} \cdot \vec{D} = 4\pi\Sigma.$$

(Eqn. 7-86)

(Using Σ for surface charge density to avoid confusion with σ – the conductivity). Note that the unit norm $\vec{n}$ is orthogonal to the conducting surface pointing into the non-conducting medium.

Similarly,

$$n \times \vec{H} = \frac{4\pi}{c} \vec{K},$$

(Eqn. 7-87)

where $\vec{K}$ is the surface current. From the junction conditions:

$$n \cdot (\vec{B} - \vec{B}_0) = 0$$

(Eqn. 7-88)

$$n \times (\vec{E} - \vec{E}_0) = 0$$

(Eqn. 7-89)

So, just outside the surface of a perfect conductor only normal E and tangential H can exist. Consider a good conductor, fields in the neighborhood of its surface must behave approximately the same as for a perfect conductor. Recall, inside a conductor fields are attenuated exponentially in length δ. So, we have a thin transitional region at the surface. From Ohm's law $J = \sigma E \Longrightarrow$ for a finite conductivity there cannot actually be a surface layer of current. Instead

$$\vec{n} \times (\vec{H} - \vec{H}_0) = 0.$$

(Eqn. 7-90)

<u>Successive approx. scheme</u>:

Just outside: only a normal E field $\vec{E} \perp$

a tangential H field $\vec{H} \parallel$ as for a perfect conductor.

Values are assumed to have been obtained from the solution of an appropriate boundary value problem. Then we use the boundary conditions and the Maxwell equations in the conductor to find the fields within the transition layer and small corrections to the fields outside. In solving the Maxwell equations within the conductor we make use of the fact that the spatial variation of the fields normal to the surface is much more rapid than variation parallel to the surface. So, we neglect all derivatives with respect to coordinates parallel to the surface compared to the normal derivative.

Tangential $H_\parallel$ outside and $\vec{n} \times (\vec{H} - \vec{H}_0) = 0$ implies the same $H_\parallel$ inside the surface.

$$\nabla \times H = \frac{4\pi}{c} J + \frac{1}{c}\frac{\partial \vec{D}}{\partial t} \quad , \quad \nabla \times E = -\frac{1}{c}\frac{\partial \vec{B}}{\partial t}$$

$$\left\{\begin{array}{l} \uparrow \text{neglecting the displacement current,} \\ \text{using } J = \sigma E \\ \text{and assuming a harmonic variation } e^{-i\omega t} \end{array}\right\}$$

We get

$$\nabla \times \vec{H}_c \simeq \frac{4\pi}{c} \sigma \vec{E}_c \quad , \quad \nabla \times E_c = \frac{i\omega}{c} \vec{B}_c = \frac{i\omega\mu_c}{c} \vec{H}_0$$

$$\left\{\begin{array}{l} \vec{n} \text{ is unit normal outward from the conductor} \\ \varepsilon \text{ is the normal coord. inward} \\ \Rightarrow \nabla \simeq -\vec{n}\dfrac{\partial}{\partial \varepsilon} \text{ neglecting the transverse derivatives} \end{array}\right.$$

Thus,

$$\left\{\begin{array}{l} \vec{E}_c \simeq \dfrac{c}{4\pi\sigma}\left\{\vec{n} \times \dfrac{\partial \vec{H}_c}{\partial \varepsilon}\right\} \\ \vec{H}_c \simeq \dfrac{ic}{\mu_c \omega}\left\{\vec{n} \times \dfrac{\partial \vec{E}_c}{\partial \varepsilon}\right\} \end{array}\right.$$

(Eqn.s 7-91)

which can be combined:

$$n \times \vec{H}_c = \frac{ic}{\mu_c \omega} n \times \left\{\vec{n} \times \frac{\partial \vec{E}_c}{\partial \varepsilon}\right\} = \frac{ic}{\mu_c \omega}\left\{\left(\vec{n} \cdot \frac{\partial \vec{E}_c}{\partial \varepsilon}\right)\vec{n} - \frac{\partial \vec{E}_c}{\partial \varepsilon}\right\}$$

$$n \times \vec{H}_c = \frac{ic}{\mu_c \omega}\left(\frac{c}{4\pi\sigma}\right)\frac{\partial}{\partial \varepsilon}\left(\vec{n} \times \frac{\partial \vec{H}_c}{\partial \varepsilon}\right)$$
$$= \frac{ic^2}{\mu_c \omega 4\pi\sigma}\frac{\partial^2}{\partial \varepsilon^2}(\vec{n} \times \vec{H}_c)$$

So,

$$\frac{\partial^2}{\partial \varepsilon^2}(\vec{n} \times \vec{H}_c) + i\mu_c \omega 4\pi\sigma(\vec{n} \times \vec{H}_c) = 0$$

or

$$\frac{\partial^2}{\partial \varepsilon^2}(\vec{n} \times \vec{H}_c) + \frac{2i}{\delta^2}(\vec{n} \times \vec{H}_c) = 0 \;\; with \quad \delta^2 = \frac{c^2 2}{4\pi\mu_c \omega\sigma}, \quad \delta = \frac{c}{\sqrt{4\pi}}\left(\frac{2}{\mu_c \omega\sigma}\right)^{1/2}$$

While

$$\vec{n} \cdot \vec{H}_c = 0.$$

So $H_{\perp} = 0$; $\frac{\partial^2 H_{\parallel}}{\partial \varepsilon^2} + \frac{2i}{\delta^2} H_{\parallel} = 0$. Consider $H_{\parallel} = Ae^{i\lambda\varepsilon}$, then

$$-\lambda^2 + \frac{2i}{\delta^2} = 0 \quad \rightarrow \quad \lambda = \frac{\sqrt{2i}}{\delta} = \frac{\sqrt{2}e^{i\frac{\pi}{4}}}{\delta} = \frac{(1+i)}{\delta}.$$

Thus,

$$H_c = \vec{H}_{\parallel} e^{-\varepsilon/\delta} e^{i\varepsilon/\delta}.$$

Using $E_c \cong \frac{-c}{4\pi\sigma} \vec{n} \times \frac{\partial \vec{H}_c}{\partial \varepsilon} = \frac{c}{4\pi\sigma} \left\{\frac{i-1}{\delta}\right\} (\vec{n} \times \vec{H}_{\parallel}) e^{-\varepsilon/\delta} e^{i\varepsilon/\delta}$, we get:

$$E_c = \sqrt{\frac{\mu_c \omega}{8\pi\sigma}} (1-i)(\vec{n} \times \vec{H}_{\parallel}) e^{-\varepsilon/\delta} e^{i\varepsilon/\delta}.$$

Just outside the surface we find a tangential E-field component:

$$E_{\parallel} \simeq \sqrt{\frac{\mu_c \omega}{8\pi\sigma}} (1-i)(\vec{n} \times \vec{H}_{\parallel}),$$

which shows a power flow into the conductor:

$$\begin{aligned} \frac{dP_{loss}}{da} &= -\frac{c}{8\pi} Re[n \cdot \vec{E} \times \vec{H}^*] \\ &= -\frac{c}{8\pi} \sqrt{\frac{\mu_c \omega}{8\pi\sigma}} Re[n \cdot (1-i)(\vec{n} \times \vec{H}_{\parallel}) \times \vec{H}_{\parallel}] \\ &= \frac{c}{8\pi} \sqrt{\frac{\mu_c \omega}{8\pi\sigma}} |H_{\parallel}|^2 \end{aligned}$$

(Eqn. 7-92)

Note, have minus sign since loss from field, and factor of $1/2$ due to time average of phases. The expression provides the time-avg. power absorbed per unit area.

Since

$$\vec{J} = \sigma \vec{E}_c = \sqrt{\frac{\mu_c \omega \sigma}{8\pi}} (1-i)(\vec{n} \times \vec{H}_{\parallel}) e^{-\varepsilon(1-i)/\delta}$$

(Eqn. 7-93)

then

$$\vec{K}_{eff} = \int_0^{\infty} J \, d\varepsilon = \left[\frac{c}{4\pi}\right] \vec{n} \times \vec{H}_{\parallel}.$$

(Eqn. 7-94)

The power loss is thus:

$$\frac{dP_{loss}}{da} = \frac{1}{2\sigma\delta}\left|K_{eff}\right|^2.$$

(Eqn. 7-95)

Note that $1/\sigma\delta$ plays the role of surface resistance.

Cylindrical Cavities and Wave Guides

Consider a hollow cylindrical wave guide. Assume a sinusoidal time dependence $e^{-i\omega t}$ for the fields inside the cylinder. Maxwell's eqn's inside the cylinder become, assuming the cylinder is filled with a uniform nondissipative medium having dielectric constant ϵ and permeability μ:

$$\left.\begin{aligned} \nabla\cdot B &= 0 \\ \nabla\cdot D &= 4\pi\rho \\ \nabla\times E &= -\frac{1}{c}\frac{\partial B}{\partial t} \\ \nabla\times H &= \frac{4\pi}{c}\vec{J} + \frac{1}{c}\frac{\partial\vec{D}}{\partial t} \end{aligned}\right\} \Rightarrow$$

$$\begin{aligned} \nabla\cdot\vec{B} &= 0 \\ \nabla\cdot\vec{E} &= 0 \\ \nabla\times E &= \frac{i\omega}{c}\vec{B} \quad \text{(no charges or currents)} \\ \nabla\times B &= -\frac{\mu\epsilon i\omega}{c}\vec{E} \end{aligned}$$

Thus,

$$\left(\nabla^2 + \mu\epsilon\frac{\omega^2}{c^2}\right)\begin{Bmatrix} E \\ B \end{Bmatrix} = 0.$$

Taking account of the cylindrical geometry, assume

$$\left.\begin{matrix} \vec{E}(x,y,z,t) \\ \vec{B}(x,y,z,t) \end{matrix}\right\} = \left\{\begin{matrix} \vec{E}(x,y)e^{\pm ikz - i\omega t} \\ \vec{B}(x,y)e^{\pm ikz - i\omega t} \end{matrix}\right.$$

(appropriate combinations give either standing or traveling waves). And we have:

$$\left[\nabla_t^2 + \left(\mu\epsilon\frac{\omega^2}{c^2} - k^2\right)\right]\begin{Bmatrix} E \\ B \end{Bmatrix} = 0.$$

For cylindrical geometry:

$\vec{E} = \vec{E}_z + \vec{E}_t$ with $\vec{E}_z = \hat{e}_3 E_z$ and $\vec{E}_t = \left(\hat{e}_3 \times \vec{E}\right)\hat{e}_3 \Rightarrow (e_3\cdot e_3)E - \left(e_3\cdot\vec{E}\right)\hat{e}_3 = E - \vec{E}_z$

Decomposing the Maxwell eqn's into transverse and parallel components:

$\nabla \cdot \vec{B} = 0 \Rightarrow \vec{\nabla}_t \cdot \vec{B}_t + \frac{\partial}{\partial z} B_z = 0$ $\quad \boxed{\vec{\nabla}_t \cdot \vec{B}_t + \frac{\partial B_z}{\partial z}}$ (v)

Similarly $\boxed{\vec{\nabla}_t \cdot \vec{E}_t + \frac{\partial E_z}{\partial z}}$ (vi)

$\nabla \times \vec{E} = \frac{i\omega}{c} \vec{B} \Rightarrow \boxed{\hat{e}_3(\nabla_t \times \vec{E}_t) = i\frac{\omega}{c} B_z}$ (i)

And $\nabla_t \mathrm{E}_z - \frac{\partial E_t}{\partial z} = i\frac{\omega}{c} \hat{e}_3 \times \vec{B}_t$ $\quad \boxed{\frac{\partial E_t}{\partial z} + i\frac{\omega}{c} \hat{e}_3 \times \vec{B}_t = \nabla_t \mathrm{E}_z}$ (ii)

$\nabla \times B = \frac{\mu\epsilon i\omega}{c} \vec{E} \Rightarrow \boxed{\hat{e}_3 \cdot (\nabla_t \times \vec{B}_t) = -\frac{\omega}{c} \mu\epsilon \mathrm{E}_z}$ (iii)

And $\nabla_t B_z - \frac{\partial B_t}{\partial z} = -\frac{i\omega}{c} \mu\epsilon \hat{e}_3 \times \vec{E}_t$

$\boxed{\frac{\partial B_t}{\partial z} - i\mu\epsilon \frac{\omega}{c} \hat{e}_3 \times \mathrm{E}_t = \nabla_t B_z}$ (iv)

If E_z and B_z are known, the transverse components of $\vec{E}$ and $\vec{B}$ are determined.

Degenerate solution: Transverse electromagnetic (TEM) wave

$\mathrm{E}_z = 0 \quad , \quad \mathrm{B}_z = 0$

(iii) (iv)

$\vec{E}_t = \vec{E}_{TEM} \Rightarrow \vec{\nabla}_t \times E_{TEM} = 0 \quad , \quad \vec{\nabla}_t \times \vec{E}_{TEM} = 0$

So, E_{TEM} is the solution of an electrostatic problem in 2D.

$k = k_0 = \frac{1}{c} \omega\sqrt{\mu\epsilon} \leftarrow$ <u>TEM mode has no cutoff frequency</u>.

(Eqn. 7-96)

$B_{TEM} = \pm\sqrt{\mu\epsilon} e_3 \times \vec{E}_{TEM}$ (iv)

TEM mode cannot exist inside an equipotential surface (then $E = 0$ everywhere from electrostatic property). It is necessary to have two or more cylindrical surfaces, etc., to support a TEM mode.

In hollow cylinders there are two types of field configurations corresponding to the longitudinal components E_z and B_z and the boundary conditions they satisfy. For a perfectly conducting cylinder the b.c.'s are

$n \times E = 0$ and $n \cdot B = 0$ at the surface

$\downarrow \qquad \downarrow$

$E_z|_s = 0 \qquad \left.\frac{\partial B_z}{\partial n}\right|_s = 0$

The two dimensional wave equation:

$$\left[\nabla_t^2 + \left(\mu\epsilon\frac{\omega^2}{c^2} - K^2\right)\right]\begin{Bmatrix} E \\ B \end{Bmatrix} = 0,$$

together with our b.c.'s specify eigenvalue problems of the usual sort. For a given ω only certain k's can occur. Since the b.c.'s on E_z and B_z are different, the eigenvalues will in general be different. The fields naturally divided into two categories:

Transverse Magnetic (TM) waves

$$B_z = 0 \text{ everywhere; b. c.} E_z|_s = 0$$

Transverse Electic (TE)waves

$$E_z = 0 \text{ everywhere; b. c.} \left.\frac{\partial B_z}{\partial n}\right|_s = 0$$

The various TM, TE, and TEM waves for a mode constitute a complete set of fields to describe an arbitrary electromagnetic disturbance in a wave guide or cavity.

Wave Guides:

For TM, $B_z = 0$ use (iv)

$$\frac{\partial B_t}{\partial z} - i\mu\epsilon\frac{\omega}{c}\hat{e}_3 \times E_t = 0$$

$$\pm ikH_t = i\epsilon\frac{\omega}{c}\hat{e}_3 \times E_t \qquad k_0 = \frac{\omega}{c}\sqrt{\mu\epsilon}$$

$$\vec{H}_t = \pm\frac{K_0}{K}\sqrt{\frac{\epsilon}{\mu}}\hat{e}_3 \times \vec{E}_t$$

$$\boxed{\vec{H}_t = \frac{\pm 1}{z}\hat{e}_3 \times \vec{E}_t \qquad z = \frac{K}{K_0}\sqrt{\frac{\epsilon}{\mu}}} \qquad (TM)$$

For TE, $E_z = 0$ use (ii)

$$\frac{\partial E_t}{\partial z} + i\frac{\omega}{c}\hat{e}_3 \times B_t = 0$$

$$\pm ikE_t + i\frac{\omega}{c}\hat{e}_3 \times B_t = 0$$

$$\pm k\hat{e}_3 \times E_t = -\frac{\omega}{c}\hat{e}_3 \times \hat{e}_3 \times B_t = -\frac{\omega}{c}[(\hat{e}_3 \cdot B_t)\hat{e}_3 - (\hat{e}_3 \cdot \hat{e}_3)B_t]$$

$$\pm k\hat{e}_3 \times E_t = \frac{\omega}{c}B_t = \frac{\omega}{c}(\mu H_t) \qquad \frac{K}{\frac{\omega}{c}\sqrt{\mu\epsilon}\sqrt{\frac{\mu}{\epsilon}}}$$

$$\vec{H}_t = \pm k \frac{c}{\omega\mu} \hat{e}_3 \times E_t = \pm \frac{K}{K_0}\sqrt{\frac{\epsilon}{\mu}} \hat{e}_3 \times E_t$$

$$\boxed{\vec{H}_t = \frac{\pm 1}{z} \hat{e}_3 \times \vec{E}_t \qquad z = \frac{K_0}{K}\sqrt{\frac{\mu}{\epsilon}}} \qquad (TE)$$

<u>TM waves: $E_z = \Psi e^{\pm ikz}$:</u>

$$\begin{cases} \dfrac{\partial E_t}{\partial z} + i\dfrac{\omega}{c} e_3 \times B_t = \nabla_t E_z \\ H_t = \dfrac{\pm 1}{z} e_3 \times E_t \Rightarrow e_3 \times B_t = \dfrac{\pm 1}{z} e_3 \times e_3 \times E_t = \dfrac{\pm\mu}{z} \vec{E}_t(-1) \end{cases}$$

$$\frac{\partial E_t}{\partial z} = \frac{d}{dz}\left(E_t(x,y)e^{\pm ikz}\right) = \pm ikE_t$$

$$\pm ikE_t = i\frac{\omega}{c}\frac{\mu}{z} E_t(-1) = \nabla_t \Psi$$

$$k_0 = \frac{\omega}{c}\sqrt{\mu\epsilon}$$

$$\pm E_t\left(ik + \frac{i\omega}{c}\frac{\mu}{\frac{k}{k_0}\sqrt{\frac{\mu}{\epsilon}}}\right) = \pm E_t\left(ik - i\left(\frac{\omega}{c}\right)^2 \frac{\sqrt{\mu\epsilon}\cdot\sqrt{\mu\epsilon}}{K}\right) = \nabla_t \Psi$$

$$E_t = \frac{\pm ik}{\left(k^2 - \left(\frac{\omega}{c}\right)^2 \mu\epsilon\right)} \nabla_t \Psi = \frac{\pm \mathrm{ik}}{\gamma^2} \nabla_t \Psi$$

$$\vec{E}(x,y,z,t) = \vec{E}(x,y)e^{\pm ikz - i\omega t} = \left(\vec{E}_t + \vec{E}_z\right)e^{\pm ikz - i\omega t}$$

<u>TE waves: $E_z = 0 \qquad H_z = \Psi e^{\pm ikz}$:</u>

$$\begin{cases} \dfrac{\partial B_t}{\partial z} - i\mu\epsilon\dfrac{\omega}{c} e_3 \times E_t = \nabla_t B_z \\ \dfrac{\partial E_t}{\partial z} + i\dfrac{\omega}{c} e_3 \times B_t = 0 \Rightarrow \pm ikE_t + i\dfrac{\omega}{c}\vec{e}_3 \times B_t = 0 \end{cases}$$

$$\pm ik\mu H_t - i\mu\epsilon\frac{\omega}{c} e_3 \times \left\{\pm\frac{\omega}{kc}\vec{e}_3 \times \vec{B}_t(-1)\right\} = \nabla_t B_z$$

$$\pm ik\mu H_t - \left(\pm\left[\frac{\omega}{c}\right]^2 i\mu\epsilon\right)\frac{1}{k}\mu H_t = \nabla_t B_z$$

$$e^{\pm ikz} H_t = \frac{\pm ik}{-k^2 + \mu\epsilon\left(\frac{\omega^2}{c^2}\right)}\left(\frac{1}{\mu}\right)\nabla_t B_z$$

$$e^{\pm ikz}H_t = \frac{\pm ik}{\gamma^2}\nabla_t H_z \qquad H_z = \Psi e^{\pm ikz}$$

$$H_t = \frac{\pm ik}{\gamma^2}\nabla_t\Psi$$

Now, consider our result: $E_z(\text{or } H_z) = \Psi e^{\pm ikz}$ for TM (or TE) $waves$
TM waves:

$$E_t = \frac{\pm ik}{\gamma^2}\nabla_t\Psi \qquad (\gamma^2 = \mu\epsilon\frac{\omega^2}{c^2} - k^2)$$

TE waves:

$$H_t = \frac{\pm ik}{\gamma^2}\nabla_t\Psi$$

The scalar function satisfies $(\nabla_t^2 + \gamma^2)\Psi = 0$, such that $\Psi|_s = 0\ (TM)$ and $\left.\frac{\partial\Psi}{\partial n}\right|_s = 0\ (TE)$ specifies an eigenvalue problem. Note that γ^2 must be non-negative because Ψ must be oscillatory in order to satisfy either b.c. on opposite sides of the cylinder. There is a spectrum γ_λ^2 of eigenvalues and corresponding solutions $\Psi_\lambda, \lambda = 1,2, \ldots$ which form an orthogonal set, with $k_\lambda^2 = \mu\epsilon\frac{\omega^2}{c^2}\gamma_\lambda^2$ for a given ω and λ, corresponding wavenumber determined. We can define a cutoff frequency:

$$\omega_\lambda = c\frac{\gamma_\lambda}{\sqrt{\mu\epsilon}}.$$

So, $k_\lambda = \frac{1}{c}\sqrt{\mu\epsilon}\sqrt{\omega^2 - \omega_\lambda^2}$. For $\omega > \omega_\lambda$ k_λ is real; waves of the λ mode can propagate in the guide. For $\omega < \omega_\lambda$ k_λ is imaginary, such λ are cutoff modes or evanescent modes.

Modes in a Rectangular Wave Guide

Consider the propagation of TE waves in the rectangular wave guide, bounded in the x-y plane by (0,0,(a,0),(a,b),(0,b); and with dielectric constant ϵ and permittivity μ.

Recall that TE has $\left.\frac{\partial\Psi}{\partial n}\right|_s = 0$ where $\begin{matrix} H_z = \Psi e^{\pm ikz} \\ E_z = 0 \end{matrix}$, also recall that $H_t = \frac{\pm ik}{\gamma^2}\nabla_t\Psi$ where Ψ satisfies $(\nabla_t^2 + \gamma^2)\Psi = 0$. So, consider the eigenvalue problem:

$$(\nabla_t^2 + \gamma^2)\Psi = 0$$

$$\left.\frac{\partial\Psi}{\partial n}\right|_S = 0$$

from which everything else follows:

$$\left(\frac{\partial^2}{\partial x^2}+\frac{\partial^2}{\partial y^2}+\gamma^2\right)\Psi = 0; \qquad \left.\frac{\partial\Psi}{\partial x}\right|_{x=0} = 0 = \left.\frac{\partial\Psi}{\partial x}\right|_{x=a}; \qquad \left.\frac{\partial\Psi}{\partial y}\right|_{x=0} = 0 = \left.\frac{\partial\Psi}{\partial y}\right|_{x=b}.$$

So want: $\frac{\partial\Psi}{\partial x}$ to behave as $\sin x$, $\frac{\partial\Psi}{\partial y}$ to behave as $\sin y$. Thus, $\Psi = H_0\cos(\alpha x)\cos(\beta y)$,

where α, β are determined by

$\sin(\alpha a) = 0 \rightarrow \ \alpha = \frac{m\pi}{a}$, and similarly, $\beta = \frac{m\pi}{b}$. So,

$$\Psi = H_0\cos\left(\frac{m\pi x}{a}\right)\cos\left(\frac{n\pi y}{b}\right),$$

where

$$\gamma_{mn}^2 = \pi^2\left(\frac{m^2}{a^2}+\frac{n^2}{b^2}\right),$$

and since $\gamma^2 = \mu\epsilon\frac{\omega^2}{c^2} - k^2$:

$$\omega_{mn} = \omega_{cutoff} = \frac{c}{\sqrt{\mu\epsilon}}\gamma_{mn} = \frac{c\pi}{\sqrt{\mu\epsilon}}\left(\frac{m^2}{a^2}+\frac{n^2}{b^2}\right)^{1/2}.$$

(Eqn. 7-97)

If $a > b$, the lowest cutoff frequency is for $m = 1, n = 0$.

$\omega_{1,0} = \frac{\pi c}{\sqrt{\mu\epsilon}a}$ corresponds to $\frac{1}{2}$ of a free-space wavelength across the guide.

Explicit fields for this mode: (choose wave traveling in positive z direction):

$$H_z = \Psi e^{ikz}e^{-i\omega t} = H_0\cos\left(\frac{\pi x}{a}\right)e^{ikz-i\omega t}$$

$$E_z = 0$$

The other components:

$H_t = \frac{ik}{\gamma^2}\nabla_t\Psi \Rightarrow t \simeq x \quad H_x\frac{ik}{\gamma^2}\frac{\partial}{\partial x}\Psi e^{ikz-i\omega t}H_0 =$
$\frac{ik}{\gamma^2}\left(-\frac{\pi}{a}\right)\sin\left(\frac{\pi x}{a}\right)e^{ikz-i\omega t}H_0$

$\gamma^2 = \frac{\pi^2}{a^2}$, so:

$$H_x = \frac{ika}{\pi}H_0\sin\left(\frac{\pi x}{a}\right)e^{ikz-i\omega t}$$

Similarly,

$$\vec{H}_t = \frac{1}{z}\hat{e}_3 \times \vec{E}_t \Rightarrow \hat{e}_3 \times \vec{H}_t = \frac{1}{\frac{\mathrm{k}_0}{\mathrm{k}}\sqrt{\frac{\mu}{\epsilon}}}(-\vec{E}_t)$$

$$\vec{E}_y = -\frac{\mathrm{k}_0}{\mathrm{k}}\sqrt{\frac{\mu}{\epsilon}}H_x\,; \qquad k^2 = k_{1,0}^2 = \mu\epsilon\frac{\omega^2}{c^2} - \frac{\pi^2}{a^2}\,; \quad \mathrm{k}_0 = \frac{\omega}{\mathrm{c}}\sqrt{\mu\epsilon}$$

$$E_y = \frac{i\omega a\mu}{\pi c}H_0 \sin\left(\frac{\pi x}{a}\right)e^{ikz-i\omega t}\,.$$

$TE_{1,0}$ mode has the lowest cutoff frequency of both TE and TM modes and so is the one used in most practical situations. For TM $E_z \propto E_0 \sin\left(\frac{m\pi x}{a}\right)\sin\left(\frac{n\pi x}{b}\right) \Rightarrow \gamma^2 = \pi^2\left(\frac{m^2}{a^2}+\frac{n^2}{b^2}\right)$ but lowest mode is $m = 1, n = 1 \Rightarrow$ So $\gamma_{TM}^2 > \gamma_{TE}^2$ by factor of $\left(1+\frac{a^2}{b^2}\right)$.

7.3.1 Transmission Lines [34]

Note the convention to use 'j' for the imaginary number instead of 'i' in places where there might be confusion with 'current' notations.

In free space, the interchange of electric and magnetic field energy gives rise to (transverse) wave propagation (at the speed of light). Similar propagation, based on the time-varying exchange between the E and B fields, occurs in directed transmission via surfaces, in dielectrics, in transmission lines, and in various hollow-pipe waveguides. In this section we examine the transmission line scenario.

The typical transmission line consists of two parallel lines. We further approximate to two parallel plate ribbons in most of the discussion, as might occur in computer bus connections or on circuit boards in general. We start with the transmission line because we can draw on basic circuit theory to help obtain a solution about the transmission properties.

If we view the transmission line as a distributed circuit, then we can consider a differential element involving a length dz of the line, across which we have an inductance Ldz (in series) and a capacitance Ldz (in parallel). This gives rise to the following changes in current and voltage across the element dz of transmission line:

$$\text{voltage change: } \frac{\partial V}{\partial z}dz = -(Ldz)\frac{\partial I}{\partial t}, \text{ and}$$

current change: $\frac{\partial I}{\partial z}dz = -(Cdz)\frac{\partial V}{\partial t}$.

Regrouping we get:

$$\frac{\partial^2 V}{\partial z^2} = \frac{1}{v^2}\frac{\partial^2 V}{\partial t^2},$$

where $v = (LC)^{-1/2}$(and similarly for I). Since D'Alembert we know the solution to the wave equation can be written as:

$$V(z,t) = f\left(t - \frac{z}{v}\right) + g\left(t + \frac{z}{v}\right),$$

for which the related $I(z,t)$must be:

$$I(z,t) = \frac{1}{Z_0}\left[f\left(t - \frac{z}{v}\right) + g\left(t + \frac{z}{v}\right)\right],$$

where $Z_0 = Lv = \sqrt{L/C}$ is known as the characteristic impedance.

Suppose our parallel plate transmission line is described with coordinates such that x is the dimension of their separation (and z is their running length, and y is their width). Assuming perfect conductors, the E-field must be normal at the boundary, a pure E_x component will do this. A pure H_y that is transverse in the z-direction then follows and we have Maxwell's equations:

$$\frac{\partial E_x(z,t)}{\partial z} = -\mu\frac{\partial H_y(z,t)}{\partial t}, \quad and$$
$$\frac{\partial H_y(z,t)}{\partial z} = -\varepsilon\frac{\partial E_x(z,t)}{\partial t}$$

Comparison with the circuit form of the answer we get: $C = \frac{b}{a}\varepsilon$, $L = \frac{a}{b}\mu$, and $v = (\varepsilon\mu)^{-1/2} = c$.

Perfect conductors transmit only transverse electromagnetic waves (TEM waves) for which there is no axial magnetic or electric field.

A transmission line typically has a 'junction', such as a resistive discontinuity, or a resistive termination. So we need to consider what happens with such boundary conditions. again, a simple circuit analysis argument clarifies. Recalling the (travelling) wave-like nature of the potential, consider a voltage wave hitting the discontinuity with voltage V^+ and having a reflection $V^-$where the overall voltage at the discontinuity, or load, is $V_L = V^+ + V^-$. Similarly for the time-varying current: $I_L = I^+ + I^-$ from which we get the transmission and reflection coefficients:

$$\rho \equiv \frac{V^-}{V^+} = \frac{R_L - Z_0}{R_L + Z_0} \quad and \quad \tau \equiv \frac{V_L}{V^+} = \frac{2R_L}{R_L + Z_0}.$$

(Eqn. 7-98)

Instantaneous power at the load is: $W^+ = I^+V^+$, while reflected power goes as:

$$W^-/W^+ = \rho^2 .$$

Decomposing into sinusoidal variations and using phasor form:

$$V = V^+e^{-j\beta z} + V^-e^{+j\beta z} \quad and \quad I = \left(\frac{1}{Z_0}\right)\left[V^+e^{-j\beta z} - V^-e^{+j\beta z}\right].$$

Note that is βz changes by 2π then the current voltage are unchanged. This being the smallest value for this to hold (not $n2\pi$), thus defines β to be related to the wavelength by:

$$\beta = \frac{2\pi}{\lambda}.$$

Let's consider the impedance seen at position $-l$ before the discontinuity:

$$Z = Z_0\left[\frac{e^{j\beta l} + \rho e^{-j\beta l}}{e^{j\beta l} - \rho e^{-j\beta l}}\right] = Z_0\left[\frac{Z_L\cos\beta l + jZ_0\sin\beta l}{Z_0\cos\beta l + jZ_L\sin\beta l}\right].$$

(Eqn. 7-99)

Transmission Line Problems

Example 7.7. ([34] 5.5b) Consider special cases of

$$Z = Z_0\left[\frac{Z_L\cos\beta l + jZ_0\sin\beta l}{Z_0\cos\beta l + jZ_L\sin\beta l}\right].$$

for when the discontinuity is: (i0 a shorted line; (ii) an open line; (iii) a half-wave line with load impedance Z_L; (iv) a quarter-wave line with load impedance Z_L.

Solution

(i) Shorted line: $Z_L = 0 \rightarrow Z_i = Z_0\left[\frac{j\sin\beta\ell}{\cos\beta\ell}\right] = jZ_0\tan\beta\,\ell.$

(ii) Open line: $Z_L \to \infty \rightarrow Z_i = Z_0\left[\frac{\cos\beta\ell}{j\sin\beta\ell}\right] = -jZ_0\cot\beta\,\ell.$

(iii) Half-wave line: $\ell = \frac{1}{2}\lambda \quad \beta\ell = \pi \rightarrow Z_i = Z_0\left[\frac{Z_L\cos\pi + jZ_0\sin\pi}{Z_0\cos\pi + jZ_L\sin\pi}\right] = Z_L$

(iv) Quarter-wave line: $\ell = \frac{1}{4}\lambda \quad \beta\ell = \pi/2 \rightarrow Z_i =$
$Z_0\left[\frac{Z_L\cos\pi/2 + jZ_0\sin\pi/2}{Z_0\cos\pi/2 + jZ_L\sin\pi/2}\right] = \frac{Z_0^2}{Z_L}.$

Example 7.8. ([34] 5.5d) Derive the expression for the current reflection coefficient: $\rho_I = I_-/I_+$, how is it related to the voltage version?

Solution
Start with:

$$V = Z_0 I_+ e^{-jBz} - z_0 I_- e^{jBz} \quad and \quad I = I_+ e^{-jBz} + I_- e^{jBz}.$$

Set $z = 0$, to get:

$$Z_L = \frac{V}{I} = \frac{Z_0 I_+ - Z_0 I_-}{I_+ + I_-}.$$

Thus,

$$\rho_I = I_-/I_+ = \frac{(Z_0 - Z_L)}{(Z_L + Z_0)} = -\rho\,.$$

The negative can be absorbed into the e^{jBz} expression: $-e^{-jBz} = e^{-jBz-\pi}$, thus the negative represents a phase difference of π rad.

Example 7.9. ([34] 5.5c) When connecting transmission lines with characteristic impedances Z_1 and Z_2 a reflection will occur if they do not have matching characteristic impedances. Show that this problem can be solved by inserting a quarter-wave length of line ($\ell = \lambda/4 \rightarrow \beta\ell = \pi/2$) with characteristic impedance $Z_* = \sqrt{Z_1 Z_2}$.

Solution
The impedance seen at the start of the quarter-length line, using the quarter length relation, has the general form derived previously:

$$Z_i = \frac{Z_0^2}{Z_L}.$$

Here we have

$$Z_i = \frac{Z_*^2}{Z_2},$$

and this must match Z_1, thus:

$$Z_1 = \frac{Z_*^2}{Z_2} \rightarrow Z_* = \sqrt{Z_1 Z_2}\ .$$

Example 7.10. A loop in a B-field is connected to a resistor R. This results in a current flowing $I(t) = V(t)/R$. Find an expression for the time-dependent torque on the loop.

Solution
$I(t) = V(t)/R$ and for a loop of area A: $V(t) = \Omega BA \cos \Omega t$

Using

$$|F_a| = iL \times B = i\ell B$$

Torque on a loop $\tau = r \times F$ → $\tau = IBA\cos\Omega t$, so we get:

$$\tau = \frac{(AB\cos\Omega t)^2\Omega}{R}.$$

7.3.2 Waveguides

Waveguides with rectangular and circular cross-section [34]

If we take the parallel plate capacitor used in the transmission line example and complete its 'enclosure' with side walls, we have a rectangular cross-section waveguide. To understand its properties we are further removed from an equivalent steady-state circuit so we go directly to an analysis using Maxwell's equations with such conducting boundaries. This reduces to a wave equation that, in turn, reduces to solving the Helmholtz equation resulting from a phasor analysis (analysis involving complex wave solutions).

As with the transmission line, the main purpose for a waveguide is guided transmission. Since the purpose is transmission of power or information, the engineering and economic refinements naturally seek to do this the most efficient (cheap) way possible. To aid in the design process of a transmission line or wave guide it is important to examine the field distributions, power transmission and the pulse dispersion properties of the guide. Practically speaking, this restricts transmission lines to the sub-millimeter range and the hollow-tube conducting guides to the millimeter and microwave ranges (1-100GHz) [34]. For pulse-based information transfer, pulse dispersion describes your bandwidth limitations, so is obviously important to find. The pulse dispersion information is encapsulated in the relation between the propagation constant k and the frequency ω, and from this relation we find the wave velocity, phase variation and attenuation [34] as will be shown.

When we speak of "guiding" waves we are describing the process of engineering a system where electromagnetic fields exchange energy with surface charges and currents such that efficient wave propagation occurs. Let's start by considering a general cylindrical system. Take the cylindrical axis to be in the z-direction. Since the enclosed region is source free (and no convection currents) we get:

$$\nabla^2 E = \mu\varepsilon \frac{\partial^2 E}{\partial t^2} \quad and \quad \nabla^2 H = \mu\varepsilon \frac{\partial^2 H}{\partial t^2},$$

with whatever boundary conditions (BC's) indicated. If we assume solutions in terms of time-harmonic waves ($e^{j\omega t - \gamma z}$), as was useful with transmission lines, we get:

$$\nabla^2 E = -k^2 E \quad and \quad \nabla^2 H = -k^2 H,$$

where $k^2 = \omega^2 \mu\varepsilon$. In our cylindrical coordinate system the three-dimensional Laplacian is split into axial ($\frac{\partial^2}{\partial z^2}$) and transverse (${\nabla_t}^2$) components:

$${\nabla_t}^2 E + \frac{\partial^2 E}{\partial z^2} = -k^2 E \rightarrow {\nabla_t}^2 E = -(k^2 + \gamma^2)E,$$

and similarly for H. The usual way to solve this type of system is to proceed in solving the above for the axial components: E_z and H_z. Then the remaining (curl-based) Maxwell equations: $\nabla \times \mathrm{E} = -\mathrm{j}\omega\mu H$ and $\nabla \times \mathrm{H} = \mathrm{j}\omega\varepsilon E$ can be used to get expression for the other components in terms of the solved E_z and H_z:

$$E_x = -\frac{1}{(k^2 + \gamma^2)}\left(\gamma \frac{\partial E_z}{\partial x} + \mathrm{j}\omega\mu \frac{\partial H_z}{\partial y}\right)$$
$$E_y = \frac{1}{(k^2 + \gamma^2)}\left(-\gamma \frac{\partial E_z}{\partial y} + \mathrm{j}\omega\mu \frac{\partial H_z}{\partial x}\right)$$
$$H_x = \frac{1}{(k^2 + \gamma^2)}\left(\mathrm{j}\omega\varepsilon \frac{\partial E_z}{\partial y} - \gamma \frac{\partial H_z}{\partial x}\right)$$
$$H_y = -\frac{1}{(k^2 + \gamma^2)}\left(\mathrm{j}\omega\varepsilon \frac{\partial E_z}{\partial x} + \gamma \frac{\partial H_z}{\partial y}\right)$$

(Eqn. 7-100)

The wave solutions above are categorized according to whether there is electric or magnetic field in the direction of propagation. There are three cases:

(i) TEM waves (transverse electromagnetic waves): $E_z = 0$ and $H_z = 0$. If this occurs, using the equations above, it would appear that all components become zero, but not so if $k^2 + \gamma^2 = 0$ as this is where the equations fail to be valid and we have the special solution (with $\gamma = \pm jk$): $E_x = E_0$ and $H_y = \frac{\gamma}{\mathrm{j}\omega\mu} E_0$. Thus:

$$E_x = E_0 \quad and \quad H_y = \pm\sqrt{\frac{\varepsilon}{\mu}} E_0 \,.$$

(ii) TM waves (transverse magnetic waves): $H_z = 0$.
(iii) TE waves (transverse electric waves): $E_z = 0$.

Let's apply the above categories to the transmission line examined earlier, that was represented as a parallel-plate capacitor ribbon in the direction of propagation made from a perfect conductor. (The edge effect is also eliminated from calculation by assuming an infinite ribbon width in many calculations, with no field change, accordingly, in the width dimension, e.g., y-direction.)

TM waves[34]
We have $H_z = 0$ and $E_z \neq 0$, thus: ${\nabla_t}^2 E_z = -(k^2 + \gamma^2)E_z$. Since $E_z = 0$ at the plate boundaries, we can easily solve to get:

$$E_z = A \sin n\pi x/a\,, \quad n = 1,2,3 \ldots,$$

and where 'a' is the plate separation. Using the equations above we then have the other EM components:

$$E_x = -\left(\frac{a\gamma}{n\pi}\right) A \cos\frac{n\pi x}{a} \qquad and \quad E_y = 0\,,$$

$$H_y = -\left(\frac{aj\omega\varepsilon}{n\pi}\right) A \cos\frac{n\pi x}{a} \qquad and \quad H_x = 0\,.$$

The different solutions $n = 1,2,3 \ldots$ correspond to the different TM 'modes', denoted TM_n.

Let's now examine the behavior of the TM modes (solutions) according to the propagation constant that can be determined (the all-important dispersion relation mentioned earlier, when it comes to pulse-based communications). We have:

$$\gamma = \sqrt{\left(\frac{n\pi}{a}\right)^2 - k^2} = \sqrt{\left(\frac{n\pi}{a}\right)^2 - \omega^2\mu\varepsilon} = j\omega\sqrt{\mu\varepsilon}\sqrt{1-(\omega_c/\omega)^2}\,,$$

where $\omega_c = \frac{n\pi}{a\sqrt{\mu\varepsilon}}$. This shows a remarkable result, for modes to be propagating, the propagation constant must be imaginary (recall 'j' is the imaginary in this notation). So, we only have propagating modal solutions if $\omega \geq \omega_c$. Clearly this provides a cutoff for propagation of modes with sufficiently high modal number but what of elimination of all such (TM modes)? This will occur for frequencies lower than $\omega_{c,1} = \frac{\pi}{a\sqrt{\mu\varepsilon}}$.

Let's use $\beta = -j\gamma$ and focus on the propagating conditions where $\omega > \omega_c$. Phase velocity is defined to be:

$$v_p = \frac{\omega}{\beta} = \frac{v}{\sqrt{1 - \left(\frac{\omega_c}{\omega}\right)^2}},$$

where $v = 1/\sqrt{\mu\varepsilon}$. Group velocity is defined to be:

$$v_g = \frac{d\omega}{d\beta} = v\sqrt{1 - \left(\frac{\omega_c}{\omega}\right)^2}.$$

Characteristic *wave* impedance is defined to be:

$$Z_{TM} = \frac{E_x}{H_y}.$$

For TEM waves this is simply:

$$Z_{TEM} = \sqrt{\frac{\varepsilon}{\mu}}.$$

For TM waves this is:

$$Z_{TM} = \frac{\beta}{\omega\varepsilon}.$$

TE waves[34]

For TE waves we have boundary condition that the normal (of the H field) must be zero, thus we have a decomposition in terms of cosines. We have $(k^2 + \gamma^2) = n\pi a$, the same relation as before for dispersion, thus all subsequent derivations will be the same, including $Z_{TE} = Z_{TM}$.

Interpretation in terms of parallel-plane modes [34]

To better understand what the TM and TE modes are describing, that is different than the TEM modes (which can always exist), let's reexamine the time-harmonic waves introduced. In the TEM propagation we found a single solution, spatially-harmonic along the cylinder, e.g., plane waves traveling along the z-axis of the cylindrical geometry. What if we considered plane waves traveling down the z-axis but at an angle, with reflection at the boundaries (due to the perfect conductor)? For such a wave to 'propagate' its multiple plane-wave reflections must be consistent. We thereby arrive at a Bragg-like reflection condition giving the same consistency relations as our TE and TM modal boundary conditions, and thus the same dispersion relation, from which everything else is the same as well. Thus, we can view the TE and TM modes as plane wave at appropriate off-axis angle for transmission.

Parallel-plate with non-ideal dielectric

A non-ideal dielectric results when, for example, convection currents are possible, causing losses. This can be captured by allowing the permittivity to be complex [34]. If we write $\varepsilon = \varepsilon' - j\varepsilon''$, the key dispersion relation becomes, for TEM waves:

$$\gamma_{TEM} = j\omega\sqrt{\mu(\varepsilon' - j\varepsilon'')} = j\omega\sqrt{\mu\varepsilon'} + \frac{1}{2}\omega\sqrt{\mu\varepsilon'}\frac{\varepsilon''}{\varepsilon'} + \cdots,$$

(Eqn. 7-101)

thus we have attenuation when we did not have it before, and the attenuation constant is given, for $\frac{\varepsilon''}{\varepsilon'}$ small, by:

$$\alpha_{TEM} \cong \frac{1}{2}\omega\sqrt{\mu\varepsilon'}\frac{\varepsilon''}{\varepsilon'}.$$

For TE and TM waves we have:

$$\gamma_{TE,TM} = \sqrt{\left(\frac{n\pi}{a}\right)^2 - \omega^2\mu(\varepsilon' - j\varepsilon'')}$$

$$= \sqrt{\left(\frac{n\pi}{a}\right)^2 - \omega^2\mu\varepsilon'}\sqrt{1 + j\omega^2\mu\varepsilon''\left[\left(\frac{n\pi}{a}\right)^2 - \omega^2\mu\varepsilon'\right]^{-1}}$$

(Eqn. 7-102)

again, for $\frac{\varepsilon''}{\varepsilon'}$ small:

$$\gamma_{TE,TM} \cong \sqrt{\left(\frac{n\pi}{a}\right)^2 - \omega^2\mu\varepsilon'}\left[1 + \frac{j\omega^2\mu\varepsilon''}{2}\left[\left(\frac{n\pi}{a}\right)^2 - \omega^2\mu\varepsilon'\right]^{-1} + \cdots\right],$$

(Eqn. 7-103)

and again, the complex contribution gives the attenuation:

$$\alpha_{TE,TM} \cong \alpha_{TEM}/\sqrt{1 - (\omega_c/\omega)^2}\ .$$

(Eqn. 7-104)

Rectangular Waveguides

Let's now consider rectangular waveguides. Upon enclosing the sides of the parallel plate transmission line examined previously we create a full enclosure and since no static E-field (non-zero) can exist inside a conductor, there is now no TEM solution. As for the TM and TE, we will find that the Laplacian can no longer dismiss one of the transverse dimensions, reducing to a simple 1D problem. Here we will have a 2D Laplacian. But with separation of variables it reduces to much the same result.

TM waves
Recall for transverse magnetic waves we have $H_z = 0$ and $E_z \neq 0$, thus: ${\nabla_t}^2 E_z = -(k^2 + \gamma^2)E_z$. Since $E_z = 0$ at the plate boundaries, we can easily solve to get:

$$E_z = A(\sin n\pi x/a)(\sin m\pi y/b), \quad n = 1,2,3 \ldots, and\ m = 1,2,3 \ldots$$

where the x-dimension of the guide is a and the y-dimension is b. Likewise, critical frequency generalizes from$\omega_c = \frac{n\pi}{a\sqrt{\mu\varepsilon}}$ to:

$$\omega_c = \frac{1}{\sqrt{\mu\varepsilon}}\left[\left(\frac{n\pi}{a}\right)^2 + \left(\frac{m\pi}{b}\right)^2\right]^{1/2} = \frac{k_c}{\sqrt{\mu\varepsilon}}.$$

(Eqn. 7-105)

From which we can write:

$$\alpha = k_c\sqrt{1 - (\omega/\omega_c)^2}\,, \quad \omega < \omega_c$$

and

$$\beta = k\sqrt{1 - (\omega_c/\omega)^2}\,, \quad \omega > \omega_c$$

with the velocities as before. The other field components for a given TM_{nm} mode are given by the equations derived previously, using the E_z solution above.

TE waves
As before with the parallel-plate transmission line we have solutions$E_z = 0$ and $H_z \neq 0$, thus: ${\nabla_t}^2 H_z = -(k^2 + \gamma^2)H_z$, and BC's as before, we have solutions:

$$H_z = B(\cos n\pi x/a)(\cos m\pi y/b), \quad n = 1,2,3 \ldots, and\ m = 1,2,3 \ldots,$$

but there is two additional solutions (that are non-trivial). Recall before that the $n = 0$ solution led to the wave vanishing so was excluded. Here we can have $n = 0$ as long as $m \neq 0$, and visa versa. Suppose a>b then the mode with the lowest cutoff frequency will be for n=1, m=0, e.g., the TE_{10} mode. In such a waveguide geometry (a>b) we see that the TE_{10} mode will dominate if the operating frequency is about 30% above the TE_{10} cutoff frequency. This will place all other modes in attenuation and will not be so close to the cutoff that large dispersive effects occur. Again, refer to the solution equations for the other components in terms of known H_z.

The dominant TE_{10} wave
For the dominant TE_{10} we have:

$$H_z = B(\cos \pi x/a)$$

$$E_y = -\frac{(j\omega\mu B)}{(\pi/a)}(\cos \pi x/a)$$
$$H_x = \frac{(j\beta B)}{(\pi/a)}(\cos \pi x/a)$$
and other three components zero. The cutoff frequency, wavelength, and wavenumber are:

$$f_c = \frac{1}{2a\sqrt{\mu\varepsilon}}, \quad \lambda_c = 2a, \quad k_c = \frac{\pi}{a}.$$

If the dielectric is imperfect, we use the complex permittivity as before, and since we have the same mathematical forms for the expansions, we have a similar result:

$$\alpha_{TE10} \cong \frac{\frac{k}{2}\frac{\varepsilon''}{\varepsilon'}}{\sqrt{1-\left(\frac{\lambda}{2a}\right)^2}}.$$

(Eqn. 7-106)

Example 7.11. ([34] 8.3f) Show the expression for energy velocity for TM_{m0} modes.

Solution

We have $V_E = \frac{W_T}{u}$ where average power flow for a width w is:

$$W_T = w\int_0^a \frac{1}{2}Re(H_x E_y^*)dx$$

Since for TM_{m0} modes we have:

$$H_x = \frac{jB}{*_c}B\sin k_c x \quad E_y = \frac{-jw\mu}{k_c}B\sin k_c x \qquad k_c = \frac{m\pi}{a}$$

then:

$$W_T = \frac{w}{2}B^2\beta w\mu\left(\frac{a}{m\pi}\right)^2\int_0^a \sin^2 k_c x dx = \frac{w}{2}B^2\beta w\mu\frac{a^2}{m^2\pi^2}\frac{a}{2}$$

The time-average energy storage per unit length is:

$$u = w\int_0^a \left\{\frac{\varepsilon}{4}\left|E_y\right|^2 + \frac{\mu}{4}[|H_z|^2 + |H_x|^2]\right\}dx = \frac{B^2 wa\mu k^2}{4k_c^2}$$

Thus,

$$V_E = \frac{\left(\frac{wa}{4}B^2\beta\omega\mu\frac{a^2}{m^2\pi^2}\right)}{\left(\frac{wa}{4}B^2k^2\mu\frac{a^2}{m^2\pi^2}\right)} = \frac{\beta\omega}{k^2} = \frac{k\sqrt{1-\left(\frac{\omega_c}{\omega}\right)^2}\cdot k/\sqrt{\mu\varepsilon}}{k^2}$$

$$= \frac{1}{\sqrt{\mu\varepsilon}}\sqrt{1-\left(\frac{\omega_c}{\omega}\right)^2}.$$

Example 7.12. ([34] 8.5a) Find average power transfer and conductor loss for a TE mode between parallel planes.

Solution

From previous problem we have:

$$W_T = \frac{w}{2}B^2\beta w\mu\frac{a^2}{m^2\pi^2}\frac{a}{2}$$

We also have for attenuation:

$$(\alpha_C)_{TE} = \frac{(W_L)_{TE}}{2(W_T)_{TE}}.$$

The current flow in both upper and lower plates is

$$|J_{sz}| = |H_z|_{x=0} = |H_z|_{x=a} = B$$

Thus,

$$(W_L)_{TM} = \frac{2wR_s}{2}|J_{sz}|^2 = \omega R_s B^2$$

Thus,

$$(\alpha_C)_{TE} = \frac{2R_s K_c^2}{a\omega\beta\mu},$$

and, using $k = \omega\sqrt{\mu\varepsilon}$, $\beta = k\sqrt{1-\left(\frac{\omega_c}{\omega}\right)^2}$, $\eta = \sqrt{\frac{\mu}{\varepsilon}}$:

$$(\alpha_C)_{TE} = \frac{2R_s}{a\eta}\frac{\left({}^{\omega_c}/_{\omega}\right)^2}{\sqrt{1-\left(\frac{\omega_c}{\omega}\right)^2}}.$$

Example 7.13. ([34 8.5c) Prove that the frequency for minimum attenuation for a TM_m mode is at $\sqrt{3}f_c$. Give the expression for the minimum attenuation and then apply it to silver conductors 2cm apart with air for dielectric, for modes 1 ,2, and 3.

Solution

We have $(\alpha_C)_{TM} = \frac{2R_S}{\eta a\sqrt{1-(w_c/w)^2}}$, $R_s = \sqrt{\frac{\pi f \mu}{\sigma}}$, *and* $2\pi f = \omega$, thus

$$(\alpha_C)_{TM} = \frac{2\sqrt{\frac{\omega\mu}{2\sigma}}}{\eta a\sqrt{1-(\omega_c/\omega)^2}}$$

Let $\frac{(\alpha_C)_{TM}}{\partial w} = 0$, then must satisfy:

$$\left(1 - \frac{\omega_C}{\omega}\right)\frac{1}{2} - \omega\frac{\omega_C^2}{\omega^3} = 0 \quad \rightarrow \quad \omega = \sqrt{3}\omega_c \quad \rightarrow \quad f = \sqrt{3}f_c$$

Substituting:

$$(\alpha_C)_{TM,min} = \sqrt{\frac{3\sqrt{3}\omega_c\mu}{\sigma\eta^2 a^2}}$$

For silver conductors $\sigma = 6.17 \times 10^7\ S/_m$ and with separation $a = .02m$ and recall:

$$k_c = \frac{m\pi}{a} \quad , \quad w_c\sqrt{\mu\varepsilon} = \frac{m\pi}{a} \quad \eta = 377,$$

thus,

$$(\alpha_C)_{TM,min} = \frac{5.04 \times 10^{-7}\sqrt{\frac{m}{2a\sqrt{\mu\varepsilon}}}}{377(.02)\sqrt{2/_3}} = \frac{.0574\sqrt{m}}{377(.02)\sqrt{2/_3}} = .00933\sqrt{m}$$

For $m = 1 (\alpha_C)_{TM,min} = 9.33 \times 10^{-3}$
For $m = 2 (\alpha_C)_{TM,min} = 1.32 \times 10^{-2}$
For $m = 3 (\alpha_C)_{TM,min} = 1.62 \times 10^{-2}$

Example 7.14. Show that the transverse components of E and H can be written as
$E_t = \frac{1}{k_C^2}[-\gamma\nabla_t E_z + j\omega\mu e_z \times \nabla_t H_z]$ and $H_t = \frac{-1}{k_C^2}[j\omega\varepsilon\hat{e}_z \times \nabla_t E_z + \gamma\nabla_t H_z]$.

Solution

Start with $E_x = \frac{-j}{k_c^2}\left(\beta\frac{\partial E_z}{\partial x} + \omega\mu\frac{\partial H_z}{\partial y}\right)$ and $E_y = \frac{j}{k_c^2}\left(-\beta\frac{\partial E_z}{\partial y} + \omega\mu\frac{\partial H_z}{\partial x}\right)$, and $\beta = -j\gamma$.

Some of the vector notation:

$$E = E_x + E_y + E_z \qquad \nabla E_z = \hat{x}\frac{\partial E}{\partial x} + \hat{y}\frac{\partial E}{\partial y} + \hat{z}\frac{\partial E}{\partial z}$$

$$E_t = E_x\hat{e}_x + E_y\hat{e}_y \qquad \nabla_t E_z = \hat{x}\frac{\partial E}{\partial x} + \hat{y}\frac{\partial E}{\partial y}$$

Substituting:

$$E_t = \frac{-j}{k_c^2}\left(-j\gamma\frac{\partial E_z}{\partial x} + \omega\mu\frac{\partial H_z}{\partial y}\right)\hat{x} + \frac{j}{k_c^2}\left(-(-j\beta)\frac{\partial E_z}{\partial y} + \omega\mu\frac{\partial H_z}{\partial x}\right)\hat{y}$$

regrouping to get the curls

$E_t = \frac{1}{k_c^2}\left[-\gamma\left(\frac{\partial E_z}{\partial x}\hat{x} + \frac{\partial H_z}{\partial y}\hat{y}\right) + j\omega\mu\left(\frac{\partial E_z}{\partial x}\hat{y} - \frac{\partial H_z}{\partial y}\hat{x}\right)\right] = \frac{1}{k_c^2}\left[-\gamma\nabla_t E_z + j\omega\mu\left(\frac{\partial E_z}{\partial x}(\hat{z}\times\hat{x}) + \frac{\partial H_z}{\partial y}(\hat{z}\times\hat{y})\right)\right]$

Thus,

$$E_t = \frac{1}{k_c^2}\left[-\gamma\nabla_t E_z + j\omega\mu\hat{z}\times\left(\frac{\partial E_z\hat{x}}{\partial x} + \frac{\partial H_z}{\partial y}\hat{y}\right)\right]$$

$$= \frac{1}{k_c^2}[-\gamma\nabla_t E_z + j\omega\mu e_z \times \nabla_t H_z].$$

Similarly,

$H_x = \frac{j}{k_c^2}\left(\omega\varepsilon\frac{\partial E_z}{\partial y} - \beta\frac{\partial H_z}{\partial x}\right)$ and $H_y = \frac{j}{k_c^2}\left(\omega\varepsilon\frac{\partial E_z}{\partial y} + \beta\frac{\partial H_z}{\partial x}\right)$ give rise to:

$$H_t = H_z\hat{e}_x + H_y\hat{e}_y$$

$$= \frac{1}{k_c^2}\left[j\omega\varepsilon\left(\frac{\partial E_z}{\partial y}\hat{e}_x - \frac{\partial E_z}{\partial x}\hat{e}_y\right) - j(-j\gamma)\left(\frac{\partial H_z}{\partial y}\hat{e}_x - \frac{\partial H_z}{\partial x}\hat{e}_y\right)\right]$$

thus

$H_t = \frac{1}{k_c^2}\left[j\omega\varepsilon\hat{e}_z \times \left(\frac{\partial E_z}{\partial y}\hat{e}_y - \frac{\partial E_z}{\partial x}\hat{e}_x\right) - \gamma\nabla_t H_z\right] = \frac{-1}{k_c^2}[j\omega\varepsilon\hat{e}_z \times \nabla_t E_z + \gamma\nabla_t H_z]$.

Example 7.15. ([34] 8.8d) Consider the maximum power carried by a $6 = GHZ\ TE_{10}$ wave is air guide 4cm wide 2cm high, taking the breakdown field in air at that frequency as $2 \times 10^6\ V/m$.

Solution
We have

$$Z_{TE} = \eta\left[1 - \left(\frac{w_c}{w}\right)^2\right]^{-1/2}$$

Thus,

$$Z_{TE} = 377\left[1 - \left(\frac{2.36 \times 10^{10}}{3.77 \times 10^{10}}\right)^2\right]^{-1/2} = 483.4\ ohms$$

We also have

$$W_T = \frac{E_0^2 ba}{4Z_{TE}}$$

Thus the max power is: $W_T = 1.655 \times 10^6 W = 1.66MW$.

7.4 Radiative Systems
7.4.1 Antenna Broadcast Systems

Fields and Radiation of a Localized Oscillating Source
By Fourier decomposition in time domain and superposition in EM we can, without loss of generality, proceed by studying:

$$\rho(\vec{x},t) = \rho(\vec{x})e^{i\omega t} \quad \text{(take real)}$$
$$\vec{J}(\vec{x},t) = \vec{J}(\vec{x})e^{i\omega t}$$

EM Potentials assumed to have the same time dependence.

In Lorentz Gauge:

$$\nabla^2 A - \frac{1}{c^2}\frac{\partial^2 A}{\partial t} = -\frac{4\pi}{c}\vec{J}.$$

Green's function analysis of:

$$\left(\nabla_x^2 - \frac{1}{c^2}\frac{\partial^2}{\partial t^2}\right)G^{(\pm)}(\vec{x},t;\vec{x}',t') = -4\pi\delta(\vec{x}-\vec{x}')\delta(t-t')$$

(Eqn. 7-107)

then yields:

$$G^{(\pm)}(\vec{x},t;\vec{x}',t') = \delta\frac{\left(t' - \left[t \mp \frac{|\vec{x}-\vec{x}'|}{c}\right]\right)}{|\vec{x}-\vec{x}'|},$$

(Eqn. 7-108)

for which we choose the negative sign (the Retarded Greens Function). By superposition:

$$\vec{A}(\vec{x},t) = \int d^3x' \int dt' \frac{J(\vec{x}',t')}{c|\vec{x}-\vec{x}'|}\delta(t' - t + |\vec{x}-\vec{x}'|/c).$$

(Eqn. 7-109)

Thus,

$\vec{A}(\vec{x},t) = \frac{1}{c}\int d^3x' \int dt' \frac{\vec{J}(\vec{x}',t')}{|\vec{x}-\vec{x}'|}\delta\left(t' + \frac{|\vec{x}-\vec{x}'|}{c} - t\right) =$
$\frac{1}{c}\int d^3x' \frac{\vec{J}(\vec{x}',t')}{|\vec{x}-\vec{x}'|}\int dt' e^{i\omega t'}\delta\left(t' + \frac{|\vec{x}-\vec{x}'|}{c} - t\right).$

The latter step formally allows integration over the delta, forcing $t' = t - \frac{|\vec{x}-\vec{x}'|}{c}$:

$$\vec{A}(\vec{x})e^{-i\omega t} = \frac{1}{c}\int d^3x' \frac{\vec{J}(\vec{x}')}{|\vec{x}-\vec{x}'|}e^{-i\omega t}e^{i\frac{\omega}{c}|\vec{x}-\vec{x}'|}$$
$$= \frac{1}{c}\int d^3x'\vec{J}(\vec{x}')\frac{e^{ik|\vec{x}-\vec{x}'|}}{|\vec{x}-\vec{x}'|}$$

(Eqn. 7-110)

where $k = \omega/c$. Since $\vec{A}$ defines the magnetic induction by $\vec{B} = \nabla \times \vec{A}$ and since:

$$\nabla \times H = \frac{4\pi}{c}\vec{J} + \frac{1}{c}\frac{\partial\vec{D}}{\partial t}$$

(with source dimension d and wavelength: $\lambda = 2\pi c/\omega$), if $d \ll \lambda$ and we consider $r \gg d$ then the current term is negligable outside the source and we have $\nabla \times B \cong \frac{1}{c}\frac{\partial E}{\partial t}$ which gives us

$$\vec{E} = \frac{i}{k}\nabla \times \vec{B}$$

Three spatial regions of interest:

Near (static) zone:	$d \ll r \ll \lambda$
Intermediate (induction) zone:	$d \ll r \sim \lambda$
Far (radiation) zone:	$d \ll \lambda \ll r$

Near zone

$r \ll \lambda \Rightarrow kr \ll 1 \Rightarrow e^{ik|\vec{x}-\vec{x}'|} \simeq 1$

So, $\lim_{kr\to 0} \vec{A}(\vec{x}) \cong \frac{1}{c}\int \frac{\vec{J}(\vec{x}')}{|\vec{x}-\vec{x}'|} d^3x'$

Using $\frac{1}{|\vec{x}-\vec{x}'|} = \sum_{\ell,m} \frac{4\pi}{2\ell+1} \quad Y_{\ell,m}(\theta,\phi)Y^*_{\ell m}(\theta',\phi')\left(\frac{r'^{\ell}}{r^{\ell+1}}\right)$, we get:

$$\vec{A}(\vec{x}) \simeq \frac{1}{c}\sum_{\ell,m} \frac{4\pi}{2\ell+1} \frac{Y_{\ell,m}(\theta,\phi)}{r^{\ell+1}} \int \vec{J}(\vec{x}')\, Y^*_{\ell m}(\theta',\phi') d^3x'$$

(Eqn. 7-111)

The near fields are quasi-stationary, oscillating harmonically as $e^{-i\omega t}$, but otherwise static in character.

<u>Far zone</u>

$r \gg \lambda \Rightarrow kr \gg 1$ cannot neglect $e^{ik|\vec{x}-\vec{x}'|}$, the exponential oscillates rapidly, in fact, and determines the behavior of the vector potential Consider $r \gg d$ we have $|\vec{x}-\vec{x}'| \simeq r - \vec{n}\cdot\vec{x}'$. We then have

$$\lim_{kr\to\infty} A(x) \cong \frac{e^{ikr}}{cr}\int d^3x' \vec{J}(\vec{x}') e^{-ik\vec{n}\cdot\vec{x}'},$$

(Eqn. 7-112)

where $\frac{e^{ikr}}{cr}$ describes an outgoing spherical wave and $\vec{J}(\vec{x}')$ has angular dependent coefficients. Corresponding $\vec{B}$ and $\vec{E}$ are transverse to the radius vector and fall off as r^{-1}. They thus correspond to radiation fields. For $d \ll \lambda$ we can expand:

$$\vec{A}(x) \simeq \frac{e^{ikr}}{cr}\sum_n \frac{(-ik)^n}{n!}\int \vec{J}(\vec{x}')\, (\vec{n}\cdot\vec{x}')^n d^3x'.$$

Considering the same development for the scalar potential:

$$\Phi(\vec{x},t) = \int d^3x' \int dt' \frac{\rho(\vec{x}',t)}{|\vec{x}-\vec{x}'|}\delta\left(t + \frac{|\vec{x}-\vec{x}'|}{c} - t\right)$$

For the electric monopole contribution $|\vec{x}-\vec{x}'| \to |x| \equiv r$. Thus,

$$\Phi_{\text{monopole}}(\vec{x},t) = \frac{q(t' = t - {}^{r}/_{c})}{r}$$

(Eqn. 7-113)

Since charge is conserved and a localized source is by definition one that does not have charge flowing into or away from it, the total charge q is independent of time. Thus the electric monopole part of the potential (and fields) of a localized source is of necessity static. The fields with harmonic time dependence $e^{-i\omega t}, \omega \neq 0$, have no monopole terms.

Electric Dipole Fields and Radiation

Consider the first term in $\vec{A}(\vec{x}) \simeq \frac{e^{ikr}}{cr} \sum_n \frac{(-ik)^n}{n!} \int J(\vec{x}')\,(\vec{n}\cdot\vec{x}')^n d^3x'$:

$$\vec{A}(\vec{x}) \simeq \frac{e^{ikr}}{cr}\int \vec{J}(\vec{x}')\,d^3x' = \frac{e^{ikr}}{cr}\left\{-\int \vec{x}\,(\vec{\nabla}'\cdot\vec{J})d^3x'\right\}$$

$$\simeq \frac{-i\omega}{cr}e^{ikr}\int \vec{x}'\,\rho(\vec{x}')d^3x'$$

(Eqn. 7-114)

Thus,

$$\vec{A}(\vec{x}) \simeq -ik\vec{p}\frac{e^{ikr}}{r} \qquad \vec{p} = \int \vec{x}'\,\rho(\vec{x}')d^3x'$$

The electric dipole fields that follow are:

$$\vec{B} = \nabla\times\vec{A} = k^2(\vec{n}\times\vec{p})\frac{e^{ikr}}{r}\left(1-\frac{1}{ikr}\right)$$

(Eqn. 7-115)

and

$$\vec{E} = \frac{i}{k}\nabla\times\vec{B} = ik\left\{-\frac{e^{ikr}}{r^2}\left(1-\frac{1}{ikr}\right)+\frac{e^{ikr}(ik)}{r}\left(1-\frac{1}{ikr}\right)\right.$$

$$\left.+\frac{e^{ikr}}{r}\left(\frac{1}{ikr^2}\right)\right\}$$

$$= ike^{ikr}\left\{\frac{ik}{r}-\frac{2}{r^2}+\frac{2}{ikr^3}\right\}(\vec{n}\times(\vec{n}\times\vec{p}))+\cdots$$

$$= k^2(\vec{n}\times\vec{p})\times\vec{n}\frac{e^{ikr}}{r^2}+2e^{ikr}\left(\frac{1}{r^3}-\frac{ik}{r^2}\right)+\cdots$$

Thus

$$\vec{E} = k^2(\vec{n}\times\vec{p})\times n\frac{e^{ikr}}{r}+[3\vec{n}(\vec{n}\cdot\vec{p})-\vec{p}]\left(\frac{1}{r^3}-\frac{ik}{r^2}\right)e^{ikr}.$$

(Eqn. 7-116)

Radiation zone:

$$\vec{B} = k^2(n\times p)\frac{e^{ikr}}{r}$$

(Eqn. 7-117)

and

$$\vec{E} = \vec{B}\times\vec{n}$$

(Eqn. 7-118)

Near zone:

$$\vec{B} = ik(n\times p)\frac{1}{r^2}$$

(Eqn. 7-119)

and

$\vec{E} = [3n(n \cdot p) - p]\frac{1}{r^3} \leftarrow$ dominates electric in near zone ($kr \ll 1$)

(Eqn. 7-120)

The time-averaged power radiated per unit solid angle by the oscillating dipole moment $\vec{p}$ is:

$$\frac{dP}{d\Omega} = \frac{c}{8\pi} Re\left[r^2 \vec{n} \cdot \vec{E} \times \vec{B}^*\right] = \frac{c}{8\pi} k^4 |(\vec{n} \times \vec{p}) \times \vec{n}|^2 = \frac{c}{8\pi} k^4 |p|^2 \sin^2\theta.$$

(Eqn. 7-121)

Thus

$$P = \frac{ck^4}{3}.$$

(Eqn. 7-122)

Consider a short, center-fed, linear antenna:

$$I(z)e^{-i\omega t} = I_0\left(1 - \frac{2|z|}{d}\right)e^{-i\omega t}$$

Using continuity

$$\nabla \cdot J = \frac{\partial \rho}{\partial t} = +i\omega\rho e^{-i\omega t} = \pm\frac{I_0 2}{d} e^{-i\omega t} \Rightarrow \rho = \pm\frac{2_i I_0}{\omega d}.$$

So,

$$P = \int_{-d/z}^{d/z} z\,\rho(z)dz = \frac{iI_0 d}{2\omega},$$

and

$$\frac{dP}{d\Omega} = \frac{I_0^2}{32\pi c}(kd)^2 \sin^2\theta.$$

Thus

$$P = \frac{I_0^2 (kd)^2}{12c} = \frac{I_0^2}{2}\left\{\frac{(kd)^2}{6c}\right\},$$

(Eqn. 7-123)

where the bracketed term is known as the radiation resistance.

So far we've discussed transmission lines and wave-guides. Let's now turn our attention to antenna radiated transmissions [34]. Even though there is no distributed circuit system to contain and direct the transmission, some directivity is possible with different antenna geometries giving rise to different field strength patterns. In designing an

antenna system and its utility for our purpose we need to know the following:
(1) The antenna field strength pattern.
(2) The maximum field strengths in the field strength pattern (to avoid dielectric breakdown and corona discharge).
(3) Radiation power output for given input voltage or current (and when taken vs input power, have radiator efficiency).
(4) The input impedance of the radiator (for matching purposes).
(5) The bandwidth given the above variables.
(6) The type of antenna.
(7) Is it an antenna array? (enables "electronic scanning").

Potentials for time-varying fields [34]
Whether time varying or not, we have $\nabla \cdot B = 0$ (no magnetic monopoles), thus the introduction of a magnetic vector potential is suggested where $B = \nabla \times A$ (since $\nabla \cdot \nabla \times A = 0$ is an identity). Substituting for B in the Maxwell equation with $\partial B/\partial t$ term we get:

$$\nabla \times (E + \partial A/\partial t) = 0 \rightarrow \quad E + \partial A/\partial t = -\nabla\Phi,$$

where for general solution we've introduced the scalar potential Φ. Using the Maxwell relation for D-field in presence of charge (electric monopole) source: $\nabla \cdot D = \rho$, we then get

$$-\nabla^2\Phi - \frac{\partial(\nabla \cdot A)}{\partial t} = \frac{\rho}{\varepsilon},$$

where ε is constant (assuming isotropy and homogeneity of dielectric, if present). Similarly, the $B = \nabla \times A$ substitution on the remaining Maxwell equation gives:

$$\nabla(\nabla \cdot A) - \nabla^2 \mathrm{A} = \mu \mathrm{J} - \mu\varepsilon\nabla(\partial\Phi/\partial t) - \mu\varepsilon(\partial^2 \mathrm{A}/\partial t^2).$$

Now, we chose $B = \nabla \times A$ because the identity $\nabla \cdot \nabla \times A = 0$ would be encountered favorably upon substitution in Maxwell's equations. We are not done with this identity, where in its other form we simply have $\nabla \times (\nabla \cdot X) = 0$. This indicates a freedom to replace A by $A + \nabla \cdot X$, etc., an realizing there still some freedom in the variables A and Φ that we've introduced, the last equation above suggests an obvious relation to adopt:

$$\nabla \cdot A = -\mu\varepsilon(\partial\Phi/\partial t),$$

which allows simplification to the following:

$$\nabla^2 \mathrm{A} - \mu\varepsilon\left(\frac{\partial^2 \mathrm{A}}{\partial t^2}\right) = -\mu \mathrm{J} \quad \textit{and} \quad \nabla^2\Phi - \mu\varepsilon\left(\frac{\partial^2\Phi}{\partial t^2}\right) = -\frac{\rho}{\varepsilon}.$$

In statics we had solution for the above in the form:

$$\Phi(\mathrm{x,y,z,t}) = \int_V \frac{\rho(\mathrm{x,y,z,t})dV}{4\pi\varepsilon r} \quad and \quad \mathrm{A}(\mathrm{x,y,z,t}) = \mu \int_V \frac{J(\mathrm{x,y,z,t})dV}{4\pi r}.$$

(Eqn. 7-124)

When dynamical we must account for the finite time of propagation of the effect of any change. We know from previous discussions that change in the EM field is comprised of waves that travel at the speed of light c. Thus it is reasonable to guess that the dynamical solution will have the form:

$$\Phi(\mathrm{x,y,z,t}) = \int_V \frac{\rho(\mathrm{x',y',z',t-r/c})dV'}{4\pi\varepsilon r} \quad and \quad \mathrm{A}(\mathrm{x,y,z,t}) = \mu \int_V \frac{J(\mathrm{x',y',z',t-r/c})dV'}{4\pi r}.$$

(Eqn. 7-125)

where the indicated potentials are known as the retarded potentials.

To be specific (using the example of [34]), let's consider the retarded potentials for the time-periodic case, using the phasor form with $e^{j\omega t}$, then the full set of Maxwell's equations reduce to:

$$B = \nabla \times A$$
$$E = -\nabla\Phi - \partial A/\partial t$$

$$\Phi(\mathrm{x,y,z,t}) = \int_V \frac{\rho(\mathrm{x',y',z'})e^{-jkr}dV'}{4\pi\varepsilon r}$$
$$\mathbf{A}(\mathrm{x,y,z,t}) = \mu \int_V \frac{J(\mathrm{x',y',z'})e^{-jkr}dV'}{4\pi r}$$
$$\nabla \cdot A = -\mathrm{j}\omega\mu\varepsilon\Phi$$

(Eqn. 7-126)

where $k = \omega/\mathrm{c}$.

Let's apply the above to the classic Hertzian dipole to start, then repeat for a general antenna. We have a wire, centered at the origin, oriented in the z-direction, of length 'h' (thus lying at –h/2 to +h/2 on the z-axis). Let's write the current variation in phasor notation: $I = I_0 e^{j\omega t}$. It is then trivial to solve the integral for 'A' (there is only a z-component because there is only a z-component of the current):

$$A_z = \mu \frac{hI_0 e^{-j\omega r/c}}{4\pi r} .$$

(Eqn. 7-127)

As noted by [34], we have the A_z as the only component in the Cartesian coordinate system. It is often convenient to express this result in spherical coordinates by a simple transformation at this juncture (prior to evaluating the fields, so that it might be determined in spherical coordinates). Thus, in spherical coordinates we have:

$$A_r = A_z \cos\theta = \mu \frac{hI_0 e^{-j\omega r/c}}{4\pi r} \cos\theta ,$$

(Eqn. 7-128)

and

$$A_\theta = -A_z \sin\theta = -\mu \frac{hI_0 e^{-j\omega r/c}}{4\pi r} \sin\theta .$$

(Eqn. 7-129)

Using the above formulae we can then get the electric and magnetic fields ($\eta = \sqrt{(\mu/\varepsilon)}$):

$$H_\phi = \frac{hI_0 e^{-j\omega r/c}}{4\pi} \left(\frac{j\omega}{rc} + \frac{1}{r^2} \right) \sin\theta ,$$

(Eqn. 7-130)

$$E_r = \frac{hI_0 e^{-j\omega r/c}}{4\pi} \left(\frac{2\eta}{r^2} + \frac{2}{j\omega\varepsilon r^3} \right) \cos\theta ,$$

(Eqn. 7-131)

$$E_\theta = \frac{hI_0 e^{-j\omega r/c}}{4\pi} \left(\frac{\eta}{r^2} + \frac{1}{j\omega\varepsilon r^3} + \frac{j\omega\mu}{r} \right) \sin\theta .$$

(Eqn. 7-132)

In the far-field zone, this simplifies to:

$$H_\phi = \frac{hI_0 e^{-j\omega r/c}}{4\pi} \left(\frac{j\omega}{rc} \right) \sin\theta ,$$

(Eqn. 7-133)

$$E_\theta = \frac{hI_0 e^{-j\omega r/c}}{4\pi} \left(\frac{j\omega\mu}{r} \right) \sin\theta = \eta H_\phi .$$

(Eqn. 7-134)

For the Poynting Vector in this simple case we have:

$$P_r = Re\left[E_\theta H_\phi^{\,*}\right] = \frac{\eta h^2 k^2 {I_0}^2}{32\pi^2 r^2} \sin^2\theta \qquad W/m^2$$

(Eqn. 7-135)

Total power:

$$W = \oint_S P \cdot dS = \int_0^{\pi} P_r 2\pi r^2 \sin\theta \, d\theta = \frac{\eta\pi h^2 {I_0}^2}{3\lambda^2}.$$

(Eqn. 7-136)

Let's now generalize this above result to currents on a general antenna when viewed in the far-zone. The observation point is chosen to be distance r from the origin. The integration volume dV' is at distance r' from the origin, and the distance from observation point to integration volume is r'', where:

$$r'' = \sqrt{r^2 + r'^2 - 2rr'\cos\varphi} \cong r - r'\cos\varphi.$$

The latter approximation holds for far-filed since $r \gg 0$ and if we have the origin on or near the observed charge, then we have $r \gg r'$ as well. So, starting with

$$\mathbf{A}(\mathrm{x}, \mathrm{y}, \mathrm{z}, \mathrm{t}) = \mu \int_{V'} \frac{\boldsymbol{J} e^{-jkr''} dV'}{4\pi r''},$$

(Eqn. 7-137)

we get the simplification:

$$\mathbf{A} \cong \frac{\mu e^{-jkr}}{4\pi r} \int_{V'} \boldsymbol{J} e^{jkr'\cos\varphi} dV'.$$

(Eqn. 7-138)

Following the notation of [34], let's define a "radiation vector" :

$$\mathbf{N} = \int_{V'} \boldsymbol{J} e^{jkr'\cos\varphi} dV', \quad thus \quad \mathbf{A} \cong \frac{\mu e^{-jkr}}{4\pi r}\mathbf{N}$$
$$= \frac{\mu e^{-jkr}}{4\pi r}\left(\hat{r}N_r + \hat{\theta}N_\theta + \hat{\phi}N_\phi\right)\big).$$

(Eqn. 7-139)

If we now consider this in the equation $\boldsymbol{B} = \nabla \times \mathbf{A}$, in spherical coordinates, dropping terms that decrease faster than 1/r, we get:

$$H_\theta = -\frac{1}{\mu}\frac{\partial}{\partial r}\left(rA_\phi\right) = \frac{jk}{4\pi r} e^{-jkr} N_\phi$$

(Eqn. 7-140)

$$H_\phi = \frac{1}{\mu r}\frac{\partial}{\partial r}(rA_\theta) = -\frac{jk}{4\pi r} e^{-jkr} N_\theta.$$

(Eqn. 7-141)

A similar analysis based off of the equation

$$E = -\frac{j\omega}{k^2}\nabla(\nabla \cdot A) - j\omega A,$$

and again to order (1/r):

$$E_\theta = -\frac{j\omega\mu}{4\pi r} e^{-jkr} N_\theta ,$$

(Eqn. 7-142)

and

$$E_\phi = -\frac{j\omega\mu}{4\pi r} e^{-jkr} N_\phi .$$

(Eqn. 7-143)

The Poynting vector is thus, asymptotically:

$$P_r = Re\left[E_\theta H_\phi{}^* - E_\phi H_\theta{}^*\right] = \frac{\eta}{8\lambda^2 r^2}\left[|N_\theta|^2 + \left|N_{\phi .}\right|^2\right].$$

(Eqn. 7-144)

The total (time average) power radiated is:

$$W = \int_0^\pi \int_0^{2\pi} P_r r^2 \sin\theta \, d\theta d\phi$$

$$= \frac{\eta}{8\lambda^2} \int_0^\pi \int_0^{2\pi} \left[|N_\theta|^2 + \left|N_{\phi .}\right|^2\right] \sin\theta \, d\theta d\phi .$$

(Eqn. 7-145)

Example 7.16. Show that a circularly polarized wave can result in the far field when a Hertzian dipole (I_0 oscillating on wire of length h) and a current loop (I on loop of radius a) are placed with shared origin and axis of symmetry, and the loop current is adjusted appropriately.

Solution

The electric and magnetic field components of the Hertzian dipole:

$$H_\phi = \frac{I_0 h}{4\pi} e^{-jkr}\left(\frac{jk}{r} + \frac{1}{r^2}\right)\sin\theta \cong \frac{I_0 h}{4\pi} e^{-jkr} \frac{jk}{r} \sin\theta$$

$$E_r = \frac{I_0 h}{4\pi} e^{-jkr}\left(\frac{2\eta}{r^2} + \frac{2}{j\omega\varepsilon r^3}\right)\cos\theta \cong 0$$

$$E_\theta = \frac{I_0 h}{4\pi} e^{-jkr}\left(\frac{j\omega\mu}{r} + \frac{1}{j\omega\varepsilon r^3} + \frac{\eta}{r^2}\right)\sin\theta \cong \frac{I_0 h}{4\pi} e^{-jkr} \frac{j\omega\mu}{r}$$

The electric and magnetic field components of the loop current:

$$E_\phi = -\frac{j\omega\mu I_a^2}{4} e^{-jkr}\left(\frac{jk}{r} + \frac{1}{r^2}\right)\sin\theta \cong \frac{\omega\mu I_a^2 e^{-jkr}}{4}\frac{k}{r}\sin\theta$$

$$H_r = \frac{j\omega\mu I_a^2}{4} e^{-jkr}\left(\frac{2}{\eta r^2} + \frac{2}{j\omega\varepsilon r^3}\right)\cos\theta \cong 0$$

$$H_\theta = \frac{j\omega\mu I_a^2}{4} e^{-jkr}\left(\frac{jw\varepsilon}{r} + \frac{1}{j\omega\varepsilon r^3} + \frac{2}{\eta r^2}\right)\sin\theta \cong \frac{\omega\mu I_a^2 e^{-jkr}}{4}\frac{\omega\varepsilon}{r}\sin\theta$$

The H_ϕ and H_θ are perpendicular and for them to be circularly polarized they need be 90° out of phase and have the same magnitude. Since one is pure imaginary, the other pure real, they must be 90° out of phase: $H_\phi = j\left(\frac{I_0 h e^{-jkr} k}{4\pi r}\right) \sin\theta \; and \; H_\theta = -\left(\frac{w^2 \mu\varepsilon I_0 a^2 e^{-jkr}}{4r}\right) \sin\theta$. The phase relation is also satisfied for the E-field: $E_\theta = j\left(\frac{I_0 h}{4\pi} e^{-jkr} \frac{\omega\mu}{r}\right) \sin\theta \; and \; E_\phi = +\left(\frac{\omega\mu I a^2 e^{-jkr} k}{4r}\right) \sin\theta$. To get their magnitudes to equal we must have:

$$I_0 = \frac{I a^2 k \pi}{h}.$$

Example 7.17. A sinusoidal current is passed thru a loop such that the current wavelength is equal to the loop circumference times 10. What current is required to radiate 100W?

Solution

We have for a loop:

$$E_\phi = \frac{-j\omega\mu I a^2}{4} e^{-jkr} \frac{jk}{r} \sin\theta$$

$$H_\theta = \frac{j\omega\mu I a^2}{4} e^{-jkr} \frac{j\omega\varepsilon}{r} \sin\theta$$

Thus,

$$P_r = \frac{1}{2} Re\left(-E_\theta H_\phi^*\right) = \frac{\omega\mu I^2 a^4 k^3 \sin^2\theta}{32 r^2}$$

For total power:

$$W = \oint_S P \cdot dS = \int_0^\pi P_r \, 2\pi r^2 \sin\theta \, d\theta = \frac{\omega\mu I^2 a^4 k^3 \pi}{16} \int_0^\pi \sin^3\theta \, d\theta$$

$$= \frac{\omega\mu I^2 a^4 k^3 \pi}{12}$$

Then,

$$W = 100 \;\rightarrow\; I = 100.7A.$$

Example 7.18. ([34] 12.4b) Previously it was claimed that $E = -\frac{j\omega}{k^2} \nabla(\nabla \cdot A) - j\omega A$, taken to order (1/r), gave rise to the formulae: $E_\theta = -\frac{j\omega\mu}{4\pi r} e^{-jkr} N_\theta$ and $E_\phi = -\frac{j\omega\mu}{4\pi r} e^{-jkr} N_\phi$. Let's now show that derivation.

Solution

We have $A = \frac{\mu e^{-jkr}}{4\pi r}(\hat{r}N_r + \hat{\theta}N_\theta + \hat{\phi}N_\phi)$ and $E = \frac{-j\omega}{k^2}\nabla(\nabla\cdot A) - j\omega A$, so let's begin by evaluating $\nabla \cdot A$ in spherical coordinates:

$$\nabla\cdot A = \frac{1}{r^2}\frac{\partial}{\partial r}(r^2 A_r) + \frac{1}{r\sin\theta}\frac{\partial}{\partial\theta}(\sin\theta\, A_\theta) + \frac{1}{r\sin\theta}\frac{\partial A_\phi}{\partial_\phi}$$

Ignoring all r-constant factors, and focusing on the dependence on r:

$$\nabla\cdot A = \frac{C_r}{r^2}\frac{\partial}{\partial r}\left(re^{-jkr}\right) + \frac{C_\theta}{r}\frac{e^{-jkr}}{r} + \frac{C_\phi}{r}\frac{e^{-jkr}}{r}$$

$$= \frac{C_r}{r^2}\left(r\cdot(-jk)e^{-jkr} + e^{-jkr}\right) + C_\theta\frac{e^{-jkr}}{r^2} + C_\phi\frac{e^{-jkr}}{r^2}$$

Recall the form of the gradient in spherical coordinates is $\nabla B = \hat{r}\frac{\partial B}{\partial r} + \hat{\theta}\frac{1}{r}\frac{\partial B}{\partial\theta} + \frac{\hat{\phi}}{r\sin\theta}\frac{\partial B}{\partial\phi}$, so we have:

$$\nabla(\nabla\cdot A) = \frac{C_{r_1}e^{-jkr}}{r} + C_{r_2}\frac{e^{-jkr}}{r^2} + \frac{C_{r_2}e^{-jkr}}{r^3}C_\theta\frac{e^{-jkr}}{r^3} + C_\phi\frac{e^{-jkr}}{r^3}$$

The only factor of $\nabla(\nabla\cdot A)$ that doesn't have a factor with a higher power than r^{-1} is due to A_r:

$$\nabla(\nabla\cdot A) \simeq \hat{r}\frac{\partial}{\partial r}\left[\frac{1}{r^2}\frac{\partial}{\partial r}\left(\frac{r^2\mu e^{-jkr}N_r}{4\pi r}\right)\right] \simeq \frac{\hat{r}\mu N_r}{4\pi}(-jk)\frac{\partial}{\partial r}\left(\frac{e^{-jkr}}{r}\right)$$

$$\simeq \frac{\hat{r}\mu N_r(-jk)^2}{4\pi}\cdot\frac{e^{-jkr}}{r}$$

Thus, we have:

$$\frac{-j\omega}{k^2}\nabla(\nabla\cdot A) \simeq \hat{r}j\omega\mu N_r\frac{e^{-jkr}}{4\pi r}$$

Thus,

$$E = \left(j\omega\mu N_r\frac{e^{-jkr}}{4\pi r}\right)\hat{r} - j\omega\mu\frac{e^{-jkr}}{4\pi r}\left(\hat{r}N_r + \hat{\theta}N_\theta + \hat{\phi}N_\phi\right)$$

$$= \frac{-j\omega\mu e^{-jkr}}{4\pi r}N_\theta\hat{\theta} - \frac{j\omega\mu e^{-jkr}N_\phi}{4\pi r}\hat{\phi}\,.$$

And, after cancellation on the E_r component, we have

$$E_\theta = \frac{-j\omega\mu e^{-jkr}}{4\pi r}N_\theta \quad and \quad E_\phi = \frac{-j\omega\mu e^{-jkr}N_\phi}{4\pi r}.$$

7.4.2 Hydrogen atom – radiative collapse

Let's consider the classical lifetime of the Bohr atom by examining the electric dipole radiation from the approximately circular motion according to the Larmor formula:

$$\frac{dU}{dt} = -\frac{2e^2a^2}{3c^3},$$

(Eqn. 7-146)

where the electron charge is – e and a is the acceleration. Let's begin by assuming non-relativistic and correcting later, so for circular orbit:

$$\frac{e^2}{r^2} = ma_r \cong m\frac{{v_\theta}^2}{r}, where\ a_r \cong a,$$

where we are using the adiabatic approximation where the orbit is nearly circular throughout). So,

$$\frac{dU}{dt} = -\frac{2}{3}\frac{{r_0}^3}{r^4}mc^3,$$

(Eqn. 7-147)

where $r_0 = \frac{e^2}{mc^2} = 2.8 \times 10^{-15}m$ is the classical electron radius. Since we also have:

$$U = -\frac{e^2}{r} + \frac{1}{2}mv^2 = -\frac{r_0}{r}mc^2,$$

(Eqn. 7-148)

upon substitution we get:

$$\dot{r} = -\frac{1}{r^2}\left(\frac{4}{3}{r_0}^2c\right) \rightarrow r^3 = {a_0}^3 - 4{r_0}^2ct.$$

(Eqn. 7-149)

Thus, the in-fall time is:

$$t_{infall} = \frac{{a_0}^3}{4{r_0}^2c} = 1.6 \times 10^{-11}s.$$

(Eqn. 7-150)

Since $|\dot{r}| \rightarrow \infty$ as $r \rightarrow \infty$, we clearly violate the non-relativistic assumption and require correction towards the end, once $|\dot{r}| > 0.1c$. If relativistic corrections are put in we have:

$$\frac{dU}{dt} = -\gamma^2\frac{2e^2a^2}{3c^3}, \qquad where\ \ \gamma = \frac{1}{\sqrt{1 - \frac{v^2}{c^2}}}.$$

We also have,

$$U = -\frac{e^2}{r} + \gamma mc^2.$$

If combined as before we now get:

$$\dot{r} = -\frac{1}{r^2}\left(\frac{4}{3}{r_0}^2 c\right)\left(1+\frac{3r_0}{2r}\right),$$

(Eqn. 7-151)

but this only indicates that in-fall is slightly faster than we have determined, towards the end. Thus, according to the classical theory (even relativistic) there should be atomic collapse, and it should be very fast. Obviously this is not what is observed, and it will take Quantum Mechanics to solve the problem (Book 4 [15]).

7.5 Exercises

(Ex. 7.1) A long slender N-turn solenoid of length L and radius b is partially filled with a cylinder of magnetic material whose permeability is μ , and with radius a (a<b). The material is coaxial with the solenoid and is of the same length. Find the expression of the inductance.

(Ex. 7.2) Show that the capacitance per unit length of two parallel cylinders, each of radius R, and whose center-to-center spacing is 2D, is given by:

$$C = \frac{\pi\epsilon_0}{\cosh^{-1}\left(\frac{D}{R}\right)} .$$

(Ex. 7.3) A cylindrical electron beam of uniform negative charge density $-\rho_0$, velocity $v = \sqrt{2eV/m}$, and radius a, is injected into a cylindrical conducting tube which is at a potential V and has a radius b. find an expression for the potential on the axis of the beam.

(Ex. 7.4) Re-derive Eqn.s 7-26 & 7-27.
(Ex. 7.5) Re-derive Eqn. 7-30.
(Ex. 7.6) Re-derive Eqn.s 7-36 & 7-37.
(Ex. 7.7) Re-derive Eqn. 7-59.
(Ex. 7.8) Re-derive Eqn. 7-68.
(Ex. 7.9) Re-derive Eqn. 7-69.
(Ex. 7.10) Re-derive Eqn. 7-70.
(Ex. 7.11) Re-derive Eqn. 7-71.
(Ex. 7.12) Re-derive Eqn. 7-76.
(Ex. 7.13) Re-derive Eqn. 7-82.
(Ex. 7.14) Re-derive Eqn. 7-91.
(Ex. 7.15) Re-derive Eqn. 7-95.
(Ex. 7.16) Re-derive Eqn. 7-96.
(Ex. 7.17) Re-derive Eqn. 7-97.

(Ex. 7.18) Re-derive Eqn. 7-98.
(Ex. 7.19) Re-derive Eqn. 7-99.
(Ex. 7.20) Re-derive Eqn. 7-100.
(Ex. 7.21) Re-derive Eqn.s 7-101, 7-102, 7-103 & 7-104.
(Ex. 7.22) Re-derive Eqn. 7-105.
(Ex. 7.23) Re-derive Eqn. 7-112.
(Ex. 7.24) Re-derive Eqn. 7-123.
(Ex. 7.25) Re-derive Eqn.s 7-128 & 7-129.
(Ex. 7.26) Re-derive Eqn.s 7-130, 7-131 & 7-132..
(Ex. 7.27) Re-derive Eqn.s 7-133 & 7-134.
(Ex. 7.28) Re-derive Eqn.s 7-135 & 7-136.
(Ex. 7.29) Re-derive Eqn.s 7-140 & 7-141.
(Ex. 7.30) Re-derive Eqn.s 7-142 & 7-143.
(Ex. 7.31) Re-derive Eqn.s 7-144 & 7-145.
(Ex. 7.32) Re-derive Eqn.s 7-149 & 7-150.

Chapter 8. The Lorentz Transform

8.1 Early hints of Special Relativity

What if particle-wave duality for matter was understood sooner? (maybe it was...)

The Lorentz transform was shown to exist for the electromagnetic wave description in 1899 [37]. In 1905 Einstein [46] indicated particle-wave duality for light (photons) to explain the photo-electric (Nobel in 19XX). If particle-wave duality extends to elementary massive particles, as well as the massless photon, then a transformation that changes one space-filling wave-field should do the same for another, whether it be massless or massive, and especially if coupled (which they are). Thus Lorentz invariance on the electromagnetic field Poynting 4-vector should be matched by invariance in any other field, thus invariance on the entire 'frame, e.g., special relativity. The requirement that matter and field be lock in-step under transformation is effectively what Einstein arrived at in 1905 with his proposal of special relativity [47]. It wasn't until the early quantum theory development in 1924 that DeBroglie [48] suggested that elementary particles should be wavelike as well (have interference phenomenon, for example), with momentum described in terms of such a wavelength by $p = \lambda/h$. So it is not until 1924 that there is the realization of particle-wave duality for all energy/matter (whether massless photon or massive electron).

What if 4-vectors were known to exist sooner?

The year 1924 was quite busy because this was also the year that Compton described the famous particle scattering experiments that demonstrated the Compton effect, which verified the special relativistic formulation combined with DeBroglie's proposed wave-particle duality. From a mathematics perspective, the theory involved a 4-vector formulation for the system energy-momentum to be conserved, i.e., it involved a 4-vector with invariance properties under Lorentz's transform. Had this experiment been done in the same year as the discovery of the Lorentz transform (1899) instead of 1924, then the concept of space-time that is implied would have evident much sooner.

Einstein is 20 years ahead of the rest
Einstein didn't have to wait until 1924 to have enough hints to discover special relativity. He was aware of prior work by Fizeau [49], which showed that velocity addition for light in moving water was not simply additive. Instead of thinking of Lorentz transformation shifting all the matter/energy in a frame, with special relativity there was transformation on the coordinate frame itself, thereby relating the frame seen at rest, say, and another in relative motion, in a transformation that wasn't simply additive (Galilean). From this perspective the discovery of 4-vector conservation under Lorentz transform then provides the basis for Einstein's 'covariant' theory of special relativity. The existence of the 4-vector (space-time) description would eventually be shown very clearly by Compton in his examination of photon-electron scattering (in 1924 [50]).

8.2 Lorentz Invariance: Space-Time; Special Relativity; Vectors and Spinors

4-vector invariance and relativistic coordinate-frame transformation
If we observe two space-time events, $\{A,B\}$, and determine their space-time separation in one frame that is at rest (without primed variables) and those in a moving frame (with primed variables), we have:

Boost in x-direction
$\Delta x' = x'_B - x'_A = \gamma(x_B - x_A) - \beta\gamma c(t_B - t_A);$
$\Delta y' = y'_B - y'_A = y_B - y_A;$
$\Delta z' = z'_B - z'_A = z_B - z_A;$
$\Delta t' = t'_B - t'_A = \gamma(t_B - t_A) - \beta\gamma(x_B - x_A)/c,$
where $\gamma = 1/\sqrt{1 - v^2/c^2}$, and $\beta = v/c$.

(Eqn.s 8-1)

The four-vector length with the above form of Lorentz transform (for a relative velocity boost in the x-direction), is invariant:
$(c\Delta t')^2 - (\Delta x')^2 - (\Delta y')^2 - (\Delta z')^2 = (c\Delta t)^2 - (\Delta x)^2 - (\Delta y)^2 - (\Delta z)^2.$

(Eqn. 8-2)

Similarly, if we consider the observed energy and momentum grouped as a four-vector, when there is a boost in the x-direction, there is the same component-level transformation:
$p_x = \gamma p_x - \beta\gamma E/c;$

$p'_y = p_y;$
$p'_z = p_z;$
$E' = \gamma E - \beta\gamma c p_x;$
with length invariant:

$$E^2 = (cp)^2 + (mc^2)^2$$

(Eqn. 8-3)

More formally, we can write the coordinate transformation under Lorentz transform as

$$x^{\mu'} = \Lambda^{\mu'}{}_{\upsilon} x^{\upsilon} ,$$

(Eqn. 8-4)

and the metric is invariant:

$$\eta = \Lambda^T \eta \Lambda.$$

(Eqn. 8-5)

If we write $\gamma = 1/\sqrt{1 - v^2/c^2} = cosh\ \varphi$, and $\gamma v = sinh\ \varphi$, then the prior x-boost transform appears as:

$$\Lambda^{\mu'}{}_{\upsilon} = \begin{pmatrix} cosh\ \varphi & -sinh\ \varphi & 0 & 0 \\ -sinh\ \varphi & cosh\ \varphi & 0 & 0 \\ 0 & 0 & 1 & 0 \\ 0 & 0 & 0 & 1 \end{pmatrix},$$

(Eqn. 8-6)

there are three such boosts. If we consider the more familiar rotation thru angle θ in the x-y plane (about the z-axis) we have the transformation:

$$\Lambda^{\mu'}{}_{\upsilon} = \begin{pmatrix} 1 & 0 & 0 & 0 \\ 0 & \cos\theta & \sin\theta & 0 \\ 0 & -\sin\theta & \cos\theta & 0 \\ 0 & 0 & 0 & 1 \end{pmatrix},$$

(Eqn. 8-7)

there are three such rotations (about the three axis), thus another three transformations. The Lorentz transform, thus, consists of six possible types, three boosts and three rotations. If we add translations (4 types) we have a total of 10 transformations that comprise the Poincare group of transforms.

The 4-vector description indicates a 4-dimensional space-time description in which these vectors can reside. With choice of Minkowski metric (Minkowski [51]), the speed of light c (in all frames) then defines a light cone structure, and thus the possible time-like 4-vectors. Under Lorentz transformation the time-like 4-vector is transformed into another time-like 4-vector.

Relativistic Doppler Effect

If we are in a Minkowski space-time (or in a 'locally flat' region that appears like Minkowski to the extent of our observation area), then we can directly derive a relativistic Doppler effect:

Consider source frame in prime and observer frame (at rest) without prime:

$$(dt')^2 = (dt)^2 - \frac{1}{c}(d\vec{x})^2 = (dt)^2\left(1 - \frac{v^2}{c^2}\right),$$

where latter holds when moving source oriented parallel to one of the coordinate axes. The direction of v is taken to be towards the observer. Thus, the time dilation is seen from the magnitude of the line element:

$$(\Delta t) = (\Delta t')\left(1 - \frac{v^2}{c^2}\right)^{-1/2}.$$

And if we consider the lightwave arriving at observer, the $\Delta\tau$ between wavecrests is:

$$\Delta\tau = \Delta t + \frac{v \cdot r}{c}\Delta t = \Delta t\left(1 - \frac{v}{c}\cos\theta\right),$$

where θ is angle from asymptotic source line (parallels line of source motion and goes through observer position) to current line. If source frequency is $\nu_s = 1/\Delta t'$ when $\Delta t'$ is one period, and observed frequency is $\nu = 1/\Delta\tau$, then we have:

$$\nu = \frac{\nu_s\left(1 - \frac{v^2}{c^2}\right)^{1/2}}{\left(1 - \frac{v}{c}\cos\theta\right)}.$$

(Eqn. 8-8)

Special cases:

(1) Source moving directly towards ($\theta = 0$): $\nu = \nu_s\sqrt{\frac{c+v}{c-v}}$.

(2) Source moving directly away ($\theta = 180$): $\nu = \nu_s\sqrt{\frac{c-v}{c+v}}$.

(3) Source passing at closest approach, motion is 'transverse' ($\theta = 90$): $\nu = \nu_s\sqrt{1 - \frac{v^2}{c^2}}$

Lorentz transform notation

More formally, we can write the coordinate transformation under Lorentz transform as

$$x^{\mu'} = \Lambda^{\mu'}{}_{\upsilon} x^{\upsilon} \,,$$

where $x^{\mu'}$ is a contravariant coordinate, and if the 4-vector length is invariant, then the metric is invariant:

$$\eta = \Lambda^T \eta \Lambda.$$

Any 4-vector entity z^{υ} that transforms as x^{υ}, is, similarly, called contravariant. This is to distinguish it from covariant 4-vectors that transform as:

$$z'_{\mu} = (\Lambda^{-1})^{\upsilon}{}_{\mu} z_{\upsilon}.$$

Note that for rotation transformations in 3D we have $R^{-1} = R^T$, so the '3-vectors' transform identically, and there is no differentiation of contravariant and covariant vectors. For the 4D Lorentz transform

$$\eta = \Lambda^T \eta \Lambda \quad \rightarrow \quad \Lambda^{-1} = \eta^{-1} \Lambda^T \eta$$

(Eqn. 8-9)

and Λ^{-1} and Λ^T are not the same.

The 4-vector reformulation of EM – implicit proof of Lorentz Invariance

If the 4-vector length is invariant under Lorentz transform, then so is the metric as indicated previously. Taken further, this invariance extends to any scalar or scalar product. The general transform properties for a full tensor generalization then follow (and are discussed at length in Ch. 9). For our purposes here, in showing that the Maxwell equations describe a Lorentz invariant theory, we must identify 4-vectors that might be formed from the electric and magnetic fields. Such 4-vectors, if grouped (inner product contraction) to form a scalar entity, will be manifestly Lorentz invariant. If such scalar entities can constructed that coincide with the Maxwell equation's, the proof of Lorentz invariance is complete by construction. Recall the component notation for the E and B fields in terms of potentials:

$$E_k = -\nabla_k \varphi - \frac{\partial A_k}{\partial t} \qquad and \qquad \epsilon_{ijk} B^k = \partial_i A_j - \partial_j A_i$$

(Eqn.s 8-10)

So far we have the 3-potential A_k and need to generalize it to a '4-potential'. Furthermore, the 3-gradient operator has a natural generalization to a 4-gradient operator (that transforms like a covariant vector). If we identify the 4-potential to be:

$$A_{\mu} = \left(-\frac{\varphi}{c}, A_k\right),$$

(Eqn. 8-11)

then (setting c=1):

$$E_k = \partial_k A_0 - \partial_0 A_k \,.$$

(Eqn. 8-12)

The forms E_k and $\epsilon_{ijk}B^k$ suggest that we examine the antisymmetric 2nd rank tensor formed from the 4-potential by the relation:

$$F_{\alpha\beta} = \partial_\alpha A_\beta - \partial_\beta A_\alpha,$$

(Eqn. 8-13)

where we see the E and B components reside in this structure:

$$\begin{matrix} & & t & x & y & z \end{matrix}$$

$$F_{\alpha\beta} = \begin{matrix} t \\ x \\ y \\ z \end{matrix} \begin{bmatrix} 0 & -E^1 & -E^2 & -E^3 \\ E^1 & 0 & B^3 & B^2 \\ E^2 & -B^3 & 0 & B^1 \\ E^3 & B^2 & -B^1 & 0 \end{bmatrix}$$

(Eqn. 8-14)

Let's consider a Lorentz transform corresponding to a boost in the x-direction (and no longer have c=1), then $\Lambda^0{}_0 = \gamma = \Lambda^1{}_1$, and $\Lambda^0{}_1 = -\gamma v/c = \Lambda^1{}_0$, and $\Lambda^2{}_2 = 1 = \Lambda^3{}_3$, with the rest zero. Let's see how the E an B components transform:

$$E'^1 = E^1 \;\; and \;\; B'^1 = B^1$$

$$E'^2 = \gamma(E^2 - vB^3) = \gamma(E^2 + (v \times B)^2)$$

$$B'^2 = \gamma\left(B^2 - \frac{v}{c}E^3\right) = \gamma\left(B^2 - \left(\frac{v}{c} \times E\right)^2\right)$$

$$E'^3 = \gamma(E^3 + (v \times B)^3)$$

$$B'^3 = \gamma\left(B^3 - \left(\frac{v}{c} \times E\right)^3\right)$$

(Eqn.s 8-15)

These are indeed the transformations consistent with the coordinate transform of Maxwell's equations (can be shown by considering an infinitesimal Lorentz boost as indicated). Since we began with a potential formulation, this is then equivalent, without loss of generality, to a proof of Lorentz invariance for the homogenous Maxwell equations. For the inhomogeneous Maxwell equations we have

$$\rho = \epsilon \nabla \cdot E = \epsilon \partial_k F^{0k} \;\; and \;\; J^i = \epsilon \partial_k F^{ik} + \epsilon \partial_0 F^{i0}$$

(Eqn.s 8-16)

thus we form the 4-vector for current to be:

$$J^\mu = (\rho, J^i),$$

which means that the inhomogeneous Maxwell equations can be written as a contravariant 4-vector expression (which is manifestly Lorentz invariant):

$$\epsilon \partial_{\upsilon} F^{\mu\nu} = \frac{1}{c} J^{\mu}.$$

(Eqn. 8-17)

For the effect of an electromagnetic field on a moving particle we have the Lorentz equation:

$$\frac{d\boldsymbol{p}}{dt} = q(\boldsymbol{E} + \boldsymbol{v} \times \boldsymbol{B}), \;\; \boldsymbol{p} = \gamma m \boldsymbol{v}.$$

(Eqn.s 8-18)

To get to a 4-vector formulation we already know that p generalizes to the 4-vector:

$$p^{\mu} = (E/c, \boldsymbol{p}).$$

(Eqn. 8-19)

On the RHS we have a $\boldsymbol{v} \times \boldsymbol{B}$ term. This suggests a tensor form involving $F^{\mu\nu}$ (to get the B^k term) and the velocity 4-vector $U^{\mu} = (\gamma c, \gamma \boldsymbol{v})$, for which we find:

$$q(\boldsymbol{E} + \boldsymbol{v} \times \boldsymbol{B}) \rightarrow \frac{q}{\gamma c} F_{j\mu} U^{\mu}$$

(Eqn. 8-20)

The 4-vector form of the Lorentz equation then becomes:

$$\frac{dv}{dt} = \frac{q}{\gamma c} F_{\nu\mu} U^{\mu}.$$

(Eqn. 8-21)

which is manifestly Lorentz invariant. Thus the Maxwell equations and the coupling of the EM Fields to charged matter are all Lorentz invariant. (Note that charge is a scalar, thus is seen to be the same in all Lorentz frames.)

8.3 Generalized Lorentz Invariance: 2x2 Hermitian and Complex Bi-Quaternion

Appearance of Lorentz-invariant Spinor Solutions in addition to Vector solutions

A mathematical subtlety that arises at this juncture that is not due to the novel three boosts introduced, but due to the three (standard) rotations. This is because the rotation group in 3D, SO(3), admits a double cover from the group SU(2), indicating two types of solutions. The relation between SO(3) and SU(2) is very important and will be covered in Sec. X, where more mathematical tools are made available to aid in the analysis. For now we focus on the double solution aspect, will this still be retained in the larger Lorentzian transform basis? We will now show that it does. We will find a new type of Lorentz invariant other than the 4-vectors already encountered that is known as a spinor. The easiest way to show that Lorentz invariance extends to a new mathematical object is to rewrite a 4-vector as a 2x2 Hermitian matrix, where V^a with components (V^0, V^1, V^2, V^3) is written as the 1-1 mapping as

$$\psi(V^a) = V^{AA'} = \begin{pmatrix} V^{00'} & V^{01'} \\ V^{10'} & V^{11'} \end{pmatrix} = \frac{1}{\sqrt{2}} \begin{pmatrix} V^0 + V^3 & V^1 + iV^2 \\ V^1 - iV^2 & V^0 - V^3 \end{pmatrix}$$

(Eqn. 8-22)

where the length of the 4-vector is related to the determinant of the matrix:

$$\det[\psi(V^a)] = \frac{1}{2}\eta_{ab}V^aV^b.$$

(Eqn. 8-21)

Thus, any transformation on the matrix that preserves the determinant, will be equivalent to a transformation that preserves the length. Since the Lorentz transform is given by the latter, then such transformations on the matrix will be equivalent to a Lorentz transformation. The matrix transformation that leaves the determinant unchanged is multiplication on the left by an element of SL(2,C) (a 2x2 matrix that is complex with unit determinant) and on the right by the Hermitian conjugate:

$$V^{AA'} \to \bar{V}^{AA'} = \mathrm{T}^A{}_B V^{BB'} \bar{\mathrm{T}}^{A'}{}_{B'}, \;\; where \; \bar{\mathrm{T}}^{A'}{}_{B'} = \overline{\mathrm{T}^A{}_B}\,.$$

(Eqn. 8-23)

Thus, there is a map from SL(2,C) to the Lorentz transforms where there is a 2-1 isomorphism. This can be seen to be directly related to the SU(2) double cover (2-1 mapping) on SO(3) as mentioned previously: consider the Lorentz transform leaving the time-like component of a chosen orthonormal tetrad invariant, what remains is the 3D rotation group SO(3). This is not simply a useful mathematical relation. We've identified the invariance under Lorentz transform as a fundamental element of the theory (giving rise to SR, etc.). ***If we see this then conveyed to an extended set of spinorial solutions, in addition to vectorial, then new forms of energy/matter are indicated for the spinorial solutions. This is precisely what is observed: matter is spinorial (spin $\pm 1/2$) and (force) fields are vectorial.***

Lorentz invariance via complex biquaternions [20-22,4]
Recall from previously that the length of a 4-vector V^a could be related to the determinant of a 2x2 complex matrix:

$$\det[\psi(V^a)] = \frac{1}{2}\eta_{ab}V^aV^b,$$

where any transformation on the matrix that preserves the determinant, will be equivalent to a transformation that preserves the length. Since the Lorentz transform is given by the latter, then such transformations on the matrix will be equivalent to a Lorentz transformation. Thus, multiplication on the left by an element of SL(2,C) (a 2x2 matrix that is complex with unit determinant) and on the right by its Hermitian conjugate:

$$V^{AA'} \to \tilde{V}^{AA'} = \mathrm{T}^A{}_B V^{BB'} \overline{\mathrm{T}}^{A'}{}_{B'}, \quad where\ \overline{\mathrm{T}}^{A'}{}_{B'} = \overline{\mathrm{T}^A{}_B},$$

will leave the determinant unchanged, thus the length invariant, thus map to a Lorentz transformation.
The map from SL(2,C) to the Lorentz transforms is a 2-1 isomorphism.

Let's now consider a similar process involving transformational invariance but instead of encoding the Lorentz transform in the form of matrix transformation invariance let's use elements of the Cayley algebras instead. Specifically, consider the following transformation:

$$q' = aqa_c^*, \quad where \quad aa_c = 1,$$

(Eqn. 8-24)

where a is a (unitary) complex bi-quaternion: $H(\mathbb{C}) \times H(\mathbb{C})$, and $q = (ct, ix, iy, iz)$. The $q' = (ct', ix', iy', iz')$ that results will correspond to a proper orthochronous Lorentz transform [20-22].

In Book 7 we see how the complex bi-quaternion transformation can be related to the complex octonion representation in the Emanator construct, thereby showing that Emanator theory is manifestly Lorentz invariant.

8.4 Spinor Math [52-55]

Let V be a vector space, V^* its dual vector space, and $V_C = V \otimes \mathbb{C}$ its complexification. Elements of V, denoted V^a, shall be referred to as "vectors." Elements of V^*, denoted V_a, shall be referred to as "covectors." The contraction mapping $V \times V^* \to \mathbb{R}$ is $(V^a, W_a) \to V^aW_a$ (Penrose index conventions begin with this, [56].) Suppose there is a Lorentzian metric, $diag(1, -1, -1, -1)$, η_{ab}, such that vectors are identified as time-like, space-like, or null, according to whether $\eta_{ab}V^aV^b$ is positive, negative or zero. The contraction mapping allows identification of vectors with covectors:

$$V^a \rightarrow V_a = \eta_{ab} V^b.$$

(Eqn. 8-25)

When a vector space is used it usually has a choice of orientation, which for 4D space amounts to a totally skew rank 4 tensor such that in a right-handed orthonormal frame $\{T^a, X^a, Y^a, Z^a\}$ satisfies:

$$\epsilon_{abcd} T^a X^a Y^a Z^a = 1,$$

(Eqn. 8-26)

where $\epsilon_{abcd} = \epsilon_{[abcd]}$ (thus $\epsilon_{abcd}\epsilon^{abcd} = -24$).

A skew-symmetric tensor of rank 2 is known as a bi-vector: F_{ab}. With the existence of the aforementioned η_{ab} and ϵ_{abcd} we can define the dual of a bivector as:

$$^*F^{ab} = -\frac{1}{2}\epsilon_{ab}{}^{cd} F_{cd},$$

(Eqn. 8-27)

where $^{**}F^{ab} = -F_{ab}$ (the eigenvalues of duality are complex).

Let's now consider the group of Lorentz transforms $L = O(1,3)$, where for element $\Lambda^a{}_b$, such that there is 4-vector length invariance:

$$(\Lambda^a{}_b Z^b)(\Lambda^c{}_d Z^d)\eta_{ac} = Z^b Z^d \eta_{bd},$$

(Eqn. 8-28)

which is accomplished by preserving the metric:

$$\Lambda^a{}_b \Lambda^c{}_d \eta_{ac} = \eta_{bd} \quad \rightarrow \quad \eta = \Lambda^T \eta \Lambda.$$

(Eqn. 8-29)

Note that in preserving the metric there are four disjoint sets of transformations according to whether $\det(\Lambda) = \pm 1$ and according to whether $\Lambda^0{}_0 < or > 0$. Since

$$\epsilon_{abcd} \Lambda^a{}_\alpha \Lambda^b{}_\beta \Lambda^c{}_\gamma \Lambda^d{}_\delta = \det(\Lambda)\, \epsilon_{\alpha\beta\gamma\delta},$$

(Eqn. 8-30)

if the transform has $\det(\Lambda) = -1$ it changes the orientation. Mostly interested in $\det(\Lambda) = 1$ and $\Lambda^0{}_0 > 0$, known as the proper orthochronous Lorentz group.

Continuing to use the notation and examples from [54]. What is the dimension of the Lorentz group? By examining infinitesimal transformations this can be determined easily. Consider $\Lambda = I + \epsilon S$ then $\eta = \Lambda^T \eta \Lambda \quad \rightarrow \quad S^T \eta + \eta S = 0$, which is 10 relations constraining S from 16 components to 6 free components, e.g., a 6 dimensional group.

Null Vectors and Spinors

Recall that

$$\det[\psi(V^a)] = \frac{1}{2}\eta_{ab}V^aV^b,$$

thus a null vector entails $\det[\psi(V^a)] = 0$, but a zero determinant means we have a matrix with rank 1, which means we can write the matrix as an outer product involving a single, two-element, complex vector, known as a spinor:

$$\psi(V^a) = V^{AA'} = \begin{pmatrix} V^{00'} & V^{01'} \\ V^{10'} & V^{11'} \end{pmatrix} = \begin{pmatrix} \alpha^0\bar{\alpha}^{0'} & \alpha^0\bar{\alpha}^{1'} \\ \alpha^1\bar{\alpha}^{0'} & \alpha^1\bar{\alpha}^{1'} \end{pmatrix} = \alpha^A\bar{\alpha}^{A'},$$

where $\bar{\alpha}^{A'} = \overline{\alpha^A}$ is the complex conjugate vector. We, thus, have spinor space S complex conjugate vector space S' and then we have the dual spaces for each of these: S^* and S'^*. Thus, SL(2,ℂ), a 2x2 complex matrix with det=1, acts on α^A a 2-element complex vector. Thus, we have in a more general example: $\bar{\alpha}^{A'B'C} = \overline{\alpha^{ABC'}}$. Of particular note is when a higher rank spinor is Hermitian.

Hermitian example: $\bar{\alpha}^{A'B'CD} = \alpha^{A'B'CD}$.

A tensor with valence n is 1-to-1 with a Hermitian spinor of rank 2n, with n contraspinor and n cospinor indices.

Since spinor space is 2D, skewing over more than two indices will give zero, and any skew two-index spinor will simply be the antisymmetric tensor ϵ_{ab} up to a complex multiple. Thus the antisymmetric tensor ϵ_{ab} will play an important role in spinor math. The determinant is also defined in terms of ϵ_{ab}. Consider for example the transformation of ϵ_{ab} when contracted with elements of $t_A{}^B$ ∈SL(2,ℂ):

$$t_a{}^b t_c{}^d \epsilon_{bd} = (\det t)\epsilon_{ac} = \epsilon_{ac}.$$

(Eqn. 8-31)

Note that ϵ_{ac} is preserved by an element of SL(2,ℂ) much like η_{ab} is preserved by the Lorentz group. From Penrose [57,58] we know that SL(2,ℂ) is 2-to-1 to the conformal group C(2) which is 1-to-1 to the proper orthochronous Lorentz group. The similar role of ϵ_{ac} in spinor length descriptions, to η_{ab} in vector length descriptions, is apparent in the relation between the associated dual spaces.

For vectors: $V^a \to V_a = \eta_{ab}V^b$, and for spinors: $\alpha^A \to \alpha_A = \alpha^B\epsilon_{BA}$, where $\epsilon^{AB}\epsilon_{CB} = \delta_C{}^A$.

Dyad

Since S is 2D, instead of a real tetrad have a complex dyad. Following the notation of [54], let's introduce the dyad (o_A,ι_A), normalized gives us:

$$\epsilon_{AB}\, o^A \iota^B = o_B \iota^B = 1.$$

(Eqn. 8-32)

We have the following relations between a normalized tetrad and a normalized dyad:

$$l^a = o^A\bar{o}^{A'}\,; \quad n^a = \iota^A\bar{\iota}^{A'}\,; \quad m^a = o^A\bar{\iota}^{A'}\,; \quad \bar{m}^a = \iota^A\bar{o}^{A'};$$

(Eqn.s 8-33)

where

$$\eta_{ab}l^a n^b = 1 = -\eta_{ab}m^a\bar{m}^b.$$

(Eqn.s 8-34)

Spinor reduction to symmetric and ϵ_{AB} factors

Spinor reduction is possible to a fully symmetric spinor with contractions associated with ϵ_{AB} factors. To see this first consider the skew contraction operation provided by

$$\epsilon_{AB}\epsilon^{CD} = \delta_A{}^C\delta_B{}^D - \delta_A{}^D\delta_B{}^C,$$

(Eqn. 8-35)

and then consider a spinor that is skew on indices CD:

$$\Phi_{...CD...} = \Phi_{...[CD]...}\ .$$

(Eqn. 8-36)

In such a case we have:

$$\epsilon_{AB}\epsilon^{CD}\Phi_{...CD...} = \epsilon_{AB}\Phi_{...C...}{}^{C}\ .$$

(Eqn. 8-37)

Thus, any indices that are skew can be contracted and replaced skew indices in a ϵ_{AB}factor. Taken to an extreme, any spinor can be reduced to a symmetric spinor, with ϵ_{AB}factors, and that symmetric tensor can always be written as a fully symmetrized outer product of valence 1 spinors (thus, a valence n spinor is constructed from n valence 1 spinors):

$$\Phi_{A...B} = \alpha_{(A} \ldots \beta_{B)}\ .$$

(Eqn. 8-38)

Each of the valence 1 spinors is a null vector and together they comprise the principle null directions of $\Phi_{A...B}$.

Tensors, Bi-vectors, and Spinors

Consider a valence 2 tensor that is mapped to a valence 4 spinor:

$$\mathrm{T}_{ab} = \mathrm{T}_{AA'BB'} = \mathrm{T}_{ABA'B'},$$

(Eqn. 8-39)

where relative order of primed indices and unprimed indices does not matter. Thus,

$$\mathrm{T}_{ab} = \mathrm{T}_{(AB)A'B'} + \frac{1}{2}\epsilon_{AB}\mathrm{T}_{C}{}^{C}{}_{A'B'}.$$

(Eqn. 8-40)

The relation can be repeated on the primed indices and we have:

$$\mathrm{T}_{ab} = \mathrm{T}_{(AB)(A'B')} + \frac{1}{2}\epsilon_{AB}\mathrm{T}_{C}{}^{C}{}_{(A'B')} + \frac{1}{2}\epsilon_{A'B'}\mathrm{T}_{(AB)C}{}^{C} + \frac{1}{4}\epsilon_{AB}\epsilon_{A'B'}\mathrm{T}_{CC'}{}^{CC'}.$$

(Eqn. 8-41)

A bivector is a skew symmetric tensor of valence two, thus having the form:

$$\mathrm{T}_{[ab]} = \frac{1}{2}\epsilon_{AB}\mathrm{T}_{C}{}^{C}{}_{(A'B')} + \frac{1}{2}\epsilon_{A'B'}\mathrm{T}_{(AB)C}{}^{C}.$$

(Eqn. 8-42)

Thus, generally speaking, a bivector T_{ab} can be written:

$$\mathrm{T}_{ab} = \Phi_{AB}\epsilon_{A'B'} + \Phi_{A'B'}\epsilon_{AB},$$

(Eqn. 8-43)

where Φ_{AB} is symmetric. Using similar derivations we can show that:

$$\epsilon_{abcd} = i(\epsilon_{AC}\epsilon_{BD}\epsilon_{A'D'}\epsilon_{B'C'} - \epsilon_{AD}\epsilon_{BC}\epsilon_{A'C'}\epsilon_{B'D'}),$$

and

$$\epsilon_{ab}\epsilon^{cd} = i(\delta_{A}{}^{C}\delta_{B}{}^{D}\delta_{A'}{}^{D'}\delta_{B'}{}^{C'} - \delta_{A}{}^{D}\delta_{B}{}^{C}\delta_{A'}{}^{C'}\delta_{B'}{}^{D'}).$$

(Eqn.s 8-44)

Spinor Examples

Example 8.1. Show that the normalization of the spinor of ϵ is -24, where:

$$\epsilon_{A\grave{A}B\grave{B}C\grave{C}D\grave{D}} = i(\epsilon_{AB}\epsilon_{CD}\epsilon_{\grave{A}\grave{C}}\epsilon_{\grave{B}\grave{D}} - \epsilon_{AC}\epsilon_{BD}\epsilon_{\grave{A}\grave{B}}\epsilon_{\grave{C}\grave{D}}.$$

Solution

$$\epsilon_{A\grave{A}B\grave{B}C\grave{C}D\grave{D}}\epsilon^{A\grave{A}B\grave{B}C\grave{C}D\grave{D}}$$
$$= -((\epsilon_{AB}\epsilon^{AB})^4 + (\epsilon_{AC}\epsilon^{AC})^4 - 2(\epsilon_{AB}\epsilon^{AC}\epsilon_{CD}\epsilon^{BD})^2)$$
$$= -(2(2)^4 - 2(2)^2)$$

Thus,

$$\epsilon_{A\grave{A}B\grave{B}C\grave{C}D\grave{D}}\epsilon^{A\grave{A}B\grave{B}C\grave{C}D\grave{D}} = -24.$$

Example 8.2. Show that if $F_{\alpha\beta}$ is null, with principle null direction along $K^\alpha = \alpha^A\overline{\alpha}^{\dot{A}}$, then

$$F_{\alpha\beta}K^\beta = 0 = {}^*F_{\alpha\beta}K^\beta\,,$$

and conclude that K^β is orthogonal to E^α and B^α.

Solution

For $F_{\alpha\beta}$ an antisymmetric tensor we can write

$$F_{\alpha\beta} \leftrightarrow F_{A\dot{A}B\dot{B}} \text{ where } F_{A\dot{A}B\dot{B}} = \epsilon_{AB}\overline{\phi}_{\dot{A}\dot{B}} + \epsilon_{\dot{A}\dot{B}}\phi_{AB}$$

With ϕ_{AB} a symmetric two index tensor. Since any symmetric n-index spinor $\phi_{A....B} = \phi_{(A....B)}$ can be written as a symmetrized product of spinors:

$$\phi_{AB} = \alpha_{(A}\beta_{B)}$$

If β is proportional to α then $\phi_{AB} = \alpha_{(A}\alpha_{B)}$ satisfies $\phi_{AB}\phi^{AB} = 0 \ \left(\Rightarrow F_{\alpha\beta}F^{\alpha\beta} = 0\right)$ and the field is called null. Here we are given that $F_{\alpha\beta}$ is null so the ϕ_{AB} associated with it (given principle null direction $\alpha^K\overline{\alpha}^K$) is $\phi_{AB} = \alpha_{(A}\alpha_{B)}$. So,

$F_{\alpha\beta}K^\beta = \left(\epsilon_{AB}\overline{\phi}_{\dot{A}\dot{B}} + \epsilon_{\dot{A}\dot{B}}\phi_{AB}\right)\alpha\overline{\alpha}^{\dot{B}} = \epsilon_{AB}\alpha_{(\dot{A}}\alpha_{\dot{B})}\alpha^B\overline{\alpha}^{\dot{B}} + \epsilon_{\dot{A}\dot{B}}\alpha_{(A}\alpha_{B)}\alpha^B\overline{\alpha}^{\dot{B}}$

$$= \epsilon_{AB}\alpha\left(\alpha_{\dot{B}}\overline{\alpha}^{\dot{B}}\right)\alpha^B + \epsilon_{\dot{A}\dot{B}}\alpha_A(\alpha_B\alpha^B)\overline{\alpha}^{\dot{B}} = 0$$

Note that $\alpha_B\alpha^B = \alpha_B\alpha_B\epsilon^{BB} = 0 \ since \ \ \epsilon^{BB} = 0$, so:

$$F_{\alpha\beta}K^\beta = 0 \Rightarrow \epsilon^\alpha = F^{0\alpha} \Rightarrow \epsilon_\alpha K^\alpha = -F_{0\alpha}K^\alpha = 0 \Rightarrow \epsilon^\alpha \perp K^\alpha\,.$$

Similarly, we have:

$${}^*F_{\alpha\beta} \leftrightarrow {}^*F_{A\dot{A}B\dot{B}} = i\left(\epsilon_{AB}\overline{\phi}_{\dot{A}\dot{B}} - \epsilon_{\dot{A}\dot{B}}\phi_{AB}\right)$$

So, ${}^*F_{\alpha\beta}K^\beta = -\left(\epsilon_{AB}\overline{\alpha}_{\dot{A}}(\overline{\alpha}_{\dot{B}}\overline{\alpha}^{\dot{B}})\alpha^B - \epsilon_{\dot{A}\dot{B}}\alpha_A(\alpha_B\alpha^B)\overline{\alpha}^{\dot{B}}\right) = 0$. Thus

$$B^\alpha = {}^*F^{\alpha 0} \Rightarrow B_\alpha K^\alpha = -{}^*F_{\alpha 0}K^\alpha = 0 \Rightarrow B^\alpha \perp K^\alpha\,.$$

Example 8.3. Show that he Weyl tensor satisfies:

$$C_{\alpha\beta\gamma\delta} + i^*C_{\alpha\beta\gamma\delta} = \epsilon_{\dot{A}\dot{B}}\epsilon_{\dot{C}\dot{D}}\Psi_{ABCD}\,,$$

Where ${}^*C_{\alpha\beta\gamma\delta} = \frac{1}{2}\epsilon_{\gamma\delta}^{\varepsilon\zeta}C_{\alpha\beta\varepsilon\zeta}$.

Solution

We know $C_{\alpha\beta\gamma\delta} = \epsilon_{\dot{A}\dot{B}}\epsilon_{\dot{C}\dot{D}}\Psi_{ABCD} + \epsilon_{AB}G_{CD}\Psi_{\dot{A}\dot{B}\dot{C}\dot{D}}$

Where $\Psi_{ABCD} = \Psi_{(ABCD)}$

Furthermore, much as in the duality notation analysis:

$$^*C_{\alpha\beta\gamma\delta} = \frac{1}{2}\epsilon^{\varepsilon\zeta}_{\gamma\delta}C_{\alpha\beta\varepsilon\zeta}$$

$$= \frac{i}{2}\left(\epsilon_{CD}\epsilon^{EF}\epsilon^{\dot{E}}_{\dot{C}}\epsilon^{\dot{F}}_{\dot{D}} - \epsilon_{\dot{C}\dot{D}}\epsilon^{\dot{E}\dot{F}}\epsilon^{E}_{C}\epsilon^{F}_{D}\right) \times \left(\epsilon_{\dot{A}\dot{B}}\epsilon_{\dot{E}\dot{F}}\Psi_{ABEF} + \epsilon_{AB}\epsilon_{EF}\overline{\Psi}_{\dot{A}\dot{B}\dot{E}\dot{F}}\right)$$

$$^*C_{\alpha\beta\gamma\delta} = \frac{i}{2}\left(2\delta^{E}_{[C}\delta^{F}_{D]}E^{\dot{E}}_{\dot{C}}C^{\dot{F}}_{\dot{D}} - 2\delta^{\dot{E}}_{[\dot{C}}\delta^{\dot{F}}_{\dot{D}]}E^{E}_{C}E^{F}_{D}\right)C_{A\dot{A}B\dot{B}E\dot{E}F\dot{F}}$$

$$= -i(\epsilon_{\dot{A}\dot{B}}\epsilon_{\dot{C}\dot{D}}\Psi_{ABCD} - \epsilon_{AB}\epsilon_{CD}\Psi_{\dot{A}\dot{B}\dot{C}\dot{D}})$$

Thus,

$$C_{\alpha\beta\gamma\delta} + i^*C_{\alpha\beta\gamma\delta} = 2\epsilon_{\dot{A}\dot{B}}\epsilon_{\dot{C}\dot{D}}\Psi_{ABCD}\,.$$

Example 8.4. Show that the Bianchi identity for the Weyl spinor has the form: $\nabla_{A\dot{A}}\Psi^{ABCD} = 0$.

Solution

The Bianchi identity is: $\nabla_{[\alpha}R_{\beta\gamma]\delta E} = 0 \Rightarrow \nabla_{[\alpha}C_{\beta\gamma]\delta E} = 0$. So,

$$\nabla_{\alpha}C^{\alpha}_{\beta\gamma\delta} = 0 \quad \Rightarrow \quad \nabla_{\delta}{}^*C^{\alpha\beta\gamma\delta} = 0 \qquad \left({}^*C_{\alpha\beta\gamma\delta} = \frac{1}{2}\epsilon^{\varepsilon\zeta}_{\gamma\delta}C_{\alpha\beta\varepsilon\zeta}\right)$$

So, $\nabla_{\alpha}\left(C^{\alpha\beta\gamma\delta} + i^*C^{\alpha\beta\gamma\delta}\right) = 0$, or $\nabla_{A\dot{A}}\left(2\epsilon^{\dot{A}\dot{B}}\epsilon^{\dot{C}\dot{D}}\psi^{ABCD}\right) = 0$, thus:

$$\nabla_{A\dot{A}}\Psi^{ABCD} = 0\,.$$

8.5 Exercises

(Ex. 8.1) Verify Eqn. 8-2.
(Ex. 8.2) Verify Eqn. 8-3.
(Ex. 8.3) Verify Eqn. 8-4.
(Ex. 8.4) Verify Eqn. 8-5.
(Ex. 8.5) Verify Eqn. 8-6.
(Ex. 8.6) Verify Eqn. 8-7.
(Ex. 8.7) Verify Eqn. 8-8.
(Ex. 8.8) Verify Eqn. 8-10.
(Ex. 8.9) Verify Eqn. 8-14.
(Ex. 8.10) Verify Eqn. 8-15.
(Ex. 8.11) Verify Eqn. 8-16.
(Ex. 8.12) Verify Eqn. 8-17.
(Ex. 8.13) Verify Eqn. 8-21.
(Ex. 8.14) Verify Eqn. 8-44.
(Ex. 8.15) Verify Eqn. 8-48.

Chapter 9. Dynamics from Geometry

In Ch. 8 we saw the fundamental aspect of the Lorentz transform. In the descriptions explored there the invariance of lengths, and scalars, under the Lorentz transform was shown to be equivalent to a having the transform leave an underlying 'flat' metric (the Minkowski metric) invariant. The obvious generalization to any space-time, would then be to consider the general Riemannian metric as starting point, with invariance under Lorentz transform. The non-constant metric then might capture dynamical aspects of the geometry, e.g., we can begin to explore General Relativity (GR). Most of the explorations of GR will be left to Book 3 [5], where more sophisticated analysis involving Manifold constructs compatible with spinor fields will be considered. In the remaining chapters (Ch. 9, 10, 11) of this book, moreover, the focus is on Gauge Fields and their geometric properties, connecting to the generalizations of the EM field, the Yang-Mills fields. This then covers the main gauge fields seen in nature, the basis of the standard model, in terms of the groups U(1) , SU(2), and SU(3), which will then be developed in Book 5 on QFT to describe the Standard Model of elementary particles and their interactions. In this chapter we continue to build out the basic theory to prepare for the more advanced issues indicated, starting with justification of the choice of generalization to the Riemannian form of the metric.

In this chapter *concrete* indices will be used (the old-fashioned approach), and definitions of fundamental aspects of tensor algebra will be developed in that context. The notes adopt the conventions and notation of Adler [59] for this initial description. In chapters 10 & 11 the mathematics switches to *abstract* indices to be more tractable. Abstract indices are described in detail in App. B in detail, as well as reformulation of the tensor algebra definitions using abstract indices.

9.1 The choice of Riemannian geometry

In studying surfaces in three dimensional Euclidean space Gauss introduced a general parametric representation of surfaces ($\{u_1, u_2\}$), where intrinsic geometric features are expressed independent of surface coordinates, including distance as

$$ds = \left(\sum_{i,k}^{2} g_{ik} du_i du_k\right)^{1/2}.$$

In 1854 Riemann [60] publishes "On the hypotheses which lie at the Foundation of Geometry." Riemann points out the restriction of Gauss to the case of two surface coordinates was not necessary and only motivated by his focus on two-dimensional surfaces in a three-dimensional space. Riemann proposed the n-dimensional, but still quadratic, generalization:

$$ds = \left(\sum_{i,k}^{n} g_{ik} du_i du_k\right)^{1/2}.$$

(Eqn. 9-1)

The quadratic form is a function of the $du_i's$, e.g. $ds = F(du_i)$, but more general would be not just quadratic and also have explicit coordinate dependencies: $ds = F(u_i, du_i)$. Such general spaces are known as Finsler spaces. It was pointed out by the famous mathematician Helmholtz, however, that most Finsler geometries can be excluded since they lack symmetries. In fact, the simple existence of systems where we rotate a "rigid body" about a fixed point require transformations to exist that are length preserving. We will now show, following a discussion from [59], that requiring distance-preserving transformations to exist requires the geometry to be Riemannian.

We assume a metric function exists in which three-parameter groups of distance-preserving transformation are possible (for rigid body). Suppose near a fixed point O we have $ds = F(dx_i)$. Introduce a local coordinate system around the "origin" O. We then have a three parameter family of transformations: $\tilde{x}_i = f_i(x_k, p_j)$ which depends on the three parameters p_j:

$$d\tilde{x}_i = \sum_{k=1}^{3} \left(\frac{\partial f_i}{\partial x_k}\right)_{x_k=0} dx_k = \sum_{k=1}^{3} \alpha_{ik}(p_j) dx_k\,,$$

and now $F(d\tilde{x}_i) = F(dx_i)$. So, to begin, consider:

$$F(x_i) = F\left(\sum_{K=1}^{3} \alpha_{iK} x_K\right) \rightarrow now\text{ use } \varepsilon's \rightarrow F(\varepsilon_i) = F\left(\sum_{K=1}^{3} \alpha_{iK} \varepsilon_K\right)$$

Given an arbitrary plane in the ε-space, there exists still a one parameter subgroup of those transformations which carries the plane into itself. Assume $\varepsilon_3 = 0$ (no loss of generality). Since $\tilde{\varepsilon}_i = \sum_{K=1}^{3} \alpha_{iK}(P_1, P_2, P_3)\varepsilon_K$ we must have: $\alpha_{31}(P_1, P_2, P_3) = 0$ and $\alpha_{32}(P_1, P_2, P_3) = 0$, thus

P_2 and P_3 can be determined as functions of the remaining free parameter $P = P_1$. We consider only $F(\varepsilon_i)$ positive definite. Now, with $\varepsilon_3 = 0$:

$$\tilde{\varepsilon}_i = \sum_{K=1}^{2} \alpha_{iK}(p)\varepsilon_K \qquad i = 1,2.$$

Let $F(\varepsilon_1, \varepsilon_2, 0) = \Phi(\varepsilon_1, \varepsilon_2, 0)$ and, so, $\Phi(\tilde{\varepsilon}_i) = \Phi(\varepsilon_i)$. We can assume without loss of generality that $p = 0$ corresponds to the identity transformation and that the matrices depend differentiably upon the parameter p:

$$\tilde{\varepsilon}_i(p) = \varepsilon_i + p\sum_{K=1}^{2} C_{iK}\varepsilon_K + O(p^2)$$

where C_{iK} is the derivative matrix of α_{iK} at $p = 0$. Now, differentiate $\Phi(\tilde{\varepsilon}_i) = \Phi(\varepsilon_i)$ with respect to p and put $p = 0$ to find the identity (also, dropping tildas):

Define the characteristic curve $\varepsilon_i(t)$ by $\frac{\partial \varepsilon_i}{\partial t} = \sum_{K=1}^{2} C_{iK}\varepsilon_K$, $i = 1,2$, then:

$$\frac{d}{dt}\Phi\varepsilon_i(t) = 0 \quad \longrightarrow \quad \sum_{i=1}^{2}\frac{\partial \Phi}{\partial \varepsilon_i}\left(\sum_{K=1}^{2} C_{iK}\varepsilon_K\right) = 0 \quad \longrightarrow \quad \Phi(\varepsilon_i) = \text{const}$$

Let $m = \min \Phi$, $M = \max \Phi$ on $\varepsilon_1^2 + \varepsilon_2^2 = 1$. Because of the homogeneity of Φ we can assert that

$$mr \leq \Phi \leq Mr \quad \text{for } \varepsilon_1^2 + \varepsilon_2^2 = r^2.$$

Thus, the curve $\Phi = a$ must be between the circles a/M and a/m, and is thereby bounded away from both zero and infinity.

The general solution to $\frac{\partial \varepsilon_i}{\partial t} = \sum_{K=1}^{2} C_{iK}\varepsilon_K$, $i = 1,2$ is:

$$\varepsilon_i = A_i e^{\lambda t} + B_i e^{\mu t}$$

(Eqn. 9-2)

with $\det\|C_{iK} - \lambda\delta_{iK}\| = 0$ etc. Thus, $\varepsilon_i = a_i \cos\lambda t + b_i \sin\lambda t$ where the integral curve is an ellipse. Instead of the parametrized form we have

$$\sum_{i,K=1}^{2} g_{iK}\varepsilon_i\varepsilon_K = 1.$$

Since $\varepsilon_3 = 0$ was arbitrary we have

$$\sum_{i,K=1}^{3} g_{iK}\varepsilon_i\varepsilon_K = \text{const}.$$

Thus, $ds^2 = \sum g_{iK}\, dx^i dx^K$, and the geometry is indeed Riemannian. Since ds^2 must be quadratic in the space differentials if the time coordinate differential *dt* equals zero, we expect that ds^2 will be quadratic as a space-time differential as well. The only new feature of the extended space time geometry will be the fact that ds^2 need to be positive definite, as can be inferred from the case of special relativity theory.

9.2 Tensor Definitions (in 4D space-time)

Consider

$$\overline{x}^j = f^j(x^0, x^1, x^2, x^3), \quad x^K = h^K(\overline{x}^0, \overline{x}^1, \overline{x}^2, \overline{x}^3)$$

defined on a Riemann space with a metric $ds^2 = \sum_{iK} g_{iK}\, dx^i dx$. Assume $g_{iK} = g_{Ki}$ has the signature (+1, -1, -1, -1) (this is a different signature convention than used elsewhere in the notes, but staying with Adler's conventions in this section).

Scalar Quantities: $ds^2 = \sum_{iK} g_{iK}\, dx^i dx^K$ has to keep the same numerical value under an arbitrary change of coordinates since it is a scalar. A scalar field is a point function in the space considered – invariant under transformation.

Contravariant Vectors: Defined analogous to distance differential 'vector'. So, for $\overline{x}^j$ and x^K are basis vectors in whatever coordinate system we have:

$$d\overline{x}^i = \sum_{j=0}^{3} \frac{\partial f^i}{\partial x^j} dx^j$$

(Eqn. 9-3)

Any set of four quantities $\varepsilon^i (i = 0,1,2,3)$ which transform according to $\overline{\varepsilon}^i = \sum_{j=0}^{3} \frac{\partial \overline{x}^i}{\partial x^j} \varepsilon^i$ forms a contravariant vector.

Covariant Vectors: Consider a scalar field φ. Consider how the four quantities $A_i = \partial\varphi/\partial x^i$ transform under a change of coordinate system from (x^i) to $(\overline{x}^i)$:

$$\frac{\partial \varphi}{\partial \overline{x}^i} = \sum_{j=0}^{3} \frac{\partial x^j}{\partial \overline{x}^i} \frac{\partial \varphi}{\partial x^j}.$$

and we have:

$$\overline{A}_i = \sum_j \frac{\partial x^j}{\partial \overline{x}^i} A_j$$

(Eqn. 9-4)

where the latter form of transformation defines a covariant vector.

Tensors:

Once we have scalars, and covariant and contravariant vectors, we have the building blocks for more general transformational objects (rank n covariant and m contravariant), known as tensors.

$\overline{\varepsilon}^i = \frac{\partial \overline{x}^i}{\partial x^j} \varepsilon^j$ for contravariant vectors (indices above)

$\overline{\eta}_i = \frac{\partial x^j}{\partial \overline{x}^i} \eta_j$ for covariant vectors (indices below)

The inner product of a covariant vector and a contravariant vector is a *scalar* invariant:

$$\overline{P} = \sum_i \overline{A}_i \overline{\varepsilon}^i = \sum_{ijK} \frac{\partial x^K}{\partial \overline{x}^i} A_K \frac{\partial \overline{x}^i}{\partial x^j} \varepsilon^j = \sum_{jK} A_K \, \varepsilon^j \delta_j^K = \sum_j A_j \, \varepsilon^j = P.$$

Consider a multilinear form P where summation is implied on indices that match when one is covariant and one contravariant -- known as a 'contraction' (the implied summation is also known as Einstein convention):

$$P = \left(T^{i_1 i_2 \cdots i_a}_{j_1, j_2 \cdots j_b}\right)\left(\varepsilon^{j_1}_{(1)} \varepsilon^{j_2}_{(2)} \cdots \varepsilon^{j_b}_{(b)}\right)\left(\eta^{(1)}_{i_1} \eta^{(2)}_{i_2} \cdots \eta^{(a)}_{i_a}\right),$$

where $T^{i_1 i_2 \cdots i_a}_{j_1, j_2 \cdots j_b}$ is a set of n^{a+b} elements (with overall rank $(a + b)$ if a tensor). T is a tensor if, for an arbitrary change of coordinates under ε and η, T transforms such that P remains unchanged (a scalar). (Thus, a tensor of rank zero is a scalar, and of rank one is either a contravariant or covariant vector. The metric g_{iK} is a second rank covariant tensor.

General transformation law of a tensor:

$$\left(\overline{T}^{i_1, \cdots i_a}_{j_1, \cdots j_b}\right)\left(\overline{\varepsilon}^{j_1}_{(1)} \cdots \overline{\varepsilon}^{j_b}_{(b)}\right)\left(\overline{\eta}^{(1)}_{i_1} \cdots \overline{\eta}^{(a)}_{i_a}\right)$$
$$= \left(T^{i_1, \cdots i_a}_{j_1 \cdots j_b}\right)\left(\varepsilon^{j_1}_{(1)} \cdots \varepsilon^{j_b}_{(b)}\right)\left(\eta^{(1)}_{i_1} \cdots \eta^{(a)}_{i_a}\right)$$

So,

$$\overline{T}^{i_1,\cdots i_a}_{j_1,\cdots j_b}\left(\frac{\partial \overline{x}^{j_i}}{\partial x^{\beta_i}}\cdots\frac{\partial \overline{x}^{j_b}}{\partial x^{\beta_b}}\right)\left(\frac{\partial x^{\alpha_1}}{\partial \overline{x}^{i_1}}\cdots\frac{\partial x^{\alpha_a}}{\partial \overline{x}^{i_a}}\right)=\overline{T}^{\alpha_1\cdots\alpha_a}_{\beta_1\cdots\beta_b}$$

or

$$\overline{T}^{K_1,\cdots K_a}_{\ell_1,\cdots \ell_b}\left(\frac{\partial \overline{x}^{K_i}}{\partial x^{\alpha_i}}\cdots\frac{\partial \overline{x}^{K_a}}{\partial x^{\alpha_a}}\right)\left(\frac{\partial x^{\beta_1}}{\partial \overline{x}^{\ell_1}}\cdots\frac{\partial x^{\beta_b}}{\partial \overline{x}^{\ell_b}}\right)=\overline{T}^{\alpha_1\cdots\alpha_a}_{\beta_1\cdots\beta_b}$$

A typical example:

$$T^{\alpha\beta}_{\gamma}=\frac{\partial \overline{x}^{\alpha}}{\partial x^{i}}\frac{\partial \overline{x}^{\beta}}{\partial x^{j}}\frac{\partial x^{K}}{\partial \overline{x}^{\gamma}}T^{ij}_{K}.$$

9.3 Tensor Algebras

Let's now consider some basic properties of tensor algebra.

Equality

A and B are equal if $A^{\alpha\beta}_{\gamma}=B^{\alpha\beta}_{\gamma}$.

Additive property

Sum of two tensors with the same number of respective indices is simply

$$A^{\alpha\beta}_{\gamma}+B^{\alpha\beta}_{\gamma}=C^{\alpha\beta}_{\gamma},$$

which is a new tensor.

Multiplicative property

The product of a tensor by a scalar is again a tensor. The product of tensors is a tensor. The tensor product $T^{\alpha\beta}_{\gamma}S^{\mu\nu}=G^{\alpha\beta\mu\nu}_{\gamma}$ is often called the outer product of the two tensors T and S. The reverse can be done, where decomposition of a tensor into a sum of vector products is done (Tensor Products of Tensors of Rank 1).

Contraction of indices

$T^{i_1 i_2\cdots i_{a-1}\sigma}_{j_1,j_2\cdots j_{b-1}\sigma}$ is a tensor of rank $a+b-2$ which one can denote $R^{i_1 i_2\cdots i_{a-1}}_{j_1,j_2\cdots j_{b-1}}$.

The Quotient Theorem

S = TA is valid if the indices count is correct. Then if any two are tensors te third must be as well.

Lowering and Raising Indices – Associated Tensors

Consider

$$T^{\alpha}_{\gamma}=g_{\gamma\beta}T^{\alpha\beta}\ \ and\ \ T_{\delta\gamma}=g_{\alpha\delta}T^{\alpha}_{\gamma}=g_{\alpha\delta}g_{\gamma\beta}T^{\alpha\beta}.$$

The second-rank tensor g_{iK} which we are using to lower indices can be chosen arbitrarily. However, once selected, it plays a central role in tensor calculus since it establishes a relation between contravariant and covariant tensors; it is called the fundamental tensor. In a metric space, such as the four-dimensional space of general relativity, it is quite natural to take for g_{iK} to be the metric tensor itself. Thus

$$g^{iK} g_{iK} = \delta^i_j.$$

Connection with Vector Calculus in Euclidean Space
Contravariant vector with components (μ^1, μ^2) $\quad \tilde{u} = \mu^1 \hat{e}_1 + \mu^2 \hat{e}_2$
Covariant vector with components (μ_1, μ_2) $\quad \mu_1 = u \cdot \hat{e}_1,\ \mu_2 = \hat{u} \cdot \hat{e}_2$

Connection between Bilinear Forms and Tensor Calculus
Let $\overline{X}_K = \sum_i a_{Ki} X_i$ and $\overline{Y}_K = \sum_i b_{Ki} Y_i$ and consider the bilinear forms: $F = \sum_i X_i Y_i$ and $\overline{F} = \sum_K \overline{X}_K \overline{Y}_K$:

$$\overline{F} = \sum_{ijK} a_{Ki} X_i b_{Kj} Y_j = \sum_{ij} \left(\sum_K a_{Ki} b_{Kj} \right) X_i Y_j$$

where

$$\overline{F} = F \text{ if } \sum_K a_{Ki} b_{Kj} = \delta_{ij}.$$

Using the transpose notation (if A^T is the transpose of a_{iK} then it is the matrix with a_{Ki}) we have our condition to be:

$$A^T B = I \quad \text{or} \quad AB^T = I$$

So, we have:

$$\begin{cases} \text{Original Matrix} & A \\ \text{Matrix contragradient to A} & B = (A^T)^{-1} \\ \text{Matrix contragradient to B} & C = (B^T)^{-1} = (B^{-1})^T = (A^T)^T = A \end{cases}$$

The contragradience relationship is an automorphism, i.e., it preserves the law of multiplication with the order of the factors. If, for example, $A = (B^T)^{-1}$ and $D = (E^T)^{-1}$ then $AD = [(BE)^T]^{-1}$.

9.4 Tensor Fields

So far transformation at a point. Now consider tensor at other points (everywhere in fact). There is now the possibility of 'moving about' and of transplanting vectors.

Vector Fields in Affine and Riemann Space
A Tensor Field is the assignment of a tensor to each point of the space.

Vector Transplantation and Affine Connections
Consider

$$\overline{\varepsilon}^i = \frac{\partial \overline{x}^i}{\partial x^K}\varepsilon^K$$

where vector $\overline{\varepsilon}^i$ is not a constant over the space since $\partial \overline{x}^i / \partial x^K$ is arbitrary. Covariance requirements on a vector field ε^K taken to be constant is then possible. Starting with a vector ε^i of constant components in an original coordinate system x^i, we see that $\overline{\varepsilon}^i$ in another coordinate system (generally) does not have constant components over the space. Let us now see how the components $\overline{\varepsilon}^i$ vary when we go from one point to a neighboring one in space along a curve parametrized with a parameter p; to do this we differentiate $\overline{\varepsilon}^i = \left(\partial \overline{x}^i / \partial x^K\right)\varepsilon^K$ with respect to p, remembering that the ε^K's are constant along the curve by assumption, thus:

$$\frac{d\overline{\varepsilon}^i}{dp} = \frac{\partial^2 \overline{x}^i}{\partial x^K \partial x^\ell}\frac{dx^\ell}{dp}\varepsilon^K = \left(\frac{\partial^2 \overline{x}^i}{\partial x^K \partial x^\ell}\frac{dx^\ell}{\partial \overline{x}^m}\frac{\partial \overline{x}^m}{dp}\frac{\partial x^K}{\partial \overline{x}^j}\right)\overline{\varepsilon}^j = \overline{\Gamma}^i_{mj}\frac{\partial \overline{x}^m}{dp}\overline{\varepsilon}^j$$

where $\overline{\Gamma}^i_{mj} = \frac{\partial^2 \overline{x}^i}{\partial x^K \partial x^\ell}\frac{dx^\ell}{\partial \overline{x}^m}\frac{\partial x^K}{\partial \overline{x}^j}$. Thus,

$$d\varepsilon^i = \Gamma^i_{mj}dx^m\varepsilon^j,$$

(Eqn. 9-5)

and this defines a general law for transplantation of the vector ε^j at the point x into the quantities $\varepsilon^i + d\varepsilon^i$ at the point $x + dx$. It is a law of affine character; that is, it has invariant structure under a linear transformation of the coordinates.

So far we have achieved vector transplantation: $d\varepsilon^i = \Gamma^i_{mj}dx^m\varepsilon^j$. Now to make it coordinate-invariant. This will force certain requirements on the Γ^i_{mj}:

To start, we have:

$$\varepsilon^i(x + dx) = \varepsilon^i + d\varepsilon^i = \varepsilon^i + \Gamma^i_{mj}dx^m\varepsilon^j.$$

Since we require $\varepsilon^i(x + dx)$ to be a vector this requires:

$$\bar{\varepsilon}^j(x+dx) = \varepsilon^i(x+dx)\left(\frac{\partial \bar{x}^j}{\partial x^i}\right)_{x+dx}$$

So,

$$\bar{\varepsilon}^j + \bar{\Gamma}^j_{ms} d\bar{x}^m \bar{\varepsilon}^s = \left(\varepsilon^i + \Gamma^i_{m\ell} dx^m \varepsilon^\ell\right)\left(\frac{\partial \bar{x}^j}{\partial x^i}\right)_{x+dx}.$$

Expanding using a Taylor series:

$$\left(\frac{\partial \bar{x}^j}{\partial x^i}\right)_{x+dx} = \left(\frac{\partial \bar{x}^j}{\partial x^i}\right)_x + \frac{\partial^2 \bar{x}^j}{\partial x^i \partial x^\eta} dx^\eta$$

Thus

$$\bar{\Gamma}^j_{ms} d\bar{x}^m \bar{\varepsilon}^s = \bar{\varepsilon}^j + \varepsilon^i \left(\frac{\partial \bar{x}^j}{\partial x^i}\right)_x + \varepsilon^i \left(\frac{\partial^2 \bar{x}^j}{\partial x^i \partial x^\eta} dx^\eta\right) + \Gamma^i_{m\ell} dx^m \varepsilon^\ell \left(\frac{\partial \bar{x}^j}{\partial x^i}\right)_x$$

$$+ \Gamma^i_{m\ell} dx^m \varepsilon^\ell \frac{\partial^2 \bar{x}^j}{\partial x^i \partial x^\eta} dx^\eta$$

And since

$$\left(\varepsilon^i + d\varepsilon^i\right) \frac{\partial^2 \bar{x}^j}{\partial x^i \partial x^\eta} dx^\eta \simeq \varepsilon^i \frac{\partial^2 \bar{x}^j}{\partial x^i \partial x^\eta} dx^\eta$$

We have

$$\bar{\Gamma}^j_{ms} d\bar{x}^m \bar{\varepsilon}^s = \Gamma^i_{m\ell} dx^m \varepsilon^\ell \left(\frac{\partial \bar{x}^j}{\partial x^i}\right) + \frac{\partial^2 \bar{x}^j}{\partial x^i \partial x^\eta} \varepsilon^i dx^\eta$$

$$= \left(\Gamma^i_{\alpha\beta}\left(\frac{\partial \bar{x}^j}{\partial x^i}\right) + \frac{\partial^2 \bar{x}^j}{\partial x^\beta \partial x^\alpha}\right) \varepsilon^\beta dx^\alpha$$

and using

$$\varepsilon^\beta dx^\alpha = \frac{\partial x^\alpha}{\partial \bar{x}^m} \frac{\partial x^\beta}{\partial \bar{x}^s} \bar{\varepsilon}^s d\bar{x}^m$$

this becomes:

$$\bar{\Gamma}^j_{ms} d\bar{x}^m \bar{\varepsilon}^s = \left(\frac{\partial \bar{x}^j}{\partial x^i} \frac{\partial x^\alpha}{\partial \bar{x}^m} \frac{\partial x^\beta}{\partial \bar{x}^s} \Gamma^i_{\alpha\beta} + \frac{\partial^2 \bar{x}^j}{\partial x^\alpha \partial x^\beta} \frac{\partial x^\alpha}{\partial \bar{x}^m} \frac{\partial x^\beta}{\partial \bar{x}^s}\right) d\bar{x}^m \bar{\varepsilon}^s$$

Thus,

$$\bar{\Gamma}^j_{ms} = \frac{\partial \bar{x}^j}{\partial x^i} \frac{\partial x^\alpha}{\partial \bar{x}^m} \frac{\partial x^\beta}{\partial \bar{x}^s} \Gamma^i_{\alpha\beta} + \frac{\partial^2 \bar{x}^j}{\partial x^\alpha \partial x^\beta} \frac{\partial x^\alpha}{\partial \bar{x}^m} \frac{\partial x^\beta}{\partial \bar{x}^s}$$

(Eqn. 9-6)

These coefficients are called "coefficients of affine connection". Note that the transformation law is inhomogeneous in the coefficient $\Gamma^i_{\alpha\beta}$. So not a tensor! This is fundamentally different from a tensor transformation law and the connections are not tensors. (Another form, generalized to gauge analysis, is in shown in Ch. 10 & 11.). Let's consider the consequence of this non-tensorial behavior:

(1) If we restrict ourselves to linear transformations of coordinates, the term $\partial^2\overline{x}^j/\partial x^\alpha \partial x^\beta$ vanishes and Γ^i_{mj} transforms like a tensor. In this special case constancy of components is therefore an acceptable criterion for a constant vector field.

(2) $\Gamma^i_{K\ell} - \overline{\Gamma}^i_{K\ell}$ transforms like a tensor since the inhomogeneous terms cancel.

(3) Certain axiomatic deductions can be made from the equation (these were first developed by Levi-Civita):
  (a) The Γ coefficients remain unsymmetric under any change of coordinates
  (b) It is impossible to find a coordinate system in which all Γ coefficients are 0 at a point.

Parallel Displacement – Christoffel Symbols

Now we focus our study on Riemann spaces. Consider the metric requirement that the scalar product of two vectors be invariant under transplantation:

$$\frac{d}{dS}\left(g_{iK}\varepsilon^i\eta^K\right) = 0 \rightarrow \quad \frac{\partial g_{iK}}{\partial x^\ell}\frac{dx^\ell}{dS}\varepsilon^i\eta^K + g_{iK}\frac{d\varepsilon^i}{dS}\eta^K + g_{iK}\varepsilon^i\frac{d\eta^K}{dS} = 0$$

Using $\frac{d\varepsilon^i}{dS} = \Gamma^i_{\ell r}\frac{dx^\ell}{dS}\varepsilon^r$ *and* $\frac{d\eta^K}{dS} = \Gamma^K_{\ell r}\frac{dx^\ell}{dS}\eta^r$ this becomes:

$$\frac{\partial g_{iK}}{\partial x^\ell}\frac{dx^\ell}{dS}\varepsilon^i\eta^K + g_{iK}\Gamma^i_{\ell r}\frac{dx^\ell}{dS}\varepsilon^r\eta^K + g_{iK}\Gamma^K_{\ell r}\frac{dx^\ell}{dS}\eta^r\varepsilon^i = 0$$

or

$$\frac{\partial g_{iK}}{\partial x^\ell}\varepsilon^i\eta^K + g_{iK}\underbrace{\Gamma^i_{\ell n}\varepsilon^n\eta^K}_{n\to i, i\to r} + \underbrace{\Gamma^K_{\ell n}\eta^n\varepsilon^i}_{n\to K, K\to r} = 0$$

with relabeling dummy indices to get:

$$\frac{\partial g_{iK}}{\partial x^\ell} = g_{rK}\Gamma^r_{\ell i} + g_{ir}\Gamma^r_{\ell K}\,. \tag{Eqn. 9-7}$$

By permuting the indices $iK\ell$ two additional equations are found and solving for Γ^r_{Ki} we get:

$$\Gamma^r_{Ki} = -\frac{1}{2}g^{\ell r}\left(\frac{\partial g_{K\ell}}{\partial x^i} + \frac{\partial g_{\ell i}}{\partial x^K} - \frac{\partial g_{iK}}{\partial x^\ell}\right).$$

(Eqn. 9-8)

Define $[iK,\ell] = \frac{1}{2}\left(\frac{\partial g_{K\ell}}{\partial x^i} + \frac{\partial g_{K\ell}}{\partial x^i} - \frac{\partial g_{iK}}{\partial x^\ell}\right)$ this is a Christoffel symbol of the first kind. Also, define $\left\{ {j \atop i\ K} \right\} = g^{j\ell}\,[iK,\ell]$ which is called the Christoffel symbol of the second kind. Thus, $\Gamma^r_{iK} = -\left\{ {r \atop i\ K} \right\}$ and the law of parallel displacement in a metric space is thus:

$$d\varepsilon^i = -\left\{ {i \atop \alpha\ \beta} \right\} dx^\alpha \varepsilon^\beta.$$

(Eqn. 9-9)

To put this into context. Consider the evolution in time of a mechanical system described by generalized coordinates $x^i(t)$, $\dot{x}^i = dx^i/dt, T = \frac{1}{2}g_{ik}\dot{x}^i\dot{x}^K$, and $V(x^i)$ which gives $F_i = -\partial V/\partial x^i$. As usual in analytical dynamics, we take $Tdt^2 = ds^2$ to define a metric on the space of the generalized coordinates, which is called configuration space. Using $L = T - V$, and $\frac{d}{dt}\left(\frac{\partial L}{\partial \dot{x}^i}\right) = \frac{\partial}{\partial x^i}$ we then get:

$$g_{iK}\ddot{x}^K + \frac{\partial g_{iK}}{\partial x^\ell}\dot{x}^\ell\dot{x}^K = \frac{1}{2}\frac{\partial g_{\ell K}}{\partial x^i}\dot{x}^\ell\dot{x}^K + F_i$$

or

$$g_{iK}\ddot{x}^K + \frac{1}{2}\left[\frac{\partial g_{iK}}{\partial x^\ell} + \frac{\partial g_{i\ell}}{\partial x^K} - \frac{\partial g_{\ell K}}{\partial x^i}\right]\dot{x}^\ell\dot{x}^K = F_i$$

and using the Christoffel notation:

$$\ddot{x}^i + \left\{ {i \atop \ell\ K} \right\}\dot{x}^\ell\dot{x}^K = F^i.$$

(Eqn. 9-10)

Geodesics in Affine and Riemann Space:
Suppose we have

$$d\varepsilon^i = \Gamma^i_{\alpha\beta}dx^\alpha\varepsilon^\beta = 0,$$

and we parametrize by q:

$$\frac{d\varepsilon^i}{dq} - \Gamma^i_{\alpha\beta}\frac{dx^\alpha}{dq}\varepsilon^\beta = 0.$$

dx^i/dq is a particular tangent vector, while a more general one is $\lambda(q)\left(\frac{dx^i}{dq}\right)$, where $\lambda(q)$ is an arbitrary function of q. In which case:

$$\frac{d}{dq}\left(\lambda(q)\frac{dx^i}{dq}\right) = \Gamma^i_{\alpha\beta}\frac{dx^\alpha}{dq}\lambda(q)\frac{dx^\beta}{dq}.$$

If we let $dp = dq/[\lambda(q)]$, then

$$\lambda(q)\frac{d}{dq}\left(\frac{dx^i}{dp}\right) = \Gamma^i_{\alpha\beta}\frac{dx^\alpha}{dp}\frac{dx^\beta}{dp}$$

and we get:

$$\frac{d^2x^i}{dp^2} - \Gamma^i_{\alpha\beta}\frac{dx^\alpha}{dp}\frac{dx^\beta}{dp} = 0$$

which define geodesic lines in an affine space. The defining equations for a geodesic in Riemann space become

$$\frac{d^2x^i}{dp^2} + \left\{\begin{matrix} i \\ \alpha\beta \end{matrix}\right\}\frac{dx^\alpha}{dp}\frac{dx^\beta}{dp} = 0.$$

(Eqn. 9-11)

Other definitions of geodesics in a Riemann space

A geodesic can also be defined as the stationary curve between points: If $S = \int_{P_0}^{P_1}\left(g_{iK}\frac{dx^i}{dp}\frac{dx^K}{dp}\right)^{1/2} dp$:

$$\delta S = \delta\int_{P_0}^{P_1}\underbrace{\left(g_{iK}\frac{dx^i}{dp}\frac{dx^K}{dp}\right)^{1/2}}_{\text{"}L\text{"}} dp = 0$$

Choose $p = s$ arc length to get:

$$\frac{d}{ds}\left(g_{iK}\frac{dx^K}{ds}\right) - \frac{\partial g_{iK}}{\partial x^i}\frac{dx^i}{ds}\frac{dx^i}{ds} = 0$$

and again we have:

$$\frac{d^2x^i}{ds^2} + \left\{\begin{matrix} i \\ \ell\,K \end{matrix}\right\}\frac{dx^\ell}{ds}\frac{dx^K}{ds} = 0.$$

9.5 Examples

Example 9.1. Describe expansion of $\boldsymbol{T}$ in a basis, where by definition: $\boldsymbol{T}(\vec{e}_\alpha, \vec{e}^\beta, \vec{e}_\gamma) = T_\alpha{}^\beta{}_\gamma$:

$$\boldsymbol{T}(\vec{e}_\alpha, \vec{e}^\beta, \vec{e}_\gamma) = T_\alpha{}^\beta{}_\gamma = T_\mu{}^\nu{}_\tau\delta^\mu_\alpha\delta^\beta_\nu\delta^\tau_\gamma = T_\mu{}^\nu{}_\tau(\vec{e}^\mu\cdot\vec{e}_\alpha)(\vec{e}^\beta\cdot\vec{e}_\nu)\cdot(\vec{e}^\tau,\vec{e}_\gamma)$$
$$= T_\mu{}^\nu{}_\tau\cdot\vec{e}^\mu\otimes\vec{e}_\nu\otimes\vec{e}^\tau(\vec{e}_\alpha, \vec{e}^\beta, \vec{e}_\gamma)$$

Thus,

$$\boldsymbol{T} = T_\mu{}^\nu{}_\tau\vec{e}^\mu\otimes\vec{e}_\nu\otimes\vec{e}^\tau,$$

where we could have used any combination of basis vectors $\vec{e}_\alpha$ and $\vec{e}^\alpha$.

Example 9.2. Consider the value of $\boldsymbol{T}$ on a set of vectors.

By definition, and using the notation of [38]:

$$\vec{A}(\vec{C}) = g(\vec{A},\vec{C}) = A^{\alpha}\vec{e}_{\alpha}(C_{\beta}\vec{e}^{\beta}) = A^{\alpha}C_{\beta}\vec{e}_{\alpha}(\vec{e}^{\beta}) = A^{\alpha}C_{\beta}\cdot\delta^{\beta}_{\alpha} = A^{\alpha}C_{\alpha}$$

So generalizing for $\boldsymbol{T}$ from Ex. 1**:**

$$\boldsymbol{T}(\vec{A},\vec{B},\vec{C}) = T^{\ \beta}_{\alpha\ \gamma}\vec{e}^{\alpha}\otimes\vec{e}_{\beta}\otimes\vec{e}^{\gamma}(\vec{A},\vec{B},\vec{C})$$

and

$$\boldsymbol{T}(\vec{A},\vec{B},\vec{C}) = T^{\ \beta}_{\alpha\ \gamma}\vec{e}^{\alpha}(\vec{A})\vec{e}_{\beta}(\vec{B})\vec{e}^{\gamma}(\vec{C})$$

$$= T^{\ \beta}_{\alpha\ \gamma}\cdot\left(\vec{e}^{\alpha}\cdot(A^{\mu}\vec{e}_{\mu})\right)\cdot\left(\vec{e}_{\beta}\cdot(B_{\nu}\vec{e}^{\nu})\right)\left(\vec{e}^{\gamma}\cdot(C^{\tau}\vec{e}_{\tau})\right)$$

$$= T^{\ \beta}_{\alpha\ \gamma}A^{\mu}B_{\nu}C^{\tau}\cdot(\vec{e}^{\alpha}\cdot\vec{e}_{\mu})(\vec{e}_{\beta}\cdot\vec{e}^{\nu})(\vec{e}^{\gamma}\cdot\vec{e}_{\tau}) = T^{\ \beta}_{\alpha\ \gamma}A^{\mu}B_{\nu}C^{\tau}\cdot\delta^{\alpha}_{\mu}\delta^{\nu}_{\beta}\delta^{\gamma}_{\tau}$$

$$= T^{\ \beta}_{\alpha\ \gamma}A^{\alpha}B_{\beta}C^{\gamma}$$

So,

$$\boldsymbol{T}(\vec{A},\vec{B},\vec{C}) = T^{\ \beta}_{\alpha\ \gamma}A^{\alpha}B_{\beta}C^{\gamma}$$

and similar calculation can be done for any choice of up and down indices.

Example 9.3. Let's show the symmetry property of the metric in this explicit notation. We have, say,
$g(\vec{A},\vec{B})$ symmetric for arbitrary $\vec{A}$ and $\vec{B}$. For this to hold we would need:
$g_{\alpha\beta} = g(\vec{e}_{\alpha},\vec{e}_{\beta}) = g(\vec{e}_{\beta},\vec{e}_{\alpha}) = g_{\beta\alpha}$,
$g^{\alpha\beta} = g(\vec{e}^{\alpha},\vec{e}^{\beta}) = g(\vec{e}^{\beta},\vec{e}^{\alpha}) = g^{\beta\alpha}$,
$g^{\beta}_{\alpha} = g(\vec{e}_{\alpha},\vec{e}^{\beta}) = g(\vec{e}^{\beta},\vec{e}_{\alpha}) = \vec{e}^{\beta}\cdot\vec{e}_{\alpha} = \delta^{\beta}_{\alpha}$,
and
$g^{\beta}_{\alpha} = g(\vec{e}^{\beta},\vec{e}_{\alpha}) = \vec{e}^{\beta}\cdot\vec{e}_{\alpha} = \delta^{\beta}_{\alpha}$.

Let's test the latter:
$g_{\alpha\beta}g^{\beta\mu} = g(\vec{e}_{\alpha},\vec{e}_{\beta})g(\vec{e}^{\beta},\vec{e}^{\mu}) = (\vec{e}_{\alpha}\cdot\vec{e}_{\beta})(\vec{e}^{\beta}\cdot\vec{e}^{\mu}) = (\vec{e}_{\beta})_{\gamma}(\vec{e}_{\beta})^{\gamma}\cdot(\vec{e}^{\beta})_{\nu}(\vec{e}^{\mu})^{\nu}$,
where $\vec{e}_{\alpha}$ is a vector for each α, and using Ex. 2. Continuing:

$$g^{\mu}_{\alpha} = g_{\alpha\beta}g^{\beta\mu} = (\vec{e}_{\alpha})_{\gamma}\cdot\vec{e}_{\beta}(\vec{e}^{\gamma})\cdot\vec{e}^{\beta}(\vec{e}_{\nu})\cdot(\vec{e}^{\mu})^{\nu} = (\vec{e}_{\alpha})_{\gamma}\cdot\delta^{\gamma}_{\beta}\cdot\delta^{\beta}_{\nu}\cdot(\vec{e}^{\mu})^{\nu}$$

$$= (\vec{e}_{\alpha})_{\beta}(\vec{e}^{\mu})^{\beta} = \vec{e}_{\alpha}(\vec{e}^{\mu}) = \delta^{\mu}_{\alpha}$$

Thus,

$$g^{\mu}_{\alpha} = \delta^{\mu}_{\alpha},$$

and the others relations are similarly verified.

Example 9.4. Let's now use the metric tensor to raise and lower indices. Here we raise an index:

$$g^{\beta\mu}T^{\alpha\ \gamma}_{\ \mu} = g^{\beta\mu}T(\vec{e}^{\alpha}, \vec{e}_{\mu}\vec{e}^{\gamma}) = g^{\beta\mu} \cdot T^{\gamma\tau\lambda}\vec{e}_{\nu}\otimes\vec{e}_{\tau}\otimes\vec{e}_{\lambda}(\vec{e}^{\alpha}, \vec{e}_{\mu}, \vec{e}^{\gamma})$$
$$= g^{\beta\mu} \cdot T^{\gamma\tau\lambda}\delta^{\gamma}_{\lambda}(\vec{e}_{\tau} \cdot \vec{e}_{\mu})\delta^{\gamma}_{\lambda}$$

thus,

$$g^{\beta\mu}T^{\alpha\ \gamma}_{\ \mu} = g^{\beta\mu} \cdot T^{\alpha\tau\gamma} \cdot g_{\tau\mu} = \delta^{\beta}_{\tau}T^{\alpha\tau\gamma} = T^{\alpha\beta\gamma}$$

Similar results can be obtained for lowering the index.

Example 9.5. Suppose we have a change of basis $L^{\alpha'}_{\mu}$ (such as from a Lorentz transform). Such transforms must satisfy by consistency:

$$\vec{e}_{\mu} = L^{\alpha'}_{\mu}\vec{e}_{\alpha'} \quad \text{and} \quad L^{\mu}_{\ \alpha'}L^{\alpha'}_{\nu} = \delta^{\mu}_{\nu}$$

What happens for transformation of $\vec{e}^{\alpha'}$:

$$L^{\mu}_{\ \alpha'}\vec{e}^{\alpha'} = L^{\mu}_{\ \alpha'}\left(\vec{e}^{\alpha'}\right)_{\nu}\vec{e}^{\nu} = L^{\mu}_{\ \alpha'}\vec{e}^{\alpha'}(\vec{e}_{\nu})\vec{e}^{\nu} = L^{\mu}_{\ \alpha'}\vec{e}^{\alpha'}\left(L^{\beta'}_{\nu}\vec{e}_{\beta'}\right)\vec{e}^{\nu}$$
$$= L^{\mu}_{\ \alpha'}L^{\beta'}_{\nu}\delta^{\alpha'}_{\beta'}\vec{e}^{\nu} = L^{\mu}_{\ \alpha'}L^{\alpha'}_{\nu}\vec{e}^{\nu}$$

Thus,

$$L^{\mu}_{\ \alpha'}\vec{e}^{\alpha'} = \delta^{\mu}_{\nu}\vec{e}^{\nu} = \vec{e}^{\mu}.$$

Similarly:

$$L^{\mu}_{\ \alpha'}L^{\beta'}_{\nu}L^{\tau}_{\ \alpha'}T^{\alpha'\ \gamma'}_{\ \beta'} = L^{\mu}_{\ \alpha'}L^{\beta'}_{\nu}L^{\tau}_{\ \alpha'}T^{\rho\ \delta}_{\ \sigma}\vec{e}^{\alpha}(\vec{e}_{\rho})\vec{e}_{\beta'}(\vec{e}^{\sigma})\vec{e}^{\gamma}(\vec{e}_{\delta}) = T^{\rho\ \delta}_{\ \sigma}\delta^{\mu}_{\rho}\delta^{\sigma}_{\nu}\delta^{\tau}_{\delta}$$
$$= T^{\mu\ \tau}_{\ \nu}.$$

Example 9.6. Consider a Null Basis in 2D Minkowski where:

$$\vec{e}_0 \cdot \vec{e}_0 = -1\ , \qquad \vec{e}_1 \cdot \vec{e}_1 = 1\ , \qquad \vec{e}_0 \cdot \vec{e}_1 = 0$$

and

$$\|g_{\alpha\beta}\| = \|\vec{e}_{\alpha} \cdot \vec{e}_{\beta}\| = \begin{pmatrix} -1 & 0 \\ 0 & 1 \end{pmatrix}, with\ inverse\ \|g^{\alpha\beta}\| = \begin{pmatrix} -1 & 0 \\ 0 & 1 \end{pmatrix}$$

and use definitions:

$$\vec{e}_{+} \equiv \vec{e}_0 + \vec{e}_1\ , \qquad \vec{e}_{-} = \vec{e}_0 - \vec{e}_1.$$

(a) Find $\vec{e}^{+}$ and $\vec{e}^{-}$.
(b) Find the metric components in the $\{+,-\}$ basis.
(c) Find $A_{+}\ and\ A_{-}$ in terms of $A^{+}\ and\ A^{-}$.
(d) Suppose $B = \vec{e}_{+}$, what are the components $B_{+}\ and\ B_{-}$.

Solution

(a) $\quad \vec{e}_{+} \cdot \vec{e}_{+} = \vec{e}_{-} \cdot \vec{e}_{-} = 0\ ,\ \vec{e}_{+} \cdot \vec{e}_{-} = \vec{e}_{-} \cdot \vec{e}_{+} = -2$
$\quad g_{++} = g_{--} = 0 \qquad\qquad g_{+-} = g_{-+} = -2$

$$\left.\begin{aligned} \vec{e}_{+} = g_{++}\vec{e}^{+} + g_{+-}\vec{e}^{-} = -2\vec{e}^{-} \Rightarrow \vec{e}^{-} = -\tfrac{1}{2}\vec{e}_{+} \\ \vec{e}_{-} = g_{--}\vec{e}^{-} + g_{-+}\vec{e}^{+} = -2\vec{e}^{+} \Rightarrow \vec{e}^{+} = -\tfrac{1}{2}\vec{e}_{-} \end{aligned}\right\} \qquad (1)$$

$$\vec{e}_0 = g_{00}\vec{e}^0 + g_{01}\vec{e}^1 = -\vec{e}^0 \Rightarrow \vec{e}^0 = -\vec{e}_0$$
$$\vec{e}_1 = g_{11}\vec{e}^1 + g_{10}\vec{e}^0 = -2\vec{e}^1 \Rightarrow \vec{e}^1 = -\vec{e}_1$$

$$\vec{e}^+ = -\frac{1}{2}\vec{e}_- = -\frac{1}{2}(\vec{e}_0 - \vec{e}_1) = \frac{1}{2}(\vec{e}^0 + \vec{e}^1)$$
$$\vec{e}^- = -\frac{1}{2}\vec{e}_+ = -\frac{1}{2}(\vec{e}_0 + \vec{e}_1) = -\frac{1}{2}(\vec{e}^0 + \vec{e}^1) = \frac{1}{2}(\vec{e}^0 - \vec{e}^1)$$

$$\vec{e}^+ = \frac{1}{2}(\vec{e}^0 + \vec{e}^1)$$
$$\vec{e}^- = \frac{1}{2}(\vec{e}^0 - \vec{e}^1)$$

(b) $\begin{Bmatrix} a & b \\ +, & - \end{Bmatrix}$: $\quad g_{ab} = 0$ if $a = b$, $g_{ab} = -2$ if $a \neq b$.

$$g_a^b = g_a^b = \delta_a^b$$
$$g^{ab} = 0 \text{ if } a = b \text{ , } g^{ab} = -\tfrac{1}{2} \text{ if } a \neq b$$

$$\begin{pmatrix} 0 & -2 \\ -2 & 0 \end{pmatrix}\begin{pmatrix} 0 & -\frac{1}{2} \\ -\frac{1}{2} & 0 \end{pmatrix} = \begin{pmatrix} 1 & 0 \\ 0 & 1 \end{pmatrix} = \delta$$

(c) $\vec{A} = A_+\vec{e}^+ + A_-\vec{e}^- = A^+\vec{e}_+ + A^-\vec{e}_-$ using relations (1)

$$= A_+\left(-\tfrac{1}{2}\vec{e}_-\right) + A_-\left(-\tfrac{1}{2}\vec{e}_+\right)$$

Since $-\frac{1}{2}A_+ = A^-$ and $-\frac{1}{2}A_- = A^+$ we have:

$A_+ = -2A^-$

$A_- = -2A^+$

(d) $B_a = g_{ab}b^b = g_{ab}\vec{e}_+(\vec{e}^b) = g_{ab}\delta^b = g_{a+}$. So, $B_+ = 0$ *and* $B_- = -2$.

Example 9.7. $\varepsilon_{ijk} = \pm[ijk]$ in an orthonormal basis, what is the antisymmetric tensor in a non-orthonormal basis with metric g?

Solution

Let $\vec{e}_{\hat{1}}, \vec{e}_{\hat{2}}, \vec{e}_{\hat{3}}$ be a right-handed orthonormal basis, so, $g_{\hat{\ell}\hat{m}} = \delta_{\ell m}$ and $\varepsilon_{\hat{\ell}\hat{m}\hat{n}} = [\ell mn]$. Let $L_a^{\hat{\ell}}$ is the transformation matrix from $\vec{e}_{\hat{\ell}}$ *to* $\vec{e}_a$. Thus, for an arbitrary basis:

$$g_{ab} = L_a^{\hat{\ell}}\, L_b^{\hat{m}}\, g_{\hat{\ell}\hat{m}}$$

where $g_{\hat{\ell}\hat{m}}$ is an orthonormal basis. We then have:

$$\underbrace{\det\|g_{ab}\|}_{g} = \left(\det\left\|L_a^{\hat{\ell}}\right\|\right)^2 \underbrace{\det\|g_{\hat{\ell}\hat{m}}\|}_{1}$$

Thus

$$g = \pm\left(\det\left\|L_a^{\hat{\ell}}\right\|\right)^2.$$

The determinant definition is now needed:

$$\varepsilon_{ijk} = \underbrace{L_i^{\hat{\ell}}\, L_j^{\hat{m}}\, L_k^{\hat{n}}\, \varepsilon_{\hat{\ell}\hat{m}\hat{n}}}_{\substack{use\ the\ det.\mathrm{def}\ of \\ \varepsilon_{\alpha\beta\gamma}A^\alpha B^\beta C^\gamma = \det\begin{vmatrix} A^1 \cdots A^N \\ B^1 \cdots B^N \\ C^1 \cdots C^N \end{vmatrix}}} = \underbrace{\det\left\|L_a^{\hat{\ell}}\right\|}_{\substack{r.h\ orthonormal\ basis \\ inset\ (-1) if\ L.h. \\ orthonornal\ basis}} \quad [ijk](\pm 1)$$

Thus

$$\sqrt{g} = \det\left\|L_a^{\hat{\ell}}\right\| = \frac{\varepsilon_{ijk}}{(\pm 1)[ijk]}$$

or simply:

$$\varepsilon_{ijk} = \pm\sqrt{g}[ijk].$$

Example 9.8.

(a) Show that $\varepsilon_{aij}\varepsilon^{bij} \equiv \delta_{ai}^{bi} = \delta_a^b\, \delta_i^i - \delta_a^i\, \delta_i^b$ and $\varepsilon_{aij}\varepsilon^{bij} = 2\delta_a^b$.

(b) Evaluate $\varepsilon_{iab}\,\varepsilon^{jki}$.

(c) Prove that $\vec{A} \times \left(\vec{B} \times \vec{C}\right) = \left(\vec{A} \cdot \vec{C}\right)\vec{B} - \left(\vec{A} \cdot \vec{B}\right)\vec{C}$.

Solution

(a) $\quad \varepsilon_{abi}\varepsilon^{jki} \equiv \delta_{ab}^{jk} = 0$ unless $\{a = j \neq b = k\}$ or $\{a = k \neq b = j\}$

$$\delta_{12}^{12} = 0 \text{ unless } i = 3 \text{ so, } \delta_{12}^{12} = \delta_{21}^{21} = 1\ ,\ \delta_{12}^{21} = \delta_{21}^{12} = -1$$
$$\delta_{13}^{13} = 0 \text{ unless } i = 2 \text{ so, } \delta_{13}^{13} = \delta_{31}^{31} = 1\ ,\ \delta_{13}^{31} = \delta_{31}^{13} = -1$$
$$\delta_{23}^{23} = 0 \text{ unless } i = 1 \text{ so, } \delta_{23}^{23} = \delta_{32}^{32} = 1\ ,\ \delta_{23}^{32} = \delta_{32}^{23} = -1$$

Thus, $\delta_{ab}^{jk} = \delta_a^j\, \delta_b^k - \delta_a^k\, \delta_b^j$ checks.

Since $\varepsilon_{aij}\varepsilon^{bij} \equiv \delta_{ai}^{bi} = \delta_a^b\, \delta_i^i - \delta_a^i\, \delta_i^b$ *and* $\sum_i \delta_a^i\, \delta_i^b = \delta_a^b$

then: $\varepsilon_{aij}\varepsilon^{bij} = 2\delta_a^b$

(b) $\quad \varepsilon_{iab}\,\varepsilon^{jki} = -\varepsilon_{bai}\varepsilon^{jki} = \varepsilon_{abi}\varepsilon^{jki} = \delta_{ab}^{jk}$

(c) $\quad \vec{A} \times \left(\vec{B} \times \vec{C}\right) = ?$

$$\left(\vec{B} \times \vec{C}\right)_i = \varepsilon_{ijk}B^j C^k = D_i$$

$$\left(\vec{A}\times\vec{D}\right)^{\ell} = \varepsilon^{\ell m i} A_m D_i = E^{\ell} = \varepsilon^{\ell m i} A_m \varepsilon_{ijk} B^j C^k = A_m B^j C^k \varepsilon_{ijk} \varepsilon^{\ell m i}$$

$$= A_m B^j C^k \varepsilon_{jki} \varepsilon^{\ell m i} = A_m B^j C^k \left(\delta_j^{\ell}\,\delta_k^{m} - \delta_j^{m}\,\delta_k^{\ell}\right)$$
$$= A_m B^{\ell} C^m - A_m B^m C^{\ell} = A_m C^m B^{\ell} - A_m B^m C^{\ell}$$
$$= (\vec{A}\cdot\vec{C})\vec{B} - (\vec{A}\cdot\vec{B})\vec{C}$$

Thus,

$$\vec{A}\times(\vec{B}\times\vec{C}) = (\vec{A}\cdot\vec{C})\vec{B} - (\vec{A}\cdot\vec{B})\vec{C}$$

Example 9.9. Compute for the upper half sphere, for vector field:

$$A = (x^2 + y^2)\frac{\partial}{\partial z}$$

for the following surface integral:

$$\int_S A d\Sigma.$$

Solution

The legs of the infinitesimal volume element are given by:

$$\frac{\partial x^i}{\partial \theta} d\theta \quad , \quad \frac{\partial x^i}{\partial y} dy.$$

Thus, the volume element is defined by:

$$d\Sigma_i = \epsilon_{ijk}\frac{\partial x^j}{\partial \theta} d\theta \,\frac{\partial x^k}{\partial y} dy.$$

Therefore, we have

$$\int_{S_2} A\cdot d\Sigma = \int \epsilon_{ijk}\, A^i \frac{\partial x^j}{\partial \theta} d\theta \,\frac{\partial x^k}{\partial y} dy$$
$$= \int \left(\epsilon_{zxy} A^2 \frac{\partial x}{\partial \theta}\frac{\partial y}{\partial y} + \epsilon_{zyx} A^2 \frac{\partial y}{\partial \theta}\frac{\partial x}{\partial y}\right) d\theta$$

$$= \int A^2 \left(r^2 \sin\theta\cos\theta\cos^2 y + r^2 \sin\theta\cos\theta\sin^2 y\right)$$
$$= \int_0^{\frac{\pi}{2}} \cos\theta\, d\theta \int_0^{2\pi} dy\, a^4 \sin^3\theta$$
$$= \frac{\pi}{2}\, a^4.$$

Example 9.10. In a selected Lorentz frame project the global law of 4-momentum conservation onto the time-basis vector of that frame to obtain a global law of energy conservation.

Solution

The global law of 4-momentum conservation is:

$$\int T^{\alpha\beta}\, d\Sigma_\beta = 0.$$

Let's write it explicitly in terms of components in a specific, but arbitrary Lorentz frame:

$$\alpha = 0 \quad \int_{\partial u} T^{00}\, d\Sigma_0 = -\int_{\partial u} T^{01} d\Sigma_1 - \int_{\partial u} T^{02}\, d\Sigma_2 - \int_{\partial u} T^{03} d\Sigma_3$$

$$\alpha = 1 \quad \int_{\partial u} T^{10}\, d\Sigma_0 = -\int_{\partial u} T^{11} d\Sigma_1 - \int_{\partial u} T^{12}\, d\Sigma_2 - \int_{\partial u} T^{13} d\Sigma_3$$

$$\alpha = 2 \quad \int_{\partial u} T^{20}\, d\Sigma_0 = -\int_{\partial u} T^{21} d\Sigma_1 - \int_{\partial u} T^{22}\, d\Sigma_2 - \int_{\partial u} T^{23} d\Sigma_3$$

$$\alpha = 3 \quad \int_{\partial u} T^{30}\, d\Sigma_0 = -\int_{\partial u} T^{31} d\Sigma_1 - \int_{\partial u} T^{32}\, d\Sigma_2 - \int_{\partial u} T^{33} d\Sigma_3$$

Let the 4-volume u be a parallelepiped with legs $\Delta t \frac{\partial}{\partial t}, \Delta x \frac{\partial}{\partial x}, \Delta y \frac{\partial}{\partial y}, \Delta z \frac{\partial}{\partial z}$. Choose an arbitrary point $t_0\,, x_0, y_0, z_0$. Then the volume of the parallelepiped is bounded by planes:

$$t = t_0,\ t = t_0 + \Delta t;$$
$$x = x_0, x = x_0 + \Delta x;$$
$$y = y_0, y = y_0 + \Delta y$$
$$z = z_0, z = z_0 + \Delta z$$

Consider now the equation corresponding to $\alpha = 0$ component. It may be written as:

$$\left.\left(\int T^{00} dxdydz\right)\right|_{t_0+\Delta t} - \left.\left(\int T^{00} dxdydz\right)\right|_{t_0} =$$
$$-\left[\left.\left(\int T^{01} dtdydz\right)\right|_{x_0+\Delta x} - \left.\left(\int T^{01} dtdydz\right)\right|_{x_0}\right]$$
$$-\left[\left.\left(\int T^{02} dtdxdz\right)\right|_{y_0+\Delta y} - \left.\left(\int T^{02} dtdxdz\right)\right|_{y_0}\right]$$
$$-\left[\left.\left(\int T^{03} dtdxdy\right)\right|_{z_0+\Delta z} - \left.\left(\int T^{03} dtdxdy\right)\right|_{z_0}\right]$$

Where for example $(\int T^{00} dxdydz)|_{t_0}$ denotes that the integration is performed over the 3-dimensional spatial volume at the time slice $t = t_0$.

The physical interpretation of the terms in this equation is straightforward:

[net change of energy content of volume V between moment t_0 and moment $t_0 + \Delta t$]
Equals
[(energy flux entering volume V through the face $x = x_0$ integrated over that face and time Δt) – (energy flux exiting the volume V through the face $x = x_0 + \Delta x$ integrated over that face and time Δt)]

Similarly for $\alpha = j$ ($j = 1,2,3$)

$$\left(\int T^{j0} dxdydz\right)\Big|_{t_0+\Delta t} - \left(\int T^{j0} dxdydz\right)\Big|_{t_0} =$$
$$-\left(\int T^{j1} dtdydz\right)\Big|_{x_0+\Delta x} - \left(\int T^{j1} dtdydz\right)\Big|_{x_0}$$
$$-\left(\int T^{j2} dtdxdz\right)\Big|_{y_0+\Delta y} - \left(\int T^{j2} dtdxdz\right)\Big|_{y_0}$$
$$-\left[\left(\int T^{j3} dtdxdy\right)\Big|_{z_0+\Delta z} - \left(\int T^{j3} dtdxdy\right)\Big|_{z_0}\right]$$

Thus,
[net change of j-th component of momentum inside volume V between $t_0$and $t_0 + \Delta t$]
Equals
[net value of the j-th component of force per unit area perpendicular to $\hat{e}_x$ integrated over faces x_0 and $x_0 + \Delta x$ and time Δt]

Similarly for $\hat{e}_x$ and $\hat{e}_z$..

Now, if we let $\Delta t, \Delta x, \Delta y, \Delta z \to 0$ the equations above will turn into the local conservation laws for temporal and spatial components. Let $\alpha = 0$:

$$\left(\int T^{00} dxdydz\right)\Big|_{t_0+\Delta t} - \left(\int T^{00} dxdydz\right)\Big|_{t_0} \to \frac{\partial T^{00}}{\partial t}\Delta t\Delta x\Delta y\Delta z$$
$$\left(\int T^{01} dtdydz\right)\Big|_{x_0+\Delta x} - \left(\int T^{01} dtdydz\right)\Big|_{x_0} \to \frac{\partial T^{01}}{\partial x}\Delta t\Delta x\Delta y\Delta z$$
$$\left(\int T^{02} dtdxdz\right)\Big|_{y_0+\Delta y} - \left(\int T^{02} dtdxdz\right)\Big|_{x_0} \to \frac{\partial T^{02}}{\partial y}\Delta t\Delta x\Delta y\Delta z$$
$$\left(\int T^{03} dtdxdy\right)\Big|_{z_0+\Delta z} - \left(\int T^{03} dtdxdy\right)\Big|_{z_0} \to \frac{\partial T^{03}}{\partial z}\Delta t\Delta x\Delta y\Delta z$$

Thus

$$\frac{\partial T^{00}}{\partial x^0} + \frac{\partial T^{0j}}{\partial x^j} = 0$$

The procedure is analogous for spatial components and yields:

$$\frac{\partial T^{j0}}{\partial x^0} + \frac{\partial T^{jk}}{\partial x^k} = 0.$$

Example 9.11. The Stress-Energy Tensor for a perfect fluid:
(a) Derive the frame-independent expression for the perfect fluid stress-energy tensor from the rest-frame components.
(b) Show that the perfect fluid the inertial mass per unit volume is isotropic and equals density plus pressure.

Solution
(a) Let's construct a rank-2 symmetric tensor based on fluid velocity $\vec{u}$:

$$T^{\alpha\beta} = Au^{\alpha}u^{\beta} + Bg^{\alpha\beta}$$

In a local rest frame, $\vec{u} = (-1,0,0,0)$, where we simply have rest-frame density and zero momentum:

$$T^{00} = A - B = \rho$$
$$T^{j0} = T^{0j} = 0$$
$$T^{jk} = B\delta^{jk} = p\delta^{jk}$$

Thus

$$B = p, A = \rho + p \Rightarrow \underline{T^{\alpha\beta} = (\rho + p)u^{\alpha}u^{\beta} + pg^{\alpha\beta}}$$

(b) Inertial mass per unit volume has form:

$$T^{00}\delta^{ij} + T^{ij} = \rho\delta^{ij} + p\delta^{ij} = (\rho + p)\delta^{ij}$$

thus isotropic.

Example 9.12. In 4D Minkowski space-time, the electromagnetic field is represented by a second-rank antisymmetric tensor $F^{\mu\nu}$ which represents a force whose change in 4-velocity (acceleration) imparted on a charge q, in 4-vector form, is then:

$$\nabla_u p = qF \rightarrow \frac{dp^{\alpha}}{d\tau} = qF^{\alpha\beta}u_{\beta}$$

(a) Compare with the Lorentz Force Law (non-relativistic scenario), to obtain the components of $F^{\mu\nu}$.
(b) Define the Dual of $F^{\mu\nu}$ by ${}^*F^{\mu\nu} = \frac{1}{2}\epsilon_{\mu\nu\alpha\beta}\, F^{\alpha\beta}$ and obtain its components.

(c) For observer with 4-velocity $\vec{u}$, show that $F(\dots,\vec{u})$ and $-\ {}^{*}F(\dots,\vec{u})$ correspond to the electric and magnetic fields seen by that observer.
(d) Two independent scalars can be constructed from $F^{\mu\nu}$: $F^{\mu\nu}F_{\mu\nu}$ and ${}^{*}F^{\mu\nu}F_{\mu\nu}$. Obtain these scalars in the E and B-fields of any observer (any Lorentz frame).

Solution
In a Lorentz basis,

$$\|g_{\mu\nu}\| = \begin{pmatrix} -1 & 0 & 0 & 0 \\ 0 & 1 & 0 & 0 \\ 0 & 0 & 1 & 0 \\ 0 & 0 & 0 & 0 \end{pmatrix}, \text{and,} \quad \|F^{\mu\nu}\| = \begin{pmatrix} 0 & E^1 & E^2 & E^3 \\ -E^1 & 0 & B^3 & -B^2 \\ -E^2 & -B^3 & 0 & B^1 \\ -E^3 & +B^2 & -B^1 & 0 \end{pmatrix}$$

Also, set $E^0 = B^0 = 0$.

(a) $* F_{\mu\nu} = \frac{1}{2}\epsilon_{\mu\nu\alpha\beta}\ F^{\alpha\beta}$ $\left[\epsilon_{1234} = 1 \text{ and} \epsilon_{\mu\nu\alpha\beta} \text{is totally antisymmetric.}\right]$
$\epsilon_{\mu\nu\alpha\beta}$ is antisymmetric in all indices $\Rightarrow \left\|* F_{\mu\nu}\right\|$ is antisymmetric.
Straightforward calculation gives:

$$* F_{12} = \frac{1}{2}(\epsilon_{1234}F^{34} + \epsilon_{1243}F^{43}) = \frac{1}{2}(B^1 + B^1) = B^1\ etc.$$

Thus

$$\left\|* F_{\mu\nu}\right\| = \begin{pmatrix} 0 & B^1 & B^2 & B^3 \\ -B^1 & 0 & E^3 & -E^2 \\ -B^2 & -E^3 & 0 & E^1 \\ -B^3 & E^2 & -E^1 & 0 \end{pmatrix},$$

and

$$\|* F^{\mu\nu}\| = \left\|g^{\mu\alpha} - F_{\alpha\beta}g^{\beta\nu}\right\| = \begin{pmatrix} 0 & -B^1 & -B^2 & -B^3 \\ B^1 & 0 & E^3 & -E^2 \\ B^2 & -E^3 & 0 & E^1 \\ B^3 & E^2 & -E^1 & 0 \end{pmatrix}$$

(b) $\vec{u} = \vec{e}_0 \Rightarrow u^0 = 1, u^1 = u^2 = u^3 = 0$. So,
$F(\dots,\vec{u})\ :\ F^{\mu\alpha}u_\alpha = F^{\mu\alpha}\vec{e}_0(\vec{e}_\alpha) = F^{\mu\alpha}g_{0\alpha} = F^{\mu 0} = E^\mu$
and
$* F(\dots,\vec{u})\ :\ F^{\mu\alpha}u_\alpha = F^{\mu\alpha}\vec{e}_0(\vec{e}_\alpha) =* F^{\mu\alpha}g_{0\alpha} =* F^{\mu 0}(-1) = -B^\mu.$

(c) $\vec{u} = \gamma\vec{e}_0 + \gamma\vec{v}, \gamma \equiv \frac{1}{\sqrt{1-\vec{v}^2}}$, $\vec{v}$ is a purely spatial vector, thus
$\vec{u}^2 = (\gamma\vec{e}_0 + \gamma\vec{v}) - (\gamma\vec{e}_0 - \gamma\vec{v}) = \gamma^2 g_{00} + \gamma^2\vec{v}^2 = \gamma^2(-1 + \vec{v}^2) = -1$
$F(\dots,\vec{u})\ :\ F^{0\alpha}u_\alpha = F^{01}u_1 + F^{02}u_2 + F^{03}u_3 = \gamma\vec{v}\cdot\vec{E}$ (since $E^0 = 0$)
Thus,

$$
\begin{aligned}
F^{j\alpha}u_\alpha &= F^{j\alpha}(\gamma\vec{e}_0 - \gamma\vec{v})_\alpha \\
&= F^{j\alpha}(\gamma g_{0\alpha} + \gamma v_\alpha) \\
&= \gamma\left[-F^{j0} + F^{j\alpha}v_\alpha\right] \\
&= \gamma\left[E^j + [jkl]v^k B^l\right] \\
&= \gamma\left(\underline{E} + \underline{v} \times \underline{B}\right)^j
\end{aligned}
$$

Thus $qF(\dots,\vec{u})$ is the Lorentz 4-force acting on partials of charge q and u velocity $\vec{u}$.

(d) $F^{\mu\nu}F_{\mu\nu} = F^{\mu\nu}g_{\mu\alpha}F^{\alpha\beta}g_{\beta\nu}$

$$
= \begin{pmatrix} 0 & E^1 & E^2 & E^3 \\ -E^1 & 0 & B^3 & -B^2 \\ -E^2 & -B^3 & 0 & B^1 \\ -E^2 & B^2 & B^1 & 0 \end{pmatrix}^{\mu\nu} \begin{pmatrix} 0 & -E^1 & -E^2 & -E^3 \\ E^1 & 0 & B^3 & -B^2 \\ E^2 & -B^3 & 0 & B^1 \\ E^3 & B^2 & -B^1 & 0 \end{pmatrix}^{\mu\nu}
$$

$$
\begin{aligned}
&= 2\left[-(E^1)^2 - (E^2)^2 - (E^3)^2 + (B^1)^2(B^2)^{2(B^3)^2}\right] \\
&= 2(\vec{B}^2 - \vec{E}^2)
\end{aligned}
$$

$$
* F^{\mu\nu}F_{\mu\nu} = \begin{pmatrix} 0 & -B^1 & -B^2 & -B^3 \\ B^1 & 0 & E^3 & -E^2 \\ B^2 & -E^3 & 0 & E^1 \\ B^2 & E^2 & -E^1 & 0 \end{pmatrix}^{\mu\nu} \begin{pmatrix} 0 & -E^1 & -E^2 & -E^3 \\ E^1 & 0 & B^3 & -B^2 \\ E^2 & -B^3 & 0 & B^1 \\ E^3 & B^2 & -B^1 & 0 \end{pmatrix}^{\mu\nu}
$$

$$
\begin{aligned}
&= 2[B^1E^1 + B^2E^2 + B^3E^3 + E^1B^1 + E^2B^2 + E^3B^3] \\
&= 4\,\vec{B}\cdot\vec{E}
\end{aligned}
$$

These are the only scalar quantities (no free indices and hence invariant under Lorentz transformations) that one can form out of the electromagnetic field tensor. (Building blocks are $F^{\mu\nu}$, $*\,F^{\mu\nu}$, $\epsilon^{\alpha\beta\gamma\delta}$, and $g^{\alpha\beta}$. Because of symmetry of $g^{\alpha\beta}$ and total antisymmetry of the rest, all other contractions vanish.

Example 9.13. Maxwell's equations can be written: ${F^{\alpha\beta}}_{;\beta} = 4\pi J^\alpha$ and ${^*F^{\alpha\beta}}_{;\beta} = 0$, where $J^\alpha$is the density-current 4-vector, and the stress-energy tensor associated with the electromagnetic field is:

$$
T^{\mu\nu} = \frac{1}{4\pi}\left(F^{\mu\alpha}{F^\nu}_\alpha - \frac{1}{4}g^{\mu\nu}F_{\alpha\beta}F^{\alpha\beta}\right).
$$

(a) Obtain $T^{\mu\nu}{}_{;\nu} = \frac{1}{4\pi}\left(\mathrm{F}^{\mu\alpha}{}_{;\nu}\mathrm{F}^{\nu}{}_{\alpha} + \mathrm{F}^{\mu\alpha}\mathrm{F}^{\mu}{}_{\alpha;\nu} - \frac{1}{2}F_{\alpha\beta}{}^{;\nu}F^{\alpha\beta}\right)$.
(b) Show $T^{\mu\nu}{}_{;\nu} = -\mathrm{F}^{\alpha\beta}J_{\beta}$.

Solution
(a) $T = T^{\mu\nu}\frac{\partial}{\partial x^{\nu}} \oplus \frac{\partial}{\partial x^{\nu}}$
The usual chain rule is valid for covariant derivatives. We can see this most easily in a local inertial basis where $\Gamma^{\alpha}_{\beta\gamma} = 0$. Using the chain rule, it is almost obvious that

$$T^{\mu\nu}{}_{;\nu} = \frac{1}{4\pi}\left(F^{\mu\nu}{}_{;\nu}F^{\mu}_{\alpha} + F^{\mu\alpha}F^{\mu}{}_{\alpha;\nu} - \frac{1}{2}\cdot g^{\mu\nu}F_{\mu\beta;\nu}F^{\alpha\beta}\right)$$
$$= \frac{1}{4\pi}\left(F^{\mu\nu}{}_{;\nu}F^{\mu}_{\alpha} + F^{\mu\alpha}F^{\mu}{}_{\alpha;\nu} - \frac{1}{2}F_{\alpha\beta}{}^{;\nu}F^{\alpha\beta}\right)$$

(b) From (a):

$$4\pi\, T^{\mu\nu}{}_{;\nu} = F^{\mu\alpha}F^{\nu}{}_{\alpha;\nu} + F^{\mu\alpha}{}_{;\beta}F^{\beta}_{\alpha} - \frac{1}{2}F_{\alpha\beta}{}^{;\nu}F^{\alpha\beta}$$
$$= F^{\mu}_{\alpha}F^{\mu\alpha}{}_{;\nu} + F^{\mu}{}_{\alpha;\beta}F^{\beta\alpha} - \frac{1}{2}F_{\alpha\beta}{}^{;\mu}F^{\alpha\beta}$$

Thus

$$F^{\mu\alpha}_{\alpha}F^{\nu\mu} \rightarrow = -F^{\mu}_{\alpha}F^{\alpha\nu}{}_{;\nu} + \frac{1}{2}\left(-F^{\mu}{}_{\alpha;\beta} + F^{\mu}{}_{\beta;\alpha} - F_{\alpha\beta}{}^{;\mu}\right)F^{\alpha\beta}$$

Therefore, we obtain

$$4\pi\, T^{\ \nu}_{\mu;\nu} = -F_{\mu\alpha}F^{\alpha\nu}{}_{;\nu} - \frac{1}{2}F^{\alpha\beta}\left(F_{\mu\alpha i\beta} + F_{\beta\mu i\alpha} + F_{\alpha\beta i\mu}\right).$$

Example 9.14. Consider the separation vector $\boldsymbol{\xi}$ between two geodesics. When describing the geodesic deviation we then have:

$$\nabla_u\nabla_u\boldsymbol{\xi} = -\boldsymbol{R}(\ldots,\boldsymbol{u},\boldsymbol{\xi},\boldsymbol{u})$$

Let's do a Newtonian analysis where $\boldsymbol{\xi}$ only has spatial components and $\tau = t$:

$$\frac{d^2\xi^j}{dt^2} = -R^j{}_{0k0}\xi^k.$$

If compared with the classic Newtonian equation for gravitational force:

$$R^j{}_{0k0} = \frac{\partial^2\Phi}{\partial x^j\partial x^k}$$

and we see that the Riemann curvature tensor is a generalization of the Newtonian tidal field. Thus, gravity is shown here to be a manifestation of space-time curvature.

Note that the radius of curvature of space-time is defined by

$$\tilde{R} = 0\left[\frac{1}{\left|R^{\alpha}_{\delta\gamma\delta}\right|^{1/2}}\right]$$

At the Earth's surface show that the radius of curvature of space-time is approximately 1 in astronomical units (the Earth-Sun distance).

Solution

$R^{j}_{0K0} = \frac{\partial^2\Phi}{\partial x^j \partial x^K}$ in a Newtonian tidal field where

$$\Phi = -\frac{MG}{r} \quad \rightarrow \quad \frac{\partial^2\Phi}{\partial r^2} = -\frac{2MG}{c^2 r^3}$$

with the latter form in geometric units. Thus

$$\tilde{R} \approx \frac{cr^{3/2}}{\sqrt{2MG}} \quad and \quad G = 6.67 \times 10^{-6}\, cm^3/g\sec^2$$

For Earth:

$$M_{\oplus} = 5.977 \times 10^{27} g$$
$$R_{\oplus} = 6.371 \times 10^{5} cm$$

Thus,

$$\tilde{R} \simeq 1.71 \times 10^{13} cm = 1.14 AU$$

For the Sun: $\tilde{R} \simeq 2.1 AU$.

Example 9.15. Let's examine the geodesic deviation on a sphere, one geodesic the equator, and another (a great circle) near it (and parallel at $\phi = 0$). Let $\vec{\varepsilon}$ be the separation vector (at $\phi = 0$ we then have $\vec{\varepsilon} = \varepsilon^{\theta}\frac{\partial}{\partial\theta}$.

Let l be the proper distance along the equatorial geodesic: $\frac{d}{dl} = u$ then denotes the tangent vector.

(a) Show that $\ell = a\phi$ along the equatorial geodesic.

(b) Show that the equatorial geodesic can be written as:

$$\frac{d^2\varepsilon^{\theta}}{d\phi^2} = -\varepsilon^{\theta} \quad and \quad \frac{d^2\varepsilon^{\phi}}{d\phi^2} = 0.$$

(c) Solve the relations in (b) subject to the initial conditions indicated.

Solution

(a) We have: $ds^2 = a^2(d\theta^2 + \sin^2\theta\, d\phi^2)$. Along the equatorial geodesic $\theta = \frac{\pi}{2}$ *and* $d\theta = 0$. Thus

$ds^2 = a^2 d\phi^2$

$$ds = a d\phi \rightarrow \int_0^{\ell} ds = \int_0^{\phi} a\, d\phi \rightarrow \ell = a\phi.$$

(b) $\nabla_{\vec{u}}\nabla_{\vec{u}}\vec{\varepsilon} = -R(\dots, \vec{u}, \vec{\varepsilon}, \vec{u})$. Let's decompose into component form:

$$\left(\varepsilon^{\alpha}_{;\beta}u^{\beta}\right)_{;\gamma}u^{\gamma} = -R^{\alpha}_{\beta\gamma\delta}u^{\beta}\varepsilon^{\gamma}u^{\delta}$$

Since $\vec{u} = a\vec{\phi}$ on the equaterial geodesic, $\vec{u} = a\sin\theta\,\vec{\phi}$ elsewhere. Now, since:

$$\vec{\varepsilon} = \varepsilon^{\theta}\hat{\theta} \rightarrow \varepsilon^{\phi} = 0$$

and we have:

$$\left(\varepsilon^{\theta}_{;\phi}u^{\phi}\right)_{;\phi}u^{\phi} = -R^{\theta}_{\phi\theta\phi}u^{\phi}\varepsilon^{\theta}u^{\phi}$$

Thus

$$\frac{d^2\varepsilon^{\theta}}{d\phi^2}(a\sin\theta)^2 = \varepsilon^{\theta}R_{\theta\phi\theta\phi} = -\varepsilon^{\theta}(a^2\sin^2\theta)$$

and we have:

$$\frac{d^2\varepsilon^{\theta}}{d\phi^2} = -\varepsilon^{\theta} \; and \; \frac{d^2\varepsilon^{\phi}}{d\phi^2} = 0$$

(c) Guess $\varepsilon^{\theta} = a\sin\phi + b\cos\phi$

When $\phi = 0, \varepsilon^{\theta} =$ the maximum separation, thus $\frac{d\varepsilon^{\theta}}{d\phi} = 0$ if an I.C., thus $a = 0$, and we get:

$$\varepsilon^{\theta} = b\cos\phi \quad and \quad \varepsilon^{\phi} = 0.$$

9.6 Exercises

(Ex. 9.1) Verify Eqn. 9-1.
(Ex. 9.2) Verify Eqn. 9-2.
(Ex. 9.3) Verify Eqn. 9-3.
(Ex. 9.4) Verify Eqn. 9-4.
(Ex. 9.5) Verify Eqn. 9-5.
(Ex. 9.6) Verify Eqn. 9-6.
(Ex. 9.7) Verify Eqn. 9-7.
(Ex. 9.8) Verify Eqn. 9-8.
(Ex. 9.9) Verify Eqn. 9-9.
(Ex. 9.10) Verify Eqn. 9-10.
(Ex. 9.11) Verify Eqn. 9-11.

Chapter 10. Gauge Fields & Geometries with Covariant Derivatives

10.1 Field Theories

The generalization from Action with point-like Lagrangian to one with field-like Lagrangian density:

$$S = \int_{t_1}^{t_2} dtL(q,\dot{q}) \quad \rightarrow \quad S = \int_{t_1}^{t_2} dt \int d^3x\mathcal{L}(\varphi(x), \partial_\mu \varphi(x))$$

(Eqn. 10-1)

and

$$\delta S = \int_{t_1}^{t_2} dt \left(\frac{\partial L}{\partial q} - \frac{d}{dt}\frac{\partial L}{\partial \dot{q}}\right) \delta q \quad \rightarrow \quad \delta S$$
$$= \int_{t_1}^{t_2} dt \int d^3x \left(\frac{\partial \mathcal{L}}{\partial \varphi(x)} - \partial_\mu \frac{\partial \mathcal{L}}{\partial \left(\partial_\mu \varphi(x)\right)}\right) \delta\varphi(x).$$

(Eqn. 10-2)

Let's now consider some of the Lagrangian densities that might be considered for various (classical) field theories and their equations of motion.

Matter-Fields with $\boldsymbol{mass} > \mathbf{0}$ (no 'matter' field that is massless)

The Lorentz Invariant Scalar Field (massive)

$$\mathcal{L} = \frac{1}{2}((\partial^\mu \varphi)(\partial_\mu \varphi) - m^2\varphi^2)$$

and

$$(\Box + m^2)\varphi(x) = 0.$$

The Complex Scalar Field (massive)

$$\varphi(x) = \frac{1}{\sqrt{2}}[a(x) + ib(x)]$$

$$\mathcal{L} = \frac{1}{2}\left((\partial^\mu a)(\partial_\mu a) - m^2a^2\right) + \frac{1}{2}\left((\partial^\mu b)(\partial_\mu b) - m^2b^2\right)$$
$$= (\partial^\mu \varphi)^*(\partial_\mu \varphi) - m^2\varphi^*\varphi$$

and

$$(\Box + m^2)\varphi(x) = 0 \quad and \quad (\Box + m^2)\varphi^*(x) = 0\,.$$

The Massive Spinor Field – The Dirac Equation

$$\mathcal{L} = \bar{\psi}(x)\big(i\gamma^\mu\partial_\mu - m\big)\psi(x),\;\; \bar{\psi} = \psi^t\gamma^0$$

and

$$\big(i\gamma^\mu\partial_\mu - m\big)\psi(x) = 0.$$

Force-Fields with $mass = 0$ and $mass > 0$

The Massless Vector Field – EM (has U(1) gauge group)

$$\mathcal{L} = -\frac{1}{4}\big(\partial_\mu A_\nu - \partial_\nu A_\mu\big)\big(\partial^\mu A^\nu - \partial^\nu A^\mu\big)$$

and if massive simply add:

$$\mathcal{L} = above + \frac{1}{2}M^2 A_\mu A^\mu$$

with

$$\partial_\nu(\partial^\nu A^\mu - \partial^\mu A^\nu) + M^2 A^\mu = 0,$$

which is known as the Proca equation in this more general form.

The Massive Electroweak Field (with SU(2) gauge group)

The massive fields in the standard model have a Yukawa potential exponential fall-off in force that is imperceptible outside an atomic range, so won't be studied until in the full quantum context in Book 5 [12].

The Massless Chromodynamics Field (with SU(3) gauge group)

The massless fields in the standard model with SU(3) leads to quark confinement and colorlessness, so has non-classical behavior from the outset so, also, won't be studied until in the full quantum context in Book 5 [12].

10.2 Overview of the Standard Model

Matter at the level of known elementary particles consists of pointlike spin-1/2, quarks and leptons. There are six quarks that are grouped as left-handed, $SU(2)_L$ doublets:

$$\binom{u}{d}_L,\; \binom{c}{s}_L,\; \binom{t}{b}_L,$$

for the up, down, charm, strange, top, and bottom quarks. Similarly, the leptons (electron, muon, tau) and their neutrinos are also $SU(2)_L$ doublets:

$$\binom{\nu_e}{e^-}_L, \binom{\nu_\mu}{\mu^-}_L, \binom{\nu_\tau}{\tau^-}_L.$$

The quarks also carry color (3 types), for the strong interaction charge for SU(3). he leptons do not feel the strong interaction (and are colorless).

There are six quarks that are grouped as right-handed, $SU(2)_R$ weak-interaction singlets:

$$(u)_R, (d)_R, (s)_R, (s)_R, (t)_R, (b)_R.$$

Likewise, there are three right-handed, $SU(2)_R$ weak-interaction singlets for the charged leptons:

$$(e^-)_R, (\mu^-)_R, (\tau^-)_R.$$

In the standard model there are no right-handed neutrinos, but it has been suggested that the right-handed neutrinos may exist, but not interact via weak interactions (or EM or strong), thus only gravitationally, and thus are a prime candidate for dark matter. Weak interactions among left-handed particles conserve electron number, muon number, and tau number. The existence of he non-weakly interacting right-handed quarks and charged lepton is seen in their strong and EM interaction scattering behavior.

The weak isospin doublets (left-handed) are part of the evidence for an $SU(2)_L$ gauge symmetry, while the three quark colors are consistent with $SU(3)_C$ and in agreement with the various baryons that are observed. Electroweak theory combines the $U(1)_Y$ weak hypercharge phase symmetry with $SU(2)_L$
to actually give the observed $U(1)_{EM}$.

From initial experiments it was not clear that neutrinos had mass (if they had mass it had to be very small). Discovery of neutrino oscillations [61], however, suggests neutrinos have mass, and if there are right-handed neutrinos they probably have mass too (Book 7 provides one of many predictions in this regard, where such massive sterile neutrinos are suggested to exist).

Early efforts realized there was a SU(2) symmetry but mistakenly attributed it to the approximate 'flavor' model for quarks, where there was initially only three quarks known: u, d, s. In considering the quark groupings $q\bar{q}$ (for the mesons), there are $3 \otimes 3 = 1 \oplus 8$, or 9 groupings predicted, and this is what is observed. Similarly, if quark groupings qqq are considered, there are $3 \otimes 3 \otimes 3 = 1 \oplus 8 \oplus 8 \oplus 10$, with 27 groupings

predicted, and again what is observed. There are two problems with the flavor model (even before expanding it to include additional flavors):
(1) It's an approximate symmetry that is largely realized due to the near degeneracy of the u, d, and s masses. The other flavors have very different masses. Thus, it is thought by some that the breaking of the SU(N) flavor symmetry (N flavors) is a consequence of the quark mass differences.
(2) The flavor model has states like "uuu" that clearly violate the Pauli exclusion principle. To resolve, if three colors were introduced then the qqq could be color neutral, as a rule, and eliminate the conflict. The $q\bar{q}$ are seen to be color neutral via anti-color Thus, a three color SU(3) exact symmetry is indicated, with three generations of elementary particles.

Sterile Neutrinos
In V-A Theory only left-handed particles participate in the weak interaction [62-64]. Since the neutrino already has no strong or electromagnetic interaction, it would be completely 'sterile'. This means that a right-handed neutrino, if it were to exist, would only be detectable gravitationally. There is the mystery of Dark Matter, so many have suggested that Dark Matter might be sterile right-handed neutrinos (with whatever masses are consistent with the galaxy structure formation simulations).

10.3 Symmetries and Conservation Laws
The association of any symmetry with a conservation law was first described by E. Noether [65,66].

Translational Symmetry
Let's consider a simple translational transformation to start:

$$x_\mu \to x'_\mu = x_\mu + a_\mu \,.$$

(Eqn. 10-3)

Then

$$\delta\mathcal{L} = \mathcal{L}[x'] - \mathcal{L}[x] = a^\mu \frac{d\mathcal{L}}{dx^\mu} = a^\mu \partial_\mu \mathcal{L}\,.$$

(Eqn. 10-4)

Suppose the Lagrangian density has the typical dependencies (participating in the variation): $\mathcal{L}(\varphi(x), \partial_\mu\varphi(x))$, we then have:

$$\delta\mathcal{L} = \frac{\partial\mathcal{L}}{\partial\varphi}\delta\varphi + \frac{\partial\mathcal{L}}{\partial(\partial_\mu\varphi)}\delta(\partial_\mu\varphi) = \frac{\partial\mathcal{L}}{\partial\varphi}a^\mu\partial_\mu\varphi + \frac{\partial\mathcal{L}}{\partial(\partial_\mu\varphi)}a^\nu\partial_\nu(\partial_\mu\varphi)$$

Thus,

$$a_\mu \partial_\nu \left[\frac{\partial \mathcal{L}}{\partial(\partial_\nu \varphi)} \partial^\mu \varphi - g^{\mu\nu} \mathcal{L}\right] = 0.$$

Thus,

$$T^{\mu\nu} = \frac{\partial \mathcal{L}}{\partial(\partial_\nu \varphi)} \partial^\mu \varphi - g^{\mu\nu} \mathcal{L}$$

(Eqn. 10-5)

is known as the stress energy tensor, and it is conserved according to:

$$\partial_\nu T^{\mu\nu} = 0.$$

(Eqn. 10-6)

T^{00} is the Hamiltonian density, thus total energy is:

$$H = \int d^3x T^{00}$$

(Eqn. 10-7)

and is a constant of the motion, as is the momentum density $T^{0\nu}$. Thus, 4-translation invariance leads to 4-momentum conservation.

10.4 Gauge Fields with desirable properties

Summary of Field Properties desired [67]

In seeking a generalization of Maxwell's formulation of EM, with its implicit Lorentz invariance, and given the spinorial Yang-Mills forms of matter from known features of elementary particles, we desire the following:

(i) External symmetries: Lorentz group at a minimum, but want Poincare (translation invariance) to whatever extent possible. Also interested in conformal group for cases where the particle mass is zero.

(ii) Internal symmetries: Want Yang-Mills gauge theories (to be discussed) that capture the known SU(2) and SU(3) features of elementary particles.

(iii) Covariance: Ability to couple to gravitation (established implicitly if working in curved space-time).

It will be shown that Gauge Theory (Ch. 10 & 11) satisfies these features. Here's a preview on gauge potentials and fields.

Gauge potentials [67]

The Lie algebra of the Lie group G is the Lie algebra g of left invariant vector fields on G. Alternatively, we could take as Lie algebra of G the tangent space G_e at the identity with Lie algebra structure induced by requiring the vector space isomorphism $\alpha: g \to G_e,\ \alpha(X): \to X(e)$ of g with G_e to be an isomorphism of the Lie algebras. For SU(n), the Lie Algebra L(G) consists of skew-Hermitian nxn matrices of trace zero.

A gauge potential is a set of functions $A_\mu(x)$ that take values in L(G), where $x = (x_1, \dots, x_\mu)$, and $\mu = 1 \dots 4$. The potential is associated with the gauge covariant derivative:

$$\nabla_\mu = \partial_\mu + A_\mu \, .$$

(Eqn. 10-8)

This operator acts on a vector function, $(f_1(x), \dots, f_m(x))$, where m is the dimension of the representation of G. If G=SU(n), then we simply have m=n using the standard representatives for SU(n). By evaluating the commutator of the derivative we get the expression for the gauge Field $F_{\mu\nu}$:

$$F_{\mu\nu} = [\nabla_\mu, \nabla_\nu] = \partial_\mu A_\nu - \partial_\nu A_\mu + [\mathrm{A}_\mu, \mathrm{A}_\nu].$$

(Eqn. 10-9)

There is non-uniqueness in the choice of functions $A_\mu(x)$ according to gauge invariance under the following transformation:

$$\nabla'_\mu = g^{-1}\nabla_\mu,$$
$$A'_\mu = g^{-1}A_\mu g + g^{-1}\partial_\mu g \, ,$$
$$\Phi' = \mathrm{g}\Phi \, ,$$
$$F_{\mu\nu}{}' = g^{-1}F_{\mu\nu}g$$

(Eqn. 10-10)

Consider $(\nabla_\mu \Phi)'$:

$$(\nabla_\mu \Phi)' = g^{-1}\nabla_\mu(\mathrm{g}\Phi) = [g^{-1}A_\mu g + g^{-1}\partial_\mu g + \partial_\mu]\Phi = (\partial_\mu + A_\mu{}')\Phi$$

(Eqn. 10-11)

Non-renormalizability of GR

If we consider GR defined by an Action principle, and using the Einstein-Hilbert Action in particular, we introduce a coupling constant of mass dimension -2. In a perturbative analysis there are then terms of divergently higher mass dimension, which then leads to impossibility of setting up the counter terms necessary for renormalization. Thus, GR can't be made renormalizable due to the odd nature of its coupling with the dimensionful parameter G. This makes the force due to GR fundamentally different from the other force fields.

10.5 Covariant Derivatives

Before moving on to the fully general gauge covariant derivative, we must first recast our description of covariant derivatives into abstract index notation (App. A.1) and understand Lie Group mathematics (Sec. 10.6). A review of tensors in this more advanced notation is in App. A.2, along with tensor definitions in terms of mapping (App. A.3) and the general formalism of Lie Derivatives (App. A.4). Using abstract index notation, let's now review some examples involving covariant derivatives including re-derivation of the Riemann tensor and other derivations in this advanced notion. In concrete indices, these derivations would take much more space, the derivations in abstract indices can be very short (and possibly more insightful for this reason). In the derivations that follow, as examples, extra steps are shown with care. Normally, many of the derivations would be even shorter.

Example 10.1. Let's start by showing that any two covariant derivatives, ∇_a and $\widetilde{\nabla}_a$ differ by a tensor C^b_{ca}:

$$(\nabla_a - \widetilde{\nabla}_a)V^b = C^b_{ca}V^c\,,$$

with C symmetric on its covariant indices: $C^b_{ca} = C^b_{ac}$. (Use abstract indices.) Then, use Leibnitz to show the analogous formula holds for any tensor:

$$(\nabla_a - \widetilde{\nabla}_a)T^{b\ldots c}_{d\ldots e}$$
$$= C^b_{fa}T^{f\ldots c}_{d\ldots e} + \cdots + C^c_{fa}T^{b\ldots f}_{d\ldots e} - C^f_{da}T^{b\ldots c}_{f\ldots e} - \cdots - C^f_{ea}T^{b\ldots c}_{d\ldots e}\,.$$

Solution

Consider $(\nabla_a - \widetilde{\nabla}_a)(V^c w_c)$, since a covariant derivative is a map between $\binom{m}{n}$ and $\binom{n}{m}$ tensor fields which satisfies, among other things: $\nabla_a f = \text{grad}\, f$ for scalars. Then, since $V^c w_c$ is a scalar we have $(\nabla_a - \widetilde{\nabla}_a)(V^c w_c) = 0$. Now, using Leibnitz: $w_c(\nabla_a - \widetilde{\nabla}_a)V^c + V^c(\nabla_a - \widetilde{\nabla}_a)w_c = 0$. Thus we find that $(\nabla_a - \widetilde{\nabla}_a)$ defines a map of the 1-form w_c to a $\binom{0}{2}$ tensor; furthermore, the map is linear. Consequently $(\nabla_a - \widetilde{\nabla}_a)$ must be a tensor of type $\binom{1}{2}$ as this is the type which maps 1-forms to $\binom{0}{2}$ tensors. Denote the tensor field by $-C^d_{ac}$:

$$(\nabla_a - \widetilde{\nabla}_a)w_c = -C^d_{ac}w_d$$

C^d_{ac} is also symmetric on its covariant indices as can be shown by considering $w_c \nabla_c f$:

$$(\nabla_a - \widetilde{\nabla}_a)\nabla_c f = \nabla_a \nabla_c f - \widetilde{\nabla}_a \nabla_c f = (\nabla_a \nabla_c - \widetilde{\nabla}_a \widetilde{\nabla}_c) f,$$

which is symmetric on a and c. So,

$$w_d(\nabla_a - \widetilde{\nabla}_a)V^d + V^c C^d_{ac} w_d = 0,$$

and $(\nabla_a - \widetilde{\nabla}_a)V^b = C^b_{ca}V^c$, so $C^b_{ca} = C^b_{ac}$.

Now, individually the covariant derivatives satisfy Leibnitz' rule as will any linear combination of them. So, the application of $(\nabla_a - \widetilde{\nabla}_a)$ to any tensor can be expressed in terms of its action on the Vector and Dual spaces of the tensor product space. Thus for contravariant indices we get a contraction as with a a dual vector: $(\nabla_a - \widetilde{\nabla}_a)w_c = -C^d_{ac}w_d$, while for covariant indices we get that of a vector: $(\nabla_a - \widetilde{\nabla}_a)V^b = C^b_{ca}V^c$. Thus,

$$(\nabla_a - \widetilde{\nabla}_a)T^{b\ldots c}_{d\ldots e} = C^b_{fa}T^{f\ldots c}_{d\ldots e} + \cdots + C^c_{fa}T^{b\ldots f}_{d\ldots e} - C^f_{da}T^{b\ldots c}_{f\ldots e} - \cdots - C^f_{ea}T^{b\ldots c}_{d\ldots e}$$

Riemann tensor

Let's now re-derive the Riemann tensor by way of the Cartan structure equations using abstract indices. The Riemann tensor is defined as follows:

$$\frac{1}{2}R^a_{bcd}V^b = \nabla_{[c}\nabla_{d]}V^a$$

(Eqn. 10-12)

Using an orthonormal frame and Cartan's formalism:

$$\nabla_k e_j = \Gamma^i_{jk} e_j$$

$$R^i_{jk\ell} = e_k \Gamma^i_{j\ell} - e_\ell \Gamma^i_{jk} + \Gamma^i_{mk}\Gamma^m_{j\ell} - \Gamma^i_{m\ell}\Gamma^m_{jk}$$

(Eqn. 10-13)

Cartan's structure equations, using the notation ω^i dual basis vectors and $[u,v]^i = u^j \frac{\partial}{\partial x^j} v^i - v^j \frac{\partial u^i}{\partial x^j}$, and have $\omega^i(e_k) = \delta^i_k = constant$ for an orthonormal frame:

$$\begin{aligned}(d\omega^i)_{jk} &= d\omega^i(e_j, e_k) = 2e^a_b \nabla_{[a} W^i_{b]} e^a_j e^b_k = (\nabla_a \omega^i_b e^a_j e^b_k - \nabla_b \omega^i_a e^a_j e^b_k) \\ &= [\nabla_j(\omega^i_b e^b_k) - \omega^i_b \nabla_j e^b_k] - [\nabla_k(\omega^i_a e^a_j) - \omega^i_a \nabla_k e^a_j] \\ &= \nabla_j\{\omega^i(e_k)\} - \nabla_k\{\omega^i(e_j)\} + \omega^i(\nabla_k e_j) - \omega^i(\nabla_j e_k) \\ &= -\omega^i(\nabla_j e_k - \nabla_k e_j) = -\omega^i(e^a_j \nabla_a e_k - e^a_k \nabla_a e_j) \\ &= -\omega^i([e_j, e_k]) = -\omega^i(c^\ell_{jk} e_\ell) = -c^i_{jk}\end{aligned}$$

Thus

$$\left(d\omega^i\right)_{jk} = -c^i_{jk}$$

(Eqn. 10-14)

If we next define connection forms: $\omega^i_j = \Gamma^i_{jk}\omega^k$ we expect $\omega_{ij} = -\omega_{ji}$ in an orthonormal frame (or any frame where the metric has constant components). Let's show this. Start with

$$\nabla_k e_j = \Gamma^i_{jk} e_j \quad \rightarrow \quad e^a_k \nabla_a e^b_j = \Gamma^i_{jk} e^b_i (e^a_k \omega^k_a) \ \rightarrow \ \nabla_a e^b_j = \Gamma^i_{jk} \omega^k_a e^b_i$$
$$\rightarrow \quad \nabla_a e^b_j = \omega^i_{ja} e^b_i \ .$$

Now have:

$$e_i \cdot \nabla_a e_j = \nabla_a\left(g_{ij}\right) - e_j \cdot \nabla_a e_i = -e_j \cdot \nabla_a e_i \ \Rightarrow \ \omega_{ij} = -\omega_{ji}$$

Notice that:

$$\left(d\omega^i\right)_{k\ell} = -c^i_{k\ell} = \Gamma^i_{k\ell} - \Gamma^i_{\ell k} = \left(\omega^i_j\right)_\ell \left(\omega^j\right)_k - \left(\omega^i_j\right)_k \left(\omega^j\right)_\ell$$

Consider

$$\omega^i_j = \omega^i_{ja} = \left(\Gamma^i_{jk}\omega^k\right)_a \ \ and \ \ \left(\omega^i_j\right)_\ell = \left(\Gamma^i_{jk}\omega^k\right)_a e^a_\ell = \Gamma^i_{jk}\omega^k_a e^a_\ell$$

So, $\left(\omega^i_j\right)_\ell \left(\omega^j\right)_k = \Gamma^i_{j\ell}\delta^j_k = \Gamma^i_{k\ell}$ and we can now write:

$$\left(d\omega^i\right)_{k\ell} = -\left(\omega^i_j \wedge \omega^j\right)_{k\ell} \ \rightarrow \ d\omega^i = -\omega^i_j \wedge \omega^j$$

(Eqn. 10-15)

Recall $\Gamma_{ijk} = \frac{1}{2}\Big(e_k g_{ij} + e_j g_{ik} - e_i g_{jk} + \underbrace{c_{kij} + c_{jik} - c_{ijk}}\Big)$, where we have $\left(\omega_{ij}\right)_k = \omega_{ija} e^a_k = \Gamma_{ijk}\omega^k_a e^a_k = \Gamma_{ijk}$ and $\left(d\omega_j\right)_{k\ell} = c_{ik\ell}$. So, In an orthonormal basis, or one where the metric is constant:

$$\left(\omega_{ij}\right)_k = \frac{1}{2}\left((d\omega_i)_{jk} + \left(d\omega_j\right)_{ki} - (d\omega_k)_{ij}\right).$$

(Eqn. 10-16)

Since the Ricci tensor

$$R^i_j = d\omega^i_j + \omega^i_k \wedge \omega^k_j$$

(Eqn. 10-17)

Implying

$$\left(R^i_j\right)_{ab} = \left(d\omega^i_j\right)_{ab} + \left(\omega^i_k \wedge k_j\right)_{ab}$$

(Eqn. 10-18)

Note that $\omega^i_j = \Gamma^i_{jk}\omega^k$, $\omega^i(e_j) = \delta^i_j$, $\left(\omega^i_j\right)_\ell = \Gamma^i_{j\ell}$, $\left(\omega^i_j\right)_a = \left(\Gamma^i_{jk}\omega^k_a\right)$ will be useful in what follows.

We have:

$$\left(d\omega_j^i\right)_{k\ell} = \left(d\omega_j^i\right)_{ab} e_k^a e_\ell^b = \nabla_{[a}\left(\omega_j^i\right)_{b]} e_k^a e_\ell^b$$
$$= e_k^a \nabla_a \left(\omega_j^i\right)_b e_\ell^b - e_\ell^b \nabla_b \left(\omega_j^i\right)_a e_k^a$$
$= \nabla_k\{\Gamma_{jm}^i \omega_b^m\} e_\ell^b - \nabla_\ell\{\Gamma_{jm}^i \omega_a^m\} e_k^a = e_k \Gamma_{j\ell}^i - e_\ell \Gamma_{jk}^i - \Gamma_{jm}^i \Gamma_{\ell k}^m + \Gamma_{jm}^i \Gamma_{k\ell}^m = e_k\left(\Gamma_{j\ell}^i\right) - e_\ell\left(\Gamma_{jk}^i\right) + 2\Gamma_{jm}^i \Gamma_{[k\ell]}^m$.

Thus,

$$\left(d\omega_j^i\right)_{k\ell} = e_k\left(\Gamma_{j\ell}^i\right) - e_\ell\left(\Gamma_{jk}^i\right) + 2\Gamma_{jm}^i \Gamma_{[k\ell]}^m$$

(Eqn. 10-19)

We also have:

$$\left(\omega_m^i \wedge \omega_j^m\right)_{k\ell} = \left(\omega_m^i \wedge \omega_j^m\right)_{ab} e_k^a e_\ell^b = \left(\omega_m^i\right)_{[a} \left(\omega_j^m\right)_{b]} e_k^a e_\ell^b$$
$$= \left(\omega_m^i\right)_k \left(\omega_j^m\right)_\ell - \left(\omega_m^i\right)_\ell \left(\omega_j^m\right)_k$$

Thus,

$$\left(\omega_m^i \wedge \omega_j^m\right)_{k\ell} = \Gamma_{mk}^i \Gamma_{j\ell}^m - \Gamma_{m\ell}^i \Gamma_{jk}^m$$

(Eqn. 10-20)

Since

$$R_{jk\ell}^i = \left(R_j^i\right)_{ab} e_k^a e_\ell^b$$

(Eqn. 10-21)

We have:

$$R_{jk\ell}^i = e_k\left(\Gamma_{j\ell}^i\right) - e_\ell\left(\Gamma_{jk}^i\right) + \Gamma_{mk}^i \Gamma_{j\ell}^m - \Gamma_{m\ell}^i \Gamma_{jk}^m + 2\Gamma_{jm}^i \Gamma_{[k\ell]}^m,$$

(Eqn. 10-22)

which is the Riemann tensor.

Example 10.2. Show that $\Gamma_{[jk]}^i = -\frac{1}{2} c_{jk}^i$.

Solution

We have $e_k^a \nabla_a e_j^b = \Gamma_{jk}^i e_i^b$, so

$$c_{jk}^i e_i^b = [\vec{e}_j, \vec{e}_k] = e_j^a \nabla_a e_k^b - e_k^b \nabla_b e_j^a = \Gamma_{jk}^i e_i^b - \Gamma_{jk}^i e_i^b = -2\Gamma_{[jk]}^i e_i^b$$

Thus,

$$\Gamma_{[jk]}^i = -\frac{1}{2} c_{jk}^i .$$

Example 10.3. Use the previous relation and $\nabla_a g_{bc} = 0$ to show that

$$\Gamma_{ijk} = \frac{1}{2}\left(e_k g_{ij} + e_j g_{ik} - e_i g_{jk} + c_{kij} + c_{jik} - c_{ijk}\right) .$$

Solution

$\nabla_a g_{bc} = 0 \ \rightarrow \ e_i^a \nabla_a\left(e_j^b e_k^c g_{bc}\right)$ and we have:

$$e_i^a\nabla_a\left(e_j^b e_k^c g_{bc}\right) = e_i^a\left(\nabla_a e_j^b\right)e_k^c g_{bc} + e_i^a\left(\nabla_a e_k^c\right)e_j^b g_{bc} + e_i^a e_j^b e_k^c\left(\nabla_a g_{bc}\right)$$

$$e_i^a\nabla_a g_{jk} = \left(e_i^a\nabla_a e_j^b\right)e_k^c g_{bc} + \left(e_i^a\nabla_a e_k^c\right)e_j^b g_{bc}$$

$$e_i g_{jk} = \Gamma_{ji}^n e_n^b e_k^c g_{bc} + \Gamma_{ki}^n e_n^c e_j^b g_{bc}$$

$$e_i g_{jk} = \Gamma_{ji}^n g_{nk} + \Gamma_{ki}^n g_{jn} = \Gamma_{kji} + \Gamma_{jki}$$

$$\begin{cases} e_i g_{jk} = \Gamma_{kji} + \Gamma_{jki} \text{ now permute indices:} \\ e_j g_{ik} = \Gamma_{kij} + \Gamma_{ikj} \\ e_k g_{ji} = \Gamma_{ijk} + \Gamma_{jik} \end{cases}$$

$$\begin{aligned} e_k g_{ij} + e_j g_{ik} - e_i g_{jk} &= \Gamma_{ijk} + \Gamma_{jik} + \Gamma_{kij} + \Gamma_{ikj} - \Gamma_{kji} - \Gamma_{jki} + \left(+\Gamma_{ijk} - \Gamma_{ijk}\right) \\ &= 2\Gamma_{ijk} + \left(\Gamma_{ikj} - \Gamma_{ijk}\right) + \left(\Gamma_{jik} - \Gamma_{jki}\right) + \left(\Gamma_{kij} - \Gamma_{kji}\right) \\ &= 2\Gamma_{ijk} - 2\Gamma_{i[jk]} + 2\Gamma_{j[ki]} - 2\Gamma_{k[ji]} = 2\Gamma_{ijk} + c_{ijk} + c_{jki} + c_{kji} \\ &= 2\Gamma_{ijk} - \left(c_{kij} + c_{jik} - c_{ijk}\right) \end{aligned}$$

So,

$$\Gamma_{ijk} = \frac{1}{2}\left(e_k g_{ij} + e_j g_{ik} - e_i g_{jk} + c_{kij} + c_{jik} - c_{ijk}\right).$$

Example 10.4. Prove the Gauss-Codacci equation:

$${}^{(3)}R_{\alpha\beta\gamma\delta} = h_\alpha^{\alpha'} h_\beta^{\beta'} h_\gamma^{\gamma'} h_\delta^{\delta'} R_{\alpha'\beta'\gamma'\delta'} + K_{\alpha\delta}K_{\beta\gamma} - K_{\alpha\gamma}K_{\beta\delta}.$$

Solution

A tensor over the tangent space of the 3-surface at a given point p is simply

$$T^{a\ldots b}_{c\ldots d} = h_e^a \cdots h_f^b h_c^g \cdots h_d^j T^{e\ldots f}_{g\ldots j},$$

where h_{ab} is the spatial metric associated with a spacetime metric g_{ab} induced by the choice of foliation $\sum_t$. (t^a a vector field on the manifold for which $t^a\nabla_a t = 1$, t the global time function that labels $\sum_t$.) Denote $D_K T^{a\ldots b}_{c\ldots d} = h_e^a \cdots h_f^b h_c^g \cdots h_d^j h_k^n \nabla_n T^{e\ldots f}_{g\ldots j}$, then:

$${}^{(3)}R^d_{abc}w_d = D_a D_b w_c - D_b D_a w_c.$$

Now we project our what is meant by $D_a D_b w_c$:

$$D_a D_b w_c = D_a\left(h_b^d h_c^e \nabla_d w_e\right)$$
$$= h_a^f h_b^g h_c^k \nabla_f\left(h_g^d h_k^e \nabla_d w_e\right)$$
$$= h_a^f h_b^g h_c^k\left\{\left(\nabla_f h_g^d\right) h_k^e \nabla_d w_e + h_g^d\left(\nabla_f h_k^e\right)\nabla_d w_e + h_g^d h_k^e \nabla_f \nabla_d w_e\right\}$$
$$= \left(h_a^f h_b^g \nabla_f h_g^d\right)\left(h_c^k h_k^e\right)\nabla_d w_e + \left(h_a^f h_c^k \nabla_f h_k^e\right)\left(h_b^g h_g^d\right)\nabla_d w_e + h_a^f\left(h_b^g h_g^d\right)\left(h_c^k h_k^e\right)\nabla_f \nabla_d w_e$$

$$h_{ab} = g_{ab} + n_a n_b \quad \rightarrow \quad h_b^g h_g^d = \left(g_b^g + n_b n^g\right)\left(g_g^d + n_g n^d\right)$$
$$= g_b^d + n_b n^d + n_b n^d + n_b n^d \underbrace{\left(n_g n^j\right)}_{-1}$$
$$= g_b^t + n_b n^d = \underline{h_b^d}$$

Thus,

$$\nabla_f h_g^d = \nabla_f\left(g_g^d + n_g n^d\right) = \left(\nabla_f n_g\right) n^d + n_g\left(\nabla_f n^d\right)$$

and we get:

$$h_a^f h_b^g \nabla_f h_g^t = g_a^f h_b^g\left(\nabla_f n_g\right) n_t^d h_a^f \underbrace{h_b^g n_g}_{0} \nabla_f n^d$$
$$= h_a^f\left(g_b^g + n_b n^g\right)\left(\nabla_f n_g\right) n^d = h_a^f \nabla_f\left(g_b^g n_g\right) n_t^d h_a^f n_b n^d \left(\frac{1}{2}\right) \nabla_f n_g$$

$$h_a^f h_b^g \nabla_f h_g^t = h_a^f\left(\nabla_f n_b\right) n^d + h_a^f n_b d^t\left(\frac{1}{2}\nabla_f\left(n^g n_g\right)\right) = \underbrace{\left(h_a^f \nabla_f n_b\right)}_{K_{ab}} n^d = K_{ab} n^d.$$

So,

$$D_a D_b w_c = h_c^e K_{ab} n^d \nabla_d w_e + h_b^d K_{ac} n^e \nabla_d w_e + h_a^f h_b^d h_c^e \nabla_f \nabla_d w_e$$

$$h_b^d n^e \nabla_d w_e = h_b^d \nabla_d\left(n^e w_e\right) - h_b^d w_e \nabla_d n^e$$
$$= -K_b^e w_e$$

Now,

$${}^{(3)}R_{abc}^d w_d = h_c^e\left(K_{ab} - K_{ba}\right) n^d \nabla_d w_e - K_{ac} K_b^e w_e + K_{bc} K_a^e w_e + \underbrace{h_a^f h_b^d}_{symmetric\ in\ L,d.} h_c^e\left(\nabla_f \nabla_d - \nabla_d \nabla_f\right) w_e$$

So

$${}^{(3)}R_{abc}^d w_d = h_a^f h_b^d h_c^e R_{fde}^g w_g - K_{ac} K_b^e w_e + K_{bc} K_a^e w_e$$

and

$w_g = h_g^j w_j$ since w_g is on Σ.

Thus,
$^{(3)}R^{d}_{abc} = h^{f}_{a}h^{d}_{b}h^{e}_{c}h^{d}_{j}R^{j}_{fgk} - K_{ac}K^{d}_{b} + K_{bc}K^{d}_{a}$

In a given coordinate system we thus have:
$^{(3)}R_{\alpha\beta\gamma\delta} = h^{\alpha'}_{\alpha}h^{\beta'}_{\beta}h^{\gamma'}_{\gamma}h^{\delta'}_{\delta}R_{\alpha'\beta'\gamma'\delta'} + K_{\alpha\delta}K_{\beta\gamma} - K_{\alpha\gamma}K_{\beta\delta}.$

Example 10.5. Show that

$$\dot{\Gamma}^{a}_{bc} = \frac{1}{2}g^{ad}\left(\nabla_{bjcd} + \nabla_{cjbd} - \nabla_{djbc}\right)$$
$$\dot{R}_{ab} = 2\nabla_{[c}\dot{\Gamma}^{c}_{b]a}$$

and

$$\dot{R} = \nabla^{a}\nabla^{b}_{jab} - \nabla_{a}\nabla^{a}_{j} - j^{ab}R_{ab}\,.$$

Solution
We have:

$$\dot{R}_{ab} = 2\nabla_{[c}\dot{\Gamma}^{c}_{b]a} = \nabla_{c}\dot{\Gamma}^{c}_{ba} - \nabla_{b}\dot{\Gamma}^{c}_{ca}$$

where $R = g^{ab}R_{ab}$ we have: $\dot{R} = \dot{g}^{ab}R_{ab} + g^{ab}\dot{R}_{ab}$, thus

$$\dot{\Gamma}^{a}_{bc} = \frac{1}{2}g^{ad}\left(\nabla_{bjcd} + \nabla_{cjbd} - \nabla_{djbc}\right)$$

Using this in $\dot{R}_{ab} = 2\nabla_{[c}\dot{\Gamma}^{c}_{b]a}$:

$$\dot{R}_{ab} = \nabla_{c}\left(\frac{1}{2}g^{cd}\left[\nabla_{bjad} + \nabla_{ajbd} - \nabla_{djba}\right]\right)$$
$$= \nabla_{b}\left(\frac{1}{2}g^{cd}\left(\nabla_{cjad} + \nabla_{ajcd} - \nabla_{djca}\right)\right)$$
$$= \frac{1}{2}g^{cd}\left[\nabla_{c}\nabla_{bjad} + \nabla_{ajbd} - \nabla_{djba} - \nabla_{b}\nabla_{cjad} + \nabla_{ajcd} - \nabla_{djca}\right]$$
$$= \frac{1}{2}\left[\nabla^{d}\nabla_{bjad} + \nabla^{d}\nabla_{ajbd} - \nabla^{d}\nabla_{djba} - \nabla_{b}\nabla^{d}_{jad} - \nabla_{b}\nabla_{a}\left(g^{cd}_{jcd}\right) + \nabla_{b}\nabla^{c}_{jca}\right]$$

and we have:

$$g^{ab}\dot{R}_{ab} = \nabla^{a}\nabla^{b}_{jab} - \nabla_{a}\nabla^{a}_{j}$$

Or,
$\dot{R} = \nabla^{a}\nabla^{b}_{jab} - \nabla_{a}\nabla^{a}_{j} + \left(\dot{g}^{\dot{a}b}\right)R_{ab}$

$\left(\dot{g}^{\dot{a}b}\right)$ =? Since $g_{ab} = g_{ac}g_{bd}g^{cd}$: $\dot{g}_{ab} = g_{ac}\dot{g}_{bd}g^{cd} + g_{ac}\dot{g}_{bd}g^{cd} + g_{ac}g_{bd}\left(\dot{g}^{\dot{c}d}\right)$.
Thus,
$-j_{ab} = -\dot{g}_{ab} = g_{ac}g_{bd}\left(g^{\dot{c}d}\right) \Rightarrow \underline{\left(g^{\dot{c}d}\right) = -j^{cd}}$

and we have the desired:

$$\dot{R} = \nabla^a \nabla^b_{jab} - \nabla_a \nabla^a_j - j^{ab} R_{ab}.$$

10.6 Lie Groups

10.6.1 Introduction

A Lie group is a group that is also a manifold, in which the operation of group multiplication is a diffeomorphism.

Left multiplication: $L_g h = gh \quad (g, h \in G)$. Then $L_g: G \to G$ is a diffeomorphism.

Right multiplication: $R_g h = hg$ is also a diffeomorphism.

The "Adjoint" map A_g is defined by $A_g h = g^{-1} hg$, which is also a diffeomorphism:

$$A_g = R_g L_{g^{-1}}$$

(Eqn. 10-23)

Given a vector ξ^a at the identity e of G the diffeomorphism L_g produces a vector $L_{g*}\xi^a$ at $L_g(e) = ge = g$. In other words, there is a vector field on G:

$$\xi^a(g) = L_{g*}[\xi^a(e)]\,,$$

(Eqn. 10-24)

which is obtained from $\xi^a(e)$ by left multiplication.

So,

$$L_{g*}\xi^a(h) = L_{g*}L_n\xi^a(e) = \left(L_g L_n\right)_*\xi^a(e) = L_{gh*}\xi^a(e) = \xi^a(gh)$$

(Eqn. 10-25)

Thus, to each tangent vector in $T_e G$ correspnds a left invariant vector field on G (i.e., left invariant $L_{g*}(\xi^a)_{field} = (\xi^a)_{field}$). It can be proven that the Lie bracket $[\vec{\xi}, \vec{\eta}]$ of two left invariant vector fields $\vec{\xi}, \vec{\eta}$ is itself a left invariant vector field. Also, if Ψ_λ is the family diffeomorphisms generated by $\vec{\xi}$ and ϕ is another 'diffeo', then $\phi\Psi_\lambda\phi^{-1}$ is the diffeo generated by $\phi_*\vec{\xi}$.

$$\vec{\xi}(x) = \frac{d}{d\lambda}\Psi_\lambda(x)\bigg|_{\lambda=0} = \frac{d}{d\lambda}\Psi_{\lambda+\alpha}(0)\bigg|_{\lambda=0} \quad where \quad \Psi_\alpha(0) = \Psi_0(x) = x$$

(Eqn. 10-26)

and

$$\phi_*[\vec{\xi}(x)] = \frac{d}{d\lambda}[\phi \circ \Psi_\lambda(x)]\bigg|_{\lambda=0}$$

(Eqn. 10-27)

At $\phi(x), \phi_*[\vec{\xi}(x)]$ is tangent to the path $\lambda \to \phi \circ \Psi_\lambda(x)$. So, $\phi_*[\vec{\xi}(\phi^{-1}(x))]$ is tangent to $\lambda \to \phi \circ \Psi_\lambda \circ \phi^{-1}(x)$ at x. Thus, $\phi_*\xi$ is tangent at x to $\lambda \to \phi \circ \Psi_\lambda \circ \phi^{-1}(x)$. Therefore, $\phi_*\vec{\xi}$ generates the diffeomorphism $\phi \circ \Psi_\lambda \circ \phi^{-1}$. Now, let $\phi = L_g$:

$$[\vec{\xi}, \vec{\eta}] = \mathcal{L}_{\vec{\xi}}\eta = -\frac{d}{d\lambda}(\Psi_{\lambda *}\vec{\eta})\bigg|_{\lambda=0} \to \phi_*[\vec{\xi}, \vec{\eta}] = \phi_*\left\{\frac{d}{d\lambda}(\Psi_{\lambda *}\vec{\eta})\right\}$$
$$= -\frac{d}{d\lambda}(\phi_*\Psi_{\lambda *}\vec{\eta}) .$$

Thus,

$$\phi_*[\vec{\xi}, \vec{\eta}] = -\frac{d}{d\lambda}[(\phi\Psi_\lambda\phi^{-1})_*\phi_*\vec{\eta}] = \mathcal{L}_{\phi_*\vec{\xi}}\phi_*\vec{\eta}$$

(Eqn. 10-28)

Since $\vec{\xi}$ and $\vec{\eta}$ are left inv. $\phi_*\vec{\xi} = \vec{\xi}, \phi_*\vec{\eta} = \vec{\eta}$ (for fields). So,

$$\phi_*[\vec{\xi}, \vec{\eta}] = \mathcal{L}_\varepsilon\vec{\eta} = [\vec{\xi}, \vec{\eta}]$$

(Eqn. 10-29)

and the bracket is proven to be left invariant as well.

The vector space T_eG can be identified with the vector space of left inv. vector fields on G. T_eG is closed under the Lie bracket justifying the name Lie *algebra*.

If $\xi^a(e)$ and $\eta^a(e)$ are vectors at $e \in G$, then $[\xi, \eta]^a(e)$ is a vector at e. Since $[\xi, \eta]$ is linear in ξ and η it defines a bilinear map $T_eG \times T_eG \to T_eG$. Thus there is an antisymenotric tensor C^a_{bc} on bc at e with

$$[\xi, \eta]^a(e) = C^a_{bc}\xi^b(e)\eta^c(e)$$

Define $C^a_{bc}(g) = L_{g*}C^a_{bc}(e)$ then C^a_{bc} is left invariant, and

$$[\xi, \eta]^a = C^a_{bc}\xi^b\eta^c \ (everywhere).$$

(Eqn. 10-30)

C^a_{bc} is known as the structure tensor of the Lie algebra, and the concrete values C^k_{ij} are the structure constants of T_eG.

Example 10.6.

(a) Prove that $w^a = [u, v]^a$ is a vector field by proving that the Leibnitz rule is satisfied.

(b) Show that in any chart: $[u, v]^k = (u^i\partial_i v^k - v^i\partial_i u^k)$.

Solution
(a) Start with:
$w(fg) = [u,v](fg) = u(v(fg)) - v(u(fg)) = u(v(f)g + fv(g)) - v(u(f)g + fu(g))$
$= u(v(f)g) + u(fv(g)) - v(u(f)g) + v(fu(g))$
$= u(v(f))g - v(u(f))g + fu(v(g)) - fv(u(g))$
$= g[u,v] + f[u,v](g)$
$w(fg) = gw(f) + fw(g)$, and the Leibnitz rule is satisfied.

(b) In a chart x: $v(f) = v^i\partial_i f$, thus :
$$[u,v](f) = u(v(f)) - v(u(f)) \rightarrow \ [u,v]^k\partial_k f = u^i\partial_i(v^j\partial_j f) - v^j\partial_j(u^i\partial_i f)$$
So,
$[u,v]^k\partial_k f = u^i(\partial_i v^j)\partial_j f + u^i\left(\partial_i(\partial_j f)\right)v^j - v^j(\partial_j u^i)\partial_i f - v^j u^i\partial_j(\partial_i f)$
$= (u^i\partial_i v^j)\partial_j f - (v^j\partial_j u^i)\partial_i f + (u^i v^j\partial_i\partial_j f - u^i v^j\partial_j\partial_i f) = (u^i\partial_i v^k - v^i\partial_i u^k)\partial_k f$
Thus,
$$[u,v]^k = (u^i\partial_i v^k - v^i\partial_i u^k)\,.$$

Example 10.7. A vector lies in a submanifold M of N if it is the tangent to the curve $c(\lambda)$ that lies in M.
(a) Prove that a vector V^a at a point $P \in M$ lies in M if and only if $V^a\nabla_a f = 0$, for all functions f on N that are constant on M.
(b) Show that if u^a and v^a are vector fields on N for which $u^a(P)$ and $v^a(P)$ lie in M when $P \in M$, then $[u,v]^a$ similarly lies in M. (Show that $[u,v]^a\nabla_a f = 0$, f constant on M.)

Solution
(a) In a given chart x^i we have $V^a\nabla_a f = V^i\partial_i f$. If the dimension of M is m and N is n then we can arrange $i = 1 \ldots m \ldots n$ such that the first m coordinates are a basis for the submanifold. Now,
$$V^i\partial_i f = V^i\partial_i f + \cdots + V^m\partial_m f + V^{m+1}\partial_{m+1} f + \cdots V^n\partial_n f$$
Now, if we specify all functions which are constant on M under f then all the derivatives $\partial_i f, \cdots, \partial_m f$ are zero, so
$$V^i\partial_i f = V^{m+1}\partial_{m+1} f + \cdots V^n\partial_n f\,.$$

If for all f this vanishes then each of the $V^{m+1}, \cdots, V^n$ must be zero, thus if $V^a\nabla_a f = 0$ for all f on N that are constant on M then V^i can only have nonzero compoents can only have nonzero compoents $i = l \ldots m$, i.e., V^a lies in M. the converse, if V^a lies in M then the $i = m + 1 \ldots n$ components must be zero and if the function f is constant on M, i.e., $\partial_i f = 0\ \forall : \in l \ldots m$, the $V^i \partial_i f = 0 \Rightarrow V^a\nabla_a f = 0$.

(b) We have $V^i\partial_i f = 0$ and $u^i\partial_i f = 0$ for the f as specified:

$$[u, v]^a\nabla_a f = [u, v]^i\partial_i f$$

$$= \left(u^j \frac{\partial v^i}{\partial x^j} - v^j \frac{\partial u^i}{\partial x^j}\right)\frac{\partial f}{\partial x^i}\begin{cases}\text{since } \dfrac{\partial f}{\partial x^i} = 0 \text{ for } i = l..m \\ \text{also } u^i \text{ and } v^i \text{ are zero for } i = m+1 \ldots n\end{cases}$$

Let $i_M = l \ldots m, i_N = m + 1 \ldots n$, then

$$[u, v]^a\nabla_a f = \left(u^{jM}\frac{\partial v^{iN}}{\partial x^{jM}} - v^{jM}\frac{\partial u^{iN}}{\partial x^{jM}}\right)\frac{\partial f}{\partial x^{iN}} = 0\,.$$

10.6.2 The general linear group $GL_n(R)$

The Lie group might be very general, such as in $GL_n(R)$. The general linear group $GL_n(R)$ is the group of inventible $n \times n$ matrices $a = \|a_j^l\|$. Any other group of matrices is a subgroup of some GL_n. Natural coordinates on $G = GL_n$ and the n^2matrix elements a_j^i of $a \in G$. The associated coordinate basis: $\partial/\partial a_j^i$ consists of vectors tangent to curves:

$$\lambda \to a(\lambda) = \|a_n^m + \lambda\delta_i^m\delta_n^j\|\,.$$

(Eqn. 10-31)

If V is tangent at e to the path $g_j^i(\lambda)$ its components along the natural coordinate basis are

$$V_j^i(e) = \frac{d}{d\lambda}g_j^i(\lambda)|_{\lambda=0}$$

(Eqn. 10-32)

where

$$V_j^i(a) = L_{a*}V_j^i(e) = \frac{d}{d\lambda}\left[a_k^i g_j^k(\lambda)\right]\Big|_{\lambda=0} = a_k^i V_j^k(e)$$

(Eqn. 10-33)

Thus, a left invariant vector field V on G has components $V_j^i(a)$ at $a \in G$ obtained from its components at the origin by left-multiplication of V_j^i by a_j^i.

Now, consider the *1* – parameter subgroup defined by

$$g(\lambda) = e^{\lambda v}$$

(Eqn. 10-34)

where v is the matrix $\|v_n^m\|$. Then,

$$\Psi_\lambda(a) = ag(\lambda)$$

(Eqn. 10-35)

is the 1 – parameter family of diffeo's generated by the left-inv. vector field $\vec{v}$:

$$\frac{d}{d\lambda} ag(\lambda)\bigg|_{\lambda=0} = a_k^i v_j^k(e) = v_j^i(a)\,.$$

(Eqn. 10-36)

Thus, the left-inv. vector fields $\vec{v}$ generate right multiplication by $e^{\lambda x}$.

For another interesting property, consider [u, v]:

$$[u, v] = \mathcal{L}_u v = \frac{d}{d\lambda} \Psi_{-\lambda} V\bigg|_{\lambda=0}\,,$$

(Eqn. 10-37)

where

$$(\Psi_{-\lambda}V)_j^i = v_n^i[\Psi_\lambda(e)]e^{-\lambda x} = v_n^i[g(\lambda)]g_j^n(-\lambda) = g_m^i(\lambda)v_n^m(e)g_j^n(-\lambda).$$

(Eqn. 10-38)

Since

$$\frac{d}{d\lambda} \Psi_{-\lambda} V_j^i = u_m^i(\lambda)v_n^m(e)g_j^n(-\lambda) - g_m^i(\lambda)v_n^m(e)u_j^n(-\lambda)$$

(Eqn. 10-39)

we have

$$[u, v]_j^i = u_m^i v_n^m \delta_j^n - \delta_m^i v_n^m u_j^n = u_m^i v_j^m - v_m^i u_j^m\,,$$

(Eqn. 10-40)

which is a simple relation between the Lie bracket [u, v] and the commutator of two matrices.

Example 10.8. In the group GL_n find the left-invariant vector fields e_j^i that agree with the vector $\frac{\partial}{\partial a_j^i}$ at the identity. Compute the structure constants $C_{jkm}^{i\ell n}$ defined by $\left[e_k^\ell, e_m^n\right] = C_{jkm}^{i\ell n} e_i^j$.

Partial Solution

Start with $\frac{\partial}{\partial a_j^i}$ is targent to $a_j^i(\lambda) = \left\| a_n^m + \lambda \delta_i^m \delta_n^j \right\|$. Rename e_j^i by v_j^i for notational convenience. So have that v_j^i is left unvariant if $v_j^i(g) = \left[L_g^* V(e)\right]_j^i$. For GL_n left multiplication corresponds to matrix multiplication. Let $v_j^i(e) = \frac{\partial}{\partial a_j^i}$, i.e., $v_j^i(e)$ is tangent to $a_j^i(\lambda)$ at $\lambda = 0$. The left-invariant vector fields v_j^i that agree with $\frac{\partial}{\partial a_j^i}$ at the identity are:

$$v_j^i(g) = \frac{d}{d\lambda} g_k^i a_j^k(\lambda)\Big|_{\lambda=0} = g_k^i v_j^k(e) = g_k^i \left(\frac{\partial}{\partial a_j^k}\right)$$

Now consider $\left[v_k^\ell, v_m^n\right] = C_{jkm}^{i\ell n} v_i^j$ to proceed. The rest is left to the reader.

Example 10.9. Prove the Jacobi identity by
(a) Showing for any three vector fields u, v, w that

$$[u,[v,w]] + [v,[w,u]] + [w,[u,v]] = 0.$$

(b) Showing that for any Lie group, the structure constant tensor satisfies: $C_{d[a}^e C_{bc]}^d = 0$.

Solution

We consider a local coordinate system, then

$$\begin{aligned}
&[u,[v,w]] + [v,[w,u]] + [w,[u,v]] \\
&= \partial_u(\partial_v\partial_w - \partial_w\partial_v) = (\partial_v\partial_w - \partial_w\partial_v)\partial_u + \partial_v(\partial_w\partial_u - \partial_u\partial_w) \\
&\quad -(\partial_w\partial_u - \partial_u\partial_w)\partial_v + \partial_w(\partial_u\partial_v - \partial_v\partial_u) - (\partial_u\partial_v - \partial_v\partial_u)\partial_w \\
&= 0
\end{aligned}$$

Using the Jacobi identity and the definition of the structure tensor:

$$\left[u^a,[v^b,w^c]\right] + \left[v^b,[w^c,u^a]\right] + \left[w^c,[u^a,v^b]\right] = 0$$

With $[X^a, Y^a]^c = C_{ab}^c X^a Y^b$ we have:

$$\left[u^a, C_{ab}^d v^b w^c\right] + \left[v^b, C_{ca}^d w^c u^a\right] + \left[w^c, C_{ab}^d u^a v^b\right] = 0$$

or

$$C_{ad}^e C_{bc}^d u^a v^b w^c + C_{bd}^e C_{ca}^d v^b w^c u^a + C_{cd}^e C_{ab}^d w^c u^a v^b = 0$$

If we add up cyclic permutations we get:

$$\begin{aligned}
&C_{ad}^e C_{bc}^d (u^a v^b w^c + v^b w^c u^a + w^c u^a v^b) \\
&\qquad + C_{bd}^e C_{ca}^d (v^b w^c u^a + w^c u^a v^b + u^a v^b w^c)
\end{aligned}$$

$$+C^e_{cd}C^d_{ab}(w^c u^a v^b + u^a v^b w^c + v^b w^c u^a) = 0$$

$$= (C^e_{ad}C^d_{bc} + C^e_{bd}C^d_{ca} + C^e_{cd}C^d_{ab})(u^a v^b w^c + v^b w^c u^a + w^c u^a v^b) = 0$$

For general u, v, w we then get: $(C^e_{ad}C^d_{bc} + C^e_{bd}C^d_{ca} + C^e_{cd}C^d_{ab}) = 0$. Consider the lower three free indices and ask if we have antisymmetric on $a \leftrightarrow b$ (yes). Same for a and c , and b and c, thus fully antisymmetric, thus have:

$$C^e_{d[a}C^d_{bc]} = 0 \,.$$

10.6.3 Rotations (SO(3) vs SU(2))

Rotations in n dimensions

The rotation group, *SO* (*n*), is the group of orientation-preserving linear maps of R^n to itself that preserve the flat metric δ_{ab}. Thus a rotation can be regarded as a tensor R^a_b on R^n with one up and one down index, mapping vectors v^a to rotated vectors $\overline{v}^a = R^a_b v^b$. In terms of the natural orthonormal frame for R^n, the components R^i_j form an $n \times n$ a matrix satisfying

$$R^T R = I,$$

where $R^{Ti}_j = R^j_i$. A matrix satisfying the above is called an *orthogonal* matrix, and the "*SO*(*n*)" means the group of special orthogonal transformations of R^n – special means having determinant one. The full orthogonal group, *O(n)* is twice as large, including all reflections.

A rotated vector has components $\hat{v}^i = R^i_j v^j$ along the original basis, $\{e_i\}$. The rotated vector is then

$$\overline{v} = e_i \overline{v}^i = e_i R^i_j v^j = \tilde{e}_i v^i.$$

(Eqn. 10-41)

The last expression is what one would naturally write down if one began with the action of *R* on the basis vectors given by:

$$\overline{e}_i = e_j R^j_i,$$

(Eqn. 10-42)

to obtain, by linearity, its action on v.

Rotations in 2 dimensions

The counterclockwise rotation by angle θ of a vector v in R^2 is described by the above equation with the matrix

$$[R^i_j(\theta)] = \begin{bmatrix} \cos\theta & -\sin\theta \\ \sin\theta & \cos\theta \end{bmatrix}.$$

(Eqn. 10-43)

That is,

$$\tilde{v}^i = R^i_j v^j \; and \; \tilde{e}_i = e_j R^j_i.$$

(Eqn. 10-44)

The group has a single parameter, θ, which runs from 0 to 2π, and the same point is labeled by θ and 2π. Thus, the manifold of the group *SO*(2) is a circle. Up to reparameterization, there is one path through the identity, namely $\theta \rightarrow R^i_j(\theta)$, and the corresponding tangent vector can be identified with the matrix $v^i_j = \frac{d}{d\theta} R^i_j \Big|_{\theta=0} = -\epsilon^i_j$, thus

$$v^i_j = \begin{bmatrix} 0 & -1 \\ 1 & 0 \end{bmatrix}.$$

(Eqn. 10-45)

The Lie algebra is then the 1-dimensional vector {rv}, r real, with the single Lie bracket, $[v, v] = 0$. (Equivalently, the structure constant tensor $C^a_{bc} = 0$.) Using the relation, $\epsilon^i_k \epsilon^k_j = \delta^i_j$, we can directly verify: exponentiating a matrix in the Lie algebra gives back a 1-parameter subgroup (here there is only one parameter, and the subgroup is the whole group).

$$e^{\theta v} = e^{-\theta\epsilon} = 1 - \theta\epsilon + \frac{1}{2!}\theta^2 + \frac{1}{3!}\theta^3\epsilon - \cdots$$
$$= \cos\theta - \epsilon\sin\theta = R(\theta)$$

The group SO(2) is isomorphic to the group U(l) of 1×1 unitary matrices: that is, U(l) is the set $\{e^{i\theta}\}$ under ordinary multiplication. The isomorphism is simply $R^i_j(\theta) \rightarrow \epsilon^{i\theta}$. There is another, larger group, that has an identical Lie algebra to that of SO(2) = U(1), namely the group R of real numbers under addition. Again, because R is a 1-dimensional group, its Lie algebra consists of the single element v tangent to the curve $\lambda \rightarrow \lambda$ through the identity at $\lambda = 0$. Two Lie groups with the same Lie algebra are locally isomorphic: a neighborhood of the identity of each group is isomorphic, but globally they arc different. In this case R is a *covering group* of SO(2) = U(1): it is a simply-connected group and there is a homomorphism ψ of R to SO(2) = U(1)

$$\psi(\lambda) = \lambda \bmod 2\pi$$

The homomorphism wraps the line infinitely many times around the circle.

Rotations in 3 dimensions

The counterclockwise rotation by angle θ about an axis along the unit vector n is described with matrix

$$[R^i_j(\theta n)] = \exp[-\epsilon^i_{jk} n^k]$$

(Eqn. 10-46)

Here n is a vector from the origin to any point on the unit sphere, and θ is any number between 0 and π. We thus have a map from a ball of radius π onto the rotation group, given by

$$\theta n \to R(\theta n).$$

(Eqn. 10-47)

Each point in the interior of the ball is mapped to a distinct point of the rotation group. Because a rotation by π about n is the same as a rotation by π about –n, diametrically opposite points on the surface of the ball are mapped to the same rotation:

$$R(\pi n) = R(-\pi n).$$

(Eqn. 10-48)

The rotation group is therefore the manifold obtained from a 3-ball by identifying diametrically opposite points of its surface; the spacy is called RP^3 , or real projective 3-space.

Example 10.10. Consider the vector fields L^a_x, L^a_y, and L^a_z that generate rotations about the axes of $\mathbb{R}^3$:

$$L_z = x\partial_y - y\partial_x, \qquad L_y = z\partial_x - x\partial_z, \qquad L_x = y\partial_z - z\partial_y \, .$$

Find their commutation relations and check that L_x commutes with $L^2 = L_x^2 + L_y^2 + L_z^2$.

Solution

$[L_x, L_y] = y\partial_x + (yz\partial_z\partial_x - yx\partial_z\partial_z - z^2\partial_x\partial_y + xz\partial_z\partial_y)$
$-x\partial_y - (zy\partial_x\partial_z - xy\partial_z\partial_z - z^2\partial_x\partial_y + xz\partial_z\partial_y) = y\partial_x - x\partial_y = -L_z$

$[L_z, L_x] = x\partial_z - z\partial_x = -L_y$

$[L_y, L_z] = z\partial_y - y\partial_z = -L_x$

If we use $x = x_1, y = x_2, z = x_3$ then $\left[L_{x_i}, L_{x_j}\right] = -L_{x_k} \;\; for \;\; (ijk) = (123)$ and all cyclic permutations. Thus,

$[L_x, L^2] = [L_x, L_x^2] + [L_x, L_y^2] + [L_x, L_z^2]$
$= L_x L_y^2 - L_y^2 L_x + L_x L_z^2 - L_z^2 L_x = (-L_z + L_y L_x)L_y - L_y(L_x L_y + L_z) + L_x L_z^2 - L_z^2 L_x$

$$= -L_zL_y - L_yL_z + (L_zL_x + L_y)L_z - L_z(L_xL_z - L_y) =$$
$$-\{L_zL_y + L_yL_z + L_yL_z - L_zL_y\}$$
$$= 0$$

Euler angles

We need Euler angles primarily to describe the relation between SO(3) and SU(2) that underlies spinors; the relation is the deeper meaning of Hamilton's spectacular discovery that multiplication in the rotation group can be written as multiplication of quaternions. We follow Goldstein's notation, but Goldstein likes to rotate coordinates instead of vectors. A rotation for us is active, the motion of a vector. The components of a counterclockwise-rotated vector relative to fixed coordinate axes are the same as the components of Goldstein's fixed (passive) vector relative to clockwise-rotated axes. So Goldstein's clockwise matrices are our counterclockwise matrices. To get our formulas, change the sign of each angle in his formulas.

A rotation can be specified by what it does to an orthonormal frame $\{e_i\}$ that starts aligned with the fixed frame i, j, k. The final location of e_3 can be specified by saying that it connects the origin to a point on the unit sphere with coordinates (θ, ϕ). The other two frame vectors are then determined up to a rotation about e_3. Then an arbitrary frame orientation can be obtained in three steps (it can of course also be gotten by a single rotation about some axis):

(i) Rotate through an angle ψ about the z-axis $R(\psi\ k)$.
(ii) Rotate through angle θ about the x-axis $R(\theta\ i)$
(iii) Rotate through angle ϕ about the z-axis $R(\phi\ k)$

Step (i) leaves e_3 fixed. Step (ii) rotates e_3 to a position in the y-z plane at an angle θ from the z-axis.
Step (iii) rotates e_3 to a final angle ϕ from the y-axis. The final frame $\bar{e}_i$ is then given by

$$\bar{e}_i = e_m R^m_i(\psi, \theta, \phi) = e_m R^m_k(\psi k) R^k_j(\theta i) R^i_i(\psi k),$$

(Eqn. 10-49)

and any vector rotated by $R^i_j(\psi, \theta, \phi)$ has components (along the fixed frame i, j, k)

$$\bar{v}^i = R^i_j(\psi, \theta, \phi) v^j.$$

(Eqn. 10-50)

Example 10.11. Examine the Lie algebra SO(3). (see [38] for further background)
(a) Show that the matrices $(K_i)_{jk} = -\epsilon_{ijk}$ satisfy the commutation relations $[K_j, K_k] = \epsilon^i_{jk} K_i$.
(b) Show that:

$$e^{K_1\theta} = \begin{pmatrix} 1 & 0 & 0 \\ 0 & \cos\theta & -\sin\theta \\ 0 & \sin\theta & \cos\theta \end{pmatrix},$$

and find the analogous

$$e^{K_2\theta} = \begin{pmatrix} \cos\theta & 0 & \sin\theta \\ 0 & 1 & 0 \\ -\sin\theta & 0 & \cos\theta \end{pmatrix} \text{ and } e^{K_3\theta} = \begin{pmatrix} \cos\theta & -\sin\theta & 0 \\ \sin\theta & \cos\theta & 0 \\ 0 & 0 & 1 \end{pmatrix}.$$

(c) Find the left-invariant vector fields $\hat{e}_1$, $\hat{e}_2$, and $\hat{e}_3$ at a point $R(\phi, \theta, \Psi)$ of SO(3) in terms of the vectors $\vec{\partial}_\theta$, etc.

Solution
(a) If $(K_i)_{jk} = -\epsilon_{ijk}$ then:

$$K_1 = \begin{pmatrix} 0 & 0 & 0 \\ 0 & 0 & -1 \\ 0 & 1 & 0 \end{pmatrix}, \quad K_2 = \begin{pmatrix} 0 & 0 & 1 \\ 0 & 0 & 0 \\ -1 & 0 & 0 \end{pmatrix}, \quad K_3 = \begin{pmatrix} 0 & -1 & 0 \\ 1 & 0 & 0 \\ 0 & 0 & 0 \end{pmatrix}$$

and

$$[K_1, K_2] = \begin{pmatrix} 0 & -1 & 0 \\ 1 & 0 & 0 \\ 0 & 0 & 0 \end{pmatrix} = K_3, [K_1, K_3] = \begin{pmatrix} 0 & 0 & -1 \\ 0 & 0 & 0 \\ 1 & 0 & 0 \end{pmatrix} = -K_2, [K_2, K_3]$$
$$= \begin{pmatrix} 0 & 0 & 0 \\ 0 & 0 & -1 \\ 0 & 1 & 0 \end{pmatrix} = K_1$$

Since $\epsilon^3_{12} = 1,\ \epsilon^2_{13} = -1,\ and\ \epsilon^1_{23} = 1$, we see that:

$$[K_j, K_k] = \epsilon^i_{jk} K_i\ .$$

(b) $e^{K_1\theta} = \sum_{n=0}^{\infty} \frac{\theta^n}{n!} K_1^n = 1 + \theta K_1 + \frac{\theta^2 K_1^2}{2} + \frac{\theta^3 K_1^3}{3!} + \frac{\theta^4 K_1^4}{4!} + \frac{\theta^5 K_1^5}{5!} + \frac{\theta^6 K_1^6}{6!} \ldots$
and since:

$$K_1^2 = \begin{pmatrix} 0 & 0 & 0 \\ 0 & -1 & 0 \\ 0 & 0 & -1 \end{pmatrix}, K_1^3 = \begin{pmatrix} 0 & 0 & 0 \\ 0 & 0 & 1 \\ 0 & -1 & 0 \end{pmatrix}, K_1^4 =$$
$$\begin{pmatrix} 0 & 0 & 0 \\ 0 & 1 & 0 \\ 0 & 0 & 1 \end{pmatrix} \quad and \quad K_1 = K_1^{(1+4n)}$$, we have:

$$e^{K_1\theta} = 1 + K_1\left(\theta - \frac{1}{3!}\theta^3 + \frac{1}{5!}\theta^5 - \cdots\right)$$
$$+ K_1^2\left(\frac{1}{2}\theta^2 - \frac{1}{4!}\theta^4 + \frac{1}{6!}\theta^6 - \cdots\right)$$
$$= K_1(\sin\theta) + \begin{pmatrix} 1 & 0 & 0 \\ 0 & (\cos\theta) & 0 \\ 0 & 0 & (\cos\theta) \end{pmatrix} = \begin{pmatrix} 1 & 0 & 0 \\ 0 & \cos\theta & -\sin\theta \\ 0 & \sin\theta & \cos\theta \end{pmatrix}$$

Similarly for $e^{K_2\theta} = \begin{pmatrix} \cos\theta & 0 & \sin\theta \\ 0 & 1 & 0 \\ -\sin\theta & 0 & \cos\theta \end{pmatrix}$ and $e^{K_3\theta} =$
$\begin{pmatrix} \cos\theta & -\sin\theta & 0 \\ \sin\theta & \cos\theta & 0 \\ 0 & 0 & 1 \end{pmatrix}$.

(c) $R(\phi, \theta, \Psi) = R(\phi K_3)R(\theta K_1)R(\Psi K_3)$ (regarding a vector as a matrix)
$\vec{\partial}_\theta = \frac{d}{d\lambda}R(\phi, \theta + \lambda, \Psi)|_{\lambda=0} = R(\phi K_3)\frac{d}{d\lambda}R[(\theta + \lambda)K_1]_{\lambda=0}R(\Psi K_3) =$
$R(\phi K_3)R(\theta K_1)K_1 R(\Psi K_3)$
So,
$K_1 R(\Psi K_3) = \begin{pmatrix} 0 & 0 & 0 \\ 0 & 0 & -1 \\ 0 & 1 & 0 \end{pmatrix}\begin{pmatrix} \cos\Psi & -\sin\Psi & 0 \\ \sin\Psi & \cos\Psi & 0 \\ 0 & 0 & 1 \end{pmatrix} =$
$\begin{pmatrix} 0 & 0 & 0 \\ 0 & 0 & -1 \\ \sin\Psi & \cos\Psi & 0 \end{pmatrix}$
and
$R^{-1}(\Psi K_3)K_1 R(\Psi K_3) = \begin{pmatrix} 0 & 0 & -\sin\Psi \\ 0 & 0 & -\cos\Psi \\ \sin\Psi & \cos\Psi & 0 \end{pmatrix} = \cos\Psi\, K_1 -$
$\sin\Psi\, K_2$.

So,
$$\vec{\partial}_\theta = R(\phi K_3)R(\theta K_1)R(\Psi K_3)[K_1\cos\Psi - K_2\sin\Psi]$$
$$= R(\phi, \theta, \Psi)K_1\cos\Psi - R(\phi, \theta, \Psi)K_2\sin\Psi$$
Thus,
$$\vec{\partial}_\theta = \hat{e}_1(R)\cos\Psi - \hat{e}_2(R)\sin\Psi ,$$
where $\hat{e}_i(R)$ is the left invariant vector field associated with $\hat{e}_i(e) = K_i$.

Similarly, we have:
$$\vec{\partial}_\phi = R(\phi K_3)K_3 R(\theta K_1)R(\Psi K_3)$$
and

$$R^{-1}(\theta K_3)K_3R(\theta K_1) = \begin{pmatrix}1 & 0 & 0\\ 0 & \cos\theta & \sin\theta\\ 0 & -\sin\theta & \cos\theta\end{pmatrix}\begin{pmatrix}0 & -1 & 0\\ 1 & 0 & 0\\ 0 & 0 & 0\end{pmatrix}\begin{pmatrix}1 & 0 & 0\\ 0 & \cos\theta & -\sin\theta\\ 0 & \sin\theta & \cos\theta\end{pmatrix}$$
$$= K_3\cos\theta + K_2\sin\theta$$

$$\vec{\partial}_\phi = R(\phi K_3)R(\theta K_1)(K_3\cos\theta + K_2\sin\theta)R(\Psi K_3)$$
$$= R(\phi,\theta,\Psi)K_3\cos\theta + R(\phi K_3)R(\theta K_1)K_2R(\Psi K_3)\sin\theta$$

$$R^{-1}(\Psi K_3)K_2R(\Psi K_3) = \begin{pmatrix}\cos\Psi & \sin\Psi & 0\\ -\sin\Psi & \cos\Psi & 0\\ 0 & 0 & 1\end{pmatrix}\begin{pmatrix}0 & 0 & 1\\ 0 & 0 & 0\\ -\cos\Psi & -\sin\Psi & 0\end{pmatrix}$$
$$= \begin{pmatrix}0 & 0 & \cos\Psi\\ 0 & 0 & -\sin\Psi\\ -\cos\Psi & \sin\Psi & 0\end{pmatrix} = K_2\cos\Psi + K_1\sin\Psi$$

$$\vec{\partial}_\phi = R(\phi,\theta,\Psi)K_3\cos\theta + R(\phi,\theta,\Psi)(K_2\cos\Psi + K_1\sin\Psi)\sin\theta$$
$$= \hat{e}_3(R)\cos\theta - \hat{e}_2(R)\cos\Psi\sin\Psi + \hat{e}_1(R)\sin\Psi\sin\theta$$

(c) $\vec{\partial}_\Psi = R(\phi K_3)R(\theta K_1)R(\Psi K_3)K_3 = R(\phi,\theta,\Psi)K_3 = \hat{e}_3$

So,

$\vec{\partial}_\Psi = \hat{e}_3$:

$\vec{\partial}_\phi = \hat{e}_3\cos\theta + \hat{e}_2\cos\Psi\sin\theta + \hat{e}_1\sin\Psi\sin\theta$

$\vec{\partial}_\theta = \hat{e}_1\cos\Psi - \hat{e}_2\sin\Psi$

are grouped to give:

$$\frac{1}{\sin\theta}\vec{\partial}_\phi = \cot\theta\,\vec{\partial}_\Psi = \hat{e}_2\cos\Psi + \hat{e}_1\sin\Psi$$
$$\hat{e}_2 = \cos\Psi\left(\frac{1}{\sin\theta}\vec{\partial}_\phi = \cot\theta\,\vec{\partial}_\Psi\right) - \sin\Psi\,\vec{\partial}_\theta$$
$$\hat{e}_1 = \sin\Psi\left(\frac{1}{\sin\theta}\vec{\partial}_\phi = \cot\theta\,\vec{\partial}_\Psi\right) - \cos\Psi\,\vec{\partial}_\theta$$

Thus:

$$\hat{e}_1 + i\hat{e}_2 = e^{-i\Psi}\vec{\partial}_\theta + \frac{ie^{-i\Psi}}{\sin\theta}\left(\vec{\partial}_\phi - \cos\theta\,\vec{\partial}_\Psi\right) \text{ and } \hat{e}_3 = \partial_\Psi.$$

Example 10.12. Check that the vector fields above reproduce the Lie algebra of so(3), e.g., verify that $[\hat{e}_j, \hat{e}_k] = \epsilon^i_{jk}\hat{e}_i$.

Solution
Starting with

$$\hat{e}_1 = \frac{\sin\Psi}{\sin\theta}\left(\vec{\partial}_\phi - \cos\theta\,\vec{\partial}_\Psi\right) + \cos\Psi\,\vec{\partial}_\theta$$

$$\hat{e}_2 = \frac{\cos\Psi}{\sin\theta}\left(\vec{\partial}_\phi - \cos\theta\,\vec{\partial}_\Psi\right) + \sin\Psi\,\vec{\partial}_\theta$$

$$\hat{e}_3 = \vec{\partial}_\Psi$$

We get:

$$\hat{e}_x = \hat{e}_1 + i\hat{e}_2 = e^{-1^{-\Psi}}\left[\partial_\theta + \frac{i}{\sin\theta}\left(\partial_\phi - \cos\theta\,\partial_\Psi\right)\right]$$

$$\hat{e}_y = \hat{e}_1 - i\hat{e}_2 = e^{i\Psi}\left[\partial_\theta + \frac{i}{\sin\theta}\left(\partial_\phi - \cos\theta\,\partial_\Psi\right)\right]$$

$$[\hat{e}_x, \hat{e}_y] = \left\{-1\frac{\cos\theta}{\sin\theta}(-i)e^{-i\Psi}\hat{e}_y - i\partial_\theta\left\{\frac{i}{\sin\theta}\left(\partial_\phi - \cos\theta\,\partial_\Psi\right)\right\}\right\}$$
$$-\left\{1\frac{\cos\theta}{\sin\theta}(-i)e^{i\Psi}\hat{e}_x - i\partial_\theta\left\{\frac{i}{\sin\theta}\left(\partial_\phi - \cos\theta\,\partial_\Psi\right)\right\}\right\}$$

$$= \frac{\cos\theta}{\sin\theta}\left(\frac{2i}{\sin\theta}\left(\partial_\phi - \cos\theta\,\partial_\Psi\right)\right) - 2i\frac{(-\cos\theta)}{\sin^2\theta}\left(\partial_\phi - \cos\theta\,\partial_\Psi\right) +$$
$$\frac{2i}{\sin\theta}(-\sin\theta)\partial_\Psi = -2i\vec{\partial}_\Psi$$

$$[\hat{e}_1 + i\hat{e}_2, \hat{e}_1 - i\hat{e}_2] = -i[\hat{e}_1, \hat{e}_2] + i[\hat{e}_2, \hat{e}_1] = -2i[\hat{e}_1, \hat{e}_2] - 2i\vec{\partial}_\Psi$$
$$[\hat{e}_1, \hat{e}_2] = \hat{e}_3$$

$$[\hat{e}_1, \hat{e}_3] = -\left[\frac{\cos\Psi}{\sin\theta}\left(\vec{\partial}_\phi - \cos\theta\,\vec{\partial}_\Psi\right) - \sin\Psi\,\vec{\partial}_\theta\right] = -\hat{e}_2$$

$$[\hat{e}_2, \hat{e}_3] = -\left[\frac{-\sin\Psi}{\sin\theta}\left(\vec{\partial}_\phi - \cos\theta\,\vec{\partial}_\Psi\right) - \cos\Psi\,\vec{\partial}_\theta\right] = -\hat{e}_1$$

So, $[\hat{e}_1, \hat{e}_2] = \hat{e}_3, [\hat{e}_2, e_3] = \hat{e}_1, [\hat{e}_3, \hat{e}_1] = \hat{e}_2$

Thus, $[\hat{e}_j, \hat{e}_k] = \epsilon^i_{jk}\hat{e}_i$.

Example 10.13. (MTW 9.13 [38]) Rotation Group Generators
Let K_l be 3×3 matrices where $(K_\ell)_{mn} = \epsilon_{\ell mn}$.
(a) Show the matrices K_1, $(K_1)^2$, $(K_1)^3$, $(K_1)^4$.
(b) Sum the series $R_x(\theta) \equiv \exp(K_1\theta) = \sum_{n=0}^{\infty}\frac{\theta^n}{n!}(K_1)^n$ and show that it is a rotation matrix for rotting angle θ about the x-axis.

(c) Show similarly for $R_z(\phi) = \exp(K_3\phi)$ and $R_y(\chi) = \exp(K_2\chi)$ for rotation ϕ around the z-axis and χ around the y-axis.
(d) Explain why $R_z(\Psi)R_x(\theta)R_z(\phi)$ defines the Euler angle coordinates for the generic element of the group SO(3).
(e) Let $\mathcal{C}$ be the curve $\mathcal{P} = R_z(t)$ through the identity matrix $\mathcal{C}(0) = \mathcal{I} \in SO(3)$. Show that its tangent, $d\mathcal{C}(0)/dt = \dot{\mathcal{C}}(0)$ does not vanish by computing $\dot{\mathcal{C}}(0)f_{12}$, where f_{12} is the function $f_{12}(\mathcal{P}) = P_{12}$ whose value is the 12 element of $\mathcal{P}$
(f) Define a vector field $\boldsymbol{e}_3$ on SO(3) by letting $\boldsymbol{e}_3(\mathcal{P})$ be the tangent (at t=0) to the curve $\mathcal{C}(t) = R_z(t)\mathcal{P}$ through $\mathcal{P}$. Show that $\boldsymbol{e}_3(\mathcal{P})$ is nowhere zero.
(g) Show that $\boldsymbol{e}_3 = \left(\frac{\partial}{\partial\Psi}\right)_{\theta\phi}$.
(h) Derive the following formulas valid for $t \ll 1$:

$$R_y(t)R_z(\Psi)R_x(\theta)R_z(\phi) = R_z(\Psi - t\cos\Psi\cot\theta)R_x(\theta - t\sin\Psi)R_z(\phi + t\cos\Psi/\sin\theta)$$

$$R_x(t)R_z(\Psi)R_x(\theta)R_z(\phi) = R_z(\Psi - t\sin\Psi\cot\theta)R_x(\theta + t\cos\Psi)R_z(\phi + t\sin\Psi/\sin\theta)$$

(i) Define $\boldsymbol{e}_1(\mathcal{P})$ and $\boldsymbol{e}_1(\mathcal{P})$ to be the tangent vectors (at t=0) to the curve $\mathcal{C}(t) = R_x(t)\mathcal{P}$ and $\mathcal{C}(t) = R_y(t)\mathcal{P}$, respectively. Show that:

$$\vec{e}_1 = \cos\Psi\frac{\partial}{\partial\theta} - \sin\Psi\left(\cot\theta\frac{\partial}{\partial\Psi} - \frac{1}{\sin\theta}\frac{\partial}{\partial\phi}\right)$$

and

$$\vec{e}_2 = \sin\Psi\frac{\partial}{\partial\theta} + \cos\Psi\left(\cot\theta\frac{\partial}{\partial\Psi} - \frac{1}{\sin\theta}\frac{\partial}{\partial\phi}\right).$$

Solution

(a) The 3×3 matrices are:

$$K_1 = \begin{pmatrix} 0 & 0 & 0 \\ 0 & 0 & 1 \\ 0 & -1 & 0 \end{pmatrix}; \quad (K_1)^2 = \begin{pmatrix} 0 & 0 & 0 \\ 0 & -1 & 0 \\ 0 & 0 & -1 \end{pmatrix}; \quad (K_1)^4 = \begin{pmatrix} 0 & 0 & 0 \\ 0 & 1 & 0 \\ 0 & 0 & 1 \end{pmatrix}; \quad (K_1)^3 = \begin{pmatrix} 0 & 0 & 0 \\ 0 & 0 & -1 \\ 0 & 1 & 0 \end{pmatrix}$$

(b) $R_x(\theta) \equiv \exp(K_1\theta) = \sum_{n=0}^{\infty}\frac{\theta^n}{n!}(K_1)^n$.
So: $R_x(\theta)$

$$= 1 + \sum_{m=0}^{\infty} \frac{\theta^{(1+4m)}}{(1+4m)!}(K_1) + \sum_{m=0}^{\infty} \frac{\theta^{(2+4m)}}{(2+4m)!}(K_1)^2 + \sum_{m=0}^{\infty} \frac{\theta^{(3+4m)}}{(3+4m)!}(K_1)^3 + \sum_{m=0}^{\infty} \frac{\theta^{(4+4m)}}{(4+4m)!}(K_1)^4$$

$$= 1 + (K_1)\left\{\sum_{m=0}^{\infty}\left[\frac{\theta^{(1+4m)}}{(1+4m)!} - \frac{\theta^{(3+4m)}}{(3+4m)!}\right]\right\} + (K_1)^2\left\{\sum_{m=0}^{\infty}\left[\frac{\theta^{(2+4m)}}{(2+4m)!} - \frac{\theta^{(4+4m)}}{(4+4m)!}\right]\right\}$$

Note that

$$\theta - \frac{\theta^3}{3!} + \frac{\theta^5}{5!} - \frac{\theta^7}{7!} + \cdots = \sin\theta$$

and

$$1 - \frac{\theta^2}{2!} + \frac{\theta^4}{4!} - \frac{\theta^6}{6!} + \frac{\theta^8}{8!} = \cos\theta$$

So,

$$R_x(\theta) = \begin{pmatrix} 1 & 0 & 0 \\ 0 & 0 & 0 \\ 0 & 0 & 0 \end{pmatrix} + \sin\theta \begin{pmatrix} 0 & 0 & 0 \\ 0 & 0 & 1 \\ 0 & -1 & 0 \end{pmatrix} + \cos\theta \begin{pmatrix} 0 & 0 & 0 \\ 0 & 1 & 0 \\ 0 & 0 & 1 \end{pmatrix} = \begin{pmatrix} 1 & 0 & 0 \\ 0 & \cos\theta & \sin\theta \\ 0 & -\sin\theta & \cos\theta \end{pmatrix}$$

(c) Similarly

$$(K_2)_{mn} = \begin{pmatrix} 0 & 0 & -1 \\ 0 & 0 & 0 \\ 1 & 0 & 0 \end{pmatrix}; (K_2)^2 = \begin{pmatrix} -1 & 0 & 0 \\ 0 & 0 & 0 \\ 0 & 0 & -1 \end{pmatrix}; (K_2)^3 = \begin{pmatrix} 0 & 0 & 1 \\ 0 & 0 & 0 \\ -1 & 0 & 0 \end{pmatrix}; (K_2)^4 = \begin{pmatrix} +1 & 0 & 0 \\ 0 & 0 & 0 \\ 0 & 0 & +1 \end{pmatrix}$$

So,

$$R_y(\chi) = \begin{pmatrix} \cos\chi & 0 & -\sin\chi \\ 0 & 1 & 0 \\ \sin\chi & 0 & \cos\chi \end{pmatrix} = \exp(K_2\chi)$$

And for

$$(K_3)_{mn} = \begin{pmatrix} 0 & 1 & 0 \\ -1 & 0 & 0 \\ 0 & 0 & 0 \end{pmatrix} \rightarrow R_z(\phi) = \begin{pmatrix} \cos\phi & \sin\phi & 0 \\ \sin\phi & \cos\phi & 0 \\ 0 & 0 & 1 \end{pmatrix} = \exp(K_3\phi)$$

(d) Consider $P = R_z(\Psi)R_x(\theta)R_z(\phi)$:
$R_z(\phi)$ is a rotation about z by ϕ taking $\vec{x}$ to $\vec{\varepsilon}$.
$R_x(\theta)$ is a rotation about " $\vec{\varepsilon}$ " by θ taking $\vec{z}$ to $\vec{\mathcal{T}}$.
$R_z(\Psi)$ is a rotation about "$\vec{\mathcal{T}}$" by Ψ taking $\vec{\varepsilon}$ to $\overrightarrow{x'}$,
A total of twelve conventions are possible in defining the Euler angles (in a R.H. coordinate system). The two most frequently used conventions differ only in choice of axis for the 2nd rotation. (Here we have the *x*-convention, the other being the *y*-convention.)

(e) $\mathcal{C}$ is the curve $\mathcal{P} = R_z(t)$ through the identity matrix $\mathcal{C}(0) = \mathcal{I} \in SO(3)$. We have

$$R_z(t) = \begin{pmatrix} \cos t & \sin t & 0 \\ -\sin t & \cos t & 0 \\ 0 & 0 & 1 \end{pmatrix} \qquad f_{12}(P) = \sin t$$

and

$$\dot{e}(t) = \frac{d}{dt}e(t) = \frac{d}{dt}[R_z(t)\ell] = \begin{pmatrix} -\sin t & \cos t & 0 \\ -\cos t & -\sin t & 0 \\ 0 & 0 & 0 \end{pmatrix}\mathcal{I} \quad with \quad \frac{d\mathcal{I}}{dt}$$

$$= 0$$

$$\dot{e}(t) = \begin{pmatrix} 0 & 1 & 0 \\ -1 & 0 & 0 \\ 0 & 0 & 0 \end{pmatrix}\mathcal{I} \neq 0$$

(f) Define the vector field $\boldsymbol{e}_3$ on SO(3) by letting $\boldsymbol{e}_3(\mathcal{P})$ be the tangent (at t=0) to the curve $\mathcal{C}(t) = R_z(t)\mathcal{P}$ through $\mathcal{P}$. Show that $\boldsymbol{e}_3(\mathcal{P})$ is nowhere zero. From previous

$$\dot{e}(0) = \begin{pmatrix} 0 & 1 & 0 \\ -1 & 0 & 0 \\ 0 & 0 & 0 \end{pmatrix} P \neq 0$$

(g) Show that $\boldsymbol{e}_3 = \left(\frac{\partial}{\partial \Psi}\right)_{\theta\phi}$.

$$\dot{e}(0) = \begin{pmatrix} 0 & 1 & 0 \\ -1 & 0 & 0 \\ 0 & 0 & 0 \end{pmatrix}\begin{pmatrix} \cos\Psi & \sin\Psi & 0 \\ -\sin\Psi & \cos\Psi & 0 \\ 0 & 0 & 1 \end{pmatrix}\begin{pmatrix} 1 & 0 & 0 \\ 0 & \cos\theta & \sin\theta \\ 0 & -\sin\theta & \cos\theta \end{pmatrix}\begin{pmatrix} \cos\phi & \sin\phi & 0 \\ -\sin\phi & \cos\phi & 0 \\ 0 & 0 & 1 \end{pmatrix}$$

which is simply

$$\dot{e}(0) = \frac{\partial}{\partial\Psi}[\mathrm{P}] \;\Rightarrow \hat{e}_3 = \left(\frac{\partial}{\partial\Psi}\right)_{\theta\phi},$$

where

$$\hat{e}_3 = \left(\frac{\partial}{\partial\Psi}\right)_{\theta\phi}$$
$$= \begin{pmatrix} -\sin\Psi & \cos\Psi & 0 \\ -\cos\Psi & -\sin\Psi & 0 \\ 0 & 0 & 0 \end{pmatrix}\begin{pmatrix} 1 & 0 & 0 \\ 0 & \cos\theta & \sin\theta \\ 0 & -\sin\theta & \cos\theta \end{pmatrix}\begin{pmatrix} \cos\phi & \sin\phi & 0 \\ -\sin\phi & \cos\phi & 0 \\ 0 & 0 & 1 \end{pmatrix}$$

(h) Show

$$R_y(t)R_z(\Psi)R_x(\theta)R_z(\phi) = R_z(\Psi - t\cos\Psi\cot\theta)R_x(\theta - t\sin\Psi)R_z(\phi + t\cos\Psi / \sin\theta)$$

$$R_x(t)R_z(\Psi)R_x(\theta)R_z(\phi) = R_z(\Psi - t\sin\Psi\cot\theta)R_x(\theta + t\cos\Psi)R_z(\phi + t\sin\Psi/\sin\theta)$$

To start we have $R_z(t)P = R_z(t)[R_z(\Psi)R_x(\theta)R_z(\phi)] = R_z(\Psi + t)R_x(\theta)R_z(\phi)$, and for $t \ll 1$:
$R_z(t)[R_z(\Psi)R_x(\theta)R_z(\phi)] =$
$=$

$$R_z(t)\begin{pmatrix} \cos\Psi & \sin\Psi & 0 \\ -\cos\Psi & \cos\Psi & 0 \\ 0 & 0 & 1 \end{pmatrix}\begin{pmatrix} 1 & 0 & 0 \\ 0 & \cos\theta & \sin\theta \\ 0 & -\sin\theta & \cos\theta \end{pmatrix}\begin{pmatrix} \cos\phi & \sin\phi & 0 \\ -\sin\phi & \cos\phi & 0 \\ 0 & 0 & 1 \end{pmatrix}$$
$$= R_z(t)\begin{pmatrix} [\cos\Psi\cos\phi - \sin\Psi\sin\phi\cos\theta] & [\cos\Psi\sin\phi + \sin\Psi\cos\phi\cos\theta] & [\sin\Psi\sin\theta] \\ [-\sin\Psi\cos\phi - \cos\Psi\sin\phi\cos\theta] & [-\sin\Psi\sin\phi + \cos\Psi\cos\theta\cos\phi] & [\cos\Psi\sin\theta] \\ \sin\theta\sin\phi & -\sin\theta\cos\phi & \cos\theta \end{pmatrix}$$

where

$$R_z(t) \cong \begin{pmatrix} 1 & 0 & 0 \\ 0 & 1 & t \\ 0 & -t & 1 \end{pmatrix}$$

$$R_x(t)R_z(\Psi)R_x(\theta)R_z(\phi) = R_z(\Psi - t\sin\Psi\cot\theta)R_x(\theta + t\cos\Psi)R_z\left(\phi + t\frac{\sin\Psi}{\sin\theta}\right)$$

Similarly, consider $R_y(t)P$ for $t \ll 1$

$$\begin{pmatrix} 11 & 0 & -t \\ 0 & 1 & 0 \\ t & 0 & 1 \end{pmatrix}P$$
$$= \begin{pmatrix} (\ldots) & (\ldots) & \sin\Psi\sin\theta - t\cos\theta \\ (\ldots) & (\ldots) & \cos\Psi\sin\theta \\ \sin\theta\sin\phi + t[\cos\Psi\cos\phi - \sin\Psi\sin\phi\cos\theta] & (\ldots) & \cos\theta + t\sin\Psi\sin\theta \end{pmatrix}$$

$\Psi \to \Psi + tA, \theta \to \theta + tB, \phi \to \phi + tC$

23 term: $\cos(\Psi + tA)\sin(\theta + tB) = \cdots$
$[\cos\Psi + tA\sin\Psi][\sin\theta + tB\cos\theta] = \cdots \qquad \cos\Psi + tA\sin\Psi\sin\theta$
$= tB\cos\theta\cos\Psi$
$A = -B\cot\theta\cot\Psi.$

13 term: $[\sin\Psi + tA\cos\Psi][\sin\theta + tB\cos\theta] = \sin\phi\sin\theta - t\cos\theta$
$tA\cos\Psi\sin\theta + tB\cos\theta\sin\Psi = -t\cos\theta$
$B = -\sin\Psi$

31 term: $\sin(\theta + tB)\cos(\phi + tC) = [\sin\theta + tB\cos\theta][\sin\phi + tC\cos\phi]$
$C = \cos\ \Psi\ /\ \sin\ \theta$
Thus,

$$R_y(x) \equiv \exp(K_2 x) = \sum_{n=0}^{\infty}\frac{x^n}{n!}(K_2)^n = 1 + \sum_{m=0}^{\infty}\frac{x^{(1+4m)}}{(1+4m)!}(K_2) + \sum_{m=0}^{\infty}\cdots$$

$$= 1 + (K_2)\left\{\sum_{m=0}^{\infty}\left[\frac{x^{(1+4m)}}{(1+4m)!} - \frac{x^{(3+4m)}}{(3+4m)!}\right]\right\} + \cdots$$

and

$$R_y(x) = \begin{pmatrix} 0 & 0 & 0 \\ 0 & 1 & 0 \\ 0 & 0 & 0 \end{pmatrix} + \sin x\begin{pmatrix} 0 & 0 & -1 \\ 0 & 0 & 0 \\ 1 & 0 & 0 \end{pmatrix} + \cos x\begin{pmatrix} 1 & 0 & 0 \\ 0 & 0 & 0 \\ 0 & 0 & 1 \end{pmatrix}$$
$$= \begin{pmatrix} \cos x & 0 & -\sin x \\ 0 & 1 & 0 \\ \sin x & 0 & \cos x \end{pmatrix}$$

So,

$R_y(t)R_z(\Psi)R_x(\theta)R_z(\phi) = R_z(\Psi - t\cos\Psi\cot\theta)R_x(\theta - t\sin\Psi)R_z(\phi + t\cos\ \Psi\ /\ \sin\ \theta\ ).$

(i) Consider $\mathcal{C}(t) = R_z(t)P = R_z(t + \Psi)R_x(\theta)R_z(\phi)$

$$\left[\frac{d}{dt}\mathcal{C}(t)\right]_{t=0} = \lim_{t\to 0}\left\{\frac{\mathcal{C}(t) - \mathcal{C}(0)}{t}\right\}$$
$$= \lim_{t\to 0}\left[\frac{R_z(t+\Psi) - R_z(\Psi)}{t}\right]R_z(\theta)R_z(\phi)$$

Could Taylor expand $R_z(t+\Psi) \sim R_z(\Psi) + t\frac{\partial R_z(\Psi)}{\partial \Psi} + \cdots$
Or, since $R_z(\Psi) = \exp(K_3\Psi)$:

$$R_z(t+\Psi) = \exp(K_3[t+\Psi]) = \exp(K_3 t)\exp(K_3\Psi)$$
$$= (1 + K_3 t)\exp(K_3\Psi) + \cdots$$

So

$$R_z(t+\Psi) = R_z(\Psi) + t\frac{\partial R_z(\Psi)}{\partial \Psi}$$

Now $\mathcal{C}(t) = R_z(\Psi - t\sin\Psi\cot\theta)$ ...

So

$$\left[\frac{d}{dt}\mathcal{C}(t)\right]_{t=0}$$
$$= \lim_{t\to 0}\left\{\frac{R_z(\Psi - t\sin\Psi\cot\theta)R_x(\theta + t\cos\Psi)R_z\left(\phi + t\frac{\sin\Psi}{\sin\theta}\right) - R_zR_xR_z}{t}\right\}$$
$$= -\sin\Psi\cot\theta\frac{\partial R_z(\Psi)}{\partial\Psi}R_x(\theta)R_z(\phi) + \cos\Psi\, R_z(\Psi)\frac{\partial R_x(\theta)}{\partial\theta}R_z(\phi)$$
$$+ \frac{\sin\Psi}{\sin\theta}R_z(\Psi)R_x(\theta)\frac{\partial R_z(\phi)}{\partial\phi}$$
$$= \left\{\cos\Psi\frac{\partial}{\partial\theta} - \sin\Psi\left(\cot\theta\frac{\partial}{\partial\Psi} - \frac{1}{\sin\theta}\frac{\partial}{\partial\phi}\right)\right\}P$$

Thus

$$\hat{e}_2 = \cos\Psi\frac{\partial}{\partial\theta} - \sin\Psi\left(\cot\theta\frac{\partial}{\partial\Psi} - \frac{1}{\sin\theta}\frac{\partial}{\partial\phi}\right).$$

Next, consider $\mathcal{C}(t) = R_y(t)P$. Thus,

$$\left[\frac{d}{dt}\mathcal{C}(t)\right]_{t=0} = \left\{-\cos\Psi\cot\theta\frac{\partial}{\partial\Psi} - \sin\Psi\frac{\partial}{\partial\theta} + \frac{\cos\Psi}{\sin\theta}\frac{\partial}{\partial\phi}\right\}P$$

and we have

$$\hat{e}_3 = -\sin\Psi\frac{\partial}{\partial\theta} - \cos\Psi\left(\cot\theta\frac{\partial}{\partial\Psi} - \frac{1}{\sin\theta}\frac{\partial}{\partial\phi}\right)$$

Now group to get

$$\hat{f}_1 = \hat{e}_2 - i\hat{e}_3 = (\cos\Psi + i\sin\Psi)\frac{\partial}{\partial\theta}$$
$$+ (-\sin\Psi + i\cos\Psi)\left(\cot\theta\frac{\partial}{\partial\Psi} - \frac{1}{\sin\theta}\frac{\partial}{\partial\phi}\right)$$

So,

$$\hat{f}_1 = e^{i\Psi}\frac{\partial}{\partial\theta} + ie^{i\Psi}(\dots)$$

Rotation about $\hat{x}$ axis $have$: $\phi = 0, \Psi = 0$ and $\hat{e}_1 = \frac{\partial}{\partial\theta}$, $\hat{e}_2 = 0$ is consistent.

Rotation about $\hat{y}$ axis $have$ $\phi = \frac{\pi}{2}, \Psi = 0$ and $\hat{e}_1 = 0$, $\hat{e}_2 = \frac{\partial}{\partial\theta}$ which is also consistent.

$$R_y(\theta)\begin{pmatrix} x \\ y \\ z \end{pmatrix} = \begin{pmatrix} \cos\theta\, x & -\sin\theta & z \\ & y & \\ \sin\theta\, x & & \cos\theta\, z \end{pmatrix}$$

This gives (h) $B = \sin\Psi \Rightarrow$ altering the signs to agree with MTW conventions.

Example 10.14. (MTW 9.14) Rotation Group Structure Constants

Use the three fields from the previous example as the basis for the manifold of the rotation group. Show the commutator coefficients for this basis are:

$$c_{\alpha\beta}{}^{\gamma} = -\epsilon_{\alpha\beta\gamma}$$

Independently of location P in the rotation group. These coefficients are also called the structure constants of the group.

Solution

Killing Vector Structure of SO(3):

$$\vec{e}_1 = \cos\Psi\frac{\partial}{\partial\theta} - \sin\Psi\left(\cot\theta\frac{\partial}{\partial\Psi} - \frac{1}{\sin\theta}\frac{\partial}{\partial\phi}\right)$$

$$\vec{e}_2 = \sin\Psi\frac{\partial}{\partial\theta} + \cos\Psi\left(\cot\theta\frac{\partial}{\partial\Psi} - \frac{1}{\sin\theta}\frac{\partial}{\partial\phi}\right)$$

$$\vec{e}_2 = \frac{\partial}{\partial\Psi}$$

Let's find the forms that correspond to $\vec{e}_1, \vec{e}_2, \vec{e}_3$ such that

$$\langle\sigma^j, \vec{e}_1\rangle = 1 \quad j = i$$
$$= 0 \quad j \neq i$$

$$\sigma^1 = A_1 d\theta + B_1 d\Psi + C d\phi$$

$$\langle\sigma^1, e_1\rangle = 1 = A_1\cos\Psi - B_1\sin\Psi\cot\theta + C_1\frac{\sin\Psi}{\sin\theta}$$

$$\langle\sigma^1, e_2\rangle = 0 = -\breve{A}_1\sin\Psi - B_1\cos\Psi\cot\theta + C_1\frac{\sin\Psi}{\sin\theta}$$

$\langle \sigma^1, e_3 \rangle = 0 = B_1$
Solving for C_1 and A_1:

$$C_1 = \sin\theta \sin\Psi$$
$$A_1 = \cos\Psi$$

So,

$$\sigma^1 = \cos\Psi\, d\theta + \sin\Psi \sin\theta\, d\phi.$$

Repeating:
$\langle \sigma^2, e_1 \rangle = 0 = -\breve{A}_2(\cos\Psi) - B_2 \sin\Psi \cot\theta + C_2 \frac{\sin\Psi}{\sin\theta}$
$\langle \sigma^2, e_2 \rangle = 1 = -\breve{A}_2 \sin\Psi - B_2 \cos\Psi \cot\theta + C_2 \frac{\sin\Psi}{\sin\theta}$
$\langle \sigma^2, e_3 \rangle = 0 = B_2$
and

$$C_2 = \sin\theta \cos\Psi$$
$$A_2 = \sin\Psi$$

So,

$$\sigma^2 = \sin\Psi\, d\theta + \sin\theta \sin\Psi\, d\phi$$

Repeating again:
$\langle \sigma^3, e_1 \rangle = 0 = A_3(\cos\Psi) - B_3 \sin\Psi \cot\theta + C_3 \frac{\sin\Psi}{\sin\theta}$
$\langle \sigma^3, e_2 \rangle = 0 = -A_3(\sin\Psi) - B_3 \cos\Psi \cot\theta + C_3 \frac{\cos\Psi}{\sin\theta}$
$\langle \sigma^3, e_3 \rangle = 0 = B_3$
and

$$C_3 = \cos\theta$$
$$A_3 = 0$$

So,

$$\sigma^3 = d\Psi + \cos\theta\, d\phi.$$

Collecting we have:

$\sigma^1 = \sin\Psi \sin\theta\, d\phi + \cos\Psi\, d\theta$
$\sigma^2 = \cos\Psi \sin\theta\, d\phi - \sin\Psi\, d\theta$
$\sigma^3 = \cos\theta\, d\phi + d\Psi$

So, the Killing metric on SO(3) is

$$K_{ab} = \delta_{ij}\sigma_a^i \sigma_b^j = \sigma_a^1 \sigma_b^1 + \sigma_a^2 \sigma_b^2 + \sigma_a^3 \sigma_b^3$$
$$= (d\Psi \cos\theta\, \mathrm{d}\phi)^2 + (d\theta^2 + \sin^2\theta\, d\phi^2)$$

10.6.4 Taub – NUT Space-Time Preliminary Review

Let's consider Taub – NUT space:

$$ds^2 = -U^{-1}dt^2 + (2\ell)^2 U(d\Psi + \cos\theta\, d\phi)^2 + (t^2 + \ell^2)(d\theta^2 + \sin^2\theta\, d\phi^2)$$

(Eqn. 10-51)

here

$$U(t) = -1 + \frac{2(mt + \ell^2)}{t^2 + \ell^2} \qquad 0 \le \Psi \le \frac{4\pi}{SU(2)}, \quad 0 \le \theta \le \pi, \quad 0 \le \phi \le 2\pi$$

(Eqn. 10-52)

Does the KV structure of SU(2) follow from that of SU(3)? Assuming that it does, regard the KV structure of Taub. This structure exists for any t = constant slice so let's look at just that:

$$d\ell^2 = a^2(d\theta^2 + \sin^2\theta\, d\phi^2) + b^2(d\Psi + \cos\theta\, d\phi)^2$$

In terms of σ^i's above:

$$d\ell^2 = a^2((\sigma^1)^2 + (\sigma^2)^2) + b^2(\sigma^3)^2$$

$$d\ell^2 \Leftrightarrow g_{ab}$$

So, $g_{ab} = a^2(K_{ab} - \sigma_a^3\sigma_b^3) + b^2(\sigma_a^3\sigma_b^3)$. Let $\Psi^a = \frac{\partial}{\partial\Psi}, \quad \Psi_a = K_{ab}\Psi^b = K_{a\Psi} = \sigma_a^3\sigma_\Psi^3 = \sigma_a^3$, then

$$g_{ab} = a^2(K_{ab} - \Psi_a\Psi_b) + b^2(\Psi_a\Psi_b)$$

Let's recap before moving on:

$$R_z(t)\underbrace{R_z(\Psi)R_x(\theta)R_z(\phi)}_{P(\Psi,\theta,\phi)} = P(\Psi + t, \theta, \phi)$$

$$R_x(t)P(\Psi,\theta,\phi) = P\left(\Psi - t\sin\Psi\cot\theta, \theta + t\cos\Psi, \phi + t\frac{\sin\Psi}{\sin\theta}\right)$$

$$R_y(t)P(\Psi,\theta,\phi) = P\left(\Psi - t\cos\Psi\cot\theta, \theta + t\sin\Psi, \phi + t\frac{\cos\Psi}{\sin\theta}\right)$$

$$V_z = \frac{d}{dt}[R_z(t)\mathcal{I}]|_{t=0} \qquad V_x = \frac{d}{dt}[R_x(t)\mathcal{I}]|_{t=0} \qquad V_y = \frac{d}{dt}[R_y(t)\mathcal{I}]\Big|_{t=0}$$

where $\mathcal{I}$ is the identity. A diffeo $\Psi: M \longrightarrow M$ maps a vector ε^a at x_0 to a vector $\Psi_x\varepsilon^a$ at $\Psi(x_0)$. So, the diffeo R_p produces a vector

$$R_{p*}\varepsilon^a \text{ at } R_p(d) = dP = P$$

so,

$$R_{p*}V_z = \frac{d}{dt}[R_z(t)P]|_{t=0}$$

We thus arrive at a Right invariant basis of vector fields for SO(3):

$\vec{V}_z$: $\vec{e}_3 = \partial/_{\partial\Psi}$

$\vec{V}_y$: $\vec{e}_2 = -\sin\Psi\,\partial/_{\partial\theta} - \cos\Psi\left(\cot\Psi\,\partial/_{\partial\Psi} - \frac{1}{\sin\theta}\,\partial/_{\partial\phi}\right)$

$\vec{V}_x$: $\vec{e}_1 = \cos\Psi\,\partial/_{\partial\theta} - \sin\Psi\left(\cot\Psi\,\partial/_{\partial\Psi} - \frac{1}{\sin\theta}\,\partial/_{\partial\phi}\right)$

where

$$e^a\sigma_b(\mathcal{I}) = \delta^a_b(\mathcal{I}) = \text{constant}$$

If we extend δ^a_b to the whole group by requiring $\delta^a_b(P) = R_{P*}\delta^a_b(\mathcal{I})$ such that δ^a_b is right invariant, then $e^a\sigma_b = \delta^a_b$ everywhere, if e_i and σ^j are the corresponding bases:

$$(e^a\sigma_b)(e_i)_a(\sigma^j)^b = \delta^a_b(e_i)_a(\sigma^j)^b = \delta^j_i$$

$$e_i\sigma^j = \delta^j_i$$

So $R_{P*}(e^a\sigma_b) = e^a\sigma_b\ ,\ R_{P*}e^a = e^a$ and

$$R_{P*}(e^a\sigma_b) = R_{P*}e^aR_{P*}\sigma_b = e^aR_{P*}\sigma_b = e^a\sigma_b$$

So $R_{P*}\sigma_b = \sigma_b$, thus the form field basis that corresponds to our right inv. basis of vector fields is a right invariant form basis for SO(3):

$\sigma^1 = \sin\Psi\sin\theta\, d\phi + \cos\Psi\, d\theta$

$\sigma^2 = \cos\Psi\sin\theta\, d\phi - \sin\Psi\, d\theta$

$\sigma^3 = \cos\theta\, d\phi + d\Psi$

Now, consider the diffeo that corresponds to left multiplication: $P(\Psi,\theta,\phi)R_z(t) = P(\Psi,\theta,\phi+t)$ where results follow by "$\Psi \leftrightarrow \phi$" from previously:

$$P(\Psi,\theta,\phi)R_x(t) = P\left(\Psi + t\frac{\sin\phi}{\sin\theta}, \theta + t\cos\phi, \phi - t\sin\phi\cot\theta\right)$$

$$P(\Psi,\theta,\phi)R_y(t) = P\left(\Psi + t\frac{\cos\phi}{\sin\theta}, \theta - t\sin\phi, \phi - t\cos\phi\cot\theta\right)$$

$$V_z = \frac{d}{dt}[\mathcal{I}R_z(t)]|_{t=0} = \frac{d}{dt}[R_z(t)]|_{t=0}$$

$$L_{P*}V_z = \frac{d}{dt}[PR_z(t)]|_{t=0}$$

So, the left inv. vector fields on SO(3) is:

$\vec{V}_z$: $\vec{e}_3 = \partial/\partial\phi$

$\vec{V}_y$: $\vec{e}_2 = -\sin\phi\,\partial/\partial\theta - \cos\phi\left(\cot\theta\,\partial/\partial\phi - \frac{1}{\sin\theta}\partial/\partial\Psi\right)$

$\vec{V}_x$: $\vec{e}_1 = \cos\phi\,\partial/\partial\theta - \sin\phi\left(\cot\theta\,\partial/\partial\phi - \frac{1}{\sin\theta}\partial/\partial\Psi\right)$

And the Left inv. form fields on SO(3):

$\sigma^1 = \sin\phi\sin\theta\, d\Psi + \cos\phi\, d\theta$

$\sigma^2 = \cos\phi\sin\theta\, d\Psi - \sin\phi\, d\theta$

$\sigma^3 = \cos\theta\, d\Psi + d\phi$

Now $d\ell^2 = a^2(d\theta^2 + \sin^2\theta\, d\phi^2) + b^2(d\Psi + \cos\theta\, d\phi)^2 = a^2((\sigma_R^1)^2 + (\sigma_R^2)^2) + b^2(\sigma_R^3)^2$

So,

$(\sigma_L^1)^2 = \sin^2\phi\sin^2\theta\,(d\Psi)^2 + 2\sin\phi\sin\theta\cos\phi\, d\Psi d\theta + \cos^2\phi\,(d\theta)^2$

$(\sigma_L^1)^2 + (\sigma_L^2)^2 = \sin^2\theta\,(d\Psi)^2 + (d\theta)^2$

$(\sigma_L^3)^2 = (d\phi + \cos\theta\, d\Psi)^2$

$$(\sigma_L^1)^2 + (\sigma_L^2)^2 + (\sigma_L^3)^2$$
$$= (d\theta)^2 + \sin^2\theta\,(d\Psi)^2 + \cos^2\theta\,(d\Psi)^2 + \cos\theta\, d\phi d\Psi$$
$$= (d\theta)^2 + (d\Psi)^2 + (d\phi)^2 + 2\cos\theta\, d\phi d\Psi$$
$$(\sigma_R^1)^2 + (\sigma_R^2)^2 + (\sigma_R^3)^2 = (d\theta)^2 + (d\Psi)^2 + (d\phi)^2 + 2\cos\theta\, d\phi d\Psi$$

And we get: $(\sigma_R)^2 = (\sigma_L)^2$. Thus, $K_{ab} = \delta_{ij}\sigma_a^i\sigma_b^i$ is both left and right inv. under SO(3):

$$d\ell^2 \to g_{ab} = a^2(K_{ab} - \sigma_{Ra}^3\sigma_{Rb}^3) + b^2\sigma_{Ra}^3\sigma_{Rb}^3$$

So, g_{ab} is right invariant under SU(2), also, g_{ab} is left invariant under the U(1) subgroup of SU(2) that fixes $\Psi_a\sigma_{Ra}^3$. The symmetry group of Taub is thus $SU(2)_R \times U(1)_L$.What are the vector fields which are generated by the right action of SU(2)?

$\Psi_\lambda(a) = ag(\lambda)$: a 1 parameter family of diffeo's which have right multiplication.

$\frac{d}{d\lambda}(\Psi_\lambda(a))\big|_{\lambda=0} = aV(identity)$: tangent to $\Psi_\lambda(a)$ should yield vector field.

$V(P) = av(identity) \Rightarrow$ left inv. vector fields.

So, to get the three K.V.'s generating the right action of SU(2), Right action → Left inv. vector fields. So, the K.V. generating the Right action of SU(2) are the left inv. K.V.'s:

$$\begin{cases} \vec{V}_\pm = e^{\mp\phi}\, \partial/\partial\theta \mp ie^{\mp i\phi}\left(\cot\theta\, \partial/\partial\phi - \frac{1}{\sin\theta}\, \partial/\partial\Psi\right) \\ \vec{V}_3 = \partial/\partial\phi \end{cases}$$

The additional K.V. generating the left U(1) action is the right inv. vector:

$$\left\{\vec{e}_3 = \partial/\partial\Psi\right.$$

Let $\partial/\partial\Psi\,(S) = 0$, then

$$\left.\begin{aligned} \vec{V}_1 &= e^{\mp i\phi}(\partial_\theta \mp \cot\theta\, \partial_\phi) \\ \vec{V}_3 &= \partial_\phi \end{aligned}\right\} \text{transitive on 2-sphere SO(3),}$$

with $\vec{e}_3 = \partial/\partial\Psi$ → transitive on 3-sphere SO(4).

So, the K.V. structure of Taub yields surfaces of transitivity which are the 3-spheres $\underline{t = \text{const}}$.

10.6.5. Quaternions and SU(2)

Rotations in 2-dimensions can be described by vectors in R^2 of unit norm, regarded as complex numbers $e^{i\phi}$, with composition of rotations given by multiplication:

$$e^{i(\phi+\theta)} = e^{i\phi}e^{i\theta}. \qquad \text{(Eqn. 10-53)}$$

The rotation group SO(2) is thereby isomorphic to U(1). Hamilton found a spectacular way to generalize this description to the 3-dimensional rotation group. The manifold of the group SO(2) is a circle, and the isomorphism identifies SO(2) with the unit circle in R^2, with a point of R^2, written in the form $t + ix$, and with multiplication specified by linearity and by the requirement $i^2 = -1$.

SO(3) is three dimensional, and, as we will see, it is almost isomorphic to the 3-sphere S^3. Hamilton suggested representing a rotation by a vector of unit norm in R^4, written in the form

$$t + xi + yj + zk.$$

(Eqn. 10-54)

Rotations about each axis are represented by

$$e^{i\alpha}, e^{j\alpha}, e^{k\alpha},$$

(Eqn. 10-55)

(although α will tum out to correspond to twice the rotation angle), where the multiplication law for two rotations about the same axis is again specified by linearity and

$$i^2 = j^2 = k^2 = -1.$$

(Eqn. 10-56)

The composition of rotations about different axes will be related to the fact that 180°-degree rotations satisfy $R(\pi i)R(\pi j) = R(\pi k)$. The law is

$$ij = k, \quad \text{and cyclic permutations,}$$
$$ji = -k, \qquad \text{and cyclic permutations.}$$

We can show that the multiplication law for quaternions is independent of the choice of basis by showing that these equations are equivalent to the relation

$$mn = -m \cdot n + m \times n,$$

for any vectors m, n constructed from *i, j, k.*

Theorem. The group of unit quaternions,

$$\{q \equiv t + xi + yj + zk, \;\; t^2 + x^2 + y^2 + z^2 = 1\},$$

is homomorphic to the rotation group with homomorphism given by

$$e^{\frac{\phi}{2}n} \to R(\phi n).$$

(Eqn. 10-57)

The map is 2–1: q and –q are mapped to the same rotation. Before we prove the theorem, note that any unit quaternion can be written in the form $e^{n\phi/2}$, with ϕ in the interval $[0, 2\pi)$. One might think that ϕ must range up to 4π, but by changing n to –n, ϕ can be replaced by $2\pi - \phi$: $e^{[(-m)\phi/2]} = e^{[m(\pi-\phi/2)]}$.

Proof. The proof is based on the second key part of the relation between rotations and quaternions – that the rotation $R(\phi n)r$ of a vector $r \in R^3$, is given by

$$R(\phi n)r = e^{\frac{\phi}{2}n} r e^{-\frac{\phi}{2}n}.$$

(Eqn. 10-58)

Because all axes are equivalent, it suffices to prove the relation for a rotation about the z-axis: $n = k$. We need only check that both sides of the equation agree on the basis vectors i, j, k. For $r = k$, the result is immediate. For $r = i$, we have

$$R(\phi k)i = \cos\phi\, i + \sin\phi\, j.$$

(Eqn. 10-59)

So,

$$e^{\frac{\phi}{2}k} r e^{-\frac{\phi}{2}k} = \left(\cos\frac{\phi}{2} + k\sin\frac{\phi}{2}\right) i \left(\cos\frac{\phi}{2} - k\sin\frac{\phi}{2}\right)$$

$$= \left(\cos^2\frac{\phi}{2} - \sin^2\frac{\phi}{2}\right) i \left(2\cos\frac{\phi}{2}\sin\frac{\phi}{2}\right) j$$

Thus

$$e^{\frac{\phi}{2}k} r e^{-\frac{\phi}{2}k} = \cos\phi\, i + \sin\phi\, j.$$

(Eqn. 10-60)

For $r = j$, the computation is essentially identical, using $jk = -kj = i$, instead of $ki = -ik = j$. To prove the Theorem, we must show that the map preserves group multiplication. But

$$e^{\frac{\phi}{2}n} e^{\frac{\psi}{2}m} r e^{-\frac{\psi}{2}m} e^{-\frac{\phi}{2}n} = R(\phi n)R(\psi m)r \Rightarrow e^{\frac{\phi}{2}n} e^{\frac{\psi}{2}m} \to R(\phi n)R(\psi m).$$

(Eqn. 10-61)

Thus the map *is* a homomorphism. Finally, two quaternions $e(m\phi/2)$ and $e(n\psi/2)$ correspond to the same rotation if and only if $R(\phi m) = R(\psi n)$. For ϕ and ψ in the range [0, 2π), the rotations agree if and only if $m = n$ and $\phi = \psi$ or $m = -n$ and $\phi =$ whichever one of the two angles $(\psi + \pi)$ or $\psi - \pi$ is in the range [0, 2π). These two possibilities correspond to the two quaternions $\pm e^{m\phi/2}$. What remains is to show that the group of unit quaternions is isomorphic to the group SU(2). One identifies Pauli-spin matrices with the unit quaternions:

$$\frac{1}{i}\vec{\sigma} \cdot n \leftrightarrow n.$$

(Eqn. 10-62)

The algebra generated by $\vec{\sigma}/i$ is identical to that generated by the quaternions, because

$$\sigma_j \sigma_k = \delta_{jk} + i\epsilon^{i}_{jk}\sigma_i \Rightarrow$$

$$m \cdot \frac{\vec{\sigma}}{i} n \cdot \frac{\vec{\sigma}}{i} = -m \cdot n + m \times n \cdot \frac{\vec{\sigma}}{i}.$$

(Eqn. 10-63)

in agreement with above results. The isomorphism between the group of quaternions and SU(2) is then

$$e^{\alpha n} \leftrightarrow e^{\alpha n \cdot \sigma / i}.$$

(Eqn. 10-64)

Note that the Lie algebra of SU(2) is identical to that of SO(3), with isomorphism

$$\frac{n}{2} \leftrightarrow n^l K_{lab}.$$

(Eqn. 10-65)

This means that locally SO(3) looks identical to SU(2): They are identical in a neighborhood of e, and left multiplication maps a neighborhood of e to a neighborhood of any other point. Since the manifold of SU(2) is the 3-spherc, SU(2) is simply-connected: any loop can be deformed to the identity. In the rotation group, however, paths $\phi \rightarrow R(\phi n)$, where ϕ runs from 0 to 2π, cannot be deformed to the identity (this is the content of the first belt trick). SU(2) is called a *covering group* of SO(3): there is a homomorphism from SU(2) to SO(3) that is locally an isomorphism. It is easy to show that any Lie group can be covered by a simply connected group, and SU(2) is this covering group (the "universal covering group") of SO(3). The additive reals are similarly the universal covering group of U(1).

Example 10.15. The homomorphism SU(2)→SO(3) is given by:

$$U = e^{i\frac{\phi}{2}\vec{n}\cdot\sigma} \rightarrow R(\psi\vec{n})\,.$$

(a) Translate the quaternion formulas into the form

$$R(\psi\vec{n})\bar{r}\cdot\sigma = U(\vec{r}\cdot\vec{\sigma})U^t$$

(b) Use the fact that $r^2 = -\det(\vec{r}\cdot\vec{\sigma})$ to show that lengths of each vector r are preserved by a map of the form

$$\bar{r}\cdot\sigma \rightarrow U(\vec{r}\cdot\vec{\sigma})U^t$$

(c) Show that the SU(2) matrix corresponding to Euler angles θ, ϕ, Ψ is

$$U = \begin{pmatrix} \alpha & \beta \\ -\beta^* & \alpha^* \end{pmatrix} \quad where \quad \begin{cases} \alpha = \cos\frac{\theta}{2}e^{i\frac{\Psi+\phi}{2}} \\ \beta = \sin\frac{\theta}{2}e^{i\frac{\Psi-\phi}{2}} \end{cases}$$

Solution

(a) The group of unit quaternions is homomorphic to the rotation group:

$$e^{\frac{\phi}{2}\vec{n}} \rightarrow R(\phi\vec{n})$$

Quaternion formulas relating the above groups explicitly are:

$$R(\phi\vec{n})r = e^{\frac{\phi}{2}\vec{n}}re^{\frac{\phi}{2}\vec{n}}$$

The group of unit quaternions is isomorphic to SU(2) where the relation can be expressed as:

$$-i\vec{\sigma}\cdot\vec{n} \leftrightarrow \vec{n}$$

Using $\vec{n} = -i\vec{\sigma}\cdot\vec{n}$ and $r = x\bar{\imath} + y\bar{\jmath} + z\bar{k} = x(-i\sigma_x\hat{\imath}) + y(-i\sigma_y\hat{\jmath}) + z(-i\sigma_z\hat{k})$

$$= (x\hat{\imath})(-i\sigma_x) + (y\hat{\jmath})(-i\sigma_y) + (z\hat{k})(-i\sigma_z)$$

$$= -i(\vec{x}\sigma_x + \vec{y}\sigma_y + \vec{z}\sigma_z)$$

$$r = -i\vec{r}\cdot\vec{\sigma}$$

Translating:

$$R(\phi\vec{n})(-i\vec{r}\cdot\vec{\sigma}) = e^{\frac{\phi}{2}(-i\vec{n}\cdot\vec{\sigma})}(-i\vec{r}\cdot\vec{\sigma})e^{\frac{\phi}{2}(-i\vec{n}\cdot\sigma)}$$

Rename $\phi \to \Psi$

$$R(\Psi\vec{n})(\vec{r}\cdot\vec{\sigma}) = e^{-i\frac{\Phi}{2}\vec{n}\cdot\vec{\sigma}}(\vec{r}\cdot\vec{\sigma})e^{-i\frac{\Phi}{2}\vec{n}\cdot\vec{\sigma}} = U(\vec{r}\cdot\vec{\sigma})U^t$$

(b) If $r^2 = -\det(\vec{r}\cdot\vec{\sigma})$ and we have a map: $\vec{r}\cdot\vec{\sigma} \to U(\vec{r}\cdot\vec{\sigma})U^t$

Then $r^2 \to -\det(V\vec{r}\cdot\vec{\sigma}U^t) = -\det U \det(\vec{r}\cdot\vec{\sigma}) \det U^t$

$$= -\det(\vec{r}\cdot\vec{\sigma}) = r^2$$

So lengths are preserved.

(c) We have $R(\Psi,\theta,\phi)r = R(\Psi k)R(\theta i)R(\phi k)r \quad with\, r = \vec{r}\cdot\vec{\sigma}.$

$$R(\Psi,\theta,\phi)r =$$

$$R(\Psi k)R(\theta i)v_i\vec{r}\cdot\vec{\sigma}u_1^t = u_3u_2u_1(\vec{r}\cdot\vec{\sigma})u_1^tu_2^tu_3^t$$

$$= u_3u_2u_1(\vec{r}\cdot\vec{\sigma})(u_3u_2u_1)^t = U(\vec{r}\cdot\vec{\sigma})U^t$$

where $u_1 = \exp\left(i\frac{\phi}{2}\vec{k}\cdot\vec{\sigma}\right)$, etc. So:

$$U = u_3u_2u_1 = \exp\left(i\frac{\Psi}{2}\underbrace{\vec{k}\cdot\vec{\sigma}}_{\sigma_z}\right)\exp\left(i\frac{\theta}{2}\underbrace{\vec{\imath}\cdot\vec{\sigma}}_{\sigma_x}\right)\exp\left(i\frac{\phi}{2}\underbrace{\vec{k}\cdot\vec{\sigma}}_{\sigma_z}\right)$$

Since σ_z is diagonal:

$$U = \begin{pmatrix} \exp\left(i\frac{\Psi}{2}\right) & 0 \\ 0 & \exp\left(-i\frac{\Psi}{2}\right) \end{pmatrix} \exp\left(i\frac{\theta}{2}\sigma_x\right) \begin{pmatrix} \exp\left(i\frac{\phi}{2}\right) & 0 \\ 0 & \exp\left(-i\frac{\phi}{2}\right) \end{pmatrix}$$

And,

$$\exp\left(i\sigma_x\frac{\theta}{2}\right) = \cos\left(\frac{\theta}{2}\right) + i\sigma_x\sin\left(\frac{\theta}{2}\right) = \begin{pmatrix} \cos\left(\frac{\theta}{2}\right) & i\sin\left(\frac{\theta}{2}\right) \\ -i\sin\left(\frac{\theta}{2}\right) & \cos\left(\frac{\theta}{2}\right) \end{pmatrix}$$

So,

$$U = \begin{pmatrix} \cos\left(\frac{\theta}{2}\right)e^{i\frac{\Psi}{2}} & \sin\left(\frac{\theta}{2}\right)e^{i\frac{\Psi}{2}} \\ -\sin\left(\frac{\theta}{2}\right)e^{-i\frac{\Psi}{2}} & \cos\left(\frac{\theta}{2}\right)e^{-i\frac{\Psi}{2}} \end{pmatrix} \begin{pmatrix} \exp\left(i\frac{\phi}{2}\right) & 0 \\ 0 & \exp\left(-i\frac{\phi}{2}\right) \end{pmatrix}$$

$$U = \begin{pmatrix} \cos\left(\frac{\theta}{2}\right) e^{i\frac{(\phi+\Psi)}{2}} & \sin\left(\frac{\theta}{2}\right) e^{i\frac{(\Psi-\phi)}{2}} \\ -\sin\left(\frac{\theta}{2}\right) e^{i\frac{(\phi-\Psi)}{2}} & \cos\left(\frac{\theta}{2}\right) e^{-i\frac{(\phi+\Psi)}{2}} \end{pmatrix}$$

$$U = \begin{pmatrix} \alpha & \beta \\ -\beta^* & \alpha^* \end{pmatrix} \Rightarrow \begin{cases} \alpha = \cos\frac{\theta}{2} e^{i\frac{\Psi+\phi}{2}} \\ \beta = \sin\frac{\theta}{2} e^{i\frac{\Psi-\phi}{2}} \end{cases}$$

Example 10.16. Find the left-invariant 1-forms w^i dual to the left-invariant vector fields e_i from previous problem. Show that the metric of problem 4.2a, $g_{ab} = 2\delta_{ij} w_a^i w_b^j$, agrees with the usual form for a homogeneous, isotropic metric on the 3-sphere (i.e., the Killing metric makes SO(3) locally isomorphic to the 3-sphere; SU(2) is globally isometric to the 3-sphere).

Solution

From previously the left invariant vector fields are:

$$e_1 + ie_2 = e^{-i\Psi}\left[\partial_\theta + \frac{i}{\sin\theta}\left(\partial_\phi - \cos\theta\, \partial_\Psi\right)\right] \text{ and } e_3 = \partial_\Psi.$$

Finding the 1-forms. Since

$$\Big(\underbrace{w^1 - iw^2}_{w}\Big)\Big(\underbrace{e_1 + ie_2}_{e}\Big) = w^1 e_1 + iw^1 e_2 - iw^2 e_1 + w^2 e_2 = 2$$

Try $w = e^{i\Psi}[d\theta - i\sin\theta\, d\phi]$, which gives $w_a e^a = 2$ and $w_a e_3^a = 0$, so works. Thus

$$w^1 - iw^2 = e^{i\Psi}[d\theta - i\sin\theta\, d\phi]$$

Similarly, guess

$w^3 = d\Psi + \cos\theta\, d\phi \qquad \Rightarrow \qquad (w^3)_a (e_3)^a = 1$

$$(w^3)_a e^a = \frac{ie^{-i\Psi}}{\sin\theta}(\cos\theta - \cos\theta) = 0$$

So, also works. Now, $g_{ab} = 2\delta_{ab} = 2\delta_{ij} w_a^i w_b^j$, so we have $\frac{1}{2} g_{ab} = (w^1 - iw^2)_a (w^1 + iw^2)_b + w_a^3 w_b^3 = d\theta^2 + \sin^2\theta\, d\phi^2 + (d\Psi + \cos\theta\, d\phi)^2$

Thus,

$$\frac{1}{2} g_{ab} = d\theta^2 + d\phi^2 + d\Psi^2 + 2\cos\theta\, d\phi d\Psi\,.$$

$w^1 - iw^2 = e^{i\Psi}(d\theta - i\sin\theta\, d\phi)$

$w^1 + iw^2 = e^{-i\Psi}(d\theta - i\sin\theta\, d\phi)$
Thus,
$w^1 = \cos\Psi\, d\theta + \sin\Psi \sin\theta\, d\phi$
$w^2 = -\sin\Psi\, d\theta + \cos\Psi \sin\theta\, d\phi$

No consider S^3 imbedded in space of higher dimension:

$$x^2 + y^2 + z^2 + w^2 = a^2$$

Transform to polar coordinates:

$$\left.\begin{aligned} x &= a\sin x \sin\theta \cos\phi \\ y &= a\sin x \sin\theta \sin\phi \\ z &= a \sin x \cos\theta \\ w &= a\cos\theta \end{aligned}\right\} dS^2 = a^2[dx^2 + \sin^2 x\,(d\theta^2 + \sin^2\theta\, d\phi^2)]$$

Want to arrive at

$$\begin{aligned} d\sigma^2 &= (d\theta^2 + \sin^2\theta\, d\phi^2) + (d\Psi + \cos\theta\, d\phi)^2 \\ &= d\theta^2 + d\Psi^2 + d\phi^2 + 2\cos\theta\, d\Psi d\phi, \end{aligned}$$

so let $a = 1$ to start. Next consider the $d\sigma^2$ term. Let
$\Psi = (\Psi' + \phi')\frac{1}{2}$ $\quad d\Psi = (d\Psi' + d\phi')\frac{1}{2}$ and $\phi = (\Psi' + \phi')\frac{1}{2}$ $\quad d\theta =$ $(d\Psi' + d\phi')\frac{1}{2}$ etc.

Then

$$\begin{aligned} d\sigma^2 &= d\theta^2 + \tfrac{1}{2}\left(d\Psi'^2 + d\phi'^2\right) + \tfrac{1}{2}\cos\theta\left(d\Psi'^2 + d\phi'^2\right) \\ &= d\theta^2 + \tfrac{1}{2}(1+\cos\theta)d\Psi'^2 + \tfrac{1}{2}(1-\cos\theta)d\phi'^2 \\ &= d\theta^2 + \cos^2\left(\tfrac{\theta}{2}\right)d\Psi'^2 + \sin^2\left(\tfrac{\theta}{2}\right)d\phi'^2 \\ &= \left(\tfrac{1}{2}d\theta^2 + \cos^2\left(\tfrac{\theta}{2}\right)d\Psi'^2\right) + \left(\tfrac{1}{2}d\theta^2 + \sin^2\left(\tfrac{\theta}{2}\right)d\phi'^2\right) \end{aligned}$$

Now, is it possible to have an x, y, such that $dx^2 + dy^2 = (\cdots)$?
Guess

$$x = \cos\left(\frac{\theta}{2}\right)\cos\left(\frac{\Psi'}{2}\right) \qquad dx$$

$$= \frac{-1}{2}\left[\sin\frac{\theta}{2}\cos\left(\frac{\Psi'}{2}\right)d\theta + \cos\frac{\theta}{2}\sin\frac{\Psi'}{2}d\Psi'\right]$$

$$y = \cos\left(\frac{\theta}{2}\right)\sin\left(\frac{\Psi'}{2}\right) \qquad dy$$

$$= -\frac{1}{2}\left[\sin\frac{\theta}{2}\sin\left(\frac{\Psi'}{2}\right)d\theta + \cos\frac{\theta}{2}\cos\frac{\Psi'}{2}d\Psi'\right]$$

after some work:

$$dx^2 + dy^2 = \frac{1}{4}\left[\sin^2\left(\frac{\theta}{2}\right)d\theta^2 + \cos^2\left(\frac{\theta}{2}\right)d\Psi'^2\right]$$

This will work when combined with a similar term from the second parenthesis in $d\sigma^2$. So, for

$$z = \sin\left(\frac{\theta}{2}\right)\cos\left(\frac{\Psi'}{2}\right) \qquad \Rightarrow \qquad dz^2 + dw^2$$

$$w = \sin\left(\frac{\theta}{2}\right)\sin\left(\frac{\Psi'}{2}\right)$$

$$= \frac{1}{4}\left[\cos^2\left(\frac{\theta}{2}\right)d\theta^2 + \sin^2\left(\frac{\theta}{2}\right)d\Psi'^2\right]$$

Thus,

$$ds^2 = dx^2 + dy^2 + dz^2 + dw^2$$

$$= \frac{1}{4}\left[d\theta^2 + \cos^2\left(\frac{\theta}{2}\right)d\Psi'^2 + \sin^2\left(\frac{\theta}{2}\right)d\phi'^2\right] = \frac{1}{4}d\sigma^2\ ,$$

and we have found the analogous coordinate transformation as with the usual S^3 polar coordinate embedding. There only remains to check that the relationship amongst the *x*, *y*, *w*, and *z* is like that of a 3-sphere in 4-dim as in the preceding S^3. Since

$$x = \cos\left(\frac{\theta}{2}\right)\cos\left(\frac{\Psi'}{2}\right) \qquad z = \sin\left(\frac{\theta}{2}\right)\cos\left(\frac{\Psi'}{2}\right)$$

$$y = \cos\left(\frac{\theta}{2}\right)\sin\left(\frac{\Psi'}{2}\right) \qquad w = \sin\left(\frac{\theta}{2}\right)\sin\left(\frac{\Psi'}{2}\right)$$

We have:

$$x^2 + y^2 + z^2 + w^2 = \cos^2\left(\frac{\theta}{2}\right) + \sin^2\left(\frac{\theta}{2}\right) = 1.$$

In conclusion, the SU(2) metric agrees with the isotropic matrix on the 3-sphere. Thus, SU(2) is globally isometric to the 3-sphere. (Both simply connected).

10.7 Exercises

(Ex. 10.1) Re-derive 10-5.
(Ex. 10.2) Verify 10-6.
(Ex. 10.3) Re-derive 10-13.
(Ex. 10.4) Re-derive 10-22.
(Ex. 10.5) Re-derive 10-28.
(Ex. 10.6) Re-derive 10-35.
(Ex. 10.7) Re-derive 10-40.
(Ex. 10.8) Complete the solution of Example 10.8

Chapter 11. Yang-Mills Gauge Fields and Gauge Covariant Derivatives

11.1 Geometry of Yang-Mills Fields

Let B be a manifold and G a group that acts freely on B (if $L_B x = x$ for any $x \in B$ then $g = e$, then 'acts freely'). Let M be the set of orbits of G, $M = B/G$. If B looks locally like a Cartesian product, $B_{local} \approx M \times G$, then B is called a principle bundle with group G. In a principle bundle the fiber space is isomorphic to the group, $F \approx G$, and he action of the group preserves the fibers. A principle fiber bundle B admits a cross section ($B = M \times G$, a Cartesian product, globally) iff (if and only if) B is 'trivial'.

Consider tangent bundles $\pi: B \to M$, where π is smooth and surjective, and for every $x \in M \ \exists \ nghb. U_x \ni \pi^{-1}(U_x)$ diffeomorphic to $U \times F$ (F is the fiber) via a diffeo that takes $\pi^{-1}(x)$ to $\{x\} \times F \ \ \forall_x \in U$ and $F = \pi^{-1}(x)$, is the fiber.

Connections

Consider the meaning of parallel transport in terms of a bundle of frames over M. Given x_0 and $\vec{e}_\alpha(x_0)$ a covariant derivative ∇ tells one how to parallel transport $\vec{e}_\alpha(x_0)$ along a given curve $x(\lambda)$, such that it is the unique solution to

$$\vec{x} \cdot \nabla \vec{e}_\alpha = 0$$

(Eqn. 11-1)

In effect ∇ tells us how to lift a path $x(\lambda)$ in M to a path $\{\vec{e}_\mu(\lambda), x(\lambda)\}$ in B through $\vec{e}_\mu(0)$ at x_0. Lifted paths of parallel transported bases are called horizontal paths in B. Without a connection, i.e., ∇, we don't know what horizontal means, but there is a natural definition of vertical paths. A vertical path satisfies $\pi[b(\lambda)] = x, \ x \text{ indep.} \, of \ \lambda$., and since $\vec{e}_\mu(\lambda) = \vec{e}_\nu(0) a^\nu_\mu(\lambda)$ this is a path in GL_n. Note: a vector ξ^c on B is vertical if $\pi_* \xi = 0$. If B is a principal bundle, then we have $b(\lambda) = bg(\lambda)$, say, where $g(\lambda)$ is a path in G. So $\dot{b}^c$ can be identified with a vector $\dot{g}^a$ at the identity in G.

Consider G to act on B on the right for definiteness, define:

$$\Psi(\mathrm{b}): \mathrm{G} \rightarrow \mathrm{B}, \quad g \longmapsto bg$$

(Eqn. 11-2)

Then $\Psi(\mathrm{b})_*$ maps $\xi^a \in \mathcal{G}$ to $\xi^c(b) = \Psi(b)^c_a \xi^a(e)$ (where $\mathcal{G}$ is the Lie algebra of G). $\Psi(\mathrm{b})_*$ is a diffeo generating right action, this then generates left inv. vector field on g which is mapped to a left inv. vector field on B, i.e., the Lie algebra structure is preserved.

Connection on bundle of Lorentz frames

Let's now consider a bundle, E, of Lorentz frames over a manifold M:

Vector on E, use:
ξ^c *with* $c, d, \ldots.$ $(I, J, \ldots$ concrete),
Vector on M, use:
ξ^α *with* $\alpha, \beta, \ldots.$ $(\mu, \nu, \ldots$ concrete),

If ξ^c is vertical then it is tangent to $\vec{e}_\nu \Lambda^\nu_\mu(\lambda)$ and can be associated with $\frac{d}{d\lambda}\Lambda^\nu_\mu(\lambda)\Big|_{\lambda=0} = \dot{\Lambda}^\nu_\mu$ an element of the Lie algebra of the Lorentz group.

Define a "Lie-algebra valued *1*–form" $\overline{\omega}^\nu_{\mu c}$ on E by

$$\overline{\omega}^\nu_{\mu c}\xi^c = \dot{\Lambda}^\nu_\mu\,, \quad \xi^c \text{ vertical}$$
$$\overline{\omega}^\nu_{\mu c}\xi^c = 0\,, \quad \xi^c \text{ horizontal}$$

Any ξ^c can be decompeted into vertical and horizontal parts: $\xi^c = h^c + v^c$. We could like to derive a relation between $\overline{\omega}$ ... and the Cartan 1–forms $\omega^\nu_{\mu\alpha}$ on M. So, what is the relation between $\omega^\nu_{\mu\alpha}$ and $\overline{\omega}^\nu_{\mu c}$?

Consider a mapping $e: M \rightarrow E$ with $x \longmapsto \left(x, \vec{e}_\mu(x)\right)$. We have a pullback: $e^*\overline{\omega}^\nu_\mu(\dot{x}) = \overline{\omega}^\nu_\mu(e_*\dot{x})$, where $\dot{x}^\alpha$ is a vector in M tangent to $x(\lambda)$, now $e_*\dot{x}$ is the vector tangent to $\vec{e}_\mu(x(\lambda))$ in E:

$$\vec{e}_\mu(x(\lambda)) = \vec{E}_\nu(\lambda)\Lambda^\nu_\mu(\lambda)$$

(Eqn. 11-3)

where $\vec{E}_\nu(\lambda)$ describes parallel-transport of $\vec{e}_\mu(0)$ along $x(\lambda)$, and $\Lambda^\nu_\mu(\lambda)$ describes a Lorentz Transform. So,

$$e_*\dot{x} = \vec{E}_\nu + \vec{e}_\nu(0)\dot{\Lambda}^\nu_\mu\,,$$

(Eqn. 11-4)

where $\vec{E}_\nu$ is the horizontal part and $\vec{e}_\nu(0)\dot{\Lambda}^\nu_\mu$ is the vertical part. Thus, $\overline{\omega}^\nu_\mu(e_*\dot{x}) = \dot{\Lambda}^\nu_\mu$ (from definition of $\overline{\omega}$). Now consider $\omega^\nu_\mu(\dot{x}) = \Gamma^\nu_{\mu\alpha}\dot{x}^\sigma$:

$$e_\nu \omega^\nu_\mu(\dot{x}) = \dot{x}^\sigma \nabla_\sigma e_\mu = \frac{D}{D\lambda} e_\mu(x(\lambda)) = \frac{D}{D\lambda}[E_\nu(\lambda)\Lambda^\nu_\mu(\lambda)]$$
$$= \left(\frac{D}{D\lambda} E_\nu(\lambda)\right)\Lambda^\nu_\mu(\lambda)\Big|_{x=0} + E_\nu(\lambda)|_{\lambda=0}\frac{d}{d\lambda}\Lambda^\nu_\mu$$

and

$$\omega^\nu_\mu(\dot{x}) = \dot{\Lambda}^\nu_\mu = \overline{\omega}^\nu_\mu(e_*\dot{x}) = e^*\overline{\omega}^\nu_\mu(\dot{x}) \,.$$

(Eqn. 11-5)

So, $\omega^\nu_\mu(\dot{x})$ is the pull-back of $\overline{\omega}^\nu_\mu$.

Connections on a Principle Fiber Bundle

Let a, b, c be Lie algebra indices

$$B \xrightarrow{\pi} M \text{ principle bundle with group G.}$$

Write a vertical path as $b(\lambda) = bg(\lambda)$, we can associate the tangent vector $\dot{b}^c$ with an element of the Lie algebra $\dot{g}^a$ in $\mathcal{G}$ (where $\mathcal{G}$ is the Lie algebra of G). The definition of connection for a Principal Fiber B essentially repeats the properties found for the Lorentz analysis:

$$\text{Denote} \qquad R_g : B \to B \qquad \forall g \in G$$
$$b \longmapsto b_g$$

Definition of a connection $\overline{A}^a_c$ on B: a Lie-algebra valued *1*–form on B for which:

(i) $\overline{A}^a_c \xi^c = \dot{g}^a$ If ξ^c is a vertical vector tangent to $b_g(\lambda)$ at b.

(ii) The subspace of horizontal vectors at each b in B, $\left\{\xi^c | \overline{A}^a_c \xi^c = 0\right\}$, is isomorphic to $T_x M$ $\left(x = \pi(b)\right)$.

(iii) If ξ^c is horizontal $R_{g*}\xi^c$ is horizontal.

Now, given a cross-section $b(x)$ of B we will get a "vector potential" A^a_α on the spacetime M:

$$A^a_\alpha = b^*_\alpha \overline{A}^a_c \ .$$

Example 11.1. Let B be a principal G-bundle over M. Let ε^a be an isovector on M, corresponding to a g-valued scalar on $\bar{\varepsilon}^a$ that satisfies:

$$\bar{\varepsilon}(bg) = g^{-1}\bar{\varepsilon}(b)g \,.$$

(a) Show that $D\bar{\varepsilon} = \text{hor}(d\bar{\varepsilon})$ is the gauge covariant derivative:

$$D_c\bar{\varepsilon}^a = d_c\bar{\varepsilon}^a + C^a_{bd}A^b_c\bar{\varepsilon}^d \,.$$

(b) Verify that $D\bar{\varepsilon}$ satisfies the defining relation for an isovector:

$$D\bar{\varepsilon}(bg) = g^{-1}D\bar{\varepsilon}(b)g \,.$$

(c) Let $b: M \to B$ be the cross-section of B for which $\bar{\varepsilon}^a(b(x)) = \varepsilon^a(x)$. Define the gauge covariant derivative $D_\alpha \varepsilon^a(x)$ by $D_\alpha \varepsilon^a(x) \equiv b_\alpha^{c*} D_c \bar{\varepsilon}^a(b(x))$, show that:

$$D_\alpha \varepsilon^a(x) = d_\alpha \varepsilon^a(x) + C_{bd}^a A_\alpha^b \varepsilon^d(x) .$$

Solution

(a) Consider a path $b(\lambda)$ in the bundle with tangent vector $\dot{b}^c$ at $\lambda = 0$. $b(\lambda)$ can be related to a horizontal path $b_{||}(\lambda)$ through $b(o)$ using $g(\lambda)$ a path through $e \in G$:

$$b(\lambda) = b_{||}(\lambda) g(\lambda)$$

If $D\bar{\varepsilon} = \text{hor}\,(d\bar{\varepsilon})$ then $\dot{b}^c D_c \bar{\varepsilon} = \frac{d}{d\lambda}\left(\bar{\varepsilon}\left(b_{11}(\lambda)\right)\right)\Big|_{\lambda=0}$

So,

$\dot{b}^c D_c \bar{\varepsilon} = \frac{d}{d\lambda}\left(\bar{\varepsilon}\left(b_{11}(\lambda)\right)\right)\Big|_{\lambda=0}$ and using $\bar{\varepsilon}(bg) = g^{-1}\bar{\varepsilon}(b)g$

$$= \frac{d}{d\lambda}\left(g(\lambda)\bar{\varepsilon}(b(\lambda))g(\lambda)^{-1}\right)\Big|_{\lambda=0}$$

$$gg^{-1} = 1$$

$$\dot{g}g^{-1} + g(\dot{g}^{-1}) = 0$$

$$= g\frac{d}{d\lambda}\bar{\varepsilon}(b(\lambda))\Big|_{\lambda=0} g^{-1} + \dot{g}\bar{\varepsilon}(b(o))e - e\bar{\varepsilon}(b(o))\dot{g}$$

$$(\dot{g}^{-1})|_{\lambda=0} = -g^{-1}\dot{g}g^{-1}|_{\lambda=0}$$

$$= -\dot{g}$$

$$= \dot{b}^c d_c \bar{\varepsilon} + \dot{g}\bar{\varepsilon} - \vec{\varepsilon}\dot{g}$$

Using the definition of connection on B: $\dot{g}^b = A_c^b \dot{b}^c$ and $[\dot{g}, \varepsilon] = C_{bd}^a \dot{g}^b \varepsilon^d$

$$\dot{b}^c D_c \bar{\varepsilon}^a = \dot{b}_c^c d_c \bar{\varepsilon}^a + C_{bd}^a \dot{g}^b \varepsilon^d = \dot{b}^c d_c \bar{\varepsilon}^a = C_{bd}^a A_c^b \dot{g}^b \bar{\varepsilon}^d$$
$$= \dot{b}^c (d_c \bar{\varepsilon}^a + C_{bd}^a A_c^b \bar{\varepsilon}^d)$$

Thus,

$$D_c \bar{\varepsilon}^a = d_c \bar{\varepsilon}^a + C_{bd}^a A_c^b \bar{\varepsilon}^d .$$

(b) Does $D\bar{\varepsilon}(bg) = g^{-1} D\bar{\varepsilon}(b) g$? We have:

$$\bar{\varepsilon}(bg) = g^{-1}\bar{\varepsilon}(b)g$$

So,

$D\bar{\varepsilon}(bg) = D(g^{-1})\bar{\varepsilon}(b)g + g^{-1}D\bar{\varepsilon}(b)g + g^{-1}\bar{\varepsilon}(b)Dg$, but, since either $g^{-1}(\lambda)$ or $g(\lambda)$ is a vertical path $D(g^{-1}) = 0, D(g) = 0$, so

$$D\bar{\varepsilon}(bg) = g^{-1}D\bar{\varepsilon}(b)g$$

(c) $D_\alpha \varepsilon^a(x) \equiv b_\alpha^{c*} D_c \bar{\varepsilon}^a\big(b(x)\big)$

$$\begin{aligned}(b_\alpha^c)^* D_c \bar{\varepsilon}^a &= (b_\alpha^c)^*[d_c \bar{\varepsilon}^a + C_{bd}^a A_c^b \bar{\varepsilon}^d] \\ &= (b_\alpha^c)^* d_c \bar{\varepsilon}^a + C_{bd}^a (b_\alpha^c)^* A_c^b \bar{\varepsilon}^d \\ &= d_\alpha \bar{\varepsilon}^a\big(b(x)\big) + C_{bd}^a A_\alpha^b \bar{\varepsilon}^d\big(b(x)\big)\end{aligned}$$

But $\bar{\varepsilon}^a\big(b(x)\big) = \varepsilon^a(x)$, so

$$D_\alpha \varepsilon^a(x) = d_\alpha \varepsilon^a(x) + C_{bd}^a A_\alpha^b \varepsilon^d(x).$$

11.2 Gauge transformations

Gauge transformations can be associated with change of cross section, i.e., a bundle automorphism. Given a connection A_c^a on a principle bundle $B \overset{\pi}{\to} M$, we want the change in A_c^a arising from an automorphism:

$$\hat{g}: B \to B$$

where

$$\hat{g}(b) = bg(b) = bg(x), \qquad x \subset \pi(b)$$

and where the latter leaves base space fixed, assumes base $\subset B$.

For space-time independent:

$$\hat{g} = R_g \;\; and \;\; \big(R_g^* A_c\big)\dot{b}^c = A_c^a\big(R_{g*}\dot{b}^c\big).$$

$\dot{b}^c$ horizontal $\to$ $R_{g*}\dot{b}^c$ horizontal $\to$ $A_c R_{g*}\dot{b}^c = 0$

$\dot{b}^c$ vertical $\to$ a tangent to $b(\lambda) = bh(\lambda)$ with $h(\lambda)$ a path through $e \in G$.

So, $R_{g*}\dot{b}^c$ tangent at bg to $bh(\lambda)g = b_g g^{-1}h(\lambda)g$. Thus $R_{g*}\dot{b}^c$ can be associated with tangent at $e \in G$ to path $g^{-1}h(\lambda)g$:

$$\frac{d}{d\lambda}[g^{-1}h(\lambda)g] = g^{-1}\dot{h}g$$

Since $g^{-1}\dot{h}g$ is a Lie algebra valued *1*-form, we need only regard the associated vector in G:

$$A_c^a\big(R_{g*}\dot{b}^c\big) = g^{-1}\dot{h}g \;\; \to \;\; R_g^* A_c = g^{-1}A_c g$$

(Eqn. 11-6)

For space-time dependent: $\hat{g}_*$ will no longer take horizontal vectors to horizontal, so the analysis becomes more involved:

$$\hat{g}^*\overline{A}_c\dot{b}^c = \overline{A}_c\hat{g}_*\dot{b}^c$$

For $\dot{b}^c$ tangent at b_0 to $b(\lambda) \to \hat{g}_*\dot{b}^c$ tangent at $bg\ to\ b(\lambda)g[b(\lambda)]$:

$$\frac{d}{d\lambda}b(\lambda)g[b(\lambda)]|_{\lambda=0} = \frac{d}{d\lambda}b(\lambda)g + \frac{d}{d\lambda}b_g(b(\lambda))\big|_{\lambda=0}$$

where $\frac{d}{d\lambda}b(\lambda)g$ is $R_{g*}\dot{b}$ and $\frac{d}{d\lambda}bg(b(\lambda))\big|_{\lambda=0} = \frac{d}{d\lambda}bg\left(g^{-1}g(b(\lambda))\right)$ and $g^{-1}g(b(\lambda))$ is a path through the identity in G.

Since $\frac{d}{d\lambda}\left(g^{-1}g(b(\lambda))\right) = g^{-1}d_cg\dot{b}^c$, we have

$$\overline{A}_c(\hat{g}_*\dot{b}^c) = \overline{A}_c(R_{g*}\dot{b}^c) + g^{-1}d_cg\dot{b}^c$$

And we get

$$\hat{g}^*\overline{A}_c = g^{-1}\overline{A}_cg + g^{-1}d_cg\,.$$

(Eqn. 11-7)

If we apply to bundle of frames with $\vec{e}'_\mu = \vec{e}_\nu\Lambda^\nu_\mu(x)$, $\omega^\nu_\mu = e^*\overline{\omega}^\nu_\mu$, $\omega^\nu_\mu(\dot{x}) = \Gamma^\nu_{\mu\alpha}\dot{x}^\alpha$, we get:

$$e'^*\overline{\omega}^\nu_{\mu c} = \Lambda^{-1}\overline{\omega}_c\Lambda + \Lambda^{-1}d_c\Lambda$$

$$\widehat{\Lambda}^*\overline{\omega}^\mu_\nu = \Lambda^\mu_\sigma\overline{\omega}^\sigma_\tau\Lambda^\tau_\nu + \Lambda^\mu_\sigma d\Lambda^\sigma_\nu$$

$\Gamma^{\mu r}_{\nu\lambda}\vec{e}^\lambda = \Lambda^\mu_\sigma\Gamma^\sigma_{\tau\varphi}\vec{e}^\varphi\Lambda^\tau_\nu + \Lambda^\mu_\sigma d_\tau\Lambda^\sigma_\nu\vec{e}^\tau$ (using $\vec{e}^\lambda = e^\tau\Lambda^\lambda_\tau$)

Thus,

$$\Gamma^\mu_{\nu\lambda} = \Lambda^\mu_\sigma\Lambda^\tau_\nu\Lambda^\varphi_\lambda\Gamma^\sigma_{\tau\varphi} + \Lambda^\mu_\sigma\Lambda^\varphi_\lambda d_\varphi\Lambda^\sigma_\nu$$

(Eqn. 11-8)

which is the law for change of Γ's under a change of orthonormal frame.

Example 11.2. (a) Let h_{ab} be a tensor on g (equivalently, a left-invariant tensor field on G). Show that $h_{ab}\varepsilon^a\eta^b$ is invariant under maps A_g. (b) Deduce from (a) and 4.7 that $g_{bc}D_\alpha\varepsilon^bD^\alpha\varepsilon^c$ is invariant under gauge transformations. (c) Extend the definition of gauge-covariant derivative to an arbitrary field, say,

$T^{a\ldots b\ \alpha\ldots\beta}_{c\ldots d\ \gamma\ldots\delta}$ with space-time and Lie-algebra indices.

Solution

(a) $A_g: \varepsilon(x) \to g(x)^{-1}\varepsilon(x)g(x)$

So $A_g: h_{ab}\varepsilon^a\eta^b \to h_{ab}g(x)^{-1}\varepsilon^a(x)g(x)g^{-1}(x)\eta^b(x)g(x)$

$$= h_{ab}g(x)^{-1}\varepsilon^a(x)\eta^b(x)g(x)$$

$$= g(x)^{-1}\left[\underbrace{h_{ab}\varepsilon^a(x)\eta^b(x)}_{group\ scalar}\right]g(x) =$$

$$[h_{ab}\varepsilon^a(x)\eta^b(x)]\underbrace{g(x)^{-1}g(x)}_{1}$$

Thus, $A_g[h_{ab}\varepsilon^a\eta^b] = h_{ab}\varepsilon^a\eta^b$

(b) $g_{bc}D_\alpha\varepsilon^b D^\alpha\varepsilon^c = g_{bc}(d_\alpha\varepsilon^b + C^b_{ed}A^e_\alpha\varepsilon^d)(d^\alpha\varepsilon^c + C^c_{ed}A^{e\alpha}\varepsilon^d)$

$= g_{bc}d_\alpha\varepsilon^b d^\alpha\varepsilon^c + g_{bc}\left[C^b_{ed}A^e_\alpha\varepsilon^d d^\alpha\varepsilon^c + C^c_{ed}A^{e\alpha}\varepsilon^d d_\alpha\varepsilon^b\right] +$
$g_{bc}C^b_{ed}A^e_\alpha\varepsilon^d C^c_{fg}A^{f\alpha}\varepsilon^g$

$= g_{bc}d_\alpha\varepsilon^b d^\alpha\varepsilon^c + [C_{ced}A^e_\alpha\varepsilon^d d^\alpha\varepsilon^c + C_{ced}A^e_\alpha\varepsilon^d d^\alpha\varepsilon^c] +$
$C_{ced}C^c_{fg}A^e_\alpha A^{f\alpha}\varepsilon^d\varepsilon^g$

$$= g_{bc}d_\alpha\varepsilon^b d^\alpha\varepsilon^c + \underbrace{(2C_{ced}A^e_\alpha)}_{B^\alpha_{cd}}(d^\alpha\varepsilon^c)\varepsilon^d + \underbrace{\left(C_{ced}C^c_{fg}A^e_\alpha A^{f\alpha}\right)}_{D_{dg}}\varepsilon^d\varepsilon^g$$

↑ $F_{bc} = g_{bc}X(d_\alpha, d^\alpha)$ ↑ ↑

↑ a tensor on $\mathbb{G}$ an α-valued tensor on $\mathbb{G}$ a tensor on $\mathbb{G}$

Thus the expression is invariant under gauge transformation from previous problem.

(c)

$$D_\mu T^{a\ldots b\ \alpha\ldots\beta}_{c\ldots d\ \gamma\ldots\delta} = \underbrace{\nabla_\mu\left(T^{a\ldots b\ \alpha\ldots\beta}_{c\ldots d\ \gamma\ldots\delta}\right)}_{\substack{\text{the usual cov.spacetime}\\ \text{derivative acting on}\\ \alpha\to\delta\text{ indices}}} + C^a_{ef}A^e_\mu T^{f\ldots b}_{c\ldots d} + \cdots + C^b_{ef}A^e_\mu T^{a\ldots f}_{c\ldots d}$$

$$-C^f_{ec}A^e_\mu T^{a\ldots b}_{f\ldots d} \cdots - C^f_{ed}A^e_\mu T^{a\ldots b}_{c\ldots f}$$

↗
Making use of $w^i = -C^i_{jk}w^j w^k$ for Cartesian basis.

11.3 Curvature

Let's now consider the Curvature.

Definition: Horizontal projection h. The tensor h^C_D is defined by $h^C_D\xi^D = \xi^C$, *if* ξ horizontal, and $h^C_D\xi^D = 0$, *if* ξ vertical.

If σ^a_C is a 1-form, its horizontal part is hor $\sigma^a_C = h^D_C \sigma^a_D$. A horizontal form thus kills vertical vectors:

$$(\text{hor } \sigma^a_C)\xi^C = 0\,, if \;\; \xi \text{ vertical}$$

Exterior derivative

Let's now define the exterior derivative of Lie algebra valued forms: Let f^a be a Lie algebra valued scalar on B and let ξ^C be a vector field on B tangent to $b(\lambda)$.

Definition: $d_C f^a$ is the 1-form given by

$$\xi^C d_C f^a = \frac{d}{d\lambda} f^a : b(\lambda)]_{\lambda=0}$$

(Since $f^a(b)$ is a vector in G, then $f^a(b_2) \cdot f^a(b_1)$ is a vector in G and the RHS is a well defined vector in G.)

Definition: If w^a_C is a 1-form $(dw)^a_{CD}$ is the two form satisfying

$$(dw)^a_{CD}\xi^C\eta^D = d_C(w^a_D\eta^D)\xi^C - d_D(w^a_C\xi^C)\eta^D - w^a_C[\xi,\eta]^C$$

In a chart b^I (and basis $\vec{e}_i$ for G), taking $\vec{\xi} = \frac{\partial}{\partial b^{I'}}$ $\vec{\eta} = \frac{\partial}{\partial b^{J'}}$ we have:

$$(dw)^i_{IJ} = \partial_I w^i_J - \partial_J w^i_J.$$

One extends to p-forms by Leibnitz: in a chart b^I, a basis $\vec{e}_i$ for the exterior derivative of a p-form is then

$$(d\sigma)^{i...j}_{IJ...K} = (p+1)\partial_{[I}\partial^{i...j}_{J...K]}$$

(you simply ignore the Lie algebra indices).

Definition: Let A^a_C be a connection on B. **The curvature** of A is a Lie algebra valued 2-form Ω^a_{CD} defined by

$$\Omega = \text{hor } dA$$

<u>TM</u>:

$$\Omega^a_{CD} = d_C A^a_D + C^a_{bc} A^b_C A^c_D$$

<u>Proof</u>: Look at $\Omega^a_{CD}\xi^C\eta^D$ for the 3 cases $\begin{cases} \xi,\eta \text{ vertical} \\ \xi,\eta \text{ horizontal} \\ \xi \text{ vertical}, \eta \text{ horizontal} \end{cases}$

Both vertical. Since $\Omega^a_{CD}\xi^C(b)\eta^D(b)$ depends only on ξ and η at b, we can pick them arbitrarily elsewhere. $\xi^C(b)$ is vertical and so corresponds to a Lie algebra vector $\overline{\xi}^a = A^a_C\xi^C$. Extend ξ^C so that it is vertical everywhere and corresponds everywhere to the same vector $\overline{\xi}^a$. Similarly let η^C correspond to $\overline{\eta}^a$ everywhere. Then$(dA)^a_{CD}\xi^C\eta^D = d_C\left(\overline{\xi}^a\right)\eta^C - d_D(\overline{\eta}^a)\xi^D - C^a_{bc}\overline{\xi}^b\overline{\eta}^C$. But because $\overline{\xi}^a$ and $\overline{\eta}^a$ are constant Lie-algebra valued scalars,

$$d\overline{\xi}^a = 0, \;\; d\overline{\eta}^a = 0.$$

Thus $(dA)^a_{CD}\xi^C\eta^D = C^a_{bc}\overline{\xi}^b\overline{\eta}^C$, and we get:

$$\Omega^a_{CD}\xi^C\eta^D = -C^a_{bc}\overline{\xi}^b\overline{\eta}^C + C^a_{bc}\overline{\xi}^b\overline{\eta}^C = 0$$

(Eqn. 11-9)

Both horizontal. We arrive at the same result, namely,

$$\Omega^a_{CD}\xi^C\eta^D = (dA)^a_{CD}\xi^C\eta^D\,,$$

(Eqn. 11-10)

but now with ξ vertical, η horizontal. We will again get $\Omega(\xi,\eta) = 0$. To see this, let ξ be the vertical vector field corresponding to a fixed $\overline{\xi}^a$ in G. Let η^C be horizontal. Then (ξ,η) is horizontal.

Proof: ξ^C generates the diffeos R_{a_λ} of B, where a_λ is the 1-parameter group tangent to $\overline{\xi}^a$ at $e \in G$. By the def of a connection, $R_{a_\lambda *}\eta$ is horizontal. Thus

$$\Omega^a_{CD} = (dA)^a_{CD}\xi^C\eta^D = 0$$

For an electromagnetic field, the curvature $\Omega_{CD} = d_C\overline{A}_D$ is a tensor on B whose pullback to M is the field tensor $F_{\mu\nu}$. [The 4 space-time dimensional $\overline{A}_C$ is, in a chart $(\eta, X^\mu), \overline{A}_I = \left(1, -A_\mu\frac{Q}{\hbar c}\right)$]. $F_{\mu\nu}$ is gauge invariant or equivalently Ω_{CD} is invariant under $A_C \to \hat{g}^*A_C$ where, as before, $\hat{g}$ is the bundle automorphism corresponding to g(x). In the general case, Ω^a_{CD} is not gauge invariant, but it transforms simply under $\hat{g}$: Write Ω' for the curvature of $\hat{g}^*A_C$.

Theorem. $\Omega'_{CD} = g^{-1}\Omega_{CD}g$

Proof. $\Omega_{IJ} = \partial_I\overline{A}_J - \partial_J\overline{A}_I + [A_I, A_J]$, where $[A_I, A_J]^a = C^a_{bc}A^b_I A^c_J$.

$$\hat{g}^*A_I = g^{-1}A_I g + g^{-1}\partial_I g$$

Thus (using $\partial_I g^{-1} = -g^{-1}\partial_I g g^{-1}$):

$$\Omega'_{IJ} = \partial_I(g^{-1}A_Jg + g^{-1}\partial_Jg) - \partial_J(g^{-1}A_Ig + g^{-1}\partial_Ig) + [g^{-1}A_Ig + g^{-1}\partial_Ig,\ g^{-1}A_Jg + g^{-1}\partial_Jg]$$
$$\Omega'_{IJ} = g^{-1}(\partial_IA_J - \partial_JA_I + [A_I, A_J])g - g^{-1}\Omega_{IJ}g.$$

The pullback $F^a_{\alpha\beta} = \psi^*(\Omega^a_{CD})$ of Ω by across section ψ is the "Yang-Mills" field associated with A^a_α.

Example 11.3. Let G be a Lie group with Lie algebra g. Recall that given an element $g\epsilon G$, there is an automorphism, $A_g: G \to G$ that leaves the identity fixed: $A_g(a) \longmapsto g^{-1}ag$.
(a) Regard G as a group of matrices and find the image under A_{g*} of an element ε^a of g.
(b) Let ζ be an element of g and let $g(\lambda) = \exp(\lambda\zeta)$. What is

$$\frac{d}{d\lambda}A_{g(\lambda)*}\varepsilon^a\ ?$$

Solution

We have $\begin{array}{l} A_g: G \to G \\ A_g(a) \longmapsto g^{-1}ag \end{array}$ where the Lie group is G and the Lie algebra is $\mathbb{G}$.

Consider $\varepsilon^a \in \mathbb{G}$ which is tangent to the curve on the group manifold $a(\lambda), a \in G$, at the identity, i.e., $a(0) = e$. Then $A_{g*}\varepsilon^a$ is $\in \mathbb{G}$ which is tangent at $g^{-1}a(\lambda)g|_{\lambda=0} = g^{-1}eg = e$ to the path $\lambda \to g^{-1}a(\lambda)g$:

$$A_{g*}\varepsilon^a = \frac{d}{d\lambda}(g^{-1}a(\lambda)g)\bigg|_{\lambda=0} = g^{-1}\dot{a}g$$

And, for G a group of matrices, where $\varepsilon^a \to \dot{a}^k_\ell = \frac{d}{d\lambda}g^{-1}a(\lambda)^k_\ell\Big|_{\lambda=0}$, we have:

$$\left(A_{g*}\varepsilon^a\right)^i_j = (g^{-1})^i_k\dot{a}^k_\ell g^\ell_j\ .$$

(b) $\zeta \in \mathbb{G}$ and $g(\lambda) = \exp(\lambda\zeta)$ ← different λ param. than above.

$$A_{g(\lambda)*}\varepsilon^a = g^{-1}\dot{a}g = \exp(-\lambda\zeta)\,\dot{a}\exp(\lambda\zeta)$$

$$\frac{d}{d\lambda}A_{g(\lambda)*}\varepsilon^a = -\zeta\exp(-\lambda\zeta)\,\dot{a}\exp(\lambda\zeta) + \exp(-\lambda\zeta)\,\dot{a}\exp(\lambda\zeta)\,\zeta$$

$$= \exp(-\lambda\zeta)\,[\dot{a},\zeta]\exp(\lambda\zeta)$$

$$= g^{-1}[\dot{a}, \zeta]g$$

and since $\zeta = \frac{d}{d\lambda} g(\lambda)\Big|_{\lambda=0} = \dot{g}$, thus, $\frac{d}{d\lambda} A_{g(\lambda)*}\varepsilon^a = g^{-1}[\dot{a}, \dot{g}]g$, and we have:

$$\left(\frac{d}{d\lambda} A_{g(\lambda)*}\varepsilon^a\right)^i_j = (g^{-1})^i_k [\dot{a}, \dot{g}]^k_\ell g^\ell_j$$

Example 11.4. Let C^a_{bc} be the structure constant tensor of a Lie algebra g.
(a) Define a left- and right- invariant metric (the Killing metric) $h_{ab} = C^d_{ae} C^e_{bd}$. Use the Jacobi identity to show that C_{abc} is totally antisymmetric, where $C_{abc} = h_{ag} C^g_{bc}$.
(b) Find h_{ab} for so(3).

Solution
(a) We have $C_{abc} = h_{ag} C^g_{bc} = C^d_{ae} C^e_{gd} C^g_{bc}$. We also have the Jacobi identity:
$C^a_{e[b} C^e_{cd]} = 0 \;\rightarrow\; C^a_{eb} C^e_{cd} = C^a_{ec} C^e_{bd} - C^a_{ed} C^e_{bc} = C^a_{e[d} C^e_{c]b}$
So,

$$C_{abc} = C^d_{ae} C^e_{g[c} C^g_{b]d} = C^d_{ae} C^e_{gc} C^g_{bd} - C^d_{ae} C^e_{gb} C^g_{cd}$$

Thus,
$C_{abc} - C_{bac} = C^e_{gc}(C^d_{ae} C^g_{bd} - C^d_{be} C^g_{ad}) - C^g_{cd}(C^d_{ae} C^e_{gb} - C^d_{be} C^e_{ga}) = 2C_{abc}$

So, $C_{abc} = -C_{bac}$, now we merely generalize this result: $C_{abc} = -C_{acb} = C_{cab} = -C_{cba}$, thus C_{abc} is antisymmetric under exchange $a \leftrightarrow b, a \leftrightarrow c, b \leftrightarrow c$, thus C_{abc} is completely antisymmetric.

(b) $SO(3)$ *has* $C^a_{bc} = \epsilon^a_{bc}$ (basis in adjoint. rep), thus $h_{ab} = C^d_{ae} C^e_{bd} = \epsilon^d_{ae} \epsilon^e_{bd} = -2\delta_{ab}$.

11.4 Parallel transport: the gauge-covariant derivative

The physical motivation for regarding $\vec{A}$ as a connection, as specifying a kind of parallel transport, came from quantum mechanics. Similar clarity regarding gauge transformations in general becomes apparent in quantum theory as well, and this will be discussed in Sec. X. Consider the Schrodinger equation for a charge Q in a magnetic field $B = \nabla \times A$ and potential well V is

$$\frac{1}{2m}\left(\frac{\hbar}{i}\nabla - \frac{Q}{c}\vec{A}\right)^2 \Psi + V\Psi = i\hbar\partial_t\Psi.$$

Suppose that V represents the potential of a box confining the particle to a region small enough that $QBR^2 \ll \hbar c$, where R is the radius of the region. If one picks an origin x_0 and writes

$$\Psi = \psi \exp\left(i\int_{x_0}^{x} A - dx\frac{Q}{\hbar c}\right),$$

then we get the corresponding "uncharged" equation:

$$-\frac{\hbar^2}{2m}\nabla^2\psi + V\psi = i\hbar\partial_t\psi.$$

(Eqn. 11-11)

Although the line integral above depends on the path from x_0 to x, for R small its value is independent of the (short) piece of the path that lies within the box. Let $\psi_E(x,t)$ be an energy eigenstate when the box is centered at $x = 0$. Suppose the box is slowly moved from x_1 to x_2, along a path $x(t)$, $V(x,t) = V_0[x - x(t)]$ with $\dot{x} \ll \left[\langle\frac{p^2}{m^2}\rangle\right]^{1/2}$. Then to first order in $\dot{x}$, $\psi_E[x - x(t), t]$ satisfies the above and when the box reaches x_2 this physical parallel transport has changed the uncharged wave function ψ from

$$\psi_E(x - x_1, t) \text{ to } \psi_E(x - x_2, t).$$

The integral equation above then implies that for the charge particle, described by Ψ, physical parallel-transport takes

$$\Psi(x - x_1, t) \text{ to } \Psi(x - x_2, t)\exp\left(i\int_{x_1}^{x_2} A - dx\frac{Q}{\hbar c}\right).$$

The line integral depends on the path from x_1 to x_2, and the fact that the box is small means that the "path" traveled by the box is well-defined – the trajectories of different points of the box are close enough to give the same change of phase. Then, under parallel-transport along a path $x(\lambda)$, the phase of a charged particle's wave function changes by

$$\exp\left(i\int A - dx\frac{Q}{\hbar c}\right).$$

The possible phases of Ψ at a point x are just the complex numbers of magnitude 1, elements $e^{i\eta}$ of the unit complex circle. The space M, together with a circle at each point x of M is a circle bundle over M; so the space of possible phases of Ψ is a circle bundle B, and the phase $\eta(x)$, defined by

$$\Psi(x) = |\Psi(x)|e^{i\eta(x)}$$

is a cross section. A gauge transformation, $A \to A + \nabla f,\ \eta \to \eta + \frac{Q}{\hbar c} f$, can, insofar as η is concerned, be regarded as a change of cross section. In which case the change in phase above becomes:

$$\frac{d\eta}{d\lambda} - \frac{Q}{\hbar c}\vec{A} - \frac{d\vec{x}}{d\lambda} = 0$$

(Eqn. 11-12)

If we define a vector field $\overline{A}_C$ on B by $\overline{A}_\eta = 1, \quad \overline{A}_i = -\frac{Q}{\hbar c} A_i$, or

$$\overline{A} = d\eta - \frac{Q}{\hbar c} A\,,$$

then parallel transport on the bundle has the standard for

$$\overline{A}_C \dot{x}^C = 0,$$

where the horizontal vector $\dot{x}^C$ has components $(\dot{\eta}, \dot{x}^i)$.

To summarize: When a charge particle is physically transported in a background magnetic field, the change in phase of the particle's wave function can be described as parallel transport on a principal U(1) bundle with connection given by the above equation.

Finally consider the result of a bundle automorphism given by $\eta'(x) = \eta(x) + \alpha(x)$. With respect to the new coordinates $(\eta', x'^i) = (\eta + \alpha, x^i)$, the components of A are given by

$$\overline{A}_{i'} = \overline{A}_i - \partial_i \alpha$$
$$\overline{A}_{\eta'} = \overline{A}_\eta .$$

The change in the physical potential A is given by

$$A' = A + \frac{\hbar c}{Q} \nabla_\alpha\,,$$

which is the gauge transformation associated with a change of phase α in the wave function of a charge particle.

11.5 Analysis of EM and Y-M Examples

Example 11.5. Show Maxwell's equations in empty space can be written in the form:

$$dF = 0 \quad and \quad d^*F = 0,$$

where

$$^*F^{ab} = \frac{1}{2} \epsilon_{abcd} F^{cd}\,.$$

Solution

$F_{\alpha\beta}$ is defined by:

$$F_{\alpha\beta} = \begin{matrix} & t & x & y & z \\ t & 0 & -E_x & -E_y & -E_z \\ x & E_x & 0 & B_z & B_y \\ y & E_y & -B_z & 0 & B_x \\ z & E_z & B_y & -B_x & 0 \end{matrix}$$

Also,

$\bar{F} = \frac{1}{2} F_{\alpha\beta} dx^{\alpha} \wedge dx^{\beta} = E_x dx \wedge dt + E_y dy \wedge dt + E_z dz \wedge dt + B_x dy \wedge dz + B_y dz \wedge dx + B_z dx \wedge dy$

and

$$\begin{aligned} d\bar{F} &= d(E_x dx \wedge dt + \cdots + B_z dx \wedge dy) \\ &= \left(\frac{\partial E_x}{\partial t} dt + \frac{\partial E_x}{\partial x} dx + \frac{\partial E_x}{\partial y} dy + \frac{\partial E_z}{\partial z} dz\right) \wedge dx \wedge dt + \cdots \\ &= \frac{\partial E_x}{\partial y} dy \wedge dx \wedge dt + \frac{\partial E_x}{\partial z} dz \wedge dx \wedge dt + \text{other E's similarly} \\ &\quad + \left(\frac{\partial B_x}{dt} dt + \frac{\partial B_x}{dx} dx\right) \wedge dy \wedge dz + \text{other B's similarly} \end{aligned}$$

$$\begin{aligned} &= dy \wedge dx \wedge dt \left(\frac{\partial E_x}{\partial y} + \frac{\partial E_y}{\partial x} - \frac{\partial B_z}{\partial t}\right) + \text{cycles} \\ &\quad + dx \wedge dy \wedge dz \left(\underbrace{\frac{\partial B_x}{\partial x} + \frac{\partial B_y}{\partial y} + \frac{\partial B_z}{\partial z}}_{\nabla \cdot B}\right) \end{aligned}$$

$= (\nabla \cdot B)(dx \wedge dy \wedge dz) + \left(\frac{\partial E_y}{\partial x} - \frac{\partial E_x}{\partial y} + \frac{\partial B_z}{\partial t}\right)(dt \wedge dx \wedge dy) + \left(\frac{\partial E_z}{\partial y} - \frac{\partial E_y}{\partial z} + \frac{\partial B_x}{\partial t}\right)(dt \wedge dy \wedge dz) + \left(\frac{\partial E_x}{\partial z} - \frac{\partial E_z}{\partial x} + \frac{\partial B_y}{\partial t}\right)(dt \wedge dz \wedge dx)$

$= (\nabla \cdot B)(dx \wedge dy \wedge dz) + \left(\nabla \times \vec{E} + \dot{\vec{B}}\right)_z (dt \wedge dx \wedge dy) + \left(\nabla \times \vec{E} + \vec{B}\right)_x (dt \wedge dy \wedge dz)$

$+\left(\nabla \times \vec{E} + \vec{B}\right)_y (dt \wedge dz \wedge dzx)$

If we make use of Maxwell's equation's:

$$\nabla \cdot B = 0$$

$$\nabla \times \vec{E} = -\dot{\vec{B}} \text{ ,}$$

we then see that:

$$d\bar{F} = 0.$$

${}^*F_{\alpha\beta}$ is defined by: ${}^*F^{ab} = \frac{1}{2}\epsilon_{abcd}F^{cd}$:

$${}^*F_{\alpha\beta} = \begin{bmatrix} 0 & B_x & B_y & B_z \\ -B_x & 0 & +E_z & -E_y \\ -B_y & -E_z & 0 & +E_x \\ -B_z & +E_y & -E_x & 0 \end{bmatrix},$$

Thus,

$${}^*\bar{F} = -B_x dx \wedge dt - B_y dy \wedge dt - B_z dz \wedge dt + E_x dz \wedge dy + E_y dx \wedge dz + E_z dy \wedge dx$$

and we have:

$$d^*\bar{F} = -\left(\frac{\partial B_x}{\partial y} dy + \frac{\partial B_x}{\partial z} dz\right) \wedge dx \wedge dt + \cdots$$

$$d^*\bar{F} = (\nabla \cdot E)(dx \wedge dy \wedge dz) + \left(\nabla \times \vec{B} + \dot{\vec{E}}\right)_z (dt \wedge dx \wedge dy) + \left(\nabla \times \vec{B} + \dot{\vec{E}}\right)_x (dt \wedge dy \wedge dz) + \left(\nabla \times \vec{B} + \dot{\vec{E}}\right)_y (dt \wedge dz \wedge dzx)$$

Now the other Maxwell relations are $\nabla \cdot E = 4\pi\rho$ and $\frac{\partial \vec{E}}{\partial t} - \nabla \times B = 4\pi J$.

So,

$$d^*F = 4\pi\rho(dx \wedge dy \wedge dz) + 4\pi J_z(dt \wedge dx \wedge dy) - 4\pi J_x(dt \wedge dy \wedge dz) - 4\pi J_y(dt \wedge dz \wedge dzx)$$

and

$$d^*F = 4\pi^*J$$

Notice $d^{2*}F = 0 \Longrightarrow \frac{\partial \rho}{\partial t} - \Delta \cdot J = 0$ the continuity equation. Inn empty space $\rho = 0, \vec{J} = 0$ and we get

$$d^*F = 0.$$

Example 11.6. Write Maxwell's equations in a background Schwarzschild space-time with coordinates (t, r, θ, φ) and metric:

$$g_{ab} = -e^{2v}dt^2 + e^{2\lambda}dr^2 + r^2(d\theta^2 + \sin^2\theta\, d\phi^2).$$

Solution

Instead of Minkowski: $g_{ab} = -dt^2 + dx^2 + dy^2 + dz^2$ and a two form $\bar{F}$ with components:

$$F_{\alpha\beta} = \begin{bmatrix} 0 & -E_x & -E_y & -E_z \\ E_x & 0 & B_z & -B_y \\ E_y & -B_z & 0 & B_x \\ E_z & B_y & -B_x & 0 \end{bmatrix},$$

we now have metric

$$g_{ab} = -e^{2v}dt^2 + e^{2\lambda}dr^2 + r^2(d\theta^2 + \sin^2\theta\, d\phi^2)$$
$$= -d\tau^2 + dR^2 + d\Theta^2 + d\Phi^2,$$

where $d\tau = e^v dt, dR - e^\lambda dr, d\Theta = rd\theta, d\Phi = r\sin\theta\, d\phi$ and the components of our two form are now written

$$F_{\alpha\beta} = \begin{bmatrix} 0 & -E_r & -E_\theta & -E_\phi \\ E_r & 0 & B_\phi & -B_\theta \\ E_\theta & -B_\phi & 0 & B_r \\ E_\phi & B_\theta & -B_r & 0 \end{bmatrix}$$

Now $d\bar{F} = 0$ is expressed by:

$$d\bar{F} = 0 \Rightarrow \begin{cases} \nabla \cdot B = 0 \\ \nabla \times \vec{E} = -\dot{\vec{B}} \end{cases}$$

modified by e^λ and e^v terms from usual spherical coordinate form. We have to modify Maxwell's eqn's for spherical coordinates $\{\hat{e}_r, \hat{e}_\theta, \hat{e}_\phi\} \Rightarrow \{e^\lambda \hat{e}_r, \hat{e}_\theta, \hat{e}_\phi\}, \frac{\partial}{\partial r} \Rightarrow e^{-\lambda}\frac{\partial}{\partial r}, \frac{\partial}{\partial \tau} = e^{-v}\frac{\partial}{\partial t}$, so:

$$d\bar{F} = 0 \Rightarrow \begin{cases} "\nabla \cdot B = 0" \\ "\nabla \times \vec{E} = -\dot{\vec{B}}" \end{cases} \Rightarrow \quad \frac{1}{r^2}e^{-\lambda}\frac{\partial}{\partial r}(r^2 B_r)$$
$$+ \frac{1}{r\sin\theta}\frac{\partial}{\partial\theta}(\sin\theta\, B_\theta) + \frac{1}{r\sin\theta}\frac{\partial}{\partial\phi}(B_\phi) = 0$$

So,

$$\hat{e}_r(e^\lambda)\frac{1}{r\sin\theta}\left[\frac{\partial}{\partial\theta}(\sin\theta\, E_\phi) - \frac{\partial E_\theta}{\partial\phi}\right]$$
$$+ \hat{e}_\theta\left[\frac{1}{r\sin\theta}\frac{\partial E_r}{\partial\phi} - \frac{e^{-\lambda}}{r}\frac{\partial}{\partial r}(rE_\phi)\right]$$
$$+ \hat{e}_\phi\frac{1}{r}\left[e^{-\lambda}\frac{\partial}{\partial r}(rE_\theta) - \frac{\partial E_r}{\partial\theta}\right]$$

$$= -e^{-v}\frac{\partial\vec{B}}{\partial t}$$

$d^*F = 0$ ∇·E=0

$$: \quad \frac{1}{r^2} e^{-\lambda} \frac{\partial}{\partial r}(r^2 E_r) + \frac{1}{r \sin\theta} \frac{\partial}{\partial \theta}(\sin\theta\, E_\theta) + \frac{1}{r \sin\theta} \frac{\partial}{\partial \phi}(E_\phi) = 0$$

While for "$\nabla \times \vec{E} = -\dot{\vec{B}}$" $\Longrightarrow$ same as above with $E \to B$ and $B \to -E$.
We get a wave equation:

$$\frac{\partial^2 \vec{E}}{\partial \tau^2} = \nabla^2 E$$

or

$$\left(e^{-v} \frac{\partial}{\partial t}\right)\left(e^{-v} \frac{\partial}{\partial t}\right)\vec{E} = \left[\frac{1}{r^2} e^{-\lambda} \frac{\partial}{\partial r}\left(r^2 e^{-\lambda} \frac{\partial}{\partial r}\right) + \frac{1}{r^2 \sin\theta} \frac{\partial}{\partial \theta}\left(\sin\theta \frac{\partial}{\partial \theta}\right) + \frac{1}{r^2 \sin^2\theta} \frac{\partial^2}{\partial \phi^2}\right]\vec{E}.$$

Example 11.7. Find the angular momentum of a dyon: an electric charge e and a magnetic charge g at rest on the z-axis. Using the relation

$$L = \frac{1}{4\pi} \int dV\, \vec{\phi} \cdot \vec{E} \times \vec{B} = \frac{1}{4\pi} \int dV\, \varepsilon_{\alpha\beta\gamma} B^\alpha \phi^\beta E^\gamma\,,$$

show that $\varepsilon_{\alpha\beta\gamma} \phi^\beta E^\gamma = -e\nabla_\alpha \cos\theta$, and use Gauss' theorem to obtain $L = \pm eg$, where he sign depends on whether e is above or below g.

Solution

We have $\vec{B} = g\frac{\vec{n}'}{(r')^2}$ $\vec{E} = e\frac{\vec{n}}{r^2}$, for starters. The electromagnetic momentum density is: $\vec{P} = \frac{(\vec{E}\times\vec{B})}{4\pi}$. Total momentum = $\int dV\, \vec{P}$. The only vector available is $\vec{R}$ so $\vec{P} \propto \vec{R}$, so total momentum = $\frac{\vec{R}}{|R|}\int dV \frac{\vec{R}}{|R|} \cdot (\vec{E} \times \vec{B})$ where $\vec{R} \cdot (\vec{n} \times \vec{n}') = 0$!, so the total momentum = 0. With the total vanishing this considerably simplifies matters, the angular momentum of the e-m field is then independent of choice of origin. So the situation to be studied above will give an unambiguous result.

Angular momentum in e-m field: $\vec{L} = \frac{1}{4\pi}\int \vec{x} \times (\vec{E} \times \vec{B}) d^3x$

Once again, the only vector valued quantity from the geometric configuration above is $\vec{R}$, so $\vec{L} \propto \vec{R}$, thus we need only consider the value of $L_z = |\vec{L}| = L$. Also, in terms of the rotational K.V. field ϕ.

$$L = \frac{1}{4\pi}\int dV\, \vec{\phi} \cdot \vec{E} \times \vec{B} = \frac{1}{4\pi}\int dV\, \varepsilon_{\alpha\beta\gamma} B^{\alpha} \dot{\phi}^{\beta} E^{\gamma}$$

So, we need to simplify $\varepsilon_{\alpha\beta\gamma}\phi^{\beta} E^{\gamma}$, and for the configuration chosen $\vec{E} = \frac{e}{r^2}\hat{r}$, so lets use spherical coordinates, for which the volume element is $r^2 \sin\theta$. Also, ϕ^{β} has only a $\beta = \phi$ component: $\phi^{\phi} = \varepsilon_{\alpha\beta\gamma}\phi^{\beta} E^{\gamma} = \varepsilon_{\alpha\beta\gamma}\phi^{\beta}\left(\frac{e}{r^2}\right) = \varepsilon_{\alpha\phi r}\left(\frac{1}{r}\right)\left(\frac{e}{r^2}\right)$ (and in the latter, only the $\alpha = \theta$ component is nonzero).

$\varepsilon_{\theta\phi r}(r^{-1})\left(\frac{e}{r^2}\right) = r^{-1} e \sin\theta = -e\nabla_{\alpha}\cos\theta$ for $\alpha = \theta$.

So, $\varepsilon_{\alpha\beta\gamma}\phi^{\beta} E^{\gamma} = -e\nabla_{\alpha}\cos\theta$ for the geometry shown.

Now consider:

$$L = \frac{1}{4\pi}\int dV\, B^{\alpha}(-e\nabla_{\alpha}\cos\theta) = \frac{1}{4\pi}\int dV\, \nabla_{\alpha}B^{\alpha}\cos\theta \quad where \quad \nabla_{\alpha}B^{\alpha} = \nabla \cdot B = 4\pi g\delta(\vec{x} - \vec{R}).$$

If g is above e on the z-axis then L = eg; if g is below then L = –eg since $\cos(\pi) = -1$. Since we knew beforehand that $\vec{L} \propto \vec{R}$ this then allows us to generalize the result to:

$$\vec{L}_{em} = eg\frac{\vec{R}}{|R|}$$

where $\vec{R}$ is the vector from e to g.

Example 11.8. Define the Poisson Bracket for electromagnetism, and show that the canonical transformations generated by the Gauss constraint,

$$\int_{\Sigma} dx\, f\nabla_{\alpha}E^{\alpha} = 0,$$

are a family of gauge transformations. Hence Σ is a spatial plane of Minkowski space and the smearing function f is smooth and vanishes at spatial infinity.

Solution
The Lagrangian for EM is:

$$L_{EM} = -\frac{1}{4}F_{\alpha\beta}F^{\alpha\beta} = -\frac{1}{4}(\partial_\alpha A_\beta - \partial_\beta A_\alpha)(\partial^\alpha A^\beta - \partial^\beta A^\alpha)$$
$$= -\frac{1}{2}(\partial_\alpha A_\beta)(\partial^\alpha A^\beta) + \frac{1}{2}(\partial_\alpha A_\beta)(\partial^\beta A^\alpha).$$

We take our canonical coordinates to be the vector potential A^α evaluated on the hypersurface Σ. For this reason we decompose it into normal and tangential parts:

$$V = -A_\alpha n^\alpha \quad and \quad {}^{(3)}A_\alpha = h_\alpha^\beta A_\beta = \vec{A}$$

where $h_{\alpha\beta} = \eta_{\alpha\beta} + n_\alpha n_\beta$.

$$\partial_\alpha A_\beta = \partial_\alpha\left(\eta_\beta^\gamma A_\gamma\right) = \partial_\alpha\left(h_\beta^\gamma - n_\beta n^\gamma\right)A_\gamma = \partial_\alpha(\vec{A} + \vec{n}V)_\beta$$

Thus

$$\eta_\alpha^\sigma \partial_\sigma = (h_\alpha^\sigma - n_\alpha n^\sigma)\partial_\sigma = h_\alpha^\sigma \partial_\sigma - n_\alpha n^\sigma \partial_\sigma = \left(\vec{\nabla} - \vec{n}\partial_{\vec{n}}\right)_\alpha$$

and

$$\partial_\alpha A_\beta = \left({}^{(3)}\nabla_\alpha - n_\alpha n^\sigma \partial_\sigma\right)\left({}^{(3)}A_\alpha + n_\beta V\right).$$

The Lagrangian density is thus:

$$L_{EM} = -\frac{1}{4}\left(\nabla_\alpha A_\beta - \nabla_\beta A_\alpha + (n_\beta \nabla_\alpha - n_\alpha \nabla_\beta)V - n_\alpha \dot{A}_\beta + n_\beta \dot{A}_\alpha\right)^2$$

$$\left(\vec{\nabla}\times\vec{A}\right)^i = \varepsilon_{ijk}\nabla^j A^k = \left(\nabla^j A^k - \nabla^k A^j\right)^i$$
$$(\nabla\times A)\cdot(\nabla\times A) = \varepsilon_{ijk}\varepsilon^\ell\left(\nabla^j A^k - \nabla^k A^j\right)(\nabla_\ell A_m - \nabla_m A_\ell)$$
$$= 2\left(\nabla^j A^k - \nabla^k A^j\right)\cdot\left(\nabla^j A^k - \nabla^k A^j\right)$$

So,

$$L_{EM} = -\frac{1}{2}\left(\vec{\nabla}\times\vec{A}\right)\cdot\left(\vec{\nabla}\times\vec{A}\right) = \frac{1}{4}(\cdots)^2$$

where

$$\left(n_\beta \dot{A}_\alpha - n_\alpha \dot{A}_\beta\right)^2 = -2\dot{\vec{A}}^2 - 2n_\beta \dot{A}_\alpha n^\alpha \dot{A}^\beta$$
$$\leftarrow \begin{cases} \dot{A}_\alpha = \dot{h}_\alpha^\beta A_\beta + h_\alpha^\beta \dot{A}_\beta \\ so\; n^\alpha \dot{A}_\alpha = \left(n^\alpha h_\alpha^\beta\right)\dot{A}_\beta = 0 \end{cases}$$

And

$$(n_\beta \dot{A}_\alpha - n_\alpha \dot{A}_\beta)(n^\beta \nabla^\alpha V - n^\alpha \nabla^\beta V) = -2\dot{A}_\alpha \nabla^\alpha V$$

Thus,

$$L_{EM} = -\frac{1}{2}(\vec{\nabla}\times\vec{A})\cdot(\vec{\nabla}\times\vec{A}) + \frac{1}{2}\left(\dot{\vec{A}} + \vec{\nabla}\mathrm{V}\right)\cdot\left(\dot{\vec{A}} + \vec{\nabla}\mathrm{V}\right).$$

So,

$$\vec{\pi} = \frac{\partial L}{\partial \dot{\vec{A}}} = \left(\dot{\vec{A}} + \vec{\nabla}\mathrm{V}\right)$$

Recall that with $\vec{\mathrm{B}} = \vec{\nabla}\times\vec{A}$, thus $\vec{\nabla}\times\vec{E} + \frac{\partial\vec{B}}{\partial t} = 0$ becomes $\vec{\nabla}\times\left(E + \frac{\partial\vec{A}}{\partial t}\right) = 0$, thus $E + \frac{\partial\vec{A}}{\partial t} = -\vec{\nabla}\mathrm{V}$, for some scalar potential V, but this is the same V given here (aside from gauge trans.) arrived at from the covariant Lagrangian. So,

$$\dot{\vec{A}} + \vec{\nabla}\mathrm{V} = -\vec{E}$$

So, the canonically conjugate momenta conjugate to $\vec{A}$ is

$$\pi_{\vec{A}} = -\vec{E}$$

Trivially $\pi_V = 0$.

Notice,

$$\begin{aligned} H_{EM} &= \vec{\pi}\cdot\dot{\vec{A}} - L_{EM} \\ &= \vec{\pi}\cdot\vec{\pi} - \vec{\pi} - \vec{\nabla}\mathrm{V} + \tfrac{1}{2}\vec{B}\cdot\vec{B} - \tfrac{1}{2}\vec{\pi}\cdot\vec{\pi} \\ &= \tfrac{1}{2}\vec{\pi}\cdot\vec{\pi} + \tfrac{1}{2}\vec{B}\cdot\vec{B} + \mathrm{V}\vec{\nabla}\cdot\vec{\pi} - \vec{\nabla}\cdot(V\vec{\pi}) \end{aligned}$$

Now consider

$$\{f,g\}_{EM} = \int_S d^3x\left(\frac{\delta f}{\delta\pi^\alpha(x)}\frac{\delta g}{\delta A_\alpha(x)} - \frac{\delta g}{\delta A_\alpha(x)}\frac{\delta f}{\delta\pi^\alpha(x)}\right)$$

$$\{f,g\}_{EM} = \int_S d^3x\left(\frac{\delta f}{\delta A_\alpha}\frac{\delta g}{\delta E^\alpha} - \frac{\delta f}{\delta E^\alpha}\frac{\delta g}{\delta A_\alpha}\right)$$

Now,

$$\left\{\int_\Sigma d^3x\, f\nabla_\alpha E^\alpha, E^\alpha(x)\right\} = \frac{\delta}{\delta A_\alpha}\left\{\int_\Sigma^{\beta_x} f\nabla_\alpha E^\alpha\right\} = 0$$

and

$$\left\{\int_\Sigma d^3x\, f\nabla_\alpha E^\alpha, A_\alpha(x)\right\} = -\frac{\delta}{\delta E^\alpha}\left\{\int_\Sigma d^3x f\nabla_\alpha E^\alpha\right\} =$$

$$-\int_\Sigma dx\, f\frac{\delta}{\delta E^\alpha}(\nabla_\alpha E^\alpha) = -\int_\Sigma dx\, f\nabla_\alpha\left(\frac{\delta E^\alpha}{\delta E^\alpha}\right)$$

$$= +\int_\Sigma dx\,\nabla_\alpha f\delta(x) = \nabla_\alpha f(x)$$

So,

$$\left\{\int_\Sigma d^3x f\nabla_\alpha E^\alpha, F(A,E)\right\} = \int dy\left[\frac{\delta P_f}{\delta \mathrm{A}_\alpha}\frac{\delta F}{\delta E^\alpha} - \frac{\delta P_f}{\delta \mathrm{E}_\alpha}\frac{\delta F}{\delta A}\right] =$$

$$\int dy\left[+\nabla_\alpha f(x)\frac{\delta F}{\delta \mathrm{A}_\alpha}\right]$$

$\{\int_\Sigma d^3xf\nabla_\alpha E^\alpha, F(A,E)\} = -\frac{d}{d\lambda}F(\Psi_\lambda A_\alpha E^\alpha)|_{\lambda=0}$
and using $\Psi_\lambda A_\alpha = A_\alpha + \nabla_\alpha f_\lambda$. Since $\int_\Sigma d^3xf\nabla_\alpha E^\alpha = 0 \Rightarrow$
$\frac{d}{d\lambda}F(\Psi_\lambda A, E) = 0$, we have:

$$F(A_\alpha, E^\alpha) = F(A_\alpha + \nabla_\alpha f, E^\alpha)$$

where

$$A_\alpha \to A_\alpha + \nabla_\alpha f$$

is a family of gauge transformations leaves F invariant.

Example 11.9. Begin with electromagnetism. Let the gauge-covariant derivative of Ψ be given by

$$D_\alpha \Psi = (\partial_\alpha - ieA_\alpha)\Psi\,.$$

Show that the result of parallel transport of Ψ around γ is given by

$$\exp\left(ie\oint_\gamma A_\alpha\, dx^\alpha\right) = \exp\left(ie\int_S F_{\alpha\beta}dS^{\alpha\beta}\right),$$

where S is any surface with boundary γ. Hint: Define $U(\lambda)\Psi(\gamma(0)) \equiv \Psi(\gamma(\lambda))$, where $\Psi(\lambda)$ is the result of parallel transporting Ψ along γ by a parameter distance λ, $\dot{\gamma}^\alpha D_\alpha \Psi = 0$. Then $U(\lambda)$ satisfies

$$\frac{dU}{d\lambda} = ie\dot{\gamma}^\alpha A_\alpha U,$$

from which it can be shown that $U(\lambda) = \exp\left(ie\int_0^\lambda A_\alpha\,\dot{\gamma}^\alpha(\lambda)d\lambda\right)$.

Solution

We have $D_\alpha\Psi = (\partial_\alpha - ieA_\alpha)\Psi$ (the form for electromagnetism), and defining $U(\lambda)\Psi(\gamma(0)) \equiv \Psi(\gamma(\lambda))$, where $\dot{\gamma}^\alpha D_\alpha\Psi = 0$. We have to start:

$$\frac{d}{d\lambda}\big(U(\lambda)\Psi(\gamma(0))\big) = \frac{dU}{d\lambda}\Psi(\gamma(0)) = \frac{d\Psi(\lambda)}{d\lambda} = \dot{\gamma}^\alpha\partial_\alpha\Psi(\gamma(\lambda))$$
$$= \dot{\gamma}^\alpha(ieA_\alpha)\Psi(\gamma(\lambda))$$

Thus,

$$\frac{dU}{d\lambda} = ie\dot{\gamma}^\alpha A_\alpha U\Psi(\gamma(0)) = ie\dot{\gamma}^\alpha A_\alpha U$$

So,

$$\frac{dU}{U} = ie\dot{\gamma}^\alpha(\lambda)A_\alpha d\lambda$$

and

$$\ln U(\lambda) = ie \int_0^\lambda A_\alpha \dot{\gamma}^\alpha(\lambda) d\lambda + C$$

If we use $U(\lambda) = \exp\left(ie \int_0^\lambda A_\alpha \dot{\gamma}^\alpha(\lambda) d\lambda\right) \Rightarrow U(0) = 1$ and $C = 0$. So, $\Psi(\gamma(\lambda))$ transported around loop. $\Psi(\gamma)$ is:

$$\Psi(\gamma) = \exp\left(ie \oint_\gamma A_\alpha\, dx^\alpha \right) \Psi(\gamma(0))$$

Furthermore, using Stokes theorem, the phase factor may be expressed as

$$\exp\left(ie \oint_\gamma A_\alpha\, dx^\alpha \right) = \exp\left(ie \int_S F_{\alpha\beta} dS^{\alpha\beta} \right).$$

Electromagnetism: charged scalar fields.

In electromagnetism, we have seen that a charge scalar field ψ on M can be regarded as a field $\overline{\psi}$ on a U(1) bundle B over M, with

$$\mathcal{L}_\eta \overline{\psi} = im\overline{\psi},$$

for a field of charge m. A cross section b: M → B corresponds to a chart (η, z) on B for which $b(x) = (0, x)$ is a cross section. The value of ψ corresponding to the $\eta = 0$ cross section is

$$\psi(x) = \overline{\psi} \circ b(x) = \overline{\psi}(0, x).$$

A vector potential A_α of the electromagnetic field is (up to a constant) the pullback to M of a connection $\overline{A}_C$ on B:

$$A_\alpha = \frac{\hbar c}{e} \pi_\alpha^C \overline{A}_C\,.$$

The gauge-covariant derivative of a field ψ is along a path $x(\lambda)$ in M is

$$\dot{x}^\alpha D_\alpha \psi = \dot{x}^\alpha \left(d_\alpha - i\frac{me}{\hbar c} A_\alpha \right) \psi,$$

and this is equivalent to the derivative of $\overline{\psi}$ along a horizontal lift of $x(\lambda)$, i.e., along a horizontal path $b_{||}(\lambda)$ over $x(\lambda)$:

$$\dot{x}^\alpha D_\alpha \psi = \frac{d}{d\lambda}\left(\overline{\psi}\left(b_{||}(\lambda) \right) \right)\Big|_{\lambda=0} = 0.$$

<u>**Proof:**</u> A horizontal lift $b_{||}(\lambda)$ safisfies $\overline{A}_C \dot{b}^C = 0$. Writing $\dot{b} = (\dot{\eta}, \dot{x})$, we have

$$\dot{\eta} + \dot{x}^\alpha \overline{A}_\alpha = 0,$$

for a horizontal path. Then the horizontal derivative of $\overline{\psi}$ is

$$\frac{d}{d\lambda}\left(\overline{\psi}\left(b_{||}(\lambda)\right)\right)\Big|_{\lambda=0} = \frac{d}{d\lambda}[\overline{\psi}(\eta(\lambda), x(\lambda))] = \dot{x}^\alpha d_\alpha\overline{\psi} + \dot{\eta}\partial_\eta\overline{\psi}$$
$$= \dot{x}^\alpha\left(d_\alpha - i\frac{me}{\hbar c}A_\alpha\right)\overline{\psi}\,.$$
In other words, the gauge-covariant derivative,
$$D_\alpha\psi = \left(d_\alpha - i\frac{Q}{\hbar c}A_\alpha\right)\psi,$$
corresponds to the horizontal derivative of $\overline{v}$,
$$D_C\psi = h_C^D(d_D\overline{v}), \quad or\ D\overline{\psi} = \text{hor } d\overline{v},$$
$$D_a v = b_a^{*D} D_C\overline{v}.$$

Isovectors

Let G be a Lie group with Lie algebra g. An isovector, ξ^a, is a g-valued scalar on a manifold M. A gauge transformation $g(x)$ acts on an isovector field ξ^i by the adjoint representation:
$$g : \xi(x + g(x)^{-1}\xi x)g(x)$$
(Eqn. 11-13)

This transformation law is equivalent to saying that ξ is the pullback, $\xi(x) = \overline{\xi}(b(x))$, to M of a g-valued scalar $\overline{\xi}^a$ on B, safistying $\overline{\xi}(bg) = g^{-1}\overline{\xi}g$. What is the gauge-covariant derivative? In analogy with electromagnetism, write
$$D_C\overline{\xi} = h_C^D\xi_D\overline{\xi} \ \text{ or } \ D\overline{\xi} = \text{hor } d\overline{\xi},$$
Thus
$$D_a\xi = b_a^{*C}D_C\overline{\xi}$$
(Eqn. 11-14)

More explicitly, let $\dot{b}^C$ be a vector tangent to $b(\lambda)$. Its horizontal projection, $b_{||}^C$ is tangent to the horizontal path $b_{||}(\lambda)$ through $b(0)$, related to $b(\lambda)$ by $b(\lambda) = b_{||}(\lambda)g(\lambda)$ with $g(\lambda)$ a path through $e \in G$. Then $\dot{g}^a = A_C^a\dot{b}^C$ and $\dot{b}^C D_C\xi$ is given by
$$\dot{b}^C D_C\overline{\xi} = \frac{d}{d\lambda}\left(\overline{\xi}\left(b_{||}(\lambda)\right)\right)\Big|_{\lambda=0} = \frac{d}{d\lambda}[\overline{\xi}(b(\lambda), g(\lambda))]$$
$$= \frac{d}{d\lambda}[g(\lambda)\overline{\xi}(b(\lambda), g^{-1}(\lambda))] = \dot{b}^C d_C\overline{\xi} + \dot{g}\overline{\xi} - \overline{\xi}\dot{g}$$
Thus,
$$\dot{b}^C D_C\overline{\xi}^a = \xi^C\left(d_C\overline{\xi}^a - c_{bd}^a A_C^d\xi^d\right),$$
using $[\dot{g}, \xi^{-1}] = c_{bd}^a\dot{g}^d\xi^i$ In other words, the horizontal derivative of an isovector on B is
$$D_C\overline{\xi}^a = d_C\overline{\xi}^a + c_{bd}^a A_C^d\overline{\xi}^d,$$

(Eqn. 11-15)

and the corresponding gauge-covariant derivative on M is then

$$D_a\xi^a = d_a\xi^a + c^a_{bd}A^d_C\xi^d.$$

(Eqn. 11-16)

Here $A^a_\alpha = b^{*C}_a A^a_C$ is (up to a constant) the Yang-Mills vector potential corresponding to the connection A^a_C.

The gauge-covariant derivative of an isovector is an isovector, because its definition looks at the value of the cross section $b(x)$ at the point x_0 where the derivative is evaluated. In other words, if one changes cross section from $b(x)$ to $\overline{b}(x) = b(z)g(x)$, the ordinary derivative of $\xi(x) = \overline{\xi}[b(x)g(x)] =$ involves the derivative dg of the gauge transformation. The gauge-covariant derivative of $\overline{\xi}$, however, is (along the direction $\dot{x}(\lambda)$) the derivative of $\overline{\xi}\left[b_{||}\left(x(\lambda)\right)g(x_0)\right]$.

Let's show directly from the definitions that $D_a\xi^a$ is an isovector, that

$$D\overline{\xi}(x) = g(x)^{-1}d\xi(x)g(x).$$

(Note that the definition of covariant derivative given above can be used for fields that transform according to arbitrary representations of the group.)

Y-M Lagrangian

Since $F^a_{\alpha\beta}$ is an isovector, $\mathcal{L} = F^a_{\alpha\beta}F^{\alpha\beta}_a$ is an isoscalar – and also a space scalar- and is therefore the natural generalization of the electromagnetic Lagrangian density $F_{\alpha\beta}F^{\alpha\beta}$:

$$I_{Y-M} - \frac{1}{4}\int dr\, F^a_{\alpha\beta}F^{\alpha\beta}_a.$$

To find the resulting field equations, note that

$$\delta F^a_{\mu\nu} = \delta\left[\partial_\mu A^a_\nu - \partial_\nu A^a_\mu + C^a_{bc}A^b_\mu A^c_\nu\right]$$
$$= \partial_\mu\delta A^a_\nu - \partial_\nu\delta A^a_\mu + C^a_{bc}A^b_\mu\delta A^c_\nu - C^a_{bc}A^b_\nu\delta A^c_\mu$$

Thus

$$\delta F^a_{\mu\nu} = D_\mu\delta A^a_\nu - D_\nu\delta A^a_\mu$$

Note that just as $\delta\Gamma^\alpha_{\beta\gamma}$ is a tensor, δA^a_α is an isovector: The inhomogeneous term in the gauge transformation law drops out if one takes the difference

$$\hat{A}^a_\alpha - A^a_\alpha,$$

of two vector potentials, so $\hat{A}^a_\alpha - A^a_\alpha$ is an isovector and

$$\delta A^a_\alpha = \frac{d}{d\lambda}A^a_\alpha(\lambda) = \lim_{\lambda\to 0}\frac{1}{\lambda}[A^a_\alpha(\lambda) - A^a_\alpha(0)],$$

is therefore an isovector. Thus

$$\delta I_{Y-M} = \int_\Omega dr\, D_\alpha \delta A^a_\beta F_a^{\alpha\beta} = \int_\Omega dr\left[D_\alpha\left(\delta A^a_\beta F_a^{\alpha\beta}\right) - \delta A^a_\beta D_\alpha F_a^{\alpha\beta}\right],$$

by Leibnitz, and since $D_\alpha = \nabla_\alpha$ when acting on an isoscalar:

$$\delta I_{Y-M} = \int_\Omega dr\, \nabla_\alpha\left(\delta A^a_\beta F_a^{\alpha\beta}\right) - \int dr\, \delta A^a_\beta D_\alpha F_a^{\alpha\beta} =$$

$$-\int_\Omega dr\, \delta A^a_\beta D_\alpha F_a^{\alpha\beta},$$

for variations δA^a_ν that vanish at the boundary $\partial\Omega$.Thus

$$\frac{\delta I_{YM}}{\delta A^a_\alpha(x)} = D_\beta F_a^{\alpha\beta}(x),$$

and the source free field equation is

$$D_\beta F_a^{\alpha\beta} = 0.$$

This is the analogue of half of Maxwell's equations. The other half, $\nabla_{[\alpha}F_{\beta\gamma]} = 0$ has the obvious analogue

$$D_{[\alpha}F^a_{\beta\gamma]} = 0$$

Which is also true. For gravity, $F^a_{\alpha\beta}$ is the Riemann tensor and $D_\beta F_a^{\alpha\beta} = 0$ is the Bianchi identity.

Proof of the identity $D_{[\alpha}F^a_{\beta\gamma]} = 0$. Start with $F^a_{\beta\gamma} = 2\nabla_{[\beta}A^a_{\gamma]} + C^a_{bc}A^b_\beta A^c_\gamma$:

$$\begin{aligned} D_\alpha F^a_{\beta\gamma} &= \nabla_\alpha F^a_{\beta\gamma} + C^a_{bc}A^b_\alpha F^c_{\beta\gamma} \\ &= 2\nabla_\alpha\nabla_{[\beta}A^a_{\gamma]} + C^a_{bc}\nabla_\alpha\left(A^b_\beta A^c_\gamma\right) \\ &\quad + C^a_{bc}A^b_\alpha\left(2\nabla_{[\beta}A^c_{\gamma]} + C^c_{de}A^d_\beta A^e_\gamma\right) \end{aligned}$$

Then

$$D_{[\alpha}F^a_{\beta\gamma]} = 2\nabla_{[\alpha}\nabla_\beta A^a_{\gamma]} + C^a_{bc}\left(\nabla_{[\alpha}A^b_\beta A^c_{\gamma]} + A^b_{[\beta}\nabla_\alpha A^c_{\gamma]}\right) + C^a_{bc}A^b_{[\alpha}2\nabla_\beta A^c_{\gamma]}$$
$$+ C^a_{bc}C^c_{de}A^b_{[\alpha}A^d_\beta A^e_{\gamma]}$$

and since $\nabla_{[\alpha}\nabla_\beta A^a_{\gamma]} = 0$, even in curved spacetime, and with cancellations:

$$D_{[\alpha}F^a_{\beta\gamma]} = C^c_{[de}C^a_{b]c}A^b_\alpha A^d_\beta A^e_\gamma = 0,$$

where the last step follows from the Jacobi Identity.

Example 11.10. For Yang-Mills Theory, $D_\alpha\Psi = (\partial_\alpha - igA_\alpha)\Psi$, where $A(\lambda)$ is matrix $A^a_{b\alpha}$, and Ψ is an isovector, Ψ^α. For different λ, the matrices $A(\lambda)$ do not commute, and we need to show that $U(\lambda)$ is given by

$$U(\lambda) = P\exp\left(ig\int_0^{\lambda} A_\alpha\,\dot{\gamma}^\alpha(\lambda)d\lambda\right).$$

So, in a power series expansion of the exponential, the matrices are to be ordered as encountered along the path, the largest values of the parameter λ being to the left.

Solution

For Yang-Mills: $D_\alpha\Psi^\alpha = \left(\partial_\alpha - igA^a_{D\alpha}(\lambda)\right)\Psi^\alpha$. Note: for different λ the matrices $A(\lambda)$ do not commute (non-Abelian theory), so we introduce the "path" ordering operator P. We have

$$\frac{d}{d\lambda}U = ig\dot{\gamma}^\alpha A_\alpha U \quad\Rightarrow\quad U(\lambda) = \exp\left(ig\int_0^{\lambda} A_\alpha\,\dot{\gamma}^\alpha(\lambda)d\lambda\right),$$

but if we wish to expand the exponential we must contend with:

$$\int_0^{\lambda} A_\alpha\,\dot{\gamma}^\alpha(\lambda)d\lambda = \int_0^{\lambda_1} A_\alpha\,\dot{\gamma}^\alpha(\lambda)d\lambda + \cdots + \int_{\lambda_N}^{\lambda} A_\alpha\,\dot{\gamma}^\alpha(\lambda)d\lambda$$
$$\simeq A^1_\alpha\Gamma^\alpha_1 + \cdots A^N_\alpha\Gamma^\alpha_N$$

where $\Gamma^\alpha_1 = \int_0^{\lambda_1}\dot{\gamma}^\alpha(\lambda)d\lambda$, etc., and $A^1_\alpha = A_\alpha(\lambda_1)$, etc. Then

$$U(\lambda) \simeq 1 + ig\int_d^{\lambda} A_\alpha\,\dot{\gamma}^\alpha(\lambda)d\lambda + \frac{1}{2}\left(ig\int_0^{\lambda} A_\alpha\,\dot{\gamma}^\alpha(\lambda)d\lambda\right)^2 + (\cdots)^3$$

where for the square and higher order terms an ordering property enters. Algebraically we have

$$\left(\int_0^{\lambda} A_\alpha\,\dot{\gamma}^\alpha(\lambda)d\lambda\right)^2 \cong (A^1_\alpha\Gamma^\alpha_1 \cdots A^N_\alpha\Gamma^\alpha_N)^2$$

with no ordering prescribed. so it could be

$$= A^1_\alpha(A^1_\alpha\Gamma^\alpha_1 \cdots \Gamma^\alpha_N)\Gamma^\alpha_1 + A^2_\alpha(A^1_\alpha\Gamma^\alpha_1 \cdots \Gamma^N_\alpha\Gamma^\alpha_N)\Gamma^\alpha_2 + \cdots$$

But this ordering doesn't account for the non-abelian character of the variable A_α. This can be resolved easily:

$$\frac{d}{d\lambda}U = ig\dot{\gamma}^\alpha A_\alpha U \quad\Rightarrow\quad U(\lambda) = \exp\left(ig\int_0^{\lambda} A_\alpha\,\dot{\gamma}^\alpha(\lambda)d\lambda\right)$$

and $U(\lambda_1) \simeq \exp(igA^1_\alpha\Gamma^\alpha_1)$ for λ_1 infinitesimal. So

$$\Psi\left(\gamma(\lambda_1)\right) = U(\lambda_1)\Psi\left(\gamma(0)\right) = \exp(igA^1_\alpha\Gamma^\alpha_1)\,\Psi\left(\gamma(0)\right).$$

Now go another infinitesimal increment from $\gamma(\lambda_1) \to \gamma(\lambda_2)$ where $\lambda_2 - \lambda_1$ is infinitesinal:

$$\Psi(\gamma(\lambda_2)) = U(\lambda_2 - \lambda_1)\Psi(\gamma(\lambda_1)) = \exp(igA_\alpha^2\Gamma_2^\alpha)\exp(igA_\alpha^1\Gamma_1^\alpha)\,\Psi(\gamma)$$

Continuing in this fashion we see that

$$P\exp\left(ig\int_0^\lambda A_\alpha\,\dot{\gamma}^\alpha(\lambda)d\lambda\right) = \exp(igA_\alpha^N\Gamma_N^\alpha)\cdots\exp(igA_\alpha^1\Gamma_1^\alpha)\,\Psi(\gamma(0)),$$

where this merely prescribes the proper ordering on the path integral under a power series expansion such matrices A_α are ordered with largest parameter values to the left. So, we have

$$U(\lambda) = \underline{P}\exp\left(ig\int_0^\lambda A_\alpha\,\dot{\gamma}^\alpha(\lambda)d\lambda\right).$$

Now, in quantum field theory ([12]), the mathematical tools to solve this already exist, using perturbation theory methods. Since

$$D_\alpha\Psi = (\partial_\alpha - igA_\alpha)\Psi = 0$$

with parallel transport along $\gamma(\lambda)$:

$$\dot{\gamma}^\alpha\partial_\alpha\Psi = ig\dot{\gamma}^\alpha A_\alpha\Psi \qquad \dot{\gamma}^\alpha\partial_\alpha \to \partial_t, \quad -gA_\alpha\dot{\gamma}^\alpha \to H_I$$

Thus

$$\partial_t\Psi = -\mathrm{i}H_I\Psi \;\; or \;\; i\frac{\partial}{\partial t}\Psi = H_I\Psi \;with\; \Psi = U(t)\Psi(\lambda(0)).$$

Thus,

$$i\frac{\partial U}{\partial t} = H_I(\mathrm{t})\mathrm{U}$$

and we can write:

$U(t) = I - i\int_{-\alpha}^{t} dt_1\, H_I(\mathrm{t}_1)\mathrm{U}(\mathrm{t}_1),$ now iterating on $\mathrm{U}(\mathrm{t}_1)$ ...

$$U(t) = I - i\int_{-\infty}^{t} dt_1\, H_I(\mathrm{t}_1) + (-\mathrm{i})^2\int_{-\infty}^{t} dt_1\int_{-\infty}^{t_1} dt_2\, H_I(\mathrm{t}_1)H_I(\mathrm{t}_2) + \cdots$$

$$+(-\mathrm{i})^{\mathrm{n}}\int_{-\infty}^{t} dt_1\int_{-\infty}^{t_1} dt_2\cdots\int_{-\infty}^{t_{n-1}} dt_n\, H_I(\mathrm{t}_1)\cdots H_I(\mathrm{t}_\mathrm{n}) + \cdots$$

Let's introduce the time ordered product:

$$U(t) = \sum_{n=0}^{\infty} \frac{(-1)^2}{n!} \int_{-\infty}^{t} dl_1 \int_{-\infty}^{t} dt_2 \cdots \int_{-\infty}^{t} dt_n \cdots T[H_I(\mathrm{t}_1) \cdots H_I(\mathrm{t_n})]$$

$$= T \exp\left[-i \int_{-\infty}^{t} dt'\, H_I(t')\right]$$

Transforming back (with bounds of integral appropriately modified for this case):

$$U(\lambda) = P \exp\left(-i \int_{0}^{\lambda} A_\alpha \dot{\gamma}^\alpha(\lambda) d\lambda\right).$$

11.6 Exercises

(Ex. 11.1) Verify Eqn. 11-3.
(Ex. 11.2) Verify Eqn. 11-4.
(Ex. 11.3) Verify Eqn. 11-5.
(Ex. 11.4) Verify Eqn. 11-6.
(Ex. 11.5) Verify Eqn. 11-7.
(Ex. 11.6) Verify Eqn. 11-8.
(Ex. 11.7) Verify Eqn. 11-9.
(Ex. 11.8) Verify Eqn. 11-10.
(Ex. 11.9) Verify Eqn. 11-11.
(Ex. 11.10) Verify Eqn. 11-12.
(Ex. 11.11) Verify Eqn. 11-13.
(Ex. 11.12) Verify Eqn. 11-14.
(Ex. 11.13) Verify Eqn. 11-15.
(Ex. 11.14) Verify Eqn. 11-16.

12. Series Outlook

We've described field dynamics in 2D, 3D, and 4D. Two-dimensional ("2D") field dynamics can be described as a complex function (that maps complex numbers to complex numbers). A

For the 3D field dynamics we did an analysis of the electromagnetic field (in 3D to start). The level of coverage began as an overview of electrostatics at the level of the graduate text Jackson [Jackson]. Some problems from Jackson Ch's 1-3 were examined closely in developing the theory itself. We then showed many more examples of ODE problems with solutions, such as for the 3D Laplacian, usually involving separation of variables. We then reviewed the famous transform, discovered by Lorentz in 1899 [37], that relates the EM field as seen by two observers differing by a relative velocity. With the existence of this transform, that brings in the time dimension along with the relative velocity, and we have effectively established a 4D theory.

From Lorentz Invariance we showed, as a point transformation, rotational invariance under SO(3) or SU(2). If Lorentz Invariance is fundamental, then we should see both forms of rotation invariance, one of vector/tensor type from SO(3), and one of spinorial type from SU(2). This is the case, as gauge fields are vectorial and matter fields are spinorial. From Lorenz Invariance as a local invariance we have the Minkowski (flat) spacetime metric, which then generalizes to the Riemannian metric (in General Relativity).

As with the point particle dynamics, for the field dynamics we showed three ways to formulate the behavior: (1) differential equation; (2) function variation (on Lagrangian); and (3) functional variation (on the Action). We saw similar dynamical limit phenomena as before, but also new phenomena, including (i) inevitable BH singularity formation (the Penrose singularity theorem); (ii) FRW Universe formation (from homogeneity and isotropy); (iii) the BH collapse singularity; (iv) the atomic collapse 'singularity'.

Unlike this book, in Book 3 [5], classical field theory with *dynamic* geometry, i.e. GR, we don't see alpha at all. Instead we see manifold constructs and the mathematics of differential geometry (and to some extent differential topology and algebraic topology). Manifold constructs are entirely encapsulated in the math background given in Book 3 (and its Appendix). An application in the area of neuromanifolds (see [24]), shows the equivalent of a geodesic path in this setting is evolution involving minimum relative entropy steps. Similar to the description of a locally flat space-time we now have a description of 'entropy' increasing/evolving according to minimum relative entropy.

In Book 2 we have established fundamental field theory concepts, carefully analyzed electromagnetism, and made an initial foray at General Relativity from a field perspective. In Book 3 [5] we will continue the description of Classical Fields and Manifolds, and in Book 5 [12] we consider Quantum Field Theory and Thermal Quantum Field Theory, while in Book 6 [6] consider Thermal Quantum Gravity (but not Quantum Gravity). Book 4 [15] describes a special case of a QG analysis (a minisuperspace analysis) where use is made of the full generally relativistic dust shell collapse analysis (Book 3), and performed a precise quantum adjoint operator analysis (Book 4 [15]). It is found that there is no QG barrier to collapse (as with the atomic nucleus), instead, the formalism appears to indicate that the geometry, in this highly constrained (symmetric) circumstance, is "pure apparatus" (see Book 4 [15] discussion on measurement theory to see the implications of this).

Appendix

A. Complex Functions

The main reference for the sections on Complex Functions to follow is Churchill & Brown [23]. Along with Jackson [32], these two texts provide extensive and critical background for details discussed.

A.1 Complex Numbers

Def.: A complex number is a two real-component number, $\mathbb{C} = \{(a, b): a, b \in \mathbb{R}\}$, with addition and multiplication defined by:

$$(a, b) + (c, d) = (a + c, b + d)$$
$$(a, b) \cdot (c, d) = (ac - bd, bc + ad)$$

One can check that $\mathbb{C}$ is a field. In particular, the additive identity is (0, 0), the multiplicative identity is (1, 0), and the multiplicative inverse of $(a, b), for[\neq (0, 0)]$, is $\left(\frac{a}{a^2+b^2}, \frac{-b}{a^2+b^2}\right)$.

Since the map $(a, 0) \longmapsto a$ is a map of $\mathbb{R}' = \{(a, 0): a \in \mathbb{R}\} \subset \mathbb{C}$ onto $\mathbb{R}$. One can further check that it is an isomorphism, and so we consider $\mathbb{R}$ as a subfield of $\mathbb{C}$.

Def.: $i = (0, 1)$.
By definition of addition and multiplication:

$$(a, b) = a + ib \qquad (a, b \in \mathbb{R}).$$

Notation: If $z = a + ib$, where $a, b \in \mathbb{R}$, we write $a = ReZ$, $b = ImZ$.

Def.: The absolute value of *z*, (<u>modulus</u> of *z*) is:

$$|z| = \sqrt{a^2 + b^2} \geq 0 \text{ where } z = a + ib$$

Def.: The <u>conjugate</u> of $z = a + ib$ is $\bar{z} = a - ib$.

Note that $z\bar{z} = a^2 + b^2 = |z|^2$, and

$$\frac{1}{z} = \frac{\bar{z}}{|z|^2}$$

(compared with formula of multiplicative inverse above, it is the same).

We can picture $\mathbb{C}$ as the plane $\mathbb{R}^2$, with the number $a + ib$ being pictured as the point (a, b). In this picture, $|z|$ is interpreted as the distance z to 0 (the origin) and, more generally, if

$$z = a + ib \text{ and } \omega = c + id$$

then

$$|z - \omega| = |(a - c) + i(b - d)| = (a - c)^2 + (b - d)^2$$
$$= \text{distance from } z \text{ to } \omega.$$

As the 2D planar geometry would suggest, there is a triangle inequality:

$$|z + \omega| \leq |z| + |\omega|$$

Proof: Since$|ReZ| \leq |Z| \; and \; |ImZ| \leq |Z|$, let's consider $z = a + ib\,, \omega = c + id$ and compute $|z + \omega|^2$ directly:

$$|z + \omega|^2 = (a + c)^2 + (b + d)^2 = a^2 + c^2 + b^2 + d^2 + 2ac + 2bd$$
$$= |z|^2 + |\omega|^2 + 2Re(z\overline{\omega}) \leq |z|^2 + |\omega|^2 + 2|z||\overline{\omega}|$$
$$\leq |z|^2 + |\omega|^2 + 2|z||\overline{\omega}|$$
$$\leq (|z| + |\omega|)^2 \quad (since\ |\overline{\omega}| = |\omega|).$$

Thus, $|z + \omega| \leq |z| + |\omega|$

Notation: For $z, \omega \in \mathbb{C}$, let's also write $d(z + \omega) = |z - \omega|$.

Corollary: If $z_1, z_2, z_3 \in \mathbb{C} \rightarrow d(z_1, z_3) \leq d(z_1, z_2) + d(z_2, z_3)$.

Proof:
Put $= z_1 - z_2\,, \omega = z_1 - z_3\,.$

A.2 Polar representation

A complex number $z = x + iy$ can be written in polar form

$$z = r(\cos\theta + i\sin\theta)$$

where $r = |z|$ and $\theta = \arctan\left(\frac{y}{x}\right)$. Further properties will be elaborated once $\cos\theta$ itself is examined (defined rigorously) in a later section.

A.3 Extended Plane, $\mathbb{C} \cup \{\infty\} = \mathbb{C}_\infty$

Let ∞ be any point not in $\mathbb{C}$ (any mathematical object disjoint to $\mathbb{C}$) and define $\mathbb{C}_\infty = \mathbb{C} \cup \{\infty\}$. We can define a "distance" on $\mathbb{C}_\infty$:

$$d(z,z') = \frac{2|z-z'|}{[(1+|z|^2)(1+|z'|^2)]^{1/2}} \qquad z,z' \in \mathbb{C}$$

and

$$d(z,\infty) = \frac{2}{(1+|z|^2)^{1/2}} \quad , \qquad z \in \mathbb{C}$$

$\left(\text{where, formally, } \lim_{z'\to\infty} d(z,z') = d(z,\infty)\right)$.

To get a picture for $\mathbb{C}_\infty$, note that infinity, $\lim_{z\to\infty} d(z,\infty) = \lim_{z\to\infty} \frac{2}{|z|} = 0$, has all the points at infinity mapped to the same point (the added point) and this wraps the plane onto a sphere.

Def.: A metric space is a pair (X,d) where X is a set and d: $X \times X \to \mathbb{R}$ is a parameter that satisfies:

1) $d(x,y) \geq 0$
2) $d(x,y) \geq 0 \Leftrightarrow x = y$
3) $d(x,y) = d(y,x)$
4) $d(x,y) \leq d(x,z) + d\{z,y\}$ (triangle inequality)

If d exists, then "X is a metric space".

Examples:

1) Let X be any set, and define
$$d(x,y) = \begin{cases} 0 & if \quad x = y \\ 1 & if \quad x \neq y \end{cases}$$

2) Let $X = \mathbb{R}$ or $\mathbb{C}$ and define $d(x,y) = |x-y|$. More generally, let $X = \mathbb{R}^n$and define:
$$d(x,y) = \sqrt{\sum_{i=1}^{n} (x_i - y_i)^2} \qquad \text{where} \qquad \begin{matrix} x = (x_1, x_2, \dots) \\ y = (y_1, y_2, \dots) \end{matrix}.$$

3) Again, let $X = R^n$ and define
$$d(x,y) = \max_{1\leq i\leq n}\{|x_i - y_i|\}$$

4) $X = R^n$ and

$$d(x,y) = \sum_{i=1}^{n} |x_i - y_i|$$

5) Let $X = \mathbb{C}_\infty = \mathbb{C} \cup \{\infty\}$
   Easily have:
   $$d(z,z') = \frac{2|z,z'|}{[(1+|z|^2)(1+|z'|^2)]^{1/2}} \quad \text{if } z,z' \in \mathbb{C}$$
   $$d(z,\infty) = \frac{2}{(1+|z|^2)^{1/2}} \quad \text{if } z \in \mathbb{C}$$
   $d(\infty,\infty) = 0$
   The difficulty is showing the triangle inequality, for this it is easiest to go back to the equivalent space (surface of sphere) via the stereographic projection.

6) Let X be any set, and let $B(x)$ denote the set of bounded, complex-valued functions on x: $f: X \to \mathbb{C}$, define $\|f\|_\infty = \sup\{|f(x)|: x \in X\}$ and
   $B(x) = \{f: X \to \mathbb{C}: \|f\|_\infty < \infty\}$
   then we can define
   $$d(f,g) = \|f-g\|_\infty$$
   where the metric is of uniform convergence where the supremum of (f-g) goes to zero.

7) Let (X,d) be a metric space and let Y be a subset of X then (Y,d) is a metric space.

Definition: The open and closed balls of radius r about $x \in X$ are respectively:

$B(x;r) = B_r(x) = \{y \in X: d(y,x) < r\}$ open ball

$\bar{B}(x;r) = \bar{B}_r(x) = \{y \in X: d(y,x) \le r\}$ closed ball

A tricky example involving a discrete metric space:
$B_{1/2}(x) = \{x\}$
$\bar{B}_{1/2}(x) = \{x\}$
but
$B_1(x) = \{x\}$

$\bar{B}_1(x) = \{\text{everything}\}$

Definition: A set $G \subset X$ (a metric space) is open if for every $x \in G$ there is $\xi > 0$ such that $B_\xi(x) \subset G$. A set $F \subset X$ is closed if its complement $F^c = X/F$ is open.

Note that it is possible to describe an open or closed set without a metric (this is an issue of topology). Also note that G can be an open set in *X* even with a boundary! The open ball only concerns itself with points in the space *X*.

Proposition: Let (X,d) be a metric space, then
(i) ϕ (empty set) and *X* are open then
(ii) If $G_1, \dots G_n$ are open then so is $\bigcap_{i=1}^n G_n$
(iii) If $\{G_i, \dots j \in J\}$ is a collection of open sets, then $\bigcup_{j\in J} G_j$ is open

Proposition: Consider the closed case.
(i) ϕ and X are closed then
(ii) If $F_1, \dots F_n$ are closed, the so is $\bigcup_{i=1}^n F_i$
(iii) If $\{F_j, : j \in J\}$ is a collection of closed sets, then $\bigcap_{j\in J} F_j$ is closed

Consider $G_n = \left(-\frac{1}{n}, \frac{1}{n}\right)$ open $\forall\, n$, then $\bigcap_{n=1}^\infty G_n = \{0\}$.

Def.: Let A be a subset of X. The interior and closure of A are respectively

$$\text{int}\, A = \dot{A} = \bigcup_{\substack{G\ open \\ G \subseteq A}} G$$

and

$$\text{closure}\, A = \bar{A} = \bigcap_{\substack{F\ closed \\ F \supset A}} F.$$

Def.: Boundary of A is $A = \bar{A} \cap \overline{(X \cap A)}$.

Note that it is possible to have $\dot{A} = \phi$ and $\bar{A} = X$, even simultaneously. $\dot{A}$ is the largest open set contained in A, and $\bar{A}$ is the smallest closed set which contains A.

Def.: $A \subset X$ is <u>dense</u> if $\bar{A} = X$. $A \subset X$ is <u>nowhere dense</u> if $\left(\dot{\bar{A}}\right) = \phi$ (where the latter is the interior of the closure).

Def.: Let (X, d) and (Y, ρ) be metric spaces, and let $f: X \to Y$ be a function. Let $a \in X, b \in Y$. We say (definition of limit):

$$\lim_{x \to a} f(x) = b$$

if for every $\xi > 0$, there is a $\delta > 0$ such that $\rho(f(x), b) < \xi$ wherever $0 < d(x, a) < \delta$.

Def.: f is continuous at a if (definition of continuity):

$$\lim_{x \to a} f(x) = f(a),$$

and f is continuous if it is continuous at every point of X.

<u>Proposition</u>: Let $f: (X, d)$ and $\to$ (Y, ρ) be a function, and let $a \in X$ and $b = f(a) \in Y$. The following are equivalent:

a) f is continuous at a if for every $\xi > 0$, $f^{-1}\left(B_\xi(b)\right)$ contains a ball centered at a.

<u>Proof:</u>

$(a \to b)$ Suppose f is continuous: Given $\xi > 0$ close $\delta > 0$ as in the definition of continuity at a. Then $f\left(B_\delta(a)\right) \subset B_\xi(b)$, so $B_\delta(a) \subset f^{-1}\left(B_\xi(b)\right)$.

$(b \to a)$: Given $\xi > 0$, let $\delta > 0$ be such that $B_\delta(a) \subset f^{-1}\left(B_\xi(b)\right)$, then δ satisfies the condition in the def. of $\lim_{x \to a}\left(f(x)\right) = f(a)$.

<u>Proposition</u>: Let $f: (X, d)$ and $\to$ (Y, ρ) be a function. The following are equivalent:

a) f is continuous

b) $f^{-1}(u)$ is open in X for every open subset U of Y.

<u>Proposition</u>: Let $f: (X, d) \to (Y, \rho)$ be a function, suppose $a \in X$ and $b = f(a) \in Y$

(a) f is continuous at a

(b) for every $\xi > 0$ $f'\left(B_\xi(b)\right)$ contains a ball with center at $a \leftarrow$ just a restatement of continuity.

Proposition: Let $f: (X, d) \rightarrow (Y, \rho)$ for a function, the following are equivalent:
(a) f is continuous
(b) $f'(u)$ is open $(in\ X)$ for every open subset U and Y

Proof
$(a \Rightarrow b)$ Let $U \subset Y$ be open, and let $G = f^{-1}(U) \subset X$. Let $a \in G$ and let $b = f(a) \in U$. Since U is open, for some $\xi > 0$, $B_\xi(b) \subset U$. Since f is continuous at a, by the previous proposition, $f^{-1}\left(B_\xi(b)\right)$ contains a ball centered at a ... G is open.
$(b \Rightarrow a)$ Given $a \in X$, let $b = f(a)$. Given $\xi > 0$, consider $B_\xi(b)$. This is an open set, so $f^{-1}\left(B_\xi(b)\right)$ is open. But $a \in f^{-1}\left(B_\xi(b)\right)$ and so $f^{-1}\left(B_\xi(b)\right)$ contains some ball around a. By the previous proposition f is continuous at a.

Proposition: Let X, Y, and Z be metric spaces and suppose $f: X \rightarrow Y$ and $g: Y \rightarrow Z$ are continuous then $gof: x \longrightarrow Z$ is continuous.

Proof: Let $u \subset z$ be open, then $(gof)^{-1}(u) = f^{-1}(g^{-1}(u))$

Proposition: Let $f: x \longrightarrow \mathbb{C}, g: X \longrightarrow \mathbb{C}$ and let $\alpha, \beta \in \mathbb{C}$. Suppose f, g are continuous. Then
(i) $\alpha f + \beta g$ is continuous
(ii) fg is continuous
(iii) f/g is continuous at every $a \in X$ such that $g(a) \neq 0$

Since constant functions are continuous, and $f: \mathbb{C} \rightarrow \mathbb{C}; z \longmapsto z$ is continuous, it follows that all polynomials are rational functions and continuous except where the denominator = 0.

A.4 Connectedness
Intuitively, a connected space is one which is "all one piece". We will make this notion precise in two different ways, and explore the relations between them.

Def.: A partition of a metric space (X, d) is a pair of nonempty subsets A and B and X such that

$$A \cap B = \phi, A \cup B = X.$$

(X, d) is connected if it admits no partition.

Proposition: A subset of $\mathbb{R}$ is connected *iff* it is an interval.

Proposition: The following are equivalent, in any metric space X.

(a) X is connected
(b) The only subsets of X which are both open and closed are ϕ and X.
(c) There is no continuous map from X onto the discrete 2-point space $\{0,1\}$

Proposition: If X is connected and $f: X \to Y$ is continuous, then $f(x)$ is connected.

Proof: Let A, B be a partition of $f(x)$, then $\{f^{-1}(A), f^{-1}(B)\}$ is a partition of X.

Corollary: If $f: X \to \mathbb{R}$ is continuous and X is connected, then $f(x)$ is an interval.

Corollary (the Intermediate value theorem): If $f: [a, b] \to \mathbb{R}$ is continuous and $f(a) \leq \xi \leq f(b)$ there is a point $C \in [a, b]$ such that $f(c) = \xi$.

Def: A path in a metric space X is a continuous map $\phi: [a, b] \to X$ for some interval $[a, b] \subset \mathbb{R}$.

By considering $\phi_1(t) = \phi(a(1 - t) + tb): [0,1] \to X$ i.e., reparameterizing the path we can simply set $[a, b] = [0,1]$

Note also that the path is not the same as the image set $\phi([a, b]) \subset X$. We can now state our second definition of "connected":

Def: A space X is path-connected if for every pair of points $x, y \in X$ there is a path $\phi([0, b]) \to X$ with $\phi(a) = x, \phi(b) = y$. ("A path is X from *x* to *y*").

Let's now explore the relationship between the notion of "connected" and path connected?

Proposition: If X is path-connected, it is connected.

Example Topologists sine curve:

Let G be the graph of $y = \sin\left(\frac{1}{x}\right)$ for $0 < x \leq \frac{1}{\pi}$
Let J be the interval joining (0, –1) and (0, 1) in $\mathbb{R}^2$
Let $X = G \cup J$.

The curve is a subset of the plane so it has a convenient metric. It is connected but not path connected. Each piece is connected, so can an open set be chosen to contain G and J? No, any open ball around J will be overlapping on G. So, this example shows that connected implies path-connected is false. To salvage something, we define:

Def: X is locally path connected at $\underline{a}(a \leq x)$ if for every open set U containing $\underline{a}$, there is an open set V such that $a \in V \subset U$ and V is path connected. X is locally path connected (LPC) if it is locally path connected at $\underline{a}$ for every $\underline{a}$. Locally path connected does not necessitate global path connectedness.

Proposition: If X is connected locally path-connected, then it is path-connected.

Proposition: If $A \subset X$ is connected and $A \subset B \subset \bar{A}$, then B is connected (In particular, $\bar{A}$ is connected).

Def: A subset C of X is called a component of X, if it is a maximal connected subset: i.e., if $C \subset C' \subset X$ and C' is connected then $C' = C$.

Example: $X = \left\{\frac{1}{n}, n = 1,2,3 \ldots\right\} \cup \{0\} \subset \mathbb{R}$.

Proposition: A component of X is closed.

Proof: If C were not closed, $C \underset{\neq}{\subset} \bar{C}$ which would still be connected, violated maximally connected notion.

Components are not necessarily open. Consider $\{0\}$ in the example earlier $\left(\frac{1}{n} \ldots \cap \{0\}\right)$.

A.5 Sequences and Completeness

Def: If $\{x_n\}$ is a sequence in a metric space (X, d) we say x_n converges to x if for every $\xi > 0$ there is $N \in \mathbb{N}$ such that $d(x_n, x) < \xi$ whenever $n > N$. (We write $x = \lim_{n \to \infty} x_n$, $or\ x_n \to x (n \to \infty)$).

Remark: The limit of a sequence, if it exists, is unique.

Def: $\{x_n\}$ is a Cauchy sequence in (X, d) if for every $\xi > 0$ there is a $N \in \mathbb{N}$ such that

$$d(x_n, x_m) < \xi \quad whenever \quad n, m > N.$$

Remark: In every metric space, every convergent sequence is a Cauchy sequence.

Def: (X, d) is complete if every Cauchy sequence converges.

Def: Let X be a metric space and $A \subset X$. A point $x \in X$ is a limit point of A if there is a sequence of distinct point $\{x_n\}$ in A with $x_n \to x(n \to \infty)$.

Remark: Intuitively, a limit point is either a point in A, or a point on the boundary of A. But not always, consider:

$$A = B_1(2) \cup \{0\} \subset \mathbb{R}^2$$

$0 \in A$, but 0 is not a limit point of A. Intuitive picture almost correct.

Proposition: Let $A \subset X$. The following are equivalent:
(a) A is closed in X ← by def. the compliment of A is open
(b) For every sequence $\{x_n\}$ in A with $x_n \to x \in X$, the $x \in A$.
(c) A contains all its limit points

Proposition: If $A \subset X$, then $A = A \cup \{x \in X : x \text{ is a limit point of A}\}$.

Recall: Metric space X is complete if every Cauchy sequence converges. We assume it is known that $(\mathbb{R}, 1 - 1)$ is complete.

Proposition: $\mathbb{R}^n$ is complete (In, particular $\mathbb{C}$ is complete).

Proposition: Let (X, d) be a complete metric space and $Y \subset X$. Then Y is complete if f Y is closed.

Consider an intersection of closed sets in $R^n \backslash \{0\}$.

$$\bigcap_{n=1}^{\infty} \overline{B_{\frac{1}{n}}(0)} = \phi \,,$$

let's use this in a theorem by Cantor.

Theorem (Cantor):(X, d) is complete iff for every sequence $\{F_n\}$ of non-empty closed sets with $F_1 \supset F_2 \supset F_3 \supset F_4 \ldots$ and $\text{diam}(F_n) \to 0$, if

$$\bigcap_{n=1}^{\infty} F_n \text{ consists of exacty one point.}$$

where $\text{diam}(A_n) = \sup\{d(x, y): x \in A, y \in A\}$.

Proof:
($\Longrightarrow$) Let $\{F_n\}$ be such a sequence of sets, for each n,choose $x_n \in F_n$. Then if $m, n > N$, both $x_m, x_n \in F_N$ and so $d(x_m, x_n) \leq \text{diam}(F_N)$, which $\to 0$ as $N \to \infty$. Therefore $\{x_n\}$ is a Cauchy sequence, and so, it converges to x, say, if it is complete. Note that $x_n \in F_N$, whenever $n \geq N$, so (F_N is closed) $x \in F_N$ for all N. $x \in \bigcap_{N=1}^{\infty} F_N \neq \phi$.

($\Longleftarrow$) Suppose the intersection property holds and let $\{x_n\}$ be a Cauchy sequence in X. Put $F_n = \{x_n, x_{n+1}, \ldots\}$. The F_n are closed, non-empty, decreasing and $\text{diam}(F_N) = \sup\{d(x_m, x_n): m, n \geq N\} \to 0$ as $N \to \infty$ (Cauchy sequence). By the condition $\exists x \in \bigcap_{N=1}^{\infty} F_N$. In particular, $x \in F_N$ for all N, and so $d(x_n, x) \subseteq \text{diam}(F_n) \to 0$ as $n \to \infty$. So $\{x_n\}$ converges to x, and X is complete.

Note that, in general, $\text{diam}(\overline{A}) = \text{diam}(A)$.

A.6 Compactness

Def: A subset K of a metric space X is compact if whenever $U = \{u_j : j \in J\}$ is a collection of open sets such that $K \subset \bigcup_{j \in J} u_j$, there is a finite sub-collection $u_{j_1} \ldots u_{j_n}$ such that $K \subset \bigcup_{k=J}^{n} u_{j_k}$. (Every open cover of K has a finite sub-cover.)

Examples

(1) $\mathbb{R}^n$ is not compact: Consider the cover by open balls $U_K = B_K(0) \leftarrow$ Open cover with no finite sub cover.

(2) Neither is $B_1(0)$: Consider $U_K = B_{1-\frac{1}{K}}(0)$, $K = \{2, \ldots \infty\}$. So $B_1(0)$ is also an open cover without a finite sub cover.

(3) $\overline{B_1(0)}$ is compact (in $\mathbb{R}^n$ at least)
(Heine-Borel Theorem)

(4) Any finite set is compact

Compact sets have some nice properties:
Def: A set A in a metric space is bounded if $A \subset B_R(x_0)$ for some $R > 0, x_0 \in x$.

Proposition: Let X be a metric space
(a) If $K \subset X$ is compact, then K is closed and bounded.
(b) If X is compact and $K \subset X$ is closed, then K is compact.

Def: K compact ($K \subset X$) if every open cover has a finite sub-cover.

Proposition:
(a) K compact $\Rightarrow$ K closed and bounded
(b) X compact, $K \subset X$ closed $\Rightarrow$ K compact

Def: A metric space is sequentially compact if every sequence in X has a convergent subsequence.

X is compact $\Leftrightarrow$ X is sequentially compact.

Def: A collection $\mathcal{F}$ of subsets of X has the finite intersection property if every finite sub-collection has non-empty intersection; if

$$F_1, F_2, \ldots F_n \in \mathcal{F} \; then \; \bigcap_{K=1}^{n} F_K \neq \phi.$$

Def: X is totally bounded if for every $\xi > 0$ there is a finite collection $x_1, \ldots, x_n$ of points of X such that

$$X = \bigcup_{K=1}^{n} B_\xi(x_K).$$

<u>Theorem</u> Let (X, d) be a metric space, the following are equivalent:
(a) X is compact
(b) Every collection f of closed subsets of X with the f.i.p has $\bigwedge_{n=1} F \neq \phi$
(c) X is complete and totally bounded
(d) Every infinite subset of X has a limit point in X
(e) X is sequentially compact.

<u>Lemma</u> ((Lebesgue covering Lemma):
Let X be sequentially compact, and let U be an open cover of X, then there exists $(\exists)$ an $\xi > 0$ such that whenever $x \in X$, there is some $u \in U$ such that $B_\xi(x) \subset U$. Assuming the Lemma, suppose X is not compact, and let $U = \{u_j \dots j \in J\}$ be an open cover with no finite sub-cover. Choose $\xi > 0$ such that for all $x \in X$ there is $u \in U$ with $B_\xi(x) \subset U$. $\{u_1\}$ does not cover X, so choose $x_2 \in X \backslash u_1$ and $u_2 \supset B_\xi(u_2)$. $\{u_1, u_2\}$ does not cover X, so choose $x_3 \in X \backslash (u_1 \cup u_2)$. Now, if $n \neq m$, say, $n > m$,
then $x_n \notin u_m$, so $x_n \notin B_\xi(x_m)$ and $d(x_n, x_m) \geq \xi$ therefore $\{x_n\}$ has no convergent subsequence, and X is not sequentially compact.

<u>Theorem</u>: X metric space, the following are equivalent:
(a) X is compact
(b) X is sequentially compact

<u>Lemma</u> (Lebesgue): Let X be sequentially compact, and let U be an open cover of X. Then $\exists \xi > 0$ such that for every $x \in X$ there is some $u \in U$ with $B_\xi(x) \subset U$.

<u>Corollary</u> (Heine – Borel Theorem): A subset of $\mathbb{R}^n$ is compact $\Leftrightarrow$ it is closed and bounded.

Def: If $A, B \subset X$, the distance from A to B is $d(A, B) = \inf\{d(x, y): x \in A, y \in B\}$

<u>Proposition</u>: Let $A, B \subset X$ and suppose A is compact, B is closed and $A \cap B = \phi$. Then $d(A, B) > 0$.

A.7 Uniform Continuity and Uniform Convergence

Def: $f: X \to Y$ is uniformly continuous if for every $\xi > 0$ there is a $\delta > 0$ such that

$$\rho\big(f(x), f(y)\big) < \xi \quad \text{whenever} \quad d(x,y) < \delta.$$

Proposition: Let X be compact, $f: X \to Y$ continuous, then f is uniformly continuous.

Def: Now let (Y, ρ) be a metric space and X a set. Suppose we are given a sequence of function $\{f_n\}$ $f_n: X \to Y$. We can define two notions of convergence.

f_n converges to f pointwise ($f_n \to f$) : f $f_n(x) \to f(x)$ for every $x \in X$, explicitely, $\forall x \in X, \forall \xi > 0 \exists N \in \mathbb{N}$ such that $\rho\big(f_n(x), f(x)\big) < \xi$ whenever $n > N$.

2nd notion: f_n converges to f uniformly ($f_n \to f$ or $f_n \rightrightarrows f$) if $\forall \xi > 0 \exists N \in \mathbb{N}$ such that $\rho\big(f_n(x), f(x)\big) < \xi$ whenever $n > N$ $\forall x \in X \leftarrow$ comes at end, important difference.

Example

$$f_n(x) = \begin{cases} 1 - nx, & 0 \leq x \leq \frac{1}{n} \\ 0 & , \frac{1}{n} < x \leq 1 \end{cases}$$

$$f_n \longrightarrow f(x) = \begin{cases} 1 & x = 0 \\ 0 & 0 < x \leq 1 \end{cases}$$

so have pointwise but not uniform convergence.

Proposition: Let $f_n: (X, d) \longrightarrow (Y, \rho)$ be continuous for each n, and suppose $f_n \longrightarrow f$, then f is continuous.

Proposition: Let $f_n: [a, b] \to (\mathbb{R})$ be integrable for each n, and $f_n \longrightarrow f$. Then f is integrable and

$$\int_a^b f_n(x)\, dx \longrightarrow \int_a^b f(x)\, dx.$$

Proof: $\left|\int_a^b f_n(x)\,dx - \int_a^b f(x)\,dx\right| \leq |f_n(x) - f(x)|dx < \xi \longrightarrow 0.$

A.8 Analytic Functions

Cauchy criterion for convergence: a numerical sequence of complex numbers $\{a_n\}_{n=1}^{\infty}$ converges iff $\forall \xi > 0 \quad \exists N \in \mathbb{N}$ such that $|a_m - a_n| < \xi$ whenever $m, n > N$. Recall that an infinite series $\sum_{K=1}^{\infty} a_K$ is said to converge if $\{s_n\}_{n=1}^{\infty}$ converges, where $s_n = \sum_{K=1}^{n} a_K$. Also, $\sum_{K=1}^{\infty} a_K$ converges absolutely if $\sum_{K=1}^{\infty} |a_K|$ converges.

Remark: If $\sum_{K=1}^{\infty} a_K$ converges, then $a_K \to 0$ as $K \to \infty$.

Proposition: Cauchy criterion for convergence of a series $\sum_{K=1}^{\infty} a_K$ converges iff $\forall \xi > 0 \quad \exists N \in \mathbb{N}$ such that

$$\sum_{K=1}^{\infty} |a_K| < \xi \quad \text{whenever} \quad n > m > N.$$

Corollary 1: If $\sum_{K=1}^{\infty} a_K$ converges absolutely, it converges.

Corollary 2 (comparison test): If $0 \leq a_K \leq b_K$ for $K = 1,2, \ldots$ and $\sum_{K=1}^{\infty} b_K$ converges, then $\sum_{K=1}^{\infty} a_K$ converges.

Proposition: (Cauchy criterion for uniform convergence)
Let X be a set, $f_n : X \to \mathbb{C}$ for $n = 1,2, \ldots$ then $\{f_n\}_{n=1}^{\infty}$ converges uniformly iff $\forall \xi > 0 \quad \exists N \in \mathbb{N}$ such that $|f_n(x) - f_m(x)| < \xi$ whenever $m, n > N$ and $x \in X$.

Proposition: (Cauchy criterion for uniform convergence of a series)
Let X be a set, $u_n : X \to \mathbb{C}$, then $\sum_{K=1}^{\infty} u_K(x)$ converges uniformly iff $\forall \xi > 0 \quad \exists N \in \mathbb{N}$ such that $|\sum_{K=1}^{n} u_K(x)| < \xi$ whenever $n > m > N$ and $x \in X$.

Corollary (Weierstrauss M-test): Let $u_n : X \to \mathbb{C}$ $(n = 1,2, \ldots)$ and suppose there are positive constants (sequence) $M_n (n = 1,2, \ldots)$ such that

(i) $|u_n(x)| \leq M_n \quad \forall x \in X, n = 1,2, \ldots$
(ii) $\sum_{n=1}^{\infty} M_n$ converges

Then $\sum_{n=1}^{\infty} u_n(x)$ converges uniformly and absolutely on X.

A.9 Power Series

Def: A power series centered at $z_0 (\in \mathbb{C})$ is a series of the form

$$\sum_{n=0}^{\infty} a_n (z - z_0)^n.$$

Example: (Geometric Series)

$$\sum_{n=0}^{\infty} z^n$$

where the partial sums are $S_n = 1 + z + z^2 + \cdots z^{n-1} = \frac{1-z^n}{1-z}$. So the geometric series converges (absolutely) if $|z| < 1$ $\{\sum_{n=0}^{\infty} |z|^n$ converges if $|z| < 1\}$, and diverges if $|z| \geq 1$.

Theorem: Let $\sum_{n=0}^{\infty} a_n (z - z_0)^n$ be a power series then $\exists R \in [0, \pm\infty]$ such that

(a) The series converges absolutely on $B_R(z_0)$ and uniformly on any sub-disk $B_r(z_0), 0 < r < R$.

(b) The series diverges at any $z \in \overline{B_R(z_0)}^C$ (i.e., if $|z - z_0| > R$)

Corollary: Suppose $R > 0$. The sum of the power series is a continuous function.

A.10 Examples

Many of the examples that follow are from midterm and final exercises given in Caltech AMa95a 1984.

Example A.1. Given $\varepsilon > 0$ find the value of N such that $n > N$ implies $|z^n| < \varepsilon$.

Solution

$|z^N| < \epsilon$ given $\epsilon > 0, |z^N| = |z|^N < \epsilon$. If $|z| < 1, |z|^n < |z|^N < \epsilon$ for $n > N$:

$N \ln|z| < \ln \epsilon$ choose $N > \frac{\ln \epsilon}{\ln|z|}$. If $|z| \geq 1, |z|^N \geq 1$ for any N $\rightarrow$ cannot be made < any arbitrary small C.

Example A.2. Prove that $1 + z + z^2 + \cdots = \frac{1}{1-z}$ if $|z| < 1$.

Solution

Let $S = \sum_{K=0}^{\infty} z^K$ A necessary conclution for S to converge is $|z|^K \rightarrow 0$ as $k \rightarrow \infty$ $\therefore$ $|z| < 1$

$$S_N = \sum_{k=0}^{N} z^N = 1 + z + z^2 + \cdots + z^N$$

So

$$\left|S_N - \frac{1}{1-z}\right| = \left|1 + z + z^2 + \cdots + z^N - \frac{1}{1-z}\right| = \frac{|z|^{N+1}}{|1-z|}$$

For given $\epsilon > 0$ let, $\frac{|z|^{N+1}}{|1-z|} < \epsilon$ *then* $|z|^{N+1} < \epsilon|1-z|$. So

$$(N+1)\ln|z| < \ln\epsilon\,|1-z| \;\rightarrow\; N+1 > \frac{\ln(\epsilon|1-z|)}{\ln|z|}$$

Choose $N > \frac{\ln(\epsilon|1-z|)}{\ln|z|}$ then for $n > N$, $\left|S_N - \frac{1}{1-z}\right| < \epsilon$

$$S_N \longrightarrow \frac{1}{1-z} \text{ as } N \longrightarrow \infty \text{ for } |z| < 1$$

So

$$\sum_{n=0}^{\infty} z^k = \frac{1}{1-z} \;, |z| < 1\,.$$

Example A.3. Use the result of **Example A.2.** to show that $a\sin\theta + a^2\sin 2\,\theta + a^3\sin 3\,\theta + \cdots = \frac{a\sin\theta}{1-2a\cos\theta+a^2}$.

Solution

$\sum_{n=0}^{\infty} a^N \sin N\theta = S$ consider $z = ae^{i\theta}$, *then* $a^N \sin N\theta = \operatorname{Im} z^N$, and

$$S = \operatorname{Im}\left(\sum_{n=0}^{\infty} a^N e^{iN\theta}\right) = \operatorname{Im}\frac{1}{1-ae^{i\theta}} \quad \text{for} \quad |a| < 1$$

Thus,

$$S = \operatorname{Im}\frac{1}{1-a\cos\theta - ai\sin\theta} = \operatorname{Im}\left(\frac{1-a\cos\theta + ai\sin\theta}{(1-a\cos\theta)^2 + (a\sin\theta)^2}\right)$$

$$= \frac{a\sin\theta}{1-2a\cos\theta + a^2}$$

Example A.4.

(a) Find $|(-i)^i|$

(b) Find the real and imaginary parts of z^z.

Solution

a) $|(-i)^i| = |e^{i\operatorname{Log}(-i)}| = |e^{i(\operatorname{Log}1 + i - \pi/2)}| = |e^{\pi/2}| = e^{\pi/2}$

b) $z^z = (x+iy)^{x+iy} = e^{(x+iy)\,\mathrm{Log}(x+iy)} = e^{(x+iy)(\mathrm{Log}\,r + i\,\mathrm{Arg}\,z)} = e^{x\,\mathrm{Log}\,r - y\,\mathrm{Arg}\,z} e^{i(y\,\mathrm{Log}\,r + x\,\mathrm{Arg}\,z)}$

or

$$z^z = r^x \cdot e^{-y\,\mathrm{Arg}\,z}[\cos(y\,\mathrm{Log}\,r + x\,\mathrm{Arg}\,z) + i\sin(y\,\mathrm{Log}\,r + x\,\mathrm{Arg}\,z)]$$

Thus,

$$Rez^z = \left(\sqrt{x^2+y^2}\right)^x e^{-y\,\mathrm{Arg}\,z} \cos\left(y\,\mathrm{Log}\sqrt{x^2+y^2} + x\,\mathrm{Arg}\,z\right)$$

$$Imz^z = \left(\sqrt{x^2+y^2}\right)^x e^{-y\,\mathrm{Arg}\,z} \sin\left(y\,\mathrm{Log}\sqrt{x^2+y^2} + x\,\mathrm{Arg}\,z\right)$$

Example A.5. $f(z) = z^{1/2} + z^{1/3}$. Construct a single-value definition of $f(z)$ such that $\mathrm{Arg}\, f(i) = 5\pi/24$.

Solution

$f(z) = z^{1/2} + z^{1/3}$ has branch points at $z = 0$. Make cut on negative real axis, so have:

$$-\pi < \theta \le \pi \quad f(z) = r^{1/2} e^{i(\theta/2 + k\pi)} + r^{1/3} e^{i\left(\theta/3 + \frac{2k\pi}{3}\right)}$$

$$f(z) = r^{1/2} e^{i\left(\frac{3\theta+6k\pi}{6}\right)} + r^{1/3} e^{i\left(\frac{2\theta+4k\pi}{6}\right)}$$

So, at $= i, r = 1\ \theta = {}^{\pi}/_{2}$:

$$f(i) = e^{i\left(\frac{3\theta+12k\pi}{12}\right)} + e^{i\left(\frac{2\theta+4k\pi}{6}\right)} = 2\cos\left(\frac{\pi}{24} - \frac{k\pi}{3}\right)\left[\cos\left(\frac{5\pi}{24} + \frac{5k\pi}{3}\right) + i\sin\left(\frac{5\pi}{24} + \frac{5k\pi}{3}\right)\right]$$

Thus

$$f(i) = 2\cos\left(\frac{\pi}{24} - \frac{k\pi}{12}\right) e^{i\left(\frac{5\pi}{24} + \frac{7k\pi}{12}\right)}$$

Since we want $\mathrm{Arg}\, f(i) = \frac{5\pi}{24} \to k = 0$, so:

$$f(i) = 2\cos\left(\frac{\pi}{24}\right) e^{i5\pi/24} \quad and \quad \boxed{\begin{array}{c} f(z) = r^{1/2} e^{i\theta/2} + r^{1/3} e^{i\theta/3} \\ -\pi < \theta \le \pi \end{array}}$$

Example A.6. Prove that all the values of $(1-i)^{\sqrt{2}i}$ lie on a straight line.

Solution

$$(1-i)^{\sqrt{2}i} = e^{\sqrt{2}i\log(1-i)} \quad \log(1-i) = \log\sqrt{2}\,\vec{\theta} i\frac{\pi}{4} + 2ki\pi$$

$$= e^{\sqrt{2}i\left(\frac{1}{2}\log 2 - i\pi/4 + 2ki\pi\right)}$$

Thus

$$(1-i)^{\sqrt{2}i} = e^{\sqrt{2}\pi/_4 - 2\sqrt{2}k\pi} \cdot e^{i\frac{\sqrt{2}}{2}\log 2} \rightarrow \text{Arg}(1-i)^{\sqrt{2}i} = \frac{\sqrt{2}}{2}\log 2$$
$$= constant.$$

where a constant Arg value means a straight line.

Example A.7. Explain why $f(z) = \bar{z}/z$ is not an analytic function of z.

Solution

$$f(z) = \frac{\bar{z}}{z} = \frac{x-iy}{x+iy} = \frac{x-iy}{x+iy}\left(\frac{x-iy}{x-iy}\right) = \underbrace{\left(\frac{x^2-y^2}{x^2+y^2}\right)}_{u} - i\underbrace{\left(\frac{2xy}{x^2+y^2}\right)}_{v}$$

To be analytic must satisfy Cauchy-Riemann relations:

$$\frac{\delta u}{\delta x} = \frac{\delta v}{\delta y},$$

but since

$$\frac{\delta u}{\delta x} = \frac{-4xy^2}{(x^2+y^2)^2} \quad and \quad \frac{\delta v}{\delta y} = \frac{-2x^3}{(x^2+y^2)^2}$$

it is clear the relations are not satisfied, thus not analytic.

Example A.8. $f(z)$ is analytic for $r > 0, -\pi < \theta < \pi$ and its real part is given by $u(r,\theta) = 1 + r^{-1/n}\cos(\theta/n)$, where n is a real integer. Find $f(z)$ given $f(1) = 2$.

Solution

Simply guess $f(z) = 1 + \left(re^{i\theta}\right)^{-1/_N}$ and verify the Cauchy-Riemann relations in polar coordinates $\left[\frac{\delta v}{\delta\theta} = r\frac{\delta u}{\delta r} \quad r\frac{\delta v}{\delta r} = -\frac{\delta u}{\delta\theta}\right]$.

Example A.9. $f(z)$ is analytic in a region that includes the unit circle $|z| = 1$. Show that for any z inside C

$$f(z) = \frac{1}{2\pi i}\oint_C f(\zeta)\left(\frac{1}{\zeta - z} - \frac{1}{\zeta - 1/\bar{z}}\right)d\zeta$$

where C is the unit circle, $\bar{z}$ = complex conjugate of z.

Solution

Have C is $|z| = 1$ and $f(z) = \frac{1}{2\pi i}\oint_C f(s)\left(\frac{1}{s-z} - \frac{1}{s-\frac{1}{\bar{z}}}\right) ds$. If

$\bar{z}$ inside $|z| = 1$ then

$\frac{1}{\bar{z}}$ is outside of $|z| = 1$. By Cauchy's formula

$$\frac{1}{2\pi i}\oint_C \frac{f(s)}{s-z} - \frac{1}{2\pi i}\oint_C \frac{f(\xi)}{s-\frac{1}{\bar{z}}} d\xi = f(z) - 0 = f(z)$$

Example A.10.

12. Find the image of the line

$$z = (1+i)t, \quad -\infty < t < \infty$$

under the mapping. $w = e^z$.

Solution

Starting with $z = (1+i)t$, $(-\infty < t < \infty)$, $\omega = e^z$ we have: $\omega = e^{(1+i)t} = e^t e^{it} = e^t[\cos t + i \sin t]$ which is a spiral.

A.11 Exercises

Exercises are largely from Caltech AMa95a Midterm and Final questions in 1985.

1. Describe the set of points $Re\ x + \text{Im}\ x \geq 1$.
2. For what complex number z does$\left|e^{-iz}\right| < 1$?
3. Sketch the mapping described by $w = az, \quad a = 2^{-1/x}(1-1)$
4. Where is the following function not continuous? Explain why.

$$f(z) = \frac{2}{z^2+1}$$

5. Does $f(x,y) = \frac{1}{2}\log(x^2 - y^2) + 1\tan^{-1}(y/x)$ satisfy the Cauchy-Riemann equation?
6. $w = u + iv$ and $W = U + iV$ are both analytic functions of z in a region. Show that if $w = \overline{W}$ in this region, then w and $\overline{W}$ are constant.

7. Is the integral

$$I = \int_0^i \frac{dz}{1-z^2}$$

independent of path? If so, state the conditions.

8. Calculate the integral

$$I = \int_0^i \frac{dz}{1 - z^2}$$

for a path along the imaginary axis.

9. Evaluate the integral

$$\oint_c \frac{\sin z}{z} dz$$

where c is the circle $|z| = 2$.

10. Evaluate the integral

$$\oint_c \frac{2z^2 + 3z - 1}{z + 1 + i} dz$$

where c is the circle $|z| = 2$.

11. Define a branch of $\log(z^2 + 1)$ that is analytic at $z = 0$ and takes the value $2\pi i$ there.

12. Show that

$$\int_0^z \left(\frac{1 - z^{n+1}}{1 - z}\right) dx = z + \frac{z^z}{2} + \cdots \frac{z^{n+1}}{n + 1}$$

13. Prove that

$$\lim_{n \to \infty} \int_0^{e^{i\theta}} \frac{z^{n+1}}{1 - z} dz = 0$$

Provided that $\theta \neq 2n\pi, n = 0, \pm 1, \pm 2, \ldots$ were the contour is the straight line from 0 to $e^{i\theta}$.

14. Use the results of the above to deduce a value for

$$\sum_{n=1}^{\infty} \frac{\cos n\,\theta}{n}, \qquad \theta \neq 2n\pi$$

15. Let $f(z)$ be analytic for $|z| \leq 1$ and $f(0) = 1$. Evaluate the integral

$$\frac{1}{2\pi i}\oint_C \left(2 + e^{i\alpha}\zeta + \frac{1}{\zeta}e^{-i\alpha}\right)\frac{f(\zeta)}{\zeta}\,d\zeta$$

to show that

$$\frac{2}{\pi\pi}\int_0^{2\pi} f(e^{i\theta})\cos^2\left(\frac{\alpha+\theta}{2}\right)d\theta = 2 + e^{-i\alpha}f'(0)$$

where C is the unit circle $|z| = 1, \alpha$ is a real constant.

16. $z = e^{-i\pi/4}$. Give all values of $|z^z|$ and $\arg(z^z)$.

17. Find the zeros of $\frac{e^z-1}{z}$. Are they simple zeros? Explain your reasoning.

18. Given $f(z) = \frac{e^z-1}{z}$, find the poles of $g(z) = \frac{f^1(z)}{f(z)}$ and evaluate the residues at these poles.

19. Prove that if $f(z) = \frac{e^z-1}{z}$, then $g(z) = f'(z)/f(z)$ is bounded on any closed contour C that does not intersect the poles of $g(z)$.

20. The contour C is the circle $|z| = 1$. Show that

$$\int_C \frac{e^z dz}{(z^2 + z - 3/4)^2} = 0.$$

21. Find constants b_n such that

$$f(x) = \frac{\sin z}{(z-2\pi)^2} = \sum_{-1}^{\infty} b_n\,(z-2\pi)^n$$

What is the significance of b_{-1}?

22. Define a single valued branch of $(1+z)^{1/z}$ which is analytic for $0 < |z| < 1$ and real when z is real and compute $f^1(0)$.

23. Given a is real with $0 < a < 1$, show that a single valued analytic branch of

$$f(z) = \log(1 - az)$$

Can be defined on a z-plane cut from $z = 1/a$ to ∞. Specify a range of $\arg(1 - az)$ that corresponds to $\arg z$ in the range $(-\pi, \pi)$.

24. Integrate $\frac{\log(1-az)}{z}$, $0 < a < 1$ around the contour $|z| = 1$ to show that

$$\frac{1}{\pi}\int_0^{\pi} \frac{\theta \sin\theta \, d\theta}{1 - 2a\cos\theta + a^2} = \frac{\log(1+a)}{a}.$$

25. Laplace's equation $\nabla^2\phi = 0$ is satisfied in a region $|z| \leq 1, \mathrm{Im}(z) \geq 0$ with

$$\phi(x,y) = 1 \text{ on } |z| = 1, y > 0$$
$$\phi(x,y) = 0 \text{ on } y = 0, |x| \leq 1$$

Find the function$\phi(x,y)$ by using the mapping

$$w = i\left(\frac{1-z}{1+z}\right).$$

26. $u(x,y)$ and $\phi(x,y)$ are two real functions of x and y that are harmonic in the same domain. Show that the product $u(x,y)\phi(x,y)$ is harmonic in the same domain if $\phi(x,y) = av(x,y)$, where $v(x,y)$ is the harmonic conjugate of $u(x,y)$ and a is a constant.

27. Solve parts (a) and (b):

(a) Show that the polynomial

$$z^6 + 4z^2 - 1$$

Has exactly two zeros in the disk $|z| < 1$.

(b) Prove that all the roots of the equation

$$z^6 - 5z^2 + 10 = 0$$

Lie in the annulus $1 < |z| < 2$.

28. Use contour integration to show that

$$\int_o^{\infty} \frac{t^{\zeta-1}dt}{1+t} = \frac{\pi}{\sin \pi\zeta}$$

Specify the limits on ζ (complex).

29. Suppose $f(z) = \frac{1}{(z^2+1)(z-2)}$

How many different series representations for $f(z)$ centered on the origin are there? Specify the regions of convergence for each series in a sketch. Do not work out the series in detail, just give the first term of each series.

30.

(a) Locate and classify the singular points in the upper half plane for the function

$$f(z) = \frac{1}{(1+z^2)\cosh\frac{\pi z}{2}}$$

(b) Compute the residues at these singular points and show that

$$2\pi i \sum \text{Residues} = 1 + \sum_{n=1}^{\infty} \frac{(-1)^{n+1}}{n(n+1)}$$

(c) State how you would use these results to evaluate

$$\int_{-\infty}^{\infty} \frac{dx}{(1+x^2)\cosh\frac{\pi x}{2}}$$

31. $f(z)$ is analytic except for poles at $b_1, b_2, \dots, b_k$, each with residues $r_1, r_2, \dots, r_k$. $|zf(z)| \to 0 \;\; as \;\; z \to \infty$. Show that

$$\sum_{n=-\infty}^{+\infty} f(n) = -\sum_{j=1}^{k} \pi r_j \cot \pi\, b_j$$

Note that no b_j is an integer. If necessary, use the residue theorem and an appropriate contour. Any previous experience you have had with the poles of $\cot \pi\, z$ may be useful also.

32.

(i) Show that the mapping $t = (i - \zeta)/(i + \zeta)$ maps the interior of the unit circle in the t-plane to the upper half of the ζ-plane.

(ii) Given that $\phi = 0$ on $|\arg t| < \pi/2$

And $\phi = 1$ on $\pi/2 < |\arg t| \le \pi$

Find the <u>conjugate harmonic</u> function to ϕ inside $|t| = 1$.

B. Differential Geometry

B.1 Abstract Indices (Penrose [52,53])

In this section abstract indices are introduced, where a vector object is denoted V^a, which is identical in meaning to $\vec{V}$ (where 'a' is not a concrete index but an abstract index). Contrast this with $\vec{V}$ being a vector in a three-dimensional Cartesian coordinate space, say, with components$\left(\vec{V}\right)_k$, where $k = 1,2, or\ 3$ corresponding to the components V_k in the x, y, and z directions. Here k is a concrete index (it can and must take on a set of values). The abstract index does not take on a set of values, in many applications, but it will take on the key property of contraction on like-indices (or Einstein summation convention) and tensor product construction when not like-indices (abstract operations inherited from the concrete index formulation). Although a notational convenience, it can be very powerful, as the application of the notation to some problems will demonstrate, and also when describing gauge field theories in Sec. X.

Definition: The real vector space V

A set **V** is a real vector space if it is an abelian group under addition (have inverse) and under scalar multiplication it is associative and distributive with multiplicative identity (but may not have inverse). Thus, we have for V^a and U^a, elements of **V**, with scalars $r, s \in \mathbb{R}$:

$$r(U^a + V^a) = rU^a + rV^a \ ,$$

$$(r + s)U^a = rU^a + sU^a, \qquad 0 + U^a = U^a$$

and

$$r(sU^a) = (rs)U^a \quad , \quad 1U^a = U^a$$

(If not specified, assume finite-dimensional.....)

(In what follows the dual basis is the roman w instead of the greek ω, to stay consistent with the use of other roman bases u, v, etc.)

Example B.1.

1.a. Let V be a finite-dimensional vector space. Show that a linear map $T: V \rightarrow V$ is determined by its action on a basis for V.

1.b. Let V be a Hilbert space with a countable basis, termed a "separable" Hilbert space. (No other Hilbert spaces are used in physics). A linear map $T: V \to V$ is bounded if $|T(v)| < K|v|$ for some fixed constant K. Show that any bounded linear map T is determined by its action on an orthonormal basis for V.

Solution

(1)(a) V is a finite-dimensional vector space. As such it is spanned by a basis of finitely many linearly independent vectors. In a given basis $\{e_i^a\}, i = 1 \dots n,$ where $n = \dim(V)$, we can express an arbitrary element of the vector space V as:

$$V^a = e_i^a V^i$$ (summation convention in effect)

A linear map $T: V \to V$ is defined by

$$T(ru^a + sw^a) = rT(u^a) + sT(w^a)$$

So, for the preceding V^a:

$$T(V^a) = T(e_i^a V^i) = V^i T(e_i^a)$$

For the case at hand $Domain(T) = V$ and $Range(T) = V$ (from $T: V \to V$) and the maps $T(e_i^a)$ will determine the mapping of an arbitrary element of V via the expansion above. So, the linear map T is determined by its action on a basis of V.

(1)(b) A Hilbert space is a linear space X together with a mapping $X \times X \to \mathbb{C}$, denoted $(x, y) \to \langle x, y\rangle$, for which:

$$\langle x, y\rangle = \overline{\langle y, x\rangle} \text{ (bar denotes complex conjugate),}$$
$$\langle \alpha x + \beta y, z\rangle = \alpha\langle x, z\rangle + \beta\langle y, z\rangle.$$

Furthermore the mapping $(x, y) \to \langle x, y\rangle$ must be strictly positive, i.e.,:

$$\langle x, x\rangle \geq 0 \;\; \forall x \in X \text{ and } \langle x, x\rangle = 0 \Rightarrow x = 0.$$

The mapping $(x, y) \to \langle x, y\rangle$ above induces a norm, thus a Hilbert space is a normed vector space. A norm induces a metric on X, in this sense a Hilbert space is a metric space. This allows us to succinctly state the final property of a Hilbert space – it is complete. Consider a basis for V, $\{e_i^a\}$, where i is the countable index of the basis for the separable Hilbert space being considered. We choose our basis to consist of linearly independent vectors (via Gram-Schmidt process) and thus have $\langle e_i^a, e_j^a\rangle = 0$ for $i \neq j$, where the norm directly induced by the mapping given above has been chosen. Again $T(V^a) = V^i T(e_i^a)$ and $\|T(V^a)\| = \|V^i T(e_i^a)\| =$

$\sum_{i=1}^{N}|V^i|^2 \langle T(e_i^a), T(e_i^a)\rangle = \sum_{i=1}^{N}|V^i|^2 \, ||T(e_i^a)||$. Now T is not simply a linear map as before, here it is bounded, so here we must check the well definedness of $T(V^a) = V^i T(e_i^a)$, i.e., is $V^i T(e_i^a)$ bounded? To recap:

$$||T(V^a)|| = \sum_{i=1}^{N}|V^i|^2 \, ||T(e_i^a)|| \, .$$

Since T is a bounded linear map: $||T(e_i^a)|| < K||e_i^a||$ for some fixed constant K. Thus, T is obviously determined by its action on an orthonormal basis for V as long as such an expansion remains bounded by $K||V||$. Since we are considering an orthonormal basis, $||e_i^a|| = 1$, thus:

$$\sum_{i=1}^{N}|V^i|^2 \, ||T(e_i^a)|| < K\sum_{i=1}^{N}|V^i|^2 \rightarrow \quad ||T(V^a)|| < K||V|| \, ,$$

where $K = \max K_i$.So, $\left\|V^i T(e_i^a)\right\| < K||V||$ if $||T(V)|| < K||V||$, thus, the statement $T(V^a) = V^i T(e_i^a)$ is well-defined when T is a bounded linear map.

B.2 Vector spaces, Tensors, p-forms

Definition: The dual vector space V* of V

Suppose V is an n-dimensional space. The set of all linear maps of **V** to ℝ is known as the dual of **V**, and is denoted **V*. V*** is also a vector space, and it is also dimension n.

.

The operative word in the description of the dual vector space in terms of a set of linear maps was that those maps be linear, in other words we have for maps σ and τ that they operate linearly on the vector space with element V^a (they preserve the underlying operations of vector addition and scalar multiplication). Thus, we have:

$$(\sigma + \tau)V = \sigma V + \tau V \quad and \quad \sigma(\tau V) = (\sigma\tau)V.$$

The vectors in **V*** are called co-vectors or 1-forms and are written with lower abstract indices, thus $\sigma \rightarrow \ \sigma_a$ and we have, shifting from map notation to abstract contraction notation:

$$\sigma(V^a) = \sigma_a V^a.$$

Note that when there were repeated indices in the concrete vector and 1-form products described previously (or within any tensor with upper and lower indices), this led to a map to the real numbers ℝ. The abstract indices are capturing this same meaning and operation since the map $\sigma(V^a)$ is to reals, thus contraction (here $\sigma_a V^a$) maps to reals. Similarly in reverse, a vector can be regarded as a linear map from V^* to R: $V^a: \sigma_a \rightarrow V^a \sigma_a$.

Example B.2. Show that a basis $\{e_i^a\}$ and its dual basis $\{w_b^i\}$ satisfy the completeness relation,

$$e_i^a w_b^i = \delta_b^a.$$

(Regard each side as a map from vector to vectors and check that the maps agree on each basis vector.)

Solution

If the map equality works on a basis and a dual basis it will work for any vector or covector by linearity, so let's check the basis etc.:

$$e_j^b(e_i^a w_b^i) = e_i^a(S_b^a) = e_j^a = e_i^a(\delta_j^i) \; iff \; S_b^a = \delta_b^a \,, where \; S_b^a = e_i^a w_b^i.$$

Similarly,

$$w_a^j(e_i^a w_b^i) = (w_a^j e_i^a) w_b^i = \delta_i^j w_b^i = w_b^j \Rightarrow S_b^a = \delta_b^a \text{ again.}$$

Thus,

$$e_i^a w_b^i = \delta_b^a.$$

Definition: Tensor

Using abstract indices it is now easy to define what a tensor is. A tensor is a multilinear map from m covectors and n vectors to $\mathbb{R}$ (e.g., the tensor has m upper indices and n lower):

$$T(\sigma_a, \dots \tau_b, u^c, \dots v^d) \rightarrow T^{a \dots b}{}_{c \dots d} \, \sigma_a \dots \tau_a \, u^c \dots v^d$$

(analogous to $\sigma(V^a) = \sigma_a V^a$).

Example B.3. Using the definitons above, show that the tensor transformation law is equivalent to the multlineararirty property when the tensor is regarded as a map. Show that if $\{f_i^a\}$ is a new basis, related to $\{e_i^a\}$, by:

$$f_i^a = a_i^m e_m^a \,,$$

then

$$T^{ij}{}_k = (a^{-1})_\ell^i (a^{-1})_m^j a_k^n T^{\ell m}{}_n$$

Solution

Denote $T^{ij}{}_k = T^{ab}{}_c w_a^i w_b^j e_k^c$; $f_i^a = a_i^m e_m^a$; $g_a^j = b_n^j w_a^n$, then

$$g_a^j f_i^a = \delta_i^j = a_i^m g_a^j e_m^a = a_i^m b_n^j (w_a^n e_m^a) = a_i^m b_m^j \,.$$

For this to be the identity, $b_m^i = (a_i^m)^{-1}$, or as a matrix, we have: $b_i^m = (a^{-1})_i^m$. Thus,

$$T^{ij}{}_{k} = T^{ab}{}_{c}[(a^{-1})^i_\ell \sigma^\ell_a][(a^{-1})^j_m \sigma^m_b][a^n_k u^c_n]$$
$$= \{T^{ab}{}_c \sigma^\ell_a \sigma^m_b u^c_n\}(a^{-1})^i_\ell (a^{-1})^j_m a^n_k$$

and we get:

$$T^{ij}{}_{k} = (a^{-1})^i_\ell (a^{-1})^j_m a^n_k T^{\ell m}{}_{n} \,.$$

Example B.4.
(a) Show that the action of a tensor on its arguments (a set of vectors or convectors), is given in terms of components along a basis by

$$T^{a\ldots b}_{c\ldots d}\sigma_a \cdots \tau_b u^c \cdots v^d = T^{i\ldots j}_{k\ldots \ell}\sigma_i \cdots \tau_j u^k \cdots v^\ell \,.$$

(b) Show that any tensor can be written as a sum of its basis tensors,

$$T^{a\ldots b}_{c\ldots d} = T^{i\ldots j}_{k\ldots \ell} e^a_i \cdots e^b_j w^k_c \cdots w^\ell_d \,.$$

and find the dimension of the vector space consisting of all tensors with m up and n down indices (where the space of vectors is d-dimensional).

Solution
(a)

$$T^{a\ldots b}_{c\ldots d}\sigma_a \cdots \tau_b u^c \cdots v^d = T^{a\ldots b}_{c\ldots d}(\sigma_i w^i_a)\cdots(\tau_j w^j_b)(u^k e^a_k)\cdots(v^\ell e^d_\ell)$$
$$= \underbrace{T^{a\ldots b}_{c\ldots d} w^i_a \cdots w^j_b e^a_k e^d_\ell}_{T^{i\ldots j}_{k\ldots \ell}} \sigma_i \cdots \tau_j u^k \cdots v^\ell$$

Thus,

$$T^{a\ldots b}_{c\ldots d}\sigma_a \cdots \tau_b u^c \cdots v^d = T^{i\ldots j}_{k\ldots \ell}\sigma_i \cdots \tau_j u^k \cdots v^\ell \,.$$

(b) Left as an exercise.

Definition: p-form
A p-form $\sigma_{a\ldots b}$ is a totally antisymmetric tensor with p covariant indices:

$$\sigma_{a\ldots b} = \sigma_{[a\ldots b]}$$

thus have $\sigma(u^a, \ldots, v^b) = \sigma([u^a, \ldots, v^b])$.

A contravariant vector at P is equivalent to the tangent to a curve through P
A contravariant vector at P in manifold M can be regarded as the tangent to a curve through P, and is sometimes defined as the equivalence class of all smooth curves that are tangent at P. Let $F = C^\infty(M)$ be the vector space of smooth scalar fields on M.

B.3 Vector and Tensor definitions by Map

Definition of a Vector at P given a Map at P

A vector V at P is a map V: f → ℝ that is linear under addition and multiplication by scalars and where that map satisfies Leibnitz rule. We can then define, in terms of a curve $c(\lambda)$ through $P = c(0)$, the linear map:

$$f \to V(f) = \frac{d}{d\lambda} f[c(\lambda)]|_{\lambda=0} \, ,$$

with existence of the a limit value at $\lambda = 0$ is due Leibnitz.

Note that a Lie bracket automatically yields a vector that satisfies the Leibnitz condition, whereas simple composition of maps $u[v(g)]$ in and of itself is a mapping for vector fields *u* and *v* that does not satisfy the Leibnitz condition. The Lie bracket, a vector arrived at from two other vectors. but that is guaranteed to satisfy Leibnitz, may prove useful later.

Definition of a Tensor at P given a Multilinear Map at P

Suppose we have manifolds M and N and $\Psi: M \to N$ is a smooth (multilinear) map. Let's consider the dual and its action mapping tensors on M to tensors on N: $\Psi_*: T(M) \to T(N)$. Let's start by considering f, a scalar on N:

$$\Psi_* V(f) = \frac{d}{d\lambda} f[\Psi \circ c(\lambda)]|_{\lambda=0} \, ,$$

where we are using $V(f) = \frac{d}{d\lambda} f[c(\lambda)]|_{\lambda=0}$ from the vector-map relation above. Consider the locally isomorphic mapping of M or N to $\mathbb{R}^m$ or $\mathbb{R}^n$. Suppose M has local coordinates x^i and N has local coordinates y^μ. Let $c^i(\lambda) = x^i(c(\lambda)) = x^i \circ c(\lambda)$, and note that

$$\Psi^\mu = (y \circ \Psi \circ x^{-1})^\mu = y^\mu \circ \Psi \circ (x)^{-1}$$

Refer to the sketch below for clarification.

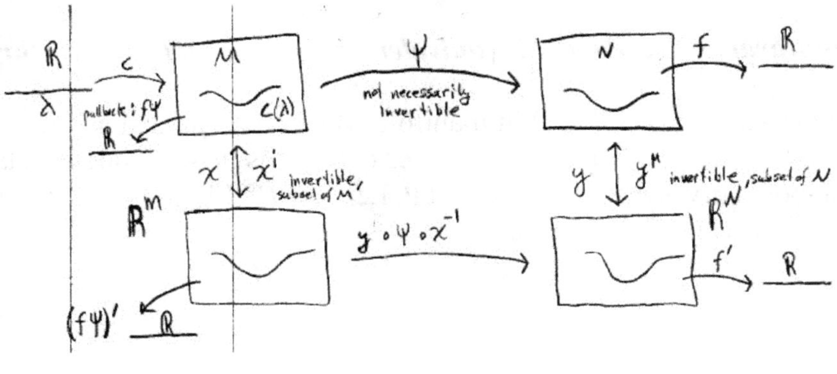

Thus,

$$(f\Psi)'[x(\alpha)] = (f\Psi)(\alpha) = f[\Psi(\alpha)] = f'[y\{\Psi(\alpha)\}]$$

or alternatively:

$$(f\Psi)'\left[x^i\right] = f'[y^\mu].$$

So can now write:

$$V(f) = \frac{d}{d\lambda} f[c(\lambda)]|_{\lambda=0}$$

and

$$V^i = \frac{d}{d\lambda} x^i[c(\lambda)]|_{\lambda=0}\,.$$

Note the use of $f = x^i$ with concrete indices (indicated by roman letter). Now let's consider the dual:

$$\Psi_* V(f) = \frac{d}{d\lambda} f[\Psi \circ \mathrm{c}(\lambda)]|_{\lambda=0}$$

and

$$(\Psi_* V)^\mu = \frac{d}{d\lambda} y^\mu[\Psi \circ \mathrm{c}(\lambda)]|_{\lambda=0}$$

Note the use of $f = y^\mu$ with abstract indices (indicated by greek letter).

$$(\Psi_* V)^\mu = \frac{d}{d\lambda}[y^\mu \circ \Psi \circ x^{-1} \circ x \circ \mathrm{c}(\lambda)]|_{\lambda=0} = \frac{d}{d\lambda}\left[\Psi^\mu\left(x \circ \mathrm{c}(\lambda)\right)\right]\Big|_{\lambda=0}\,,$$

thus,

$$(\Psi_* V)^\mu = \frac{\partial \Psi^\mu}{\partial c^i(\lambda)} \left.\frac{\partial c^i(\lambda)}{\partial \lambda}\right|_{\lambda=0} = \frac{\partial \Psi^\mu}{\partial x^i} \left.\frac{\partial x^i[c(\lambda)]}{\partial \lambda}\right|_{\lambda=0} = \frac{\partial \Psi^\mu}{\partial x^i} V^i\,.$$

Thus,

$$(\Psi_* V)^\mu = \frac{\partial \Psi^\mu}{\partial x^i} V^i\,.$$

From this we see that Ψ_* is a linear map from vectors at P to vectors at $\Psi(\mathrm{P})$ and as such may be regarded as an object with one covariant index at P and one contravariant index at $\Psi(\mathrm{P})$:

$$\Psi_* \simeq \Psi_a^\alpha \Rightarrow \Psi_* V^a = \Psi_a^\alpha V^a = U^\alpha$$

This is sometimes describes as Ψ "drags along" a vector V^a from M to N (this terminology becomes clearer in later sections when we discuss Lie Derivatives). In general Ψ will not be invertible and one cannot associate a vector on N with a unique vector on M. Co-vectors, however, are naturally "pulled back" by Ψ from M to N. The function $f: N \longrightarrow \mathbb{R}$ is pulled back to $(f\Psi): M \longrightarrow \mathbb{R}$, where $(f\Psi)[M] = f[\Psi(\mathrm{M})]$, and the co-vector $\nabla_\alpha f$ is pulled back to: $\Psi^*\nabla_\alpha f = \nabla_a(f\Psi)$. This means:

$$\nabla_i(f\Psi) = \frac{d}{dx^i}\left[(f\Psi)'(x(\alpha))\right] = \frac{d}{dx^i}[(f\Psi)(\alpha)] = \frac{d}{dx^i}\left[f(\Psi(\alpha))\right]$$

$$= \frac{d}{dx^i}\left[f'(y \circ \Psi(\alpha))\right] = \frac{d}{dx^i}\left[f'(y \circ \Psi \circ x^{-1} \circ x(\alpha))\right]$$

$$= \frac{d}{dx^i}\left[f'(y \circ \Psi \circ x^{-1}(x))\right] = \frac{\partial f'}{\partial y^\mu}\,\frac{\partial y^\mu\left(\Psi \circ x^{-1}(x(\alpha))\right)}{\partial x^i}$$

$$= \frac{\partial f'}{\partial y^\mu}\,\frac{\partial \Psi^\mu(x)}{\partial x^i}$$

Thus

$$\nabla_i(f\Psi) = \partial_\mu f\,\frac{\partial \Psi^\mu}{\partial x^i}.$$

So, co-vectors W_a at $\Psi(\mathrm{P})$ are pulled back by the same object Ψ_a^α to covectors at P.

$$(\Psi_* V)^\mu = \frac{\partial \Psi^\mu}{\partial x^i} V^i \quad \rightarrow \quad \Psi_* V^a = \Psi_a^\alpha V^a = U^\alpha$$

$$\nabla_i(f\Psi) = \frac{\partial \Psi^\mu}{\partial x^i}\partial_\mu f \quad \rightarrow \quad \Psi^* W_a = \Psi_a^\alpha W_\alpha = \sigma_a$$

Generalizing -- contravariant tensors on M are dragged to N:

$$\Psi_* T^{\alpha\ldots\beta} = \Psi_a^\alpha \ldots \Psi_b^\beta T^{a\ldots b}$$

and covariant tensors are pulled back to covariant tensors at P:

$$\Psi^* T_{a\ldots b} = \Psi_a^\alpha \ldots \Psi_b^\beta T_{\alpha\ldots\beta}\,.$$

A smooth isomorphism Ψ from a manifold M to itself is called a diffeomorphism.

$$(\Psi T)^{ab}{}_c(P) = \Psi_e^a \Psi_f^b (\Psi^{-1})_c^g T^{ef}{}_g\left(\Psi^{-1}(P)\right)$$

Thus, given a tensor field T on M, Ψ produces a new tensor field (ΨT).

Smooth vector field ⇔ integral curves

Consider V^a, a vector field on M, and $c(\lambda)$, a curve whose tangent vector $U^a(\lambda)$ coincide at each point of c with the value of V^a at that point:

$$U^a(\lambda) = V^a(c(\lambda)) \rightarrow \frac{d}{d\lambda}c^i(\lambda) = V^i(c(\lambda))$$

Any system of differential equations can be cast in this form, whence, by the existence and uniqueness theorems for ODE's of order N we have the following theorem.

Theorem: Let V^a be a vector field on M, then there is a maximal integral curve $c(\lambda)$ (cannot be extended to a longer integral curve) passing through each point of M and $c(\lambda)$ is unique. Thus, any vector field generates a *1*-parameter group of diffeomorphisms. The orbit of a point M under the action of the group is an integral curve of the vector field.

B.4 Lie derivatives

Suppose U^a, V^a are two vector fields. A family of diffeomorphisms Ψ_λ are generated by U^a and drag V^a:

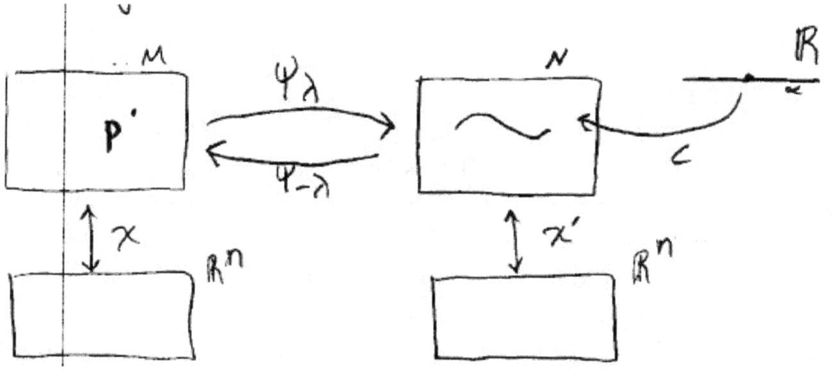

Note that $\Psi_{-\lambda} \bullet c(0) = P$:

$$(\Psi_{-\lambda} V^i)(P) = \frac{d}{d\alpha} x^i[\Psi_{-\lambda} \circ c(\alpha)]|_{\alpha=0}$$

$$= \frac{d}{d\alpha}\left[x^i \circ \Psi_{-\lambda} \circ (x')^{-1} \circ (x') \circ c(\alpha)\right]\Big|_{\alpha=0}$$

$$(\Psi_{-\lambda} V^i)(P) = \frac{d}{d\alpha}\left[\Psi^i_{-\lambda}\left(x^i \circ c(\alpha)\right)\right]\Big|_{\alpha=0} = \frac{\partial \Psi^i_{-\lambda}}{\partial (x^j)'} \frac{\partial (x^j)'}{\partial \alpha}\Bigg|_{\alpha=0}$$

$$= \left(\frac{\partial}{\partial (x^j)'} \Psi^i_{-\lambda}\right)\Bigg|_{\alpha=0} V^{j'}(\Psi_\lambda(P))$$

$$\frac{d}{d\lambda}(\Psi_{-\lambda} V^i)(P) = \frac{\partial}{\partial (x^j)'}\left(\frac{\partial \Psi^i_{-\lambda}}{\partial \lambda}\right) V^{j'} + \frac{\partial \Psi^i_{-\lambda}}{\partial (x^j)'} \frac{\partial V^{j'}(\Psi_\lambda(P))}{\partial \lambda}$$

Substituting

$$\frac{d}{d\lambda}\Psi^i_{-\lambda}(P) = \frac{d}{d\lambda}x^i(\Psi_\lambda[c(0)]) = -U^i(P)$$

and

$$\frac{\partial x^i}{\partial x^j}\frac{\partial x^j(\Psi_\lambda(P))}{\partial\alpha\partial\lambda} = \frac{\partial V^i(P)}{\partial\alpha} = \frac{\partial V^i}{\partial x^j}\frac{\partial x^j}{\partial\alpha} = U^j\frac{\partial}{\partial x^j}V^i$$

We get:

$$\frac{d}{d\lambda}(\Psi_{-\lambda}V^i)(P) = -\frac{\partial U^i}{\partial x^j}V^j + U^j\frac{\partial V^i}{\partial x^j} = [U,V]^i .$$

Thus, diffeomorphisms are symmetries of the metric iff (if and only if) V^α Lie-derives the metric:

$$\Psi_\lambda g_{\alpha\beta} = 0 \to \mathcal{L}_V g_{\alpha\beta} = 0 \quad \to \quad V^\alpha\ is\ a\ Killing\ vector\ field.$$

If V^α is a Killing Vector one can always choose a chart for which a given vector field has components $V^\mu = \delta^\mu_0$. A linear map S from tensor fields $T^{a\ldots b}_{c\ldots d}$ to $\mathbb{R}$ is itself a tensor field iff S is linear under multiplication of T by a scalar field: $S(fT) = fS(T)$. A linear map from tensor fields to tensor fields is itself a tensor field iff it is linear under multiplication by scalar fields. Consider the tensor field T^a_{bc}, it can be regarded as a multilinear map taking one covector field and two vector fields to $\mathbb{R}$; or as a linear map taking one vector field to a tensor field with two covariant indices; or as a linear map taking tensor fields with two up indices to vector fields; or .....

Example B.5. Show that in a chart, the components of the Lie derivative have the form:

$$\pounds_u T^{i\ldots j}_{k\ldots\ell} = u^m\partial_m T^{i\ldots j}_{k\ldots\ell} + T^{i\ldots j}_{m\ldots\ell}\partial_k u^m + \cdots T^{i\ldots j}_{k\ldots m}\partial_\ell u^m - T^{m\ldots j}_{k\ldots\ell}\partial_m u^i - \cdots - T^{i\ldots m}_{k\ldots\ell}\partial_m u^j$$

Solution

Definition of Lie derivative: $\pounds_u T^{a\ldots b}_{c\ldots d} = \frac{d}{d\lambda}\{(\Psi_{-\lambda}T)^{a\ldots b}_{c\ldots d}\}_{\lambda=0}$

Notice $\pounds_v(f) = \frac{d}{d\lambda}\{\Psi_{-\lambda}f\}|_{\lambda=0}$. If we let the integral curves of V^a be denoted by $c(\lambda)$:

$$\pounds_v(f) = \frac{d}{d\lambda}f(c(\lambda))\big|_{\lambda=0} = V(f)$$

Also, by choosing a chart:

$$\pounds_u V^a = \frac{d}{d\lambda}\{(\Psi_{-\lambda})_* V^a\}_{\lambda=0}$$

Since

$$[(\Psi_{-\lambda})_* V']^i(p) = \frac{d}{d\alpha} x^i[\Psi_{-\lambda} \circ c(\alpha)]_{\alpha=0}$$
$$= \frac{d}{d\alpha}\left[x^i \circ \Psi_{-\lambda} \circ (x')^{-1} \circ (x') \circ c(\alpha)\right]\Big|_{\alpha=0}$$

Thus,

$$[(\Psi_{-\lambda})_* V']^i(p) = \frac{d}{d\alpha}\left[\Psi^i_{-\lambda} x' \circ c(\alpha)\right] =$$
$$\left.\frac{\partial \Psi^i_{-\lambda}(x^j)'}{\partial (x^j)'} \frac{\partial (x^j)'}{\partial \alpha}\right|_{\alpha=0}$$

So, $(\pounds_u V)^i = \frac{d}{d\lambda}\left\{\frac{\partial \Psi^i_{-\lambda}(x')}{\partial (x^j)'} \frac{\partial (x^j)'}{\partial \alpha}\right\}_{\substack{\alpha=0 \\ \lambda=0}}$

Now $\left.\frac{d\Psi^i_{-\lambda}(P)}{d\lambda}\right|_{\substack{\lambda=0 \\ \alpha=0}} = \frac{d}{d\lambda}\, x^i\,(\Psi_{-\lambda}[c(o)])|_{\substack{\lambda=0 \\ \alpha=0}} = -u^i(P)$

And $\frac{d}{d\lambda}\left(\frac{\partial\left[(x^j)'(\Psi_\lambda(P))\right]}{\partial \alpha}\right)_{\substack{\lambda=0 \\ \alpha=0}} = \left.\frac{\partial (v^j)}{\partial \alpha}\right|_{\alpha=0} = \left.\frac{\partial (v^j)'}{\partial x^i} \frac{\partial x^i}{\partial \alpha}\right|_{\alpha=0} = u^i \frac{\partial (v^j)'}{\partial x^i}$

Thus

$$(\pounds_u v)^i = \left[\frac{\partial}{\partial (x^j)'} - \left(u^i(P)\right)\right]\left(\frac{\partial (x^j)'}{\partial \alpha}\right) + \left(\frac{\partial x^i}{\partial (x^j)'}\right)\left(u^k \frac{\partial (v^j)'}{\partial x^k}\right)$$
$$= u^j \frac{\partial v^i}{\partial x^j} - v^j \frac{\partial u^i}{\partial x^j} = [u, v]^i$$

And

$$\pounds_u v = [u, v]^a\,.$$

By choosing a convenient coord. system this can be done much more simply. Using $\pounds_v(f) = v(f)$ and $\pounds_u v = [u, v]^a$ and Leibnitz' rule one can generalize to an arbitrary tensor. So, we have:

$$(\pounds_u v)^i = u^j \partial_j v^i - v^j \partial_j u^i \quad and \qquad (\pounds_u f) = u^j \partial_j f$$

Need $\pounds_u w$ where w is a l-form. So consider Leibnitz rule:

$$\pounds_u(w_a v^a) = w_a \pounds_u v^a + v^a \pounds_u w_a$$

In a chart:

$$[\pounds_u(w_a v^a)] = w_i u^j \partial_j v^i - w_i v^j \partial_j u^i + v^i(\pounds_u w_a)_i$$

Also

$$[£_u(w_i v^i)] = u^j \partial_j (w_i v^i) = w_i u^j \partial_j v^i + v^i u^j \partial_j w_i$$

So

$$v^i (£_u w_a)_i = w_i v^j \partial_j u^i + v^i u^j \partial_j w_i = v^i (u^j \partial_j w_i + w_j \partial_i u^j)$$

Thus,

$$(£_u w)_i = u^j \partial_j w_i + w_j \partial_i u^j \ .$$

Now a general tensor can be expressed as a sum of tensor products of vectors and l-forms, so applying the Lie derivative and using the Leibnitz property one gets:

$$\left(£_u T^{a...b}_{c...d}\right)^{i'...j'}_{k'...\ell'} = £_u \{T^{i...j}_{k...\ell} e^a_i \cdots e^b_j w^k_c \cdots w^\ell_d\}^{i'...j'}_{k'...\ell'} = \left(£_u T^{i'...j'}_{k'...\ell'}\right) +$$
$$T^{i'...j'}_{k'...\ell'} (£_u e^a_i)^{i'} + \cdots T^{i'...j'}_{k'...\ell} \left(£_u w^\ell_d\right)_{\ell'}$$
$$= u^m \partial_m T^{i'...j'}_{k'...\ell'} + T^{i...j}_{k...\ell} \left\{u^m \partial_m e^{i'}_{(i)} - e^m_{(j)} \partial_m u^i\right\} + \cdots +$$
$$T^{i...j}_{k...\ell} \left\{u^m \partial_m w^{(\ell)}_{\ell'} + w^{(\ell)}_m \partial_{\ell'} u^m\right\}$$
$$= u^m \partial_m T^{i'...j'}_{k'...\ell'} - T^{m...j'}_{k'...\ell'} \partial_m u^{i'} \cdots + T^{i'...j'}_{k'...m} \partial_{\ell'} u^m$$

Dropping primes we get the result:

$$£_u T^{i...j}_{k...\ell} = u^m \partial_m T^{i...j}_{k...\ell} + T^{i...j}_{m...\ell} \partial_k u^m + \cdots T^{i...j}_{k...m} \partial_\ell u^m - T^{m...j}_{k...\ell} \partial_m u^i - \cdots$$
$$- T^{i...m}_{k...\ell} \partial_m u^j \ .$$

Example B.6. Show that if ∇_a is the covariant derivative associated with any metric, the Lie derivative can be written in the form:

$$£_u T^{a...b}_{c...d} = u^e \nabla_e T^{a...b}_{c...d} + T^{a...b}_{e...d} \nabla_c u^c + T^{a...b}_{c...e} \nabla_d u^e - T^{e...b}_{c...d} \nabla_e u^a$$
$$- T^{a...e}_{c...d} \nabla_e u^b$$

Solution

$$£_u T^{a...b}_{c...d} = \left(£_u T^{i...j}_{k...\ell}\right) e^a_i \cdots e^b_j \, w^k_c \cdots w^\ell_d$$
$$= u^m \partial_m \left(T^{a...b}_{c...d}\right) + T^{a...b}_{m...d} (w^k_c \partial_k) u^m + \cdots - T^{a...m}_{c...d} \left(e^b_j \partial_m u^j\right)$$
$$= u^e \nabla_e T^{a...b}_{c...d} + T^{a...b}_{e...d} \nabla_c u^c + T^{a...b}_{c...e} \nabla_d u^e - T^{e...b}_{c...d} \nabla_e u^a -$$
$$T^{a...e}_{c...d} \nabla_e u^b$$

This can be done since ∇_a, the covariant derivative associated with any metric reduces to $\partial_i = e^a_i \nabla_a$ when mapped onto a given chart (def. of cov. deriv. on manifold), locally this mapping "e^a_i" is diffeomorphic so $\nabla_a = (e^a_i)^{-1} \partial_i = w^i_a \partial_i$. Since $e^e_m w^m_e = 1$

Example B.7. Let $T^{\alpha\beta}$ be a symmetric tensor (such as the energy-momentum tensor). Show that $T^{\alpha\beta}$ is divergence-free ($\nabla_\beta T^{\alpha\beta} = 0$) then the current $J^\alpha = T^{\alpha\beta}V_\beta$ associated with the Killing Vector V_β is conserved ($\nabla_\alpha J^\alpha = 0$).

Solution

A vector V^β is a Killing Vector (K. V.) $if\ \nabla_\alpha V_\beta = -\nabla_\beta V_\alpha$. We have $\nabla_\alpha J^\alpha = (\nabla_\alpha T^{\alpha\beta})V_\beta + T^{\alpha\beta}\nabla_\alpha V_\beta = \frac{1}{2}(T^{\alpha\beta} + T^{\beta\alpha})\nabla_\alpha V_\beta = \frac{1}{2}(T^{\alpha\beta}\nabla_\alpha V_\beta + T^{\alpha\beta}\nabla_\beta V_\alpha)$

$\nabla_\alpha J^\alpha = \frac{1}{2}\left(T^{\alpha\beta}\nabla_\alpha V_\beta + T^{\alpha\beta}(-\nabla_\alpha V_\beta)\right) = 0$

Thus, $\nabla_\alpha J^\alpha = 0$, and the current is conserved.

Example B.8. What conserved quantities correspond to the integral over a space-like plane in Minkowski space,

$$Q = \int \mathrm{n}_\alpha J^\alpha dV,$$

where J^α is the current associated with translations, rotations, boosts.

Partial (nonmathematical) Solution:

4 translations: timelike K.V. field yields conservation of energy while spacelike K.V. field yields conservation of momentum.
3 rotations: K.V. field yields conservation of angular momentum.
3 boosts → K.V. yields Lorentz transformation along given axis.

Example B.9. Show that the expression for the exterior derivative:

$$(d\sigma)_{ab\ldots c} = (p+1)\nabla_{[a}\sigma_{b\ldots c]}$$

is independent of connection. Repeat for Lie derivative:

$$\pounds_u T^{a\ldots b}_{c\ldots d} = u^e\nabla_e T^{a\ldots b}_{c\ldots d} + T^{a\ldots b}_{e\ldots d}\nabla_c u^e + \cdots + T^{a\ldots b}_{c\ldots e}\nabla_d u^e - T^{e\ldots b}_{c\ldots d}\nabla_e u^a - \cdots - T^{a\ldots e}_{c\ldots d}\nabla_e u^a .$$

Solution

We have: $(d\sigma)_{ab\ldots c} = (p+1)\nabla_{[a}\sigma_{b\ldots c]}$ where σ is a p-form. This is independent of connection, i.e., choice of covariant derivative, if

$$\nabla_{[a}\sigma_{b\ldots c]} = \widetilde{\nabla}_{[a}\sigma_{b\ldots c]}$$

Also note:

$$\nabla_{[a}\sigma_{b\ldots c]} = 0 \quad if \quad C^f_{[ab}\sigma_{f\ldots c]} = 0 .$$

Since C^f_{ab} is symmetric on its covariant indices antisymmetrization over them will give zero. Thus $\nabla_{[a}\sigma_{b...c]} = \tilde{\nabla}_{[a}\sigma_{b...c]}$, thus the exterior derivative is independent of connection.

Let's now examine

$$\pounds_u T^{a...b}_{c...d} = u^e\nabla_e T^{a...b}_{c...d} + T^{a...b}_{e...d}\nabla_c u^e + \cdots + T^{a...b}_{c...e}\nabla_d u^e - T^{e...b}_{c...d}\nabla_e u^a - \cdots - T^{a...e}_{c...d}\nabla_e u^a$$

Consider the difference of $\pounds_u T^{a...b}_{c...d}$ defined above by covariant derivative ∇: $\pounds_u^{\nabla}$ and that using $\tilde{\nabla}$: $\pounds_u^{\tilde{\nabla}}$, then

$$\begin{aligned}
&(\pounds_u^{\nabla} - \pounds_u^{\tilde{\nabla}})T^{a...b}_{c...d} \\
&= u^e(\nabla_e - \tilde{\nabla}_e)T^{a...b}_{c...d} + T^{a...b}_{e...d}(\nabla_c - \tilde{\nabla}_c)u^e + \cdots + T^{a...b}_{c...e}(\nabla_d - \tilde{\nabla}_d)u^e \\
&\qquad -T^{e...b}_{c...d}(\nabla_e - \tilde{\nabla}_e)u^a - \cdots - T^{a...e}_{c...d}(\nabla_e - \tilde{\nabla}_e)u^b \\
&= u^e\{C^a_{fe}T^{f...b}_{c...d} + \cdots + C^b_{fe}T^{a...f}_{c...d} - C^f_{ce}T^{a...b}_{c...d} - \cdots - C^f_{de}T^{a...b}_{c...f}\} \\
&+T^{a...b}_{e...d}C^e_{cg}u^g + \cdots + T^{a...b}_{c...e}C^e_{dg}u^g - T^{e...b}_{c...d}C^a_{eg}u^g - \cdots - T^{a...e}_{c...d}C^b_{eg}u^g \\
&= u^e\{C^a_{fe}T^{f...b}_{c...d} + \cdots + C^b_{fe}T^{a...f}_{c...d} - C^f_{ce}T^{a...b}_{f...d} - \cdots - C^f_{de}T^{a...b}_{c...f}\} \\
&\qquad +u^e\{-C^a_{fe}T^{f...b}_{c...d} - \cdots - C^b_{fe}T^{a...f}_{c...d} + C^f_{ce}T^{a...b}_{f...d} + \cdots + C^f_{de}T^{a...b}_{c...f}\} \\
&= 0
\end{aligned}$$

So, the Lie derivative is also independent of connection.

Example B.10. Prove the following key identities relating Lie derivatives and exterior derivatives. For any p-form $w_{a...c}$ and vector u_a,

$$d\pounds_u w = \pounds_u dw \quad and \quad \pounds_u w = u \cdot dw + d(u \cdot w)\,.$$

Solution

Start with:

$$\pounds_u w_{a...c} = \frac{d}{d\lambda}(\Psi^*_{-\lambda}w_{a...c})_{\lambda=0}$$

We then have

$$d\pounds_u w_{a...c} = \frac{d}{d\lambda}(d\Psi^*_{-\lambda}w_{a...c})_{\lambda=0} = \frac{d}{d\lambda}\left(\Psi^*_{-\lambda}(dw)\right)_{\lambda=0} = \pounds_u dw$$

Thus,

$$d\pounds_u w = \pounds_u dw\,.$$

This step uses the relation $d(\Psi^*_{-\lambda}w) = \Psi^*_{-\lambda}(dw)$ which I will now prove using

$$\Psi^*(\mathrm{d}f) = \mathrm{d}(f\Psi).$$

Going to a coordinate frame where

$$w_{a\ldots c} = w_{i\ldots k} dx_a^i \wedge \cdots \wedge dx_c^k$$

We have

$$dw_{\alpha a\ldots c} = d(w_{i\ldots k})_\alpha \wedge dx_a^i \wedge \cdots \wedge dx_c^k = \left(\frac{\partial w_{i\ldots k}}{\partial x^n}\right) dx_\alpha^n \wedge dx_a^i \cdots \wedge dx_c^k$$

Now,

$\Psi^*(dw_{\alpha a\ldots c}) = \Psi^*\left\{\frac{\partial w_{i\ldots k}}{\partial x^n} dx_\alpha^n \wedge dx_a^i \cdots \wedge dx_c^k\right\} = \left(\frac{\Psi^* \partial w_{i\ldots k}}{\partial x^n}\right)\{(\Psi^* dx_\alpha^n) \wedge (\Psi^* dx_a^i) \wedge \cdots \wedge (\Psi^* dx_c^k)\}$

$$= \frac{w_{i\ldots k}(\Psi(P))}{\partial \Psi(x^n)} d\Psi(x^n)_\alpha \wedge d\Psi(x^i)_{\mathrm{a}} \wedge \cdots \wedge d\Psi(x^k)_{\mathrm{c}}$$

Let $y - \Psi(x)$

$$\Psi^*(dw_{\alpha a\ldots c}) = \frac{\partial w_{i\ldots k}(\Psi(P))}{\partial y^n} dy_\alpha^n \wedge dy_a^i \wedge \cdots \wedge dy_c^k$$
$$= d(\Psi^* w_{a\ldots c})$$

The proof is independent of the coordinates used in the definition due to the properties of the exterior derivative on form fields:

$w_{a\ldots c} = w_{i\ldots k} dx_a^i \wedge \cdots \wedge dx_c^k = w_{i'\ldots k'} dx_a^{i'} \wedge \cdots \wedge dx_c^{k'}$

$dw = dw_{i'\ldots k'} \wedge dx_a^{i'} \wedge \cdots \wedge dx_c^{k'} \left(\frac{\partial x^i}{\partial x^{i'}} \cdots \frac{\partial x^k}{\partial x^{k'}} A_{i\ldots k}\right) \wedge dx_a^{i'} \cdots \wedge dx_c^k = dw_{i\ldots k} \wedge dx_a^i \wedge \cdots \wedge dx_c^k$

Consider $\pounds_u w = \frac{d}{d\lambda}(\Psi_{-\lambda}^* w_{a\ldots c})_{\lambda=0}$ and choosing a chart:

$$\frac{d}{d\lambda}(\Psi_{-\lambda}^* w_{a\ldots c})(P) := \frac{d}{d\lambda}\left(\Psi_{-\lambda}^*[w_{i\ldots k}(P) dx_a^i \wedge \cdots \wedge dx_c^k]\right)$$
$$= \frac{d}{d\lambda}\left(w_{i\ldots k}(\Psi_{-\lambda}(P)) d(\Psi_{-\lambda} x^i)_a \wedge \cdots d(\Psi_{-\lambda} x^k)_c\right)$$
$$= \frac{dw_{i\ldots k}(\Psi_{-\lambda}(P))}{d\lambda} \wedge d(\Psi_{-\lambda} x^i)_a \wedge \cdots d(\Psi_{-\lambda} x^k)_c +$$
$$+ w_{i\ldots k}(\Psi_{-\lambda}(P)) d\left(\frac{d}{d\lambda}\Psi_{-\lambda} x^i\right)_a \wedge \cdots \wedge (\Psi_{-\lambda} x^k)_c + \cdots$$
$$+ w_{i\ldots k}(\Psi_{-\lambda}(P)) d(\Psi_{-\lambda} x^i)_a \wedge \cdots \wedge \left(\frac{d}{d\lambda}\Psi_{-\lambda} x^k\right)_c$$

Also note: $\left.\frac{d(w_{i\ldots k})}{d\lambda}\right|_{\lambda=0} = \left.\frac{d(w_{i\ldots k})}{d\Psi_{-\lambda}(x)} \frac{d\Psi_{-\lambda}(x)}{d\lambda}\right|_{\lambda=0} = \left.\frac{(dw)m_{i\ldots k}}{d\Psi_{-\lambda}(x)} \frac{d\Psi_{-\lambda}(x)}{d\lambda}\right|_{\lambda=0}$
$= (p+1) u^m \nabla_{[m} w_{i\ldots k]} = u^m \nabla_m w_{[i\ldots k]} + u^m \nabla_i w_{[m\ldots k]}$

So, in a chart:

$$\frac{d}{d\lambda}(\Psi^*_{-\lambda}w_{a\ldots c})(P)\bigg|_{\lambda=0}$$

$$= u^m\nabla_m w_{[i\ldots k]}dx^i_a \wedge \cdots \wedge dx^k_c + u^m\nabla_i w_{[m\ldots k]}dx^i_a \wedge \cdots \wedge dx^k_c$$

$$+w_{i\ldots k}\{d(u^m\delta^i_m) \wedge \cdots \wedge dx^k_t + \cdots + dx^i \wedge d(u^m\delta^k_m)\}$$

$$= u \cdot dw + \left(u^m\nabla_\ell w_{mj\ldots k}dx^\ell_a + w_{mj\ldots k}du^m\right)dx^i_b \cdots dx^k_c$$

$$= u \cdot dw + \nabla_\ell\left(u^m w_{mj\ldots k}\right)dx^\ell_a \wedge dx^i_a \wedge \cdots dx^k_c$$

Thus

$$£_u w = u \cdot dw + d(u \cdot w)$$

Since the chart was arbitrary the result holds in general.

C. Kaluza-Klein Theory

We have seen that the electromagnetic and Yang-Mills fields are both described by Lagrangians constructed from the curvature F of a connection $\overline{A}$. The connection is a 1-form defined on a bundle over spacetime, and a natural way to attempt to unify gravity with Yang-Mills theory is to regard the bundle as real, to suppose that spacetime has more than four dimensions. The extra dimensions would not be seen, at least not in the usual way, if they were topologically compact and had length much smaller than, say, the size of a proton.

So, to unify Yang-Mills theory and gravity, one would like to say that the action is now the gravitational action $\int R\, dr$ evaluated in a higher dimensional space. This would require that the universe from a state in which all directions were comparable in size (early Big Bang Homogeneity) to the present situation with some dimensions enormously longer than others. We will see that such an evolution is in fact a solution to the field equations. The shrinking of the small dimensions is presumed to stop at about the Planck length (worst case), and the universe then settles to a kind of ground state, in which the extra dimensions have the high symmetry one expects of a wave function corresponding to the lowest energy state of a system. One key coincidence makes this scenario attractive: If one assumes simply that the higher dimensions are (i) symmetric and (ii) there is a 4+d dimensional geometry with a d-dimensional group G of isometries, then the action $\int R\, dr$ breaks up into a sum of two parts. One is the 4-dim Einstein action, and the other is the action of the Yang-Mills fields corresponding to the group G. Let's now examine some of the key claims made thus far, starting with the indication that such small dimensions is a solution to the Einstein Field equations.

Kasner in d+4 Dimensions

Let us begin with the Detweiler-Chodos solution describing a 5-dimensional universe with one contracting dimension and three expanding dimensions. It is just the Kasner solution in five dimensions. The prior derivation of the Kasner Ricci tensor did not explicitly involve the dimension of the spacetime. The geometry with metric

$$ds^2 = -dt^2 + \sum_{i=1}^{d} \left(\frac{t}{t_o}\right)^{2P_i} (dy^i)^2 = -dt^2 + \sum t^{2P_i} (dx^i)^2$$

(where $x^i = y^i / t_o^{P_i}$) has Ricci tensor components: $R_0^0 = \frac{1}{t^2} \sum P_i^2 - P_i$, and

$$R_1^1 = \sum_{j=1} \frac{p_1 p_j}{t^2} + \frac{p_1(p_1 - 1)}{t^2} = \frac{p_1}{t^2}(p - p_1 + p_1 - 1) = \frac{p_1}{t^2} = (p-1),$$

where $p = \sum p_i$. Similarly for $R_2^2 = \frac{p_2}{t^2}(p-1)$, ... and $R_1^1 = 0, R_2^2 = 0$, ..., are all satisfied if

$$p = 1$$

$R_0^0 = 0$ then implies

$$\sum p_i^2 = 1$$

(there is again the trivial solution $p_i = 0$, all i). Detweiler and Chodos set d – 4, $p_1 = p_2 = p_3$ for spatial homogeneity. Then

$$p_4 = 1 - 3p_1, \;\; p_4^2 = 1 - 3p_1^2$$

With solution

$$p_1 = \frac{1}{2}, \quad p_4 = -\frac{1}{2}.$$

Then

$$ds^2 = -dt^2 + \frac{t}{t_o} \sum_{i=1} (dx^i)^2 + \frac{t_o}{t}(dx^4)^2$$

The 5-dimensional Kaluza-Klein Theory

Let's construct a metric g_{CD} on a U(1) bundle B over the spacetime M for which the 5-dimensional Einstein action $\int R \, dV$ is the sum of the 4-dimensional Einstein action and the 4-dimensional Maxwell action. The metric will be made from the spacetime metric $g_{\alpha\beta}$ and a connection $\overline{A}_C$ whose curvature is proportional to the electromagnetic field. The task is not difficult: One requires that g_{CD} make horizontal vectors perpendicular to vertical vectors, where as usual ξ^C is horizontal if $\xi^C\overline{A}_C = 0$. Then a horizontal vector ξ^C can be identified with the vector $\xi^\alpha = \pi^\alpha_C\xi^C$ on M and assigned the same length. Vertical vectors γ^C are multiples of the generator η^C of the U(1) action (η^C is tangent at b to the path $\eta \to be^{i\eta}$), and by assigning a constant length ℓ to the vector field η^C, the length of every vertical vector is fixed. The resulting metric is simply:

$$g_{CD} = \ell^2\overline{A}_C\overline{A}_D + {}^4\overline{g}_{CD}$$

where ${}^4\overline{g}_{CD}$ is the pullback of $g_{\alpha\beta}$ to B:

$${}^4\overline{g}_{CD} = \pi^\alpha_C\pi^\beta_D g_{\alpha\beta}.$$

Proof: For u^C horizontal $g_{CD}u^Cu^D = {}^4\overline{g}_{CD}u^Cu^D - g_{\alpha\beta}u^\alpha u^\beta$. For u^C horizontal, v^C vertical, $\pi^\alpha_C v^C = 0, \overline{A}_C u^C = 0 \to g_{CD}u^Cv^D = 0$. For u^C vertical, $v^C = k\eta^C,\ g_{CD}v^Cv^D = k^2\ell^2$. In a chart (η, x^μ), we then have:

$$ds^2 = \ell^2\left(d\eta + \overline{A}_\mu dx^\mu\right)^2 + g_{\mu\nu}dx^\mu dx^\nu,$$

where $\overline{A}_\mu$ is proportional to the vector potential A_μ. Write $\ell\overline{A}_\mu = kA$ (the "ℓ" is there to make $d\omega^5 = kF$).

Curvature

An orthonormal basis (ω^μ) of 1-forms on M can be pulled back to 1-forms $\overline{\omega}^\mu := \pi^*\omega^\mu$ on B. We will then use as a basis $\widehat{\omega}^I$ on B the 1-forms

$$\widehat{\omega}^5 = \ell\overline{A}, \qquad \widehat{\omega}^\mu = \overline{\omega}^\mu.$$

We need the relation

$$d\pi^*\omega = \pi^*d\omega^I,$$

which is valid for any map $\pi: B \to M$ and any p-form ω on M.

Proof: For 0-forms (scalars) f, $\pi^*df = d(f\cdot\pi) = d(\pi^*f)$. Any p-form is a sum of terms of the form $\omega = f\ dg \wedge \ldots \wedge dh$. So

$$\pi^*d\omega = \pi^*[dz\cdot dg \wedge \ldots \wedge dh] = \pi^*df \wedge \pi^*[dg \wedge \ldots \wedge dh]$$
$$= d\pi^*f \wedge \pi^*[dg \wedge \ldots \wedge dh]$$

Thus

$$\pi^*d\omega = d\pi^*[fdg \wedge \ldots \wedge dh] = d\pi^*\omega.$$

Using Cartan's first structure equation:

$$d\widehat{\omega}^5 = \left(\ell d\overline{A}_\mu - k\overline{F}\right) = \frac{1}{2}kF_{\mu\nu}\overline{\omega}^\mu \wedge \overline{\omega}^\nu$$

and

$$d\omega^5 = -d\widehat{\omega}^5 \wedge \overline{\omega}^\mu$$

Since

$$d\widehat{\omega}^\mu = \pi^* d\omega^\mu = -\pi^*\left(\omega^\mu_\nu \wedge \omega^\nu\right) = -\overline{\omega}^\mu_\nu \wedge \overline{\omega}^\nu = -\omega^\mu_I \wedge \omega^I$$
$$= -\overline{\omega}^\mu_\nu \wedge \overline{\omega}^\nu + \left(\widehat{\omega}^\mu_{\nu.5} - \widehat{\omega}^\mu_{5.\nu}\right)\overline{\omega}^\nu \wedge \widehat{\omega}^5$$

Then

$$\widehat{\omega}^5_\mu = \frac{1}{2} k F_{\mu\nu} \overline{\omega}^\nu \;\; and \;\; \widehat{\omega}^\mu_\nu = \overline{\omega}^\mu_\nu - \frac{1}{2} k F^\mu_\nu \omega^5.$$

The Riemann tensor now follows from the second structure equation.

$$R^I_J = d\omega^I_J + \omega^I_K \wedge \omega^K_J$$

$\widehat{R}^\mu_\nu$:

$$d\widehat{\omega}^\mu_\nu = d\overline{\omega}^\mu_\nu - \frac{1}{2} K \partial_\sigma F^\mu_\nu \omega^\sigma \wedge \widehat{\omega}^5 - \frac{1}{4} K^2 F_{\mu\nu} F_{\alpha\tau} F^\sigma \wedge \overline{\omega}^\tau$$

$$\widehat{\omega}^\mu_I \wedge \widehat{\omega}^I_\nu = \overline{\omega}^\mu_\sigma \wedge \overline{\omega}^\sigma_\nu - \frac{1}{2} \cdot K F^\mu_\sigma \overline{\omega}^5 \wedge \overline{\omega}^\sigma_\nu - \frac{1}{2} K \overline{\omega}^\mu_\sigma \wedge F^\sigma_\nu \widehat{\omega}^5$$
$$- \frac{1}{2} K^2 F^\mu_\sigma F_{\nu\tau} \overline{\omega}^\sigma \wedge \overline{\omega}^\tau$$

So,

$$\widehat{R}^\mu_\nu = \bar{R}^\mu_\nu + \frac{1}{8} k^2 \left(F^\mu_\tau F_{\nu\sigma} - F^\mu_\sigma F_{\nu\tau} - 2F^\mu_\nu F_{\sigma\tau}\right)\overline{\omega}^\sigma \wedge \overline{\omega}^\tau$$
$$- \frac{1}{8} k \left[\partial_\sigma F^\mu_\nu \overline{\omega}^\sigma - F^\mu_\tau \overline{\omega}^\tau_\nu + F^\tau_\nu \overline{\omega}^\mu_\tau\right] \wedge \widehat{\omega}^5$$

$$= \bar{R}^\mu_\nu + \frac{1}{8} k^2 \left(F^\mu_\tau F_{\nu\sigma} - F^\mu_\sigma F_{\nu\tau} - 2F^\mu_\nu F_{\sigma\tau}\right)\overline{\omega}^\sigma \wedge \overline{\omega}^\tau - \frac{1}{2} K^2 F^\mu_\sigma F_{\nu\tau} \overline{\omega}^\sigma \wedge \widehat{\omega}^5$$

Similarly:

$\widehat{R}^5_\sigma$:

$$d\widehat{\omega}^5_\sigma = -\frac{1}{2} K \partial_\mu F_{\sigma\nu} \overline{\omega}^\mu \wedge \overline{\omega}^\nu - \frac{1}{2} K F_{\sigma\nu} \overline{\omega}^\nu_\mu \wedge \overline{\omega}^\mu$$

$$\widehat{\omega}^5_\mu \wedge \widehat{\omega}^\mu_\sigma = \frac{1}{2} K F_{\mu\nu} \overline{\omega}^\nu \wedge \left(\overline{\omega}^\mu_\sigma - \frac{1}{2} K F^\mu_\sigma \widehat{\omega}^5\right.$$
$$= -\frac{1}{2} K \, F_{\mu\nu} \, \widehat{\omega}^\mu_\sigma \, \wedge \overline{\omega}^\nu - \frac{1}{4} K^2 \, F^\mu_\sigma \, F_{\mu\nu} \, \overline{\omega}^\nu \wedge \widehat{\omega}^5$$

$$\widehat{R}^5_\sigma = \frac{1}{2} K \left[\partial_\mu F_{\sigma\nu} \overline{\omega}^\mu - F_{\mu\nu} \, \overline{\omega}^\mu_\sigma - F_{\sigma\mu} \, \overline{\omega}^\mu_\nu\right] \wedge \overline{\omega}^\nu - \frac{1}{2} K^2 \, F^\mu_\sigma \, F_{\mu\nu} \, \overline{\omega}^\nu \wedge \widehat{\omega}^5$$
$$= \tfrac{1}{2} \nabla_\sigma F_{\sigma\nu} \overline{\omega}^\mu \wedge \overline{\omega}^\nu + \tfrac{1}{4} K^2 \, F^\mu_\sigma \, F_{\mu\nu} \, \widehat{\omega}^5 \wedge \overline{\omega}^\nu$$
$$= \tfrac{1}{4} \nabla_\sigma F_{\mu\nu} \overline{\omega}^\mu \wedge \overline{\omega}^\nu + \tfrac{1}{4} K^2 \, F^\mu_\sigma \, F_{\mu\tau} \, \widehat{\omega}^5 \wedge \overline{\omega}^\tau$$

Then

$$R^\mu_{\nu\sigma\tau} = R^\mu_{\nu\sigma\tau} + \tfrac{1}{4} K^2 \left(F^\mu_\tau \, F_{\nu\sigma} - F^\mu_\sigma \, F_{\nu\tau} - 2F^\mu_\nu \, F_{\sigma\tau}\right)$$

and

$R^5_{\sigma\mu\nu} = R_{\mu\nu\zeta\delta} = \frac{1}{2}K\nabla_\sigma F_{\mu\nu}$

Thus,
$R^5_{\sigma 5\tau} = \frac{1}{4}K^2 F^{\mu}_{\sigma}\, F_{\mu\tau}$ and the Ricci tensor becomes:

$$R_{\mu\nu} = R^{\sigma}_{\mu\sigma\nu} + R^5_{\mu 5\nu} = R_{\mu\nu} + \frac{1}{4}K^2\left(F^{\sigma}_{\mu}\, F_{\nu\sigma} - 2F^{\sigma}_{\nu}\, F_{\sigma\nu}\right) + \frac{1}{4}K^2 F^{\mu}_{\sigma}\, F_{\nu\tau}$$

$$= R_{\mu\nu} - \frac{1}{2}K^2 F^{\sigma}_{\mu}\, F_{\nu\sigma}$$

and

$$R_{\mu 5} = R^{\sigma}_{\mu\sigma 5} = R^5_{\sigma\mu} = \frac{1}{2}K\nabla_\nu F^{\nu}_{\mu}$$

and

$$R_{55} = \frac{1}{4}K^2 F^{\sigma\tau}F_{\sigma\tau}$$

We can then write

$$\dot{R} = R - \frac{1}{2}K^2 F_{\sigma\tau}F^{\sigma\tau} + \frac{1}{4}K^2 F^{\sigma\tau}F_{\sigma\tau} = R - \frac{1}{4}K^2 F_{\sigma\tau}F^{\sigma\tau}$$

With $K^2 = \frac{16\pi G}{C^4}$ and $L = 2\pi\ell$ is the length of the 5-dim fiber. The Einstein –Maxwell Action then results:

$$I = \frac{c^4}{16\pi GL}\int d^5\tau R = \frac{c^4}{16\pi G}\int dr\left(R - \frac{1}{4}K^2 F_{\alpha\beta}F^{\alpha\beta}\right)$$

$$= \int dr\left(\frac{c^4}{16g}R - \frac{1}{4}F_{\alpha\beta}F^{\alpha\beta}\right).$$

Note that the above action does not allow the 5-metric to be varied arbitrarily. Einstein-Maxwell is a “minisuperspace” theory – a subtheory of 5-dim gravity, with a symmetry imposed by hand (the principal fiber bundle indicated by the dimensional contraction and decoupling).

Length of Fiber

In the bundle picture of electromagnetism, the constant relating $\bar{A}_\mu$ to the physical vector potential A_μ was fixed by requiring parallel transport of phase to agree with horizontal motion in the bundle. Equivalently, the wave function if a charged particle with charge η is to be regarded as a scalar ψ on B which has the behavior $e^{im\eta}$ along a fiber:

$$\pounds_{\vec{\eta}}\psi = im\psi.$$

Since $e_\mu = \partial_\mu - \frac{K}{\ell} A_\mu \partial_\eta$ is the horizontal lift to B of ∂_μ on M, our requirement that $e_\mu \psi = \left(\partial_\mu - im\frac{e}{\hbar c}\right)\psi$ implies $\frac{K}{\ell} = \frac{e}{\hbar c}$. Since K is fixed, we must choose

$$\ell = 8\pi^{3/2}\sqrt{\frac{\hbar c}{e^2}}\ell_{P\ell} = 520\ell_{P\ell}$$

where the Planck's length $\ell_{P\ell} = \left(\frac{G\hbar}{e^2}\right)^{1/2} = 1.63 \times 10^{-33} cm.$

n-dimensional Kaluza-Klein Theory

We want now to generalize the Kaluza-Klein theory to higher dimensions in such a way that the higher dimensional Einstein action breaks up into the sum of 4-dimensional Einstein plus Yang-Mills. We again regard the real spacetime as a principle bundle, this time acted on by the group G of the Yang-Mills theory, and endowed with the Yang-Mills connection $\bar{A}$. The prior argument now implies a metric of the form

$$g_{CD} = q_{ab}\,\bar{A}^a_C\,\bar{A}^b_D - {}^4\bar{g}_{CD}$$

Where q_{ab} is a left-invariant metric on G. the fibers are then homogeneous spaces – spaces on which G acts transitively as a group of isometries, and the natural vertical vector fields (the field η^C on B associated with elements η^a of G) are Killing vectors of B with metric g_{CD}. For the computation of R, we will need only the fact that a left-invariant metric gives a left-invariant (i.e., constant) Ricci scalar R_G. The computation of the Ricci tensor for homogeneous spaces is useful in cosmology, however, and after the Kaluza-Klein computation it will be shown.

Curvature

We can write

$$g_{CD} = g_{ij}\bar{A}^i_C\,\bar{A}^j_D + \eta_{\mu\nu}\bar{\omega}^\mu_C\,\bar{\omega}^\nu_D.$$

where g_{ij} is a (constant) component of the group metric with respect to a basis e^a_i for the Lie algebra ϑ, and where

$$\bar{\omega}^\mu = \pi^*\omega^\mu$$

If e^C_i is the vertical vector field associated with e^a_i, then because $\bar{A}$ is a connection form

$$A^a_C\,e^C_i = e^a_i,$$

and because we have chosen the basis (e^a_i) for G,

$$A_C^j\, e_i^C = \delta_i^j.$$

If $\bar{e}_\mu^C$ is the horizontal lift of e_μ^α on M to B, then $\varpi^\mu(\vec{e}_\nu) = \delta_\nu^\mu$ and $A_C^i\, e_\mu^C = 0$. Thus $(\bar{e}_\mu, e_i)$ is the basis dual to $[\varpi^\mu, \bar{A}^i]$. The computation now parallels that for electromagnetism, starting with writing the expression for hor $d\bar{A} = k\bar{F}$ (using Cartan):

$$d\bar{A}^i = -\widehat{\omega}_j^i \wedge \bar{A}^j - \widehat{\omega}_\mu^i \wedge \varpi^\mu.$$

Now to compute the terms on the left and match up with those on the right::

$$d\bar{A}^i(\bar{e}_\mu, \bar{e}_\nu) = \text{hor } d\bar{A}^i(\bar{e}_\mu, \bar{e}_\nu) = k\bar{F}^i(\bar{e}_\mu, \bar{e}_\nu) = kF_{\mu\nu}^i$$

Thus

$d\bar{A}^i(\bar{e}_j, \bar{e}_\mu) = -A^i([\bar{e}_j, \bar{e}_\mu]) = 0$, since $R_{g*}\bar{e}_\mu = \bar{e}_\mu$. and $d\bar{A}^i(\bar{e}_j, \bar{e}_k) = -A^i([\bar{e}_j, \bar{e}_k]) = -C_{jk}^i$.

Thus

$$d\bar{A}^i = -\frac{1}{2}C_{jk}^i\bar{A}^j \wedge \bar{A}^k + \frac{1}{2}\, k\, F_{\mu\nu}^i\varpi^\mu \wedge \varpi^\nu.$$

For the RHS we have

$$d\varpi^\mu = -\widehat{\omega}_\gamma^\mu \wedge \omega^\nu - \widehat{\omega}_i^\mu \wedge \bar{A}^i$$

So compute:

$$d\varpi^\mu = \pi^* d\omega^\mu - \varpi_\nu^\mu \wedge \varpi^i$$

with solution

$$\widehat{\omega}_j^i = -\frac{1}{2}C_{jk}^i A^k$$

and

$$\widehat{\omega}_\mu^i = \frac{1}{2}\, k\, F_{\mu\nu}^i\varpi^\nu$$

So

$$\widehat{\omega}_\nu^\mu = \varpi_\nu^\mu - \frac{1}{2}\, k\, F_{i\nu}^\mu \bar{A}^i,$$

where g_{ij} is used to lower the group index.

Curvature form R_J^I

The computation of R_ν^μ is nearly identical to that for electromagnetism.. Apart from adding the index I to $F_\nu^{i\mu}$, there are additional terms in $d\widehat{\omega}^\mu$, that convert ∇F to DF, the gauge-covariant derivative:

$d\widehat{\omega}_\nu^\mu = d\varpi_\nu^\mu - \frac{1}{2}\, k\partial_l F_{i\nu}^\mu \omega^l \wedge \bar{A}^i + \frac{1}{4}kF_{i\nu}^\mu C_{jk}^i \bar{A}^j \wedge \bar{A}^k - \frac{1}{4}k^2 F_{\mu\nu}^i F_{i\sigma\tau}\varpi^\sigma \wedge \varpi^\tau$

$$R^{\mu}_{\nu} = R^{\mu}_{\nu} + \frac{1}{8}k^2\left(F^{i\mu}_{\tau}\, F_{i\nu\sigma} - F^{i\mu}_{\sigma}\, F_{i\nu\tau} - 2F^{i\mu}_{\nu}\, F_{i\sigma\tau}\right)\bar{\omega}^{\sigma} \wedge \bar{\omega}^{\tau}$$
$$- \frac{1}{2}k\, D_{\sigma}F^{\mu}_{i\nu}\bar{\omega}^{\sigma} \wedge \bar{A}^{\tau}$$

The computation of R^{i}_{σ} is similar:

R^{i}_{σ}:

$$d\hat{\omega}^{i}_{\sigma} = \frac{1}{2}k\partial_{I}F^{i}_{\sigma J}\omega^{I} \wedge \omega^{J} - \frac{1}{2}kF^{i}_{\sigma\nu}\bar{\omega}^{\nu}_{\mu} \wedge \bar{\omega}^{\mu}$$
$$\hat{\omega}^{i}_{\mu} \wedge \hat{\omega}^{\mu}_{\sigma} = -\frac{1}{2}kF^{j}_{\mu\nu}\bar{\omega}^{\mu}_{\sigma} \wedge \bar{\omega}^{\nu} - \frac{1}{4}k^2F^{i\mu}_{\sigma}F_{j\mu\nu}\bar{\omega}^{\nu}_{\mu} \wedge \bar{\omega}^{\mu}$$
$$\hat{\omega}^{i}_{j} \wedge \hat{\omega}^{j}_{\sigma} = -\frac{1}{2}C^{i}_{jk}\bar{A}^{k} \wedge \frac{1}{2}kF^{j}_{\sigma\mu}\bar{\omega}^{\mu}$$
$$\rightarrow R^{i}_{\sigma} = \frac{1}{4}\, D_{\sigma}F^{i}_{\mu\nu}\bar{\omega}^{\mu} \wedge \bar{\omega}^{\nu} + \frac{1}{4}k^2F^{i\mu}_{\sigma}F_{j\mu\tau}\bar{A}^{j} \wedge \bar{\omega}^{\tau}.$$

Finally, there are the additional components R^{i}_{j} of the curvature 2-form that are not present in the 5-dimensional case:

$$R^{i}_{j} = d\hat{\omega}^{i}_{j} + \hat{\omega}^{i}_{k} \wedge \hat{\omega}^{k}_{j} = d\hat{\omega}^{i}_{j} + \hat{\omega}^{i}_{k} \wedge \hat{\omega}^{k}_{j} + \hat{\omega}^{i}_{\mu} \wedge \hat{\omega}^{\mu}_{j}$$
$$= -\frac{1}{4}R^{i}_{Gjk\ell}\bar{A}^{k} \wedge \bar{A}^{\ell} - \frac{1}{4}k^2F^{i\sigma}_{\mu}F_{j\sigma\nu}\bar{\omega}^{\mu} \wedge \bar{\omega}^{\nu},$$

where $R^{a}_{G\,bcd}$ is the Riemann tensor on G (with constant components) corresponding to the left-invariant metric g_{ab}. Because $R^{i}_{j\,\mu\nu}$ does not contribute to R, the only new term in the action is a constant,

$R_G = R^{ij}_{ij}$:

$$R = R - \frac{1}{4}k^2F^{i}_{\sigma\tau}F^{\sigma\tau}_{i} + R_G$$

Thus

$$I = I_{EINSTEIN-MAXWELL} - \Lambda \int dr$$

where the cosmological constant Λ is proportional to $-R_G$.

Kaluza-Klein monopole

The Kaluza-Klein monopole [68,69] for magnetic monopoles in flat space, where we see a magnetic monopole as a connection on a twisted U(1) bundle, is examined next.

Magnetic monopole

For a charged particle moving in the field of a magnetic charge g, the Schrodinger equation cannot be written because there is no globally defined vector potential $\vec{A}$ with $\nabla \times \vec{A} = \vec{B}$. (This is immediate from Green's theorem: $g = \frac{1}{4\pi}\oint \vec{B} \cdot d\vec{S} = \frac{1}{4\pi}\oint \nabla \times \vec{A} \cdot d\vec{S} = 0$). Following Dirac, physicists have described monopoles by means of a vector potential that is singular along a ray ("string"). Because the physical field is singular only at the position of the monopole, the string singularity has no meaning, and its position is correspondingly arbitrary – i.e., gauge dependent. As we will see, one <u>can</u> define a connection from $\bar{A}_C$ on a U(1) bundle B over $R^3 \cdot \{\text{point}\}$ (the point is the monopole's position), and the lack of well defined $\vec{A}$ on R^3 reflects the fact that the bundle is twisted: Because B is not a trivial bundle it has no global cross section. Thus the pullback $A_a = -\frac{\hbar c}{Q} x_a^C \bar{A}_C$ of $\bar{A}_C$ by the cross-section x is only defined locally. Let us first write down vector potentials $A_a^\pm$ that are regular everywhere except on half of the z-axis.

$$\nabla \times A = B = g\frac{\hat{r}}{r^2} \to \varepsilon^{r\theta\phi}\left(\partial_\theta A_\phi - \partial_\phi A_\theta\right) = \frac{g}{r^2}$$

Thus,

$$\partial_\theta A_\phi - \partial_\phi A_\theta = g \sin\theta.$$

Solutions that retain axial symmetry about z-axis then have the form

$$A_r = A_\theta = 0, \quad A_o = g(K - \cos\theta).$$

A vector potential regular on the + z-axis is $A_a^+ = g(1-\cos\theta)\nabla_a\phi$ and one that is regular on the – z-axis is $A_a^- = -g(1+\cos\theta)\nabla_a\phi$.

We want to bundle with a connection $\bar{A}_C$ for which A_a^+ is the pullback by a cross section x^+ defined on $R^3 - \{-z\ axis\}$, and A_a^- is the pullback by a cross section x^- defined on $R^3 - \{+z\ axis\}$. In terms of coordinates (η, x) for the bundle, we can write $x^\pm(x) = (\eta^\pm(x), x)$. Then $x_a^{+C}\bar{A}_C - x_a^{-C}\bar{A}_C = \nabla_a(\eta^+ - \eta^-)$, and $A_a^+ - A_a^- = -\frac{\hbar c}{Q}\nabla_a(\eta^+ - \eta^-) = 2g\nabla_a\phi$.

Thus

$$\eta^+ - \eta^- = -\frac{2gQ}{\hbar c}\phi + \text{constant}.$$

Because $\phi + 2\pi$ and ϕ refer to the same point,

$$\exp[i(\eta^+ - \eta^-)]|_\phi = \exp[i(\eta^+ - \eta^-)]|_{\phi+2\pi} \to \exp\left(i\frac{2gQ}{\hbar c}2\pi\right) = 1$$

$$\to gQ = \frac{n}{2}\hbar$$

(the Dirac-Saha quantization condition). This condition is equivalent to requiring that the angular momentum of the g, Q system be quantized. The orbital angular momentum is always an integral multiple of $\hbar$, but the

Coulomb fields of g and Q contribute angular momentum gQ. This implies that systems of electric and magnetic charges can have half-integral spin, even though the individual particles are spinless.

Let us now look at the structure of our U(1) bundle. The cross section $x^{\pm}$ correspond to charts for the bundle over, say, the upper and lower half of the 3-space. The bundle is locally a product of space with a circle (with U(1)), where the zero of the U(1) at x can be chosen as the point $\eta^{+}(x)$, for x in the upper half space. Globally, however, the bundle is not trivial: Where the charts overlap in the form

$$\eta^{+} - \eta^{-} = -n\phi + \text{constant.}$$

Say that circles over the bottom half-space are glued to the circles over the top half-space after a relative rotation by an angle that varies from 0 to 2πn. It is a bundle with n twists.

Here is a more elegant description of the 1-twist bundle – call it $B_{1/2}$. $B_{1/2}$ is 4-dimensional, with base space $M = R^3 - (0)$ and can be identified with $C^2 - (0)$. A point ξ in $B_{1/2}$ is then a pair (ξ^0, ξ^1) of complex numbers, a 2-component spinor. An element $u = \exp(i\alpha)$ in U(1) acts on $B_{1/2}$ by multiplication:

$$u = \xi \to u\xi$$

To define the projection $\pi: B_{1/2} \to M$, write $r = \left(\bar{\xi}_I \xi^I\right) = (\xi^0)^2 + (\xi^1)^2$, and represent ξ by the Euler angles of the element U^I_J of SU(2) that rotates $\sqrt{r}\,\delta^I_0$ to ξ^I.

$$\xi^0 = \sqrt{r}\, e^{\frac{i}{2}(\psi+\phi)} \cos\frac{\theta}{2}$$

$$\xi^1 = \sqrt{r}\, i e^{\frac{i}{2}(\psi+\phi)} \cos\frac{\theta}{2}$$

Then $\pi: \xi \to (r, \theta, \phi)$. The $\sqrt{r}$ allows the projection π from $C^2 \to R^3$ to be written in the alternative form $\xi \to r^i = \sigma^i_{IJ} \bar{\xi}^I \xi^J$, where $r^1 = r \sin\theta \cos\phi$, etc.

Restricted to an r = constant sphere, the bundle is the 3-sphere S^3 and the projection to the two sphere with coordinates θ, ϕ is called the Hopf fibration. The isomorphism $S^3 \approx SU(2)$ provided the projection $\pi: SU(3) \to SU(2)/U(1) \approx S^2$, and S^3 is a circle bundle over S^2 with one twist. When regarded as a bundle in this way, the unit sphere S^3 has a natural connection: the unique connection that is constructed from the group elements alone and which is invariant under the action of SU(2), namely

$$\hat{A} = \frac{1}{i}(\bar{\xi}_I d\xi^I).$$

(Because ξ^I is an SU(2)-spinor, $\bar{\xi}_I d\xi^I$ is an SU(2) scalar). With the Euler angles as coordinates:

$$\hat{A} = d\psi + \cos\theta\, d\phi.$$

To Check:

$$\bar{\xi}^0 d\xi^0 = e^{-\frac{i}{2}(\psi+\phi)} \cos\frac{\theta}{2}\left[\frac{i}{2}(d\psi + d\phi)\cos\frac{\theta}{2} - \frac{1}{2}\sin\frac{\theta}{2}\, d\theta\right] e^{\frac{i}{2}(\psi+\phi)}$$

$$\bar{\xi}^1 d\xi^1 = e^{\frac{i}{2}(-\psi+\phi)} \sin\frac{\theta}{2}\left[\frac{i}{2}(d\psi - d\phi)\sin\frac{\theta}{2} + \frac{1}{2}\cos\frac{\theta}{2}\, d\theta\right] e^{\frac{i}{2}(\psi-\phi)}$$

So

$$\bar{\xi}_I d\xi^I = \frac{i}{2}\left(\cos^2\frac{\theta}{2} + \sin^2\frac{\theta}{2}\right) d\psi + \frac{i}{2}\left(\cos^2\frac{\theta}{2} - \sin^2\frac{\theta}{2}\right) d\phi = \frac{i}{2}(d\psi + \cos\theta\, d\phi).$$

Finally, we can extend the connection to the full $B_{1/2}$, writing

$$\bar{A}_{1/2} = \frac{2\mathrm{Im}(\bar{\xi}_I d\xi^I)}{(\bar{\xi}_I \xi^I)}.$$

Because the contribution from dr to $\bar{\xi}_I d\xi^I$ is purely real, $\bar{A}$ retains its form. Then $\bar{F} = dA = -\sin\theta\, d\theta d\phi$, the magnetic coulomb field. The cross section $x(x) = (\psi = 0, x)$ pulls $\bar{A}_{1/2}$ back to the singular potential $A = \cos\theta\, d\phi$. But $\bar{A}_{1/2}$ in the form given above is manifestly regular on $B_{1/2}$.

Example C.1. Consider the Kaluza-Klein monopole with metric:

$$ds^2 - dt^2 + \left(1 + \frac{1}{\rho}\right)^2 dr^2 + r^2 d\Omega^2 + \left(\frac{2r}{1+\rho}\right)^2 (d\Psi + \cos\theta\, d\phi)^2$$

where $\rho^2 = 1 + r^2$. Show that the solution is regular at r=0 by showing that the spatial metric agrees to $0(r^2)$ with the ordinary metric of a 3-sphere. Use coordinates w,x,y,z associated with elements of α, β of SU(2) by:

$$w + iz = r\cos\frac{\theta}{2} e^{\frac{i}{2}(\phi+\Psi)} = \alpha$$

$$x + iy = -ir\sin\frac{\theta}{2} e^{\frac{i}{2}(\Psi-\phi)} = r\beta$$

Solution

We have:

$$dw + idz = \frac{\alpha}{r}dr + \frac{\alpha}{\cos\frac{\theta}{2}}\left(-\frac{1}{2}\sin\frac{\theta}{2}\right)d\theta + \alpha\left(\frac{i}{2}(d\phi + d\Psi)\right)$$

thus

$$dw^2 + dz^2 = \left[\left(\frac{dr}{r} - \frac{\frac{1}{2}\sin\frac{\theta}{2}}{\cos\frac{\theta}{2}}d\theta\right)^2 + \left(\frac{i}{2}(d\phi + d\Psi)\right)^2\right]|\alpha|^2$$

similarly

$$dx + idy = \frac{\gamma}{r}dr + \frac{\frac{1}{2}\cos\frac{\theta}{2}d\theta}{\sin\frac{\theta}{2}}\gamma + \gamma\left(\frac{i}{2}(d\Psi - d\phi)\right)$$

thus

$$dx^2 + dy^2 = \left[\left(\frac{dr}{r} - \frac{\frac{1}{2}\cos\frac{\theta}{2}}{\sin\frac{\theta}{2}}d\theta\right)^2 + \left(\frac{i}{2}(d\Psi - d\phi)\right)^2\right]|\alpha|^2$$

and combiing:

$$dx^2 + dy^2 + dz^2 + dw^2$$

$$= \left(r^2\cos^2\frac{\theta}{2}\right)\left[\frac{dr^2}{r^2} + \frac{1}{4}\tan^2\frac{\theta}{2}d\theta^2 - \frac{\tan\frac{\theta}{2}}{r}drd\theta\right]$$

$$+ \left(r^2\sin^2\frac{\theta}{2}\right)\left[\frac{dr^2}{r^2} + \frac{1}{4}\cot^2\frac{\theta}{2}d\theta^2 - \frac{\cot\frac{\theta}{2}}{r}drd\theta\right]$$

$$+ \left(r^2\cos^2\frac{\theta}{2}\right)\frac{1}{4}(d\phi^2 + 2d\phi d\Psi + d\Psi^2)$$

$$+ \left(r^2\sin^2\frac{\theta}{2}\right)\frac{1}{4}(d\phi^2 + 2d\phi d\Psi + d\Psi^2)$$

$$= dr^2 + r^2\left(\frac{1}{4}\right)d\theta^2 + r^2\left(\frac{1}{4}\right)(d\phi^2 + d\Psi^2)$$

$$+ r^2\left(\frac{1}{2}\right)\frac{\left(\cos^2\frac{\theta}{2} - \sin^2\frac{\theta}{2}\right)}{\cos\theta}d\phi d\Psi$$

$$= dr^2 + \frac{1}{4}r^2[d\theta^2 + d\phi^2 + d\Psi^2 + 2\cos\theta\, d\phi d\Psi]$$

$$\hookrightarrow d\phi^2 = (\cos^2\theta + \sin^2\theta)d\phi^2$$

$$= dr^2 + \frac{1}{4}r^2[d\theta^2 + \sin^2\theta\, d\phi^2 + (d\Psi + \cos\theta\, d\phi)^2]$$

$$= dr^2 + \frac{1}{4}r^2[d\Omega^2 + (d\Psi + \cos\theta\, d\phi)^2]$$

Now, consider the Kaluza-Klein monopole metric give near $r = 0: \rho = 1 + O(r^2)$

$$ds^2 = -dt^2 + \left(1 + \frac{1}{1 + O(r^2)}\right)^2 dr^2 + r^2 d\Omega^2$$

$$+ \left(\frac{2r}{2 + O(r^2)}\right)^2 (d\Psi + \cos\theta \, d\phi)^2$$

$$= -dt^2 + (4 + O(r^2))dr^2 + r^2 d\Omega^2 r^2 1 + O(r^2)(d\Psi + \cos\theta \, d\phi)^2$$

Thus, to order r^2 the spatial metric $= \frac{1}{4}(dx^2 + dy^2 + dz^2 + dw^2)$ given above. Since spatial metric agrees with that on a 3-sphere as r → 0 then by the regularly of the metric on a 3-sphere we know that the K-K monopole is also regular.

D. Advanced Ordinary Differential Equations (ODE's)

Many of the examples and Exercises in this appendix are from class notes from Caltech AMa101c 1987.

Example D.1. Consider

$$r^2y'' + ry' + \lambda y = 0, \qquad y(1) = 0, y(2) = 0.$$

Let $y = x^r \to y' = rx^{r-1} \to y'' = r(r-1)x^{r-2}$, and we get:

$$r^2 = -\lambda \quad \to \quad r = \pm i\sqrt{\lambda}\,.$$

Thus, we have for general solution:

$$y = A\, sin(\sqrt{\lambda}\log x) + B\cos(\sqrt{\lambda}\log x).$$

The B,C, at x=1 forces B=0, while $y(2) = 0$ implies

$$sin(\sqrt{\lambda}\log 2) = 0 \to \sqrt{\lambda}\log 2 = n\pi \to \lambda =$$

$\lambda_n = (\frac{n\pi}{\log 2})^2$

The corresponding eigenfunction:

$$u_n(x) = sin\,(\frac{n\pi}{\log 2}\log x)\,.$$

We need an orthonormal set:

$$\phi_n(x) = \frac{u_n(x)}{\sqrt{x}\int_1^2 \frac{u_n{}^2(x)}{x}d_x}$$

then, $f(x) = \sum_{n=0} A_n\phi_n(x)$.

Example D.2. In this problem we consider the maximum overshoot that occurs at a discontinuity when representing the function by way of a Fourier series. Suppose we have a discontinuity at x_0, let's shift our origin and make $x_0 = 0$. Now renormalizing the discontinuity and centering it on the x-axis (so the standard step function profile at the origin with value -1 when less than zero and +1 when greater than zero). The power series of $f(x)$="step function" (this is a local analysis relevant to the overshoot calculation) is:

$$f(x) = \frac{4}{\pi}(\sin x + \frac{\sin 3x}{3} + \frac{\sin 5x}{5} + \cdots).$$

Recall the identity:

$$cos\, y + cos\, 3y + \cdots cos(2m-1)y = \frac{1}{2}\frac{sin\, 2my}{sin\, y}.$$

We can avail ourselves of this identity if we integrate both sides (and mult. by $4/\pi$):

$$s_{2m-1}(x) = \frac{4}{\pi}\left(\sin x + \frac{\sin 3x}{3} + \cdots\right) = \frac{2}{\pi}\int_0^x \frac{\sin 2my}{\sin y}dy.$$

Thus,

$$s'_{2m-1}(x) = \frac{2}{\pi}\frac{\sin 2mx}{\sin x}$$

Extrema one at sin(2mx)=0, or $x = \frac{4\pi}{2m}$, and considering the sin(x), term k=1correspond to the greatest maximum. So,

$$s_{2m-1}\left(\frac{\pi}{2m}\right) = \frac{2}{\pi}\int_0^{\frac{\pi}{2m}} \frac{\sin 2my}{\sin y}dy \to \frac{2}{\pi}\int_0^{\pi} \frac{\sin t}{t}\, dt \ \text{as}\ m \to \infty.$$

Since $\frac{2}{\pi}\int_0^{\pi} \frac{\sin t}{t}\, dt \approx 1.179$, the overshoot goes as $(0.179)[f(0^+) - f(0^-)]$. The maximum error is thus 18%. Transforming the origin back to x_0, the maximum overshoot value is $.08949[f(x_0^+) - f(x_0^-)]$ which is also:

$$s\left(x_0 + \frac{\pi}{2m}\right) - f(x_0^+).$$

Thus

$$s_{2m-1}\left(x_0 + \frac{\pi}{2m}\right) - f(x_0^+) \sim .08949[f(x_0^+) - f(x_0^-)]$$

The maximum overshoot is at position $x_0 + \frac{\pi}{2m}$. Using 2k=2m-1, or $m = k + 1/2$, the maximum overshoot is at:

$$x = x_0 + \frac{\pi}{k + 1/2}.$$

Example D.3. Solve $L = \frac{d^2}{dx^2} - \frac{2d}{dx} + 2\,,\ \ L * u = 0$.

Solution

$u^2 - 2u' + 2u = 0$

$u = Ae^{(1+j)x} + Be^{(i-j)x}$

$u(o) = A + B = 0$

$u = Ae^x(e^{jx} - e^{-jx}) = 2Ae^x sinx$

$u' = 2Ae^x[cosx + sinx]$

$u'(\pi/4) = 2\sqrt{2}\, e(\pi/4)A = 0$

$so, A = 0, thus\ u = 0\ \ trivial\ ....$

$G''\left(\frac{x}{\epsilon}\right) - 2G'^{\left(\frac{x}{\epsilon}\right)} + 2G\left(\frac{x}{\epsilon}\right) = \delta(x - E)$

Soln.

$$\begin{cases} G\left(\frac{x}{\in}\right) = A, e^{(1+i)x} + A_2 e^{(1-i)x} & x < E \\ G\left(\frac{x}{\in}\right) = B, e^{(1+i)x} + B_2 e^{(1-i)x} & x > E \end{cases}$$

$$\underset{8 \to 10^+}{lin} \left[\frac{\partial G}{\partial x}\Big|_{x=E+3} - \frac{\partial G}{\partial x}\Big|_{x=\in-3}\right] = 1$$

$$B_1 - A_1 = \frac{-e^{1-i)\in}}{w[e^{1+i)\in}, e^{1-i)\in}]} \ ; \ B_2 - A_2 = \frac{e^{1-i)\in}}{w[e^{1+i)\in}, e^{1-i)\in}]}$$

Choose $A_1 = A_2 = 0$

$$G\left(\frac{x}{\in}\right) = \left\{\frac{-e^{1-i)\in}\, e^{1+i)\in} + e^{1+i)\in} e^{1-i)\in}}{w[e^{1+i)\in}, e^{1-i)\in}]}\right\} \qquad x \geq E$$

$$w[e^{1+i)\in}, e^{1-i)\in}] = \begin{vmatrix} e^{1+i)\in} & e^{1-i)\in} \\ (1+i)e^{1+i)\in} & (1-i)e^{1-i)\in} \end{vmatrix} = (1-i)e^{2E} - (1+i)e^{2E} = -2ie^{2E}$$

$$G\left(\frac{x}{E}\right) = \left\{\frac{-e^{1-i)\in}\, e^{1+i)x} + e^{1+i)\in} e^{1-i)x}}{-2ie^{2E}} \qquad \begin{matrix} x \geq E \\ x < E \end{matrix}\right.$$

$H\left(\frac{x}{E}\right)$ *simply equals the* $G\left(\frac{x}{E}\right)$ *term will all* $1 \to -1$:

$$H\left(\frac{x}{E}\right) = \begin{cases} \dfrac{-e^{1-i)\in}\, e^{1+i)\in} + e^{1+i)\in} e^{1-i)\in}}{w[e^{1+i)\in}, e^{1-i)\in}]} & x \geq E \\ O\prime & x < E \end{cases}$$

$$w[e^{-1+i)\in}, e^{-1-i)\in}] = \begin{cases} \dfrac{-e^{1-i)x}\, e^{1+i)\in} + e^{1+i)x} e^{1-i)\in}}{-2ie^{2x}} & x \geq E \\ O & x < 0 \end{cases}$$

$$= \begin{cases} \dfrac{-e^{1-i)x}\, e^{1+i)\in} + e^{1+i)x} e^{1-i)\in}}{-2i} & x \geq E \\ O & x < 0 \end{cases}$$

$$= \begin{cases} \dfrac{-e^{(-1-i)x}\, e^{1+i)\in} + e^{-1+i)x} e^{(-1-i)\in}}{-2ie^{-2E}} & x \geq E \\ O & x < 0 \end{cases}$$

Thus G(E/x)=H(x/E)

Example D.4. Solve

$$\frac{d^2\Psi}{dx^2} + \left(K^2 - \lambda u(x)\right)\Psi = 0 \qquad \Psi(0) = \Psi(L) = 0 \qquad 0 \leq x \leq L$$

Solution

$$\frac{d^2}{dx^2} G\left(\frac{x}{E}\right) + k^2 = -\delta(x - E)$$

So, $\Psi(x) = -\lambda \int_0^L Gk(x \mid E) u(x_0) \Psi(x_0) dx_0$

$$Gk\left(x \,\middle|\, E\right) = \underset{P}{\Sigma} \frac{\emptyset p(x) \emptyset p(\Sigma)}{kp^2 - k^2}$$

$$G_k\left(\frac{X}{\in}\right) = \frac{1}{ksin(AL)}\begin{Bmatrix} Sin(kx)sin[K(L-\in)] \\ Sin(ky)sin[K(L-\in)] \end{Bmatrix} \qquad \begin{matrix} x \le x_0 \\ x \ge x_0 \end{matrix}$$

$$\Psi(x) = \lambda \sum_p \int_0^l \frac{\emptyset p(\in)u(\in)\Psi(\in)d\in\emptyset p(x_0)dx_0}{k^2-kp^2}$$

$$\Psi_n = \emptyset_n + \lambda \sum_{p\neq n} \int_0^l \frac{\emptyset p(\in)u(\in)\emptyset\gamma p(x)}{k^2-kp^2}$$

For zeroth approximation to 4n

$$\Psi_n(x) + \lambda \sum_{P\neq n}\left(\frac{Upn}{K^2-KP^2}\right)\emptyset p$$

$$u_{pn} = \int_0^l \emptyset p(\in)u(\in)\emptyset_n(\in)dE$$

$$\Psi_n^{(2)} = \emptyset_n(x) + \lambda \sum_{p\neq n}\frac{(upn)}{K^2-K_p}\emptyset p + \lambda^2 \sum_{pq\neq n}\left[\frac{upquqn}{(K^2-kp^2)(k^2-kq^2)}\right]\emptyset p$$

$$(K^2)^{(1)} = Kn^2 + \lambda u_{nn}$$

$$(K^2)^{(1)} = Kn^2 + \lambda u_{nn}$$

$$(K^2)^{(1)} = K_n^2 + \lambda u_n n$$

$$(K^2)^{(1)} = K_n^2 + \lambda u_n n + \lambda^2 \sum_{p\neq n}\frac{u_n pupn}{k_n^2-kp^2}$$

Thus;

$$\Psi_n^{(i)} = \emptyset_n(x) + \lambda \sum_{p\neq n}\left(\frac{Upn}{k_n^2-kp^2}\right)$$

$$\Psi_n^{(2)} = \emptyset_n(x) + \lambda \sum_{p\neq n}\left(\frac{Upn}{k_n^2-\lambda u_n n-kp^2}\right)\emptyset p(x) +$$

$$\lambda^2 \sum_{pq\neq n}\left[\frac{UpqUqp}{(K_n^2-k_n^2)(K_n^2\mp K_q^2}\right]\emptyset p$$

Example D.5. Construct the Green's function $g\left(\frac{x}{\in};\lambda\right)$ *which satisfies.*

$$-\frac{d^2g}{dx^2} - \lambda g = \delta(x-E) \qquad 0 < x, \ \in < l$$

$$\frac{dg}{dx}\left(\frac{0}{E};\lambda\right) = \frac{dg}{dx}\left(\frac{l}{E};\lambda\right) = 0$$

Show that g has simple poles at $\lambda = \frac{n^2\pi^2}{l^2}$ $with\, n = 0,1,2,\ldots$ By integrating g along a large circle in the λ plane obtain the bilinear expansion for the δ function. Obtain the solution of

$$-\frac{d^2u}{dx^2} - \lambda u = f(x) \qquad 0 < x, E < l \quad u'(0) = 0 \qquad u'(l) = 0$$

As an eigenfunction expansion. Find the consistency condition if $\lambda = \lambda_k$. What is the bilinear series for $g(x|E;\lambda)$? Discuss the equation with inhomogeneous boundary conditions.

Solution

$$\frac{-d^2g}{dx^2} - \lambda g = \delta(x - E) \qquad O < x,\ E < l$$
$$\frac{d^2g}{dx^2} - \lambda g = 0$$
$$g = Ae^{-kx}$$
$$g' = -KAe^{-Kx}$$
$$g'' = -\alpha^2 Ae^{-Kx}$$
$$A\lambda = +\alpha^2 A$$
$$\sqrt{\lambda} = \alpha$$
$$g = Ae^{-i\sqrt{\lambda x}} + Be^{\sqrt{\lambda}x}$$
$$g = \begin{cases} aSin\sqrt{\lambda}x + bcos\sqrt{\lambda}x & 0 < x < E \\ Csin\sqrt{\lambda}x + dcos\sqrt{\lambda}x & E < x < l \end{cases}$$
$$\frac{dg}{dx}\left(\frac{O}{E}\right)\lambda = \frac{dg}{dx}\left(\frac{O}{E}\right)$$
$$a\sqrt{\lambda}\,cos\sqrt{\lambda}(O) - b\sqrt{\lambda}\,cos\sqrt{\lambda}(O)^0 = 0$$
$$c\sqrt{\lambda}\,cos\sqrt{\lambda}(l) - d\sqrt{\lambda}\,cos\sqrt{\lambda}(l)^0 = 0$$
$$c = dtan\sqrt{\lambda}\,l$$
$$\frac{-dg}{dx}\left(\frac{X}{E};\lambda\right)|^{+}_{x=E}\frac{dg}{dx}\left(\frac{x}{E}\right);\ \lambda)|_{x=E^+} = -1$$
$$+b\sqrt{\lambda}\,sin\sqrt{\lambda}\,E + c\sqrt{\lambda}\,cos\sqrt{\lambda}\,E + d\sqrt{\lambda}\,sin\sqrt{\lambda}\,E = -1$$
$$+b\sqrt{\lambda}\,sin\sqrt{\lambda}\,E + d\sqrt{\lambda}\,tan\sqrt{\lambda}\,l\,cos\sqrt{\lambda}\,E + \sqrt{\lambda}\,sin\sqrt{\lambda}\,E = -1$$
$$b = \frac{-1}{\sqrt{\lambda}\,sin\sqrt{\lambda E}} - d(tan\sqrt{\lambda E}.Cot\sqrt{\lambda E} - 1$$
Continuity: $g(\frac{x}{E};\ \lambda|_{x=E-} = g(\frac{x}{E};\ \lambda|_{x=E^T}$
$$bcos\sqrt{\lambda E} = dtan\sqrt{\lambda l}\,sin\sqrt{\lambda E} + dcos\sqrt{\lambda E}$$
$$b = d(tan\sqrt{\lambda l}tan\sqrt{\lambda E} + 1$$
$$dtan\sqrt{\lambda l}tan\sqrt{\lambda E} + 1 = \frac{-1}{\sqrt{\lambda}sin\sqrt{\lambda E}} - d(tan\sqrt{\lambda l}\cdot cos\sqrt{\lambda E} - 1$$
$$d\left(tan\sqrt{\lambda l}\left[tan\sqrt{\lambda E} + cot\sqrt{\lambda E}\right]\right) = \frac{-1}{\sqrt{\lambda}sin\sqrt{\lambda E}}$$
$$d = \frac{-1}{tan\sqrt{\lambda E}}\left(\frac{cos\sqrt{\lambda E}\cdot sin\sqrt{\lambda E}}{\sqrt{\lambda}sin\sqrt{\lambda E}}\right)$$
$$d = \frac{-cos\sqrt{\lambda E}}{\sqrt{\lambda}tan\sqrt{\lambda l}}$$
$$C = dtan\sqrt{\lambda} = \frac{Cos\sqrt{\lambda E}}{\sqrt{\lambda}} = C$$
Both b and d l low up when tan$tan\sqrt{\lambda l} = 0$ or when $\sqrt{\lambda}l = tan^{-1}(0) = n\pi \qquad n = 0,1,2,\ldots$
$$\lambda = \left(\frac{n\pi}{l}\right)^2 \qquad n = 0,1,2,\ldots$$
Consider first $O < x < E \qquad n = 0,1,2,\ldots$
$$y = \frac{1}{\sqrt{\lambda}}\left(sin\sqrt{\lambda E} + \frac{cos\sqrt{\lambda E}}{tan\sqrt{\lambda E}}\right)cos\sqrt{\lambda E} \qquad let\lambda = \mathbb{Z}\ for\ integral$$

$\phi g = \phi - \frac{1}{\sqrt{Z}}\left(sin\sqrt{Z}E + \frac{cos\sqrt{ZE}}{tan\sqrt{ZE}}\right) cos\sqrt{ZE}$

$\phi g = 2\pi i \cdot \sum residues = 2\pi i\left(\sum_{n=1}^{\infty} \frac{2cos\frac{n\pi E}{l}}{n\pi} cos\frac{n\pi x}{l} + \cdots\right)$

$\frac{d}{dZ} cos\sqrt{Z}Ecos\sqrt{Z}x|_{Z=0} = \cos(\sqrt{Z}E)\left(\frac{-sin\sqrt{Z}x \cdot x}{2\sqrt{Z}}\right) + \cos(\sqrt{Z}x)\left(\frac{-sin\sqrt{Z}E \cdot E}{2\sqrt{Z}}\right)$

$\phi g = 2\pi i\left(\frac{1}{l} + \sum_{n=1}^{\infty} \frac{2}{l} \cos\left(\frac{n\pi}{l}\right) \cdot cos\left(\frac{n\pi x}{l}\right)\right)$

$\delta(x - E) = \frac{1}{l} + \sum_{n=1}^{\infty} cos\left(\frac{n\pi E}{l}\right) cos\left(\frac{n\pi x}{l}\right)$

$let\ u(x, E) = \sum_{n=1}^{\infty} b_n(E) cos\frac{n\pi x}{l}$ $\quad what\ satisfies\ u'(0) = 0, u'(l) = 0$

$u' = -\sum_{n=1}^{\infty} b_n(E) sin\frac{n\pi x}{l} \cdot \frac{n\pi}{l}$ $\qquad \sqrt{\lambda} = \frac{n\pi}{l}$

$u'' = -\sum_{n=1}^{\infty} b_n(E) cos\frac{n\pi x}{l} \cdot \left(\frac{n\pi}{l}\right)^2$

$= \sum_{n=1}^{\infty} cos\frac{n\pi x}{l} \cdot b_n(E) \cdot \left[\left(\frac{n\pi}{l}\right)^2 - \lambda\right] = f(x)$

Multiply by $cos\frac{n\pi x}{l} \cdot$ $\quad and\ integrate\ \int_0^l (m)\, dx$

$\sum_{n=1}^{\infty} \int_0^l \cos\left(\frac{n\pi x}{l}\right) \cos\left(\frac{n\pi x}{l}\right) . b_n(E)\left[\left(\frac{n\pi}{l}\right)^2\right] dx = \int_0^l f(x) cos\frac{n\pi x}{l} dx$

$\frac{l}{2}\delta(n - m)$

$\frac{l}{2}\delta b_m(E)\left[\left(\frac{m\pi}{l}\right)^2 - \lambda\right] = \int_0^l f(x) cos\frac{n\pi x}{l} dx$

Shift …. To the more familiar n

$b_n(E) = \frac{1}{\epsilon\frac{l}{2}\left[\left(\frac{n\pi}{l}\right)^2\right] - \lambda} \int_0^l f(x) cos\frac{n\pi x}{l} dx$

Consistency condition, $\lambda = \lambda_k$

$u(x, E) = \sum_{n=1}^{\infty}\left(\frac{1}{\frac{l}{2}\left[\left(\frac{n\pi}{l}\right)^2 - \lambda\right]} \int_0^l f(x) cos\frac{n\pi x}{l} dx\right) cos\frac{n\pi x}{l}$

The consistency condition of $\lambda = \lambda_k$ $\quad is$

$\int_0^l f(x) cos\frac{k_a x}{l} dx = 0$

$g\left(\frac{x}{E}; \lambda\right) = \sum_{n=0}^{\infty} \frac{\cos\left(\frac{n\pi}{l}\right) cos\left(\frac{n\pi}{l}\right)}{\left[\left(\frac{n\pi}{l}\right)^2 - \lambda\right]} x \left[\frac{2}{El}\right]$

Inhomogeneous boundary conditions will introduce a weighing function into the $\int_0^l f(\cos\quad) \frac{n\pi}{l} w(x) dx$

Example D.6. Consider the eigenvalue problem.

$$-\frac{d^2\emptyset}{dx^2}-\lambda\emptyset=0 \qquad 0<x<l \qquad \emptyset(0)=\emptyset(l) \qquad \emptyset'(0)=\emptyset'(l)$$

Show that the eigenvalues are $\lambda_n=\frac{4n^2\pi^2}{l^2}$ $n=0,1,2,\ldots$, and that each eigenvalue except $\lambda=0$ is degenerate (that is, the eigenvalue has multiplicity greater than one). Show that the normalized eigenfunctions can be written either as

$$\emptyset_0=\sqrt{1/l} \qquad \emptyset_n^{(1)}(x)=\sqrt{2/l}\,sin\frac{2n\pi x}{l} \qquad \emptyset_n^{(2)}(x)=\sqrt{2/l}\,cos\frac{2n\pi x}{l}$$

Or

$$\Psi_0(x)=\sqrt{1/l} \qquad \Psi_n^{(1)}=\sqrt{1/l}\exp(i2n\pi x) \qquad \Psi_n^{(2)}(x)=\sqrt{1/l}\,exp\,(i2n\pi x/l)$$

Construct $g(x|E;\lambda)$ directly from the differential equation which satisfies and by integrating g along a large circle in the complex λ-plane show.

$$\delta(x-E)=\frac{1}{l}+\sum_{n=1}^{\infty}\frac{2}{l}\left(sin\frac{2n\pi x}{l}\;sin\left(\frac{2n\pi E}{l}\right)+cos\frac{2n\pi x}{l}\;cos\frac{2n\pi E}{l}\right)$$

Solution

$$\frac{-d^2g}{dx}-\lambda g=0 \quad \emptyset(0)=\emptyset(l),\ \emptyset'(O)=\emptyset'(l) \qquad 0<x<l$$

$$g=\begin{cases}acos\sqrt{\lambda}x+bsin\sqrt{\lambda}x & 0<x<E\\ Ccos\sqrt{\lambda}x+dsin\sqrt{\lambda}x & E<x<l\end{cases}$$

$$a=Ccos\sqrt{\lambda}x+dcos\sqrt{\lambda}x \qquad \emptyset(O)=\emptyset(l)$$

$$+b=-Csin\sqrt{\lambda}x+dcod\sqrt{\lambda}x \qquad \emptyset(O)=\emptyset(l)$$

$$acos\sqrt{\lambda}E+bsin\sqrt{\lambda}E=Ccos\sqrt{\lambda}E \qquad \text{continuity}$$

$$C\sqrt{\lambda}sin\sqrt{\lambda}E-d\sqrt{\lambda}cos\sqrt{\lambda}E+\left(B\sqrt{\lambda}Ecos-A\sqrt{\lambda}\,sin\sqrt{\lambda}E\right)=Jump$$

$$\begin{matrix}\\ \\ sin\sqrt{\lambda}E\\ cos\sqrt{\lambda}E\end{matrix}\begin{bmatrix}1&0&-cos\sqrt{\lambda}l&sin\sqrt{\lambda}l\\0&1&sin\sqrt{\lambda}l&-cos\sqrt{\lambda}l\\cos\sqrt{\lambda}E&sin\sqrt{\lambda}E&-cos\sqrt{\lambda}E&-sin\sqrt{\lambda}E\\-sin\sqrt{\lambda}E&+cos\sqrt{\lambda}E&-sincos\sqrt{\lambda}E&-cos\sqrt{\lambda}E\end{bmatrix}\begin{bmatrix}a\\b\\c\\d\end{bmatrix}=\begin{bmatrix}0\\0\\0\\\frac{1}{\sqrt{\pi}}\end{bmatrix}$$

$$sin\sqrt{\lambda}E\begin{bmatrix}1&0&-cos\sqrt{\lambda}l&sin\sqrt{\lambda}l\\0&1&sin\sqrt{\lambda}l&-cos\sqrt{\lambda}l\\cos\sqrt{\lambda}E&sin\sqrt{\lambda}E&-cos\sqrt{\lambda}E&-sin\sqrt{\lambda}E\\0&1&0&-1\end{bmatrix}\begin{bmatrix}a\\b\\c\\d\end{bmatrix}=\begin{bmatrix}0\\0\\0\\\frac{1}{\sqrt{\lambda}cos\sqrt{\lambda}E}\end{bmatrix}$$

$$\begin{bmatrix}1&0&-cos\sqrt{\lambda}l&sin\sqrt{\lambda}l\\0&1&sin\sqrt{\lambda}l&-cos\sqrt{\lambda}l\\1&0&-1&0\\0&1&0&-1\end{bmatrix}\begin{bmatrix}a\\b\\c\\d\end{bmatrix}=\begin{bmatrix}0\\0\\-\frac{sin\sqrt{\lambda}E}{\sqrt{x}cos\sqrt{\lambda}E}\\\frac{1}{\sqrt{\lambda}cos\sqrt{\lambda}E}\end{bmatrix}$$

$$\begin{bmatrix} 0 & 0 & (1-\cos\sqrt{\lambda}l) & \sin\sqrt{\lambda}l \\ 0 & 0 & \sin\sqrt{\lambda}l & (1-\cos\sqrt{\lambda}l) \\ 1 & 0 & -1 & 0 \\ 0 & 1 & 0 & -1 \end{bmatrix}\begin{bmatrix} a \\ b \\ c \\ d \end{bmatrix} = \begin{bmatrix} \frac{\sin\sqrt{\lambda}E}{\sqrt{\lambda}\cos\sqrt{\lambda}E} \\ \frac{1}{\sqrt{x}\cos\sqrt{\lambda}E} \\ -\frac{\sin\sqrt{\lambda}E}{\sqrt{\lambda}\cos\sqrt{\lambda}E} \\ \frac{1}{\sqrt{\lambda}\cos\sqrt{\lambda}E} \end{bmatrix} = \begin{bmatrix} A \\ B \\ C \\ D \end{bmatrix}$$

$$(1-\cos\sqrt{\lambda}l)a + \sin\sqrt{\lambda}l\, b = A$$
$$\sin\sqrt{\lambda}l\, a + (1-\cos\sqrt{\lambda}l)b = B$$
$$a = \frac{B-(1-\cos\sqrt{\lambda}l)b}{\sin\sqrt{\lambda}l}$$
$$(1-\cos\sqrt{\lambda}l)\left(\frac{B-(1-\cos\sqrt{\lambda}l)b}{\sin\sqrt{\lambda}l}\right) + \sin\sqrt{\lambda}l = A$$
$$b\left((\sin\sqrt{\lambda}l - \frac{(1-\cos\sqrt{\lambda}l)^2}{\sin\sqrt{\lambda}l}\right) = A - \left(\frac{(1-\cos\sqrt{\lambda}l)B}{\sin\sqrt{\lambda}l}\right)$$
$$b\left((\sin^2\sqrt{\lambda}l - \frac{(1-2\cos\sqrt{\lambda}l)+\cos^2\sqrt{\lambda}l}{\sin\sqrt{\lambda}l}\right) = A - \left(\frac{(1-\cos\sqrt{\lambda}l)B}{\sin\sqrt{\lambda}l}\right)$$
$$b = \frac{A\sin\sqrt{\lambda}l}{2\cos\sqrt{\lambda}l-\cos^2\sqrt{\lambda}l} - \frac{(1-\cos\sqrt{\lambda}l)B}{\sin\sqrt{\lambda}l}\cdot\frac{\sin\sqrt{\lambda}l}{2\cos\sqrt{\lambda}l-\cos^2\sqrt{\lambda}l}$$
$$= \frac{A\tan\sqrt{\lambda}l}{(2-\cos\sqrt{\lambda}l)} - \frac{B(1-\cos\sqrt{\lambda}l)}{\cos\sqrt{\lambda}l(2-\cos\sqrt{\lambda}l)}$$
$$a = \frac{-\sin\sqrt{\lambda}l(l-E)+\sin\sqrt{\lambda}E}{2\sqrt{\lambda}(1-\cos\sqrt{\lambda}l)}$$
$$b = \frac{-\cos\sqrt{\lambda}(l-E)+\cos\sqrt{\lambda}E}{2\sqrt{\lambda}(1-\cos\sqrt{\lambda}l)}$$
$$c = \frac{-\sin\sqrt{\lambda}(l+E)+\sin\sqrt{\lambda}E}{2\sqrt{\lambda}(1-\cos\sqrt{\lambda}l)}$$
$$d = \frac{\cos\sqrt{\lambda}(l+E)+\cos\sqrt{\lambda}E}{2\sqrt{\lambda}(1-\cos\sqrt{\lambda}l)}$$
$$\emptyset g\left(\frac{x}{E};\lambda\right)dx = 2\sqrt{\lambda}i\sum residues = 2\sqrt{\lambda}i\delta(x-E)$$

Example D.7. Show that the general solution of
$-u'' - \lambda = \exp(-x) \quad 0 < x < \infty \quad u(0) = 0$
Is $u(x) = A\sin(\sqrt{\lambda}x) + \frac{\cos\sqrt{\lambda}x - \exp(-x)}{1+\lambda}$
Hence the system does not have a solution in $l_2^c(0,\infty)$ for real positive λ. Show that this conclusion is also valid if λ = 0. Find the unique l_2 solution if λ is not in $(0,\infty)$.

Solution

For real positive λ we have an $A^2Sin^2(\sqrt{\lambda}x)$ term when we calculate $u(x)u^*(x)$ thus there will exist a term $\int_0^\infty \frac{A^2}{2}dx$ and thus is not less than ∞, hence.

The system does not have a solution in $l_2^{(c)}(0,\infty)$ for real positive λ.

If λ=0: $u(x) = 1 - \exp(-x)$, once again we have a constant term $\int_0^\infty 1dx$ what also allows exp.

Let $\lambda = -\alpha$ let λ be complex! i.e. study the general case

$u(x) = Asin(i\sqrt{\alpha}x) + \frac{\cos(i\sqrt{\alpha}) - \exp(-x)}{1-\alpha}$

$= Aisinh(\sqrt{\alpha}x) + \frac{\cosh(\sqrt{\alpha}) - \exp(-x)}{1-\alpha}$

$u(x)u^*(x) = \left[\frac{cosh(\sqrt{\alpha}) - exp\,(-x)}{1-\alpha}\right] - A^2Sinh^2(\sqrt{\alpha}x)$

In order that $\int_0^\infty u(x)u^*(x)\lambda < \infty$ the increasing exponential terms must cancel:

$$\frac{cosh^2(\sqrt{\alpha}x}{(1-\alpha)^2} - A^2Sinh^2(\sqrt{\alpha}x) = 0$$

$tanh^2(\sqrt{\alpha}x) = \frac{1}{A^2(1-\alpha)^2}$; $\tanh(\sqrt{\alpha}x) = \frac{1}{A(1-\alpha)}$

Thus,

$$\int_0^\infty \frac{1}{(1-\alpha)^2}\left[cosh^2(\sqrt{\alpha}x) - 2\exp(-x)\cosh(\sqrt{\alpha}x) + \exp\,(-2x)\right] - A^2sinh^2(\sqrt{\alpha}x)dx$$

$$\int_0^\infty \frac{1}{(1-\alpha)^2}\left[\left(\frac{e^{\sqrt{\alpha}x} + e^{-\sqrt{\alpha}x}}{2}\right)^2 - 2e^{-x}\left(\frac{e^{\sqrt{\alpha}x} + e^{-\sqrt{\alpha}x}}{2}\right) - A^2\left(\frac{e^{\sqrt{\alpha}x} + e^{-\sqrt{\alpha}x}}{2}\right)^2\right]dx$$

Assuming $\alpha < 1$

$\frac{1}{(1-\alpha)^2}\left(\frac{e^{\sqrt{\alpha}x}+2}{2}\right) - A^2\left(\frac{e^{\sqrt{\alpha}x}-2}{2}\right) = 0$

$e^{2\sqrt{\alpha}x} + 2 - A^2(1-\alpha)^2(e^{2\sqrt{\alpha}x} - 2) = 0$

$e^{2\sqrt{\alpha}x}(1 - [A(1-\alpha)]\,) + 2(1 + [A(1-\alpha)]^2 = 0$

$A(1-\alpha) = i$

$(1-\alpha) = \frac{i}{A}$

$\alpha = 1 - \frac{i}{A}$

Example D.8. Consider

$$\frac{-d}{dx}\left(x\frac{du}{dx}\right) - \lambda\frac{u(x)}{x} = 0 \quad 0 < x < 1$$

$$-(xg')' - \frac{\lambda g}{x} = \delta(x - \varepsilon)$$

$$g(x|\varepsilon;\lambda) = \{^{Ax^{-i\sqrt{\lambda}}}_{B\left(x^{i\sqrt{\lambda}} - x^{-i\sqrt{\lambda}}\right) \quad \varepsilon<x<1}$$

Solution

Continuity: $Ax^{-i\sqrt{\lambda}} = B\left(x^{i\sqrt{\lambda}} - x^{-i\sqrt{\lambda}}\right)\ I_\varepsilon$

Jump: $-i\sqrt{\lambda}\,A\frac{x^{-i\sqrt{\lambda}}}{x} = Bi\sqrt{\lambda}\frac{x^{i\sqrt{\lambda}}}{x} + Bi\sqrt{\lambda}\frac{x^{-i\sqrt{\lambda}}}{x} - 1I\varepsilon$

$$A\varepsilon^{-i\sqrt{\lambda}} = B(\varepsilon^{i\sqrt{\lambda}} - \varepsilon^{-i\sqrt{\lambda}})$$

$$-i\sqrt{\lambda}\,A\frac{\varepsilon^{-i\sqrt{\lambda}}}{\varepsilon} = Bi\sqrt{\lambda}\,\frac{\varepsilon^{i\sqrt{\lambda}}}{\varepsilon} + Bi\sqrt{\lambda}\,\frac{\varepsilon^{-i\sqrt{\lambda}}}{\varepsilon} - 1$$

$$[^{\varepsilon^{-i\sqrt{\lambda}}}_{-i\sqrt{\lambda}\,\varepsilon^{-i\sqrt{\lambda}}} \qquad ^{\varepsilon^{-i\sqrt{\lambda}} - \varepsilon^{i\sqrt{\lambda}}}_{(-i\sqrt{\lambda})(\varepsilon^{i\sqrt{\lambda}} + \varepsilon^{-i\sqrt{\lambda}})}]|\ ^A_B| = |^0_1|$$

$$g(x|\varepsilon;\lambda) = \begin{cases} \frac{i}{2\sqrt{\lambda}} x^{-i\sqrt{\lambda}}(\varepsilon^{i\sqrt{\lambda}} - \varepsilon^{-i\sqrt{\lambda}}) \\ \frac{i}{2\sqrt{\lambda}} \varepsilon^{-i\sqrt{\lambda}}(x^{i\sqrt{\lambda}} - x^{-i\sqrt{\lambda}}) \end{cases}$$

$$g(x|\varepsilon;\lambda) = \frac{i}{2\sqrt{x}}\, x_<^{-i\sqrt{\lambda}}(x_<^{i\sqrt{\lambda}} - x_<^{-i\sqrt{\lambda}})$$

also have:

$$\frac{1}{2\pi i}\oint g d\lambda = -x\delta(x - \lambda)$$

$$g+= \frac{i}{2|\lambda|^{1/2}} x_<^{-1|\lambda|^{1/2}}(x_<^{i|\lambda|^{1/2}} - x_<^{-i|\lambda|^{1/2}})$$

$$g-= \frac{i}{-2|\lambda|^{1/2}} x_<^{i|\lambda|^{1/2}}(x_<^{-1|\lambda|^{1/2}} - x_<^{i|\lambda|^{1/2}})$$

$$[g] = g_+ - g_- = \frac{i}{2|\lambda|^{1/2}}(x_<^{-i|\lambda|^{1/2}} - x_<^{i|\lambda|^{1/2}})(x_<^{i|\lambda|^{1/2}} - x_<^{-i|\lambda|^{1/2}})$$

$$= \frac{1}{2|\lambda|^{1/2}}\Big[x^{-i|\lambda|^{1/2}}\varepsilon^{i|\lambda|^{1/2}} - x^{i|\lambda|^{1/2}}\varepsilon^{i|\lambda|^{1/2}} - x^{-i|\lambda|^{1/2}}\varepsilon^{-i|\lambda|^{1/2}} + x^{i|\lambda|^{1/2}}\varepsilon^{-i|\lambda|^{1/2}}\Big]$$

Let $\lambda = v^2$

$$x\delta(x - \varepsilon) = \int_0^\infty \frac{1}{2v} v\left[x^{-iv}\varepsilon^{iv} - x^{iv}\varepsilon^{iv} - x^{-iv}\varepsilon^{-iv} + x^{iv}\varepsilon^{-iv}\right]dv$$

$$= \int_0^\infty \frac{1}{2}\left[x^{-iv}\left(\varepsilon^{iv} - \varepsilon^{-iv}\right) - x^{iv}\left(\varepsilon^{iv} - \varepsilon^{-iv}\right)\right]dv$$
$$= \frac{2}{\pi}\int_0^\infty Sin\ (\log x^v)\ Sin(\log \varepsilon^v)dv$$

Multiply by f(x) and integrals from 0 to 1:

$\int_0^1 f(x)d(x-\varepsilon) = \int_0^1 \frac{1}{x} f(x)dx \int_0^\infty \frac{2}{\pi} Sin(v\log x) Sin\ (v\log\varepsilon)dv$

Let $F_s^\wedge(v) = \int_0^1 \frac{1}{x} Sin(v\log x) f(x)dx$

Then $f(\varepsilon) = \frac{2}{\pi}\int_0^\infty F_s^\wedge\ (v) Sin(v\log x)dv$

Exercises

The Exercises that follow are largely drawn from class notes for Caltech AMa95bc 1985 and from midterm and finals problems in 1985.

Exercise D.1. Solve

$$L = \frac{d^2}{dx^2} + 4\frac{d}{dx} - 3$$

Thus,

$$p_0 y'' + p_1 y' + p_2 y = 0,\ \ with\ p_0 = 1, p_1 = 4, p_2 = -3.$$

Exercise D.2. If $f(t) = \begin{cases} (1-t^2)^{v-1/2} & for\ t<1 \quad for\ v>0 \\ 0 & for\ t>1 \end{cases}$,

expand the cosine term in $F_c(\lambda)$ the Fourier cosine transform and use term by term integration show.

$F_c(\lambda) = 2^{v-1/2}T(v+1/2)\lambda^{-v}J_v(\lambda)$

Exercise D.3. A strip of metal occupies the region $-b \le y \le b, -\infty < x < \infty$. An imposed temperature profile $\theta(x)$ is a function of x alone. As the consequence of this temperature distribution certain stresses are set up in the strip. The stresses are given by the various second derivatives of a function $\emptyset(x,y)$ satisfying

$$\emptyset_{xxxx} + 2\emptyset_{xxyy} + \emptyset_{yyyy} = -E\alpha\theta_{xx}$$

Where the constants α *and* E are the linear coefficient of thermal expansion and the elastic modulus respectively. The boundary conditions on $\emptyset$ are

$$\emptyset(x, \pm b) = \emptyset_y(x, \pm b) = 0$$

Let $\theta = \theta_0 \exp(Ex)$ $for\ x < 0\ and\ \theta = 0\ for\ x > 0\ where\ E\ is\ a\ small\ positive\ constant.$ (eventually allowed to approach 0). Take a Fourier transform in x and show that.

$$\emptyset = \frac{-E\alpha\theta_0}{\pi}\int_{-\infty}^{\infty} \exp(-i\lambda x)\frac{1}{i\lambda^3}[A\cosh\lambda y + B\lambda y\sinh\lambda y - 1/2]d\lambda$$

$$A = \frac{\sinh b\lambda + \lambda b\cosh b\lambda}{2\lambda b + \sinh 2\lambda b}$$

And

$$B = -\frac{\sinh b\lambda}{2\lambda b + \sinh 2\lambda b}$$

Obtain the explicit form of the longitudinal stress given by$\tau = \emptyset_{yy}$.
To get numerical values of τ we can use numerical integration which is difficult or we can try to sum the residues. Determine approximately where the poles of the denominator are. What curve is asymptotic to their locations? If we approximate the denominator by $\sinh 2\lambda b$ alone the residues may be summed explicitly. Show that this leads to.

$$\tau = E\alpha\theta_0\left[\frac{\exp(\pi x/b) + \cos(\pi x/b}{2\cos(\pi x/b) + 2\cos h(\pi x/b}\right] - 1 + \frac{1}{\pi}arctan\frac{2\exp(\pi x/2b)\cos(\pi x/b}{\exp(\pi x/b) - 1} - \frac{y}{2b}\left[\frac{\sinh(\pi x/2b)\sin(\pi x/2b)}{\cos^2(\pi x/2b) + sinh^2(\pi x/2b)}\right]$$

For $x < 0$. (The function arctan is an angle in the second quadrant).

Exercise D.4. The Jacobi elliptic function $\theta_3(u, q)$ is defined by the series

$\theta_3(u,q) = 1 + 2\sum_{n=1}^{\infty} q^{n^2}\cos(2nu) \qquad |q| \leq 1$

When q is small this representation is a useful way of computing the elliptic function but when q is close to 1 convergence is slow. Show that

$$\theta_3(u, \exp(-1/2\alpha^2)) = \frac{\sqrt{2\pi}}{\alpha}\sum_{m=-\infty}^{\infty} exp[(-2/\alpha^2)(u + m\pi)^2]$$

And indicate why this is a more convenient for $|q|$ close to 1.

Exercise D.5. Consider the problem of finding a Green's function $g(x, y'E, n)$ for the infinite strip region $-\infty < x < \infty \quad 0 \leq y \leq l$ satisfying

$$g_{xx} + g_{yy} = \delta(x - E)\delta(y - n)$$

With $g = 0$ for $y = 0$ or $y = l$. Take a Fourier transform in x to yield

$$G = \frac{-\exp(+i\lambda E)}{\sqrt{2\pi}\lambda sinh\lambda l}\begin{cases} sinh\lambda(l-n)sinh\lambda y & y < n \\ sinh\lambda(l-n)sinh\lambda n & y < n \end{cases}$$

Use the inversion theorem and sum residues to give.

$$g = \frac{1}{4\pi} ln\left[\frac{\cosh(\pi/l)(E-x) - \cos(\pi/l)(n-y)}{\cosh(\pi/l)(E-x) - \cos(\pi/l)(n+y)}\right]$$

Expand the argument of the lnfunction in the form of an infinite product so as to show how this result can be interpreted in terms of the method of images in which the strip is replaced by the whole plane with positive unit sources at $(E, 4l+n), (E, 2l+n), (E, n), ((E, 2l+n)\ldots$ The lines $y = l, y = 0$ lines of symmetry where the effects of all sources cancel.

Exercise D.6. Consider the following

$$\emptyset_{xx} - \frac{1}{c^2}\emptyset_{tt} = p(x,t)$$
$$\emptyset(x,0) = 0$$
$$\emptyset_t(x,0) = 0$$
$$p(x,t) = \delta(x)\delta(t)$$

Take a double Fourier transform and show that the displacement resulting from a concentrated impulse is a square wave.

Exercise D.7. Make an educated guess for a solution of the following equations and check your guess by substitution.

(i) $w'(x) = x$
(ii) $w'(x) + w(x) = e^x$
(iii) $w'(x) + w(x) = 1$
(iv) $w'(x) + (a/x)w(x) = 0$
(v) $w'(x) + (a/x)w(x) = x^3$
(vi) $w''(x) + w'(x)/x - w(x)/x^2 = 0$

Exercise D.8. An aircraft autopilot detects both the displacement ζ from course and the rate of displacement $d\zeta/dt$. The signal q given to the flight controls is given by

$$q = -Q\zeta + A\frac{d\zeta}{dt}.$$

However, the flight controls generate a plane displacement change in course that is given by

$$\zeta = -Bq - P\frac{dq}{dt}.$$

(1)

Show that the plane will be always out of control unless

$$AB > PQ.$$

Suppose that $AP = \frac{1}{4}(AB + PQ)^2$, what does this imply for the control of, in addition, condition (1) is satisfied?

Exercise D.9. Find the solution of

$$w''(z) + w(z) = \sec z$$

For which

$$w(0) = 0$$
$$w'(0) = 1.$$

Exercise D.10. Investigate the nature of the point at infinity for the equation

$$z^4 w''(z) - w(z) -- = 0$$

Find two linearly independent solutions valid about the point at infinity. Where do these solutions cease to be analytic?

Exercise D.11. Consider the Fourier sine and cosine series of

$$f(x) = x \sin x \qquad (0 \le x \le \pi).$$

(a) Before you determine these series, predict the asympototic order of magnitude of the Fourier coefficients b_n and a_n, for large n, in these half-range expansions. Give necessary explanations for your prediction.

(b) Then determine the two series representations for $f(x)$.

Exercise D.12. What is the smallest N such that the N-term partial sum of the Fourier series of $f(x) = |x|$ in $-\pi \le x \le \pi$ approximates the function with root-mean-square error < 0.05? That is, with $(E_N^*)^{1/2} < 0.05$, where

$$E_N^* = \int_{-\pi}^{\pi} \left\{ f(x) - \left[\frac{a_o}{2} + \sum_{n=1}^{N} a_{2n-1} \cos(2n-1)\, x \right] \right\}^2 dx,$$

$a_{n's}$ being the pertinent Fourier coefficients.

Exercise D.13.

(a) Determine the Fourier cosine series of the function

$$f(x) = \cos vx\,, \quad 0 \le x < \pi, \qquad v \ne 0, \pm 1, \pm 2, \ldots. \tag{1}$$

(b) From the resulting series deduce that for non-integral v,

$$\frac{\pi}{\sin \nu x} = \frac{1}{\nu} + \sum_{n=1}^{\infty}(-1)^n \left(\frac{1}{\nu-n} + \frac{1}{\nu+n}\right),$$

(2)

$$\pi \cot \nu\pi - \frac{1}{\nu} = \sum_{n=1}^{\infty} \left(\frac{1}{\nu-n} + \frac{1}{\nu+n}\right).$$

(3)

(c) If the above "partial fraction expansions" of $\csc \nu\pi$ and $\cot \nu\pi$ are regarded as functions of v, what must you show to justify termwise integration with respect to v from $\nu = -a$ to a, for some $a (> 0)$? What is the largest value of "a" beyond which the result from termwise integration may not be true?

(d) If justified, integrate (3) from $\nu = 0$ to z (Does the left-hand side of (3) tend to zero as $\nu \to 0$?) and show that

$$\frac{\sin \pi z}{\pi z} = \prod_{n=1}^{\infty}\left(1 - \frac{z^2}{n^2}\right) \qquad (|z| < 1),$$

Where $\prod_1^{\infty} u_n$ indicates the infinite product $u_1 u_2 u_3$ …..

Exercise D.13. Apply the Laplace transform method to calculate the response $y(t)$ of a damped harmonic oscillator to a forcing function $f(t)$ described by

$$y'' + 2\alpha y' + \omega_o^2 y = f(t) \qquad (t > 0)$$

Under the initial conditions $y(0) = y'(0) = 0$. (α, ω_o are real positive constants.)

(a) Consider $f(t)$ to be such that its Laplace transform $\tilde{f}(s)$ exists, but otherwise arbitrary, and solve for arbitrary ω_o and α.

(b) Investigate the limiting case of $\omega_o = \alpha$.

(c) Find the large time asymptotic behavior of $y(t)$ for the special case of $\alpha = 0$ and $f(t) = f_o \sin \omega_o t (t > 0$, and $f = 0$ for $t < 0)$.

Exercise D.14. The diffusion of heat, in time t, along an infinitely long (insulated) metal bar satisfies the partial differential equation

$$\frac{\partial u}{\partial t} = \frac{\partial^2 u}{\partial x^2} \qquad (-\infty < x < \infty, t > 0).$$

Given the initial condition $u(x, 0) = f(x)$, find $u(x, t)$ for $t > 0$ by applying the Fourier transform method with respect to x, assuming $f(x)$ to be arbitrary but its Fourier transform exists. Do you need to state some assumptions regarding the integral defining the Fourier transform of $u(x, t)$ in order to proceed?

(a) Carry out the inversion integral, using the convolution theorem or otherwise, to give the solution in the form of a single-integral representation for arbitrary $f(x)$.
(b) Deduce the solution when $f(x)$ is the Dirac delta function.

Exercise D.15.
(a) Find the Green's function $G(x,\xi)$ which satisfies the

DE $\quad G_{xx} + \lambda^2 G = \delta(x-\xi) \qquad (0 \le x, \xi \le 1).$

BC's $\quad G_x(0,\xi) = 0 \text{ and } G_x(1,\xi) = 0.$

By using the method of direct integration and the jump conditions of G across the point $x = \xi$. Here λ is a real-valued constant parameter. For what particular values of λ does G not exist?
(b) Now consider the problem of vibration of a free-ended string subject to forces varying harmonically in time.

$$u_{xx} - \frac{1}{c^2} u_{tt} = f(x)e^{-i\omega t} \qquad (0 \le x \le 1).$$

$$u_x(0,t) = u_x(1,t) = 0 \qquad (t > 0).$$

Disregard any specific initial conditions for the moment and seek a solution of the form $u(x,t) = \phi(x)\exp(i\omega t)$, where ϕ satisfies

$$\phi_{xx} + \lambda^2\phi = f(x) \qquad (\lambda = \omega/c),$$

And $\hat{\omega}$ is such that $G(x,\xi;\lambda)$ exists. Determine $u(x,t)$ for arbitrary $f(x)$ by making nuse of the Green's function. Describe a method by which you can have the solution satisfy given initial conditions.

Exercise D.16. Given that $u(x,t)$ is C^2 for (x,t) lying in a domain D and satisfies $u_{tt} - c^2 u_{xx} = 0$ in D, show that the values of u at the four points PQRS which are the vertices of a parallelogram in D formed by four arbitrary characteristic lines $\xi = x - ct = \xi_1, \xi = \xi_2, \eta = x + ct = \eta_1, and\ \eta = \eta_2$, satisfy the relation

$$u_P + u_R = u_Q + u_S.$$

Exercise D.17. A string, fixed at one end at $x = 0$, is set in motion from rest by a prescribed displacement at the free end at $x = 1$ so we have (with the wave velocity c normalized to unity)

DE $\quad u_{tt} - u_{xx} = 0 \qquad (0 < x < 1, t > 0)$

IC's $\quad u(x,0) = 0, \qquad u_t(x,0) = 0 \qquad (0 \le x \le 1)$

BC's $\quad u(0,t) = 0, \qquad u(1,t) = f(t) \qquad (t > 0).$

(a) Show that by change of variables $u = v(x,t) + w(x,t)$, where $w(x,t) = xf(t)$, v satisfies an inhomogeneous wave equation and subject to homogeneous boundary conditions

$$v(0,t) = 0 \qquad \text{and} \qquad v(1,t) = 0 \qquad (t > 0)$$

Give the differential equation and initial conditions for $v(x,t)$.

(b) Solve the new problem for v by a method of your choice, assuming $f(t)$ to be twice continuously differentiable but otherwise arbitrary.

(c) Investigate the case when $f(t) = \sin \pi t$ and discuss the physical significance of your result.

Exercise D.18. Solve the following radiation problem (of waves into the half space $x > 0$):

DE $\quad u_{tt} = u_{xx} \quad (0 < x < \infty, t > 0),$

BC's $\quad u(0,t) = F(t), \quad u(\infty,t) = 0, \quad (0 \leq t < \infty),$

IC's $\quad u(x,0) = 0, \quad u_t(x,0) = 0, \quad (0 < x < \infty).$

(a) By the simplest method (including direct applications of basic principles) you can devise, or by applying the Laplace transform with respect to t, or the Fourier sine transform with respect to x.

(b) Show that the total wave energy radiated away from the origin is at the rate equal to $[dF/dt]^2$. Also show that this is equal to the rate of workign by the external force in maintaining the motion $F(t)$ at $x = 0$.

Exercise D.19. For free vibration of a string fixed at both ends, with its displacement $u(x,\tau)$ satisfying

DE $\quad u_{xx} = u_{\tau\tau} \quad (\tau = ct > 0,\ 0 \leq x \leq \pi),$

DC $\quad u(o,\tau) = u(\pi,\tau) = 0 \quad (\tau > 0),$

And arbitrary initial conditions, show that $u(x,\tau)$ has the basic properties

$$u(x,\tau) = -u(\pi - x, \tau + \pi) = u(x, \tau + 2\pi) \qquad (0 \leq x \leq).$$

So u is 2π-periodic in τ for fixed x and is antisymmetric about $x = \pi/2$ over every half-period of π in τ.

Exercise D.20. Consider the problem of heat conduction through a solid slab (having homogeneous thermal constants c, ρ, k) occupying the first quadrant $x \geq 0,\ y \geq 0$, with the temperature distribution $T(x,y,t)$ satisfying

DE $\quad T_t = k(T_{xx} + T_{yy})$

IC $\quad T(x, y, 0) \equiv 0 \qquad (x \geq 0,\ y \geq 0)$

BC $\quad T = T_o H(t) \qquad$ On the two sides $x = 0 (y \geq 0)$ and $y = 0 (x \geq 0)$

Where $H(t)$ is the Heaviside step function, T_o is a positive constant.

(a) Find the solution for this corner problem [Hint: Try the separation of variables of the form $T(x, y, t) = X(x, t)Y(y, t)$ and make use of the solution to Rayleigh's problem.]

(b) Is the total heat flux into the cornered solid less than that into Rayleigh's semi-infinite slab, other things being equal? By how much?

$$\left[\text{You may need:} \qquad \int_0^\infty \operatorname{erfc}(\eta)\, d\eta = 1/\sqrt{\pi}.\right]$$

(c) Find the Duhamel's formula corresponding to the arbitrary $BC{:}\, T = f(t)$ on the two side surfaces.

Exercise D.21. Let $u(x, y)$ be twice continuously differentiable in the unit circle $\left(0 \leq r = (x^2 + y^2)^{1/2} < 1\right)$ and continuous in $0 \leq r \leq 1$, and let u satisify

$$u_{xx} + u_{yy} + 4u_x - 4u = 0 \qquad (0 \leq r < 1)$$

With BC: $u(r, \theta) = \exp(\zeta\theta - 2\cos\theta) \quad (r = 1, 0 < \theta < 2\pi)$

Where $\theta = \tan^{-1}(y/x)$ *and* ζ is a real (or complex) parameter.

(a) By using the transformation $u = e^{ax}(x, y)$ find the constant a so that the DE for v will not have v_x term. Before you solve for u find the value of u at the origin in the simplest way you can devise.

(b) Then solve for u and express your result in an integral or series representation.

Exercise D.22. The potential problem on a hemispherical shell of unit radius has its potential $V(\theta, \phi)$ satisfying (θ being the spherical polar angle)

DE $\quad \sin\theta \frac{\partial}{\partial\theta}\left[\sin\theta \frac{\partial V}{\partial\theta}\right] + \frac{\partial^2 V}{\partial\phi^2} = 0 \qquad \left(0 \leq \theta \leq \frac{\pi}{2},\ 0 \leq \phi < 2\pi\right)$

BC $\quad V\left(\frac{\pi}{2}, \phi\right) = f(\phi) \qquad (0 < \phi < 2\pi)$

By the separation of variables $V(\theta, \phi) = \Theta(\theta)\Phi(\phi)$ show that the general solution is

$$V(\theta,\phi) = a_o + \sum_{n=1}^{\infty}\left(\tan\frac{\theta}{2}\right)^n [a_n \cos n\,\phi + b_n \sin n\,\phi]$$

And that for the given BC,

$$V(\theta,\phi) = \frac{1}{2\pi}\int_o^{2\pi} \frac{f(\alpha)(1-\rho^2)d\alpha}{1-2\rho\cos(\phi-\alpha)+\rho^2} \qquad \left(\rho = \tan\frac{\theta}{2}\right)$$

[This formula for the shell is the analogue of the Poisson formula for the circular disc.]

References

[1] Winters-Hilt, S. Classical Mechanics and Chaos. (Physics Series: "Physics from Maximal Information Emanation" Book 1.)
[2] Mandelbrot, Benoît (1982). The Fractal Geometry of Nature. W H Freeman & Co.
[3] Feigenbaum, M. J. (1976). "Universality in complex discrete dynamics" (PDF). Los Alamos Theoretical Division Annual Report 1975–1976.
[4] Winters-Hilt, S. Emanation, Emergence, and Eucatastrophe. (Physics Series: "Physics from Maximal Information Emanation" Book 7.)
[5] Winters-Hilt, S. The Dynamics of Manifolds. (Physics Series: "Physics from Maximal Information Emanation" Book 3.)
[6] Winters-Hilt, S. Thermal & Statistical Mechanics, and Black Hole Thermodynamics. (Physics Series: "Physics from Maximal Information Emanation" Book 6.)
[7] *Faraday, Michael (1839). Experimental Researches in Electricity, vols. i. and ii.*
[8] Maxwell, James C. (1865). "A dynamical theory of the electromagnetic field". *Philosophical Transactions of the Royal Society of London.* **155**: 459–512.
[9] Hunt, Bruce J. (1991). *The Maxwellians.*
[10] M. Gunaydin, F. Gursey, Quark statistics and octonions, PRD, 1974.
[11] Conway, J.H. and D.A. Smith, On Quaternions and Octonions: their geometry, arithmetic, and symmetry, A K Peters, Wellesley, Massachusetts, 2005.
[12] Winters-Hilt, S. Quantum Field Theory and the Standard Model. (Physics Series: "Physics from Maximal Information Emanation" Book 5.)
[13] Winters-Hilt, S. Informatics and Machine Learning: from Martingales to Metaheuristics. (2021) Wiley.
[14] Doran, C. and A. Lasenby. Geometric Algebra for Physicsists. CUP 2013.
[15] Winters-Hilt, S. Quantum Mechanics, Path Integrals, and Algebraic Reality. (Physics Series: "Physics from Maximal Information Emanation" Book 4.)

[16] Penrose, Roger (1965), "Gravitational collapse and space-time singularities", Phys. Rev. Lett., 14 (3): 57.
[17] Winters-Hilt S. Emanator Theory is shown to be an optimal Martingale process at the fractal edge of chaos, where the Gravitational constant G is hypothesized to be a multiscale fractal coupling parameter. Advanced Studies in Theoretical Physics, 2023.
[18] Winters-Hilt, S. "Machine-Learning based sequence analysis, bioinformatics & nanopore transduction detection". ISBN: 978-1-257-64525-1. (2011).
[19] Hawking, S. W. (1992). "Chronology protection conjecture". *Phys. Rev. D.* **46** (2): 603.
[20] P.R. Girard. The Quaternion group and modern physics. Eur. J. Phys. 5 (1984): 25-32.
[21] Synge, J.L. Quaternions, Lorentz Transformations and the Conway-Dirac-Eddington Matrices.
[22] Cailler, C. 1917. Archs. Sci. Phys. Nat. ser. 4, 44 p. 237.
[23] Brown, W. B. and R. V. Churchill. Complex Variables and applications. McGraw-Hill Science, 2003.
[24] Caves, Carlton M.; Fuchs, Christopher A.; Schack, Ruediger (2002-08-20). "Unknown quantum states: The quantum de Finetti representation". Journal of Mathematical Physics. 43 (9): 4537–4559.
[25] Woodhouse, N.M.J. Introduction to Analytical Dynamics. Springer, 2nd Edition. 2009.
[26] Chaichian, M. and A.P. Demichev. Path Integrals with Generalized Grassmannian Variables. CBPF-NF-022/95.
[27] Fetter, A.L and J.D Walecka, Theoretical Mechanics of Particles and Continua, Dover (2003).
[28] Landau, L.D. and E.M. Lifshitz. Fluid Mechanics. 2013.
[29] Liepmann, H.W. and A. Roshko. Elements of Gasdynamics. Dover. 2002.
[30] Demidov, S. S. (2005). *Treatise on the differential calculus*. ISBN 978-0080457444.
[31] Thomson, William (Lord Kelvin) 1869 'On vortex motion.' Transactions of the Royal Society of Edinburgh, 25, 217–260.
[32] Jackson, J.D. Classical Electrodynamics, 2nd Edition. Wiley 1975.
[33] Landau, Lev D.; Lifshitz, Evgeny M. (1971). The Classical Theory of Fields. Vol. 2 (3rd ed.). Pergamon Press.
[34] Ramo, S., J.R. Whinnery, and T. VanDuzer. Fields and Waves in Communication Electronics. John Wiley and sons, 1994.
[35] Purcell, E. M. Electricity and Magnetism. McGraw-Hill Science, 1984.

[36] Heaviside, O. and E. Whittaker. Electromagnetic Theory. Ams Chelsea Publishing. 2003.
[37] Lorentz, Hendrik Antoon (1899), "Simplified Theory of Electrical and Optical Phenomena in Moving Systems" , *Proceedings of the Royal Netherlands Academy of Arts and Sciences*, **1**: 427–442.
[38] Misner, Charles W., Thorne, K. S., & Wheeler, J. A. Gravitation. Princeton University Press, 2017. ISBN: 9780691177793.
[39] Robinson, Abraham (1963), Introduction to model theory and to the metamathematics of algebra, Amsterdam: North-Holland, ISBN 978-0-7204-2222-1, MR 0153570
[40] Robinson, Abraham (1966), Non-standard analysis, Princeton Landmarks in Mathematics (2nd ed.), Princeton University Press, ISBN 978-0-691-04490-3, MR 0205854
[41] R. D. Richtmyer (1978), Principles of Advanced Mathematical Physics Vol. 1 & 2, Springer-Verlag, New York.
[42] Green, G (1828). *An Essay on the Application of Mathematical Analysis to the Theories of Electricity and Magnetism*. Nottingham, England.
[43] Bender, C.M. and S.A. Orszag. Advanced Mathematical Methods for Scientists and Engineers: Asymptotic Methods and Perturbation Theory. Springer. 1999.
[44] Jarlskog, G.; Jönsson, L.; Prünster, S.; Schulz, H. D.; Willutzki, H. J.; Winter, G. G. (1 November 1973). "Measurement of Delbrück Scattering and Observation of Photon Splitting at High Energies". *Physical Review D*. American Physical Society (APS). **8** (11): 3813–3823.
[45] Euler, H. and B. Kockel, Naturwiss. 23, 246 (1935).
[46] Einstein, A. "On a heuristic point of view concerning the production and transformation of light" (Ann. Phys., Lpz 17 132-148).
[47] Einstein, A. (1905) "*Zur Elektrodynamik bewegter Körper*", *Annalen der Physik* 17: 891; English translation On the Electrodynamics of Moving Bodies by George Barker Jeffery and Wilfrid Perrett (1923).
[48] de Broglie, Louis Victor. "On the Theory of Quanta" (PDF). *Foundation of Louis de Broglie* (English translation by A.F. Kracklauer, 2004. ed.).
[49] Fizeau, H. (1851). "Sur les hypothèses relatives à l'éther lumineux". Comptes Rendus. 33: 349–355.
[50] Compton, Arthur H. (May 1923). "A Quantum Theory of the Scattering of X-Rays by Light Elements". Physical Review. 21 (5): 483–502.

[51] Minkowski, H. (1908). *"Die Grundgleichungen für die elektromagnetischen Vorgänge in bewegten Körpern"* . Nachrichten der Gesellschaft der Wissenschaften zu Göttingen, Mathematisch-Physikalische Klasse: 53–111. English translation: "The Fundamental Equations for Electromagnetic Processes in Moving Bodies". In: The Principle of Relativity (1920), Calcutta: University Press, 1–69.
[52] Penrose, R., W. Rindler (1984) Volume 1: Two-Spinor Calculus and Relativistic Fields, Cambridge University Press, United Kingdom.
[53] Penrose, R., W. Rindler (1986) Volume 2: Spinor and Twistor Methods in Space-Time Geometry. Cambridge University Press, United Kingdom.
[54] Huggett, S.A. and K.P. Tod. An Introduction to Twistor Theory. CUP 1994.
[55] Steane, A.M. Relativity Made Relatively Easy: Volume 1. OUP 2012.
[56] Penrose, R. (1968) in *Battelle Recontres*, eds. C.M. deWitt and J.A. Wheeler, Benjamin, New York.
[57] Penrose, R. (1955) Math. Proc. Camb. Phil. Soc. 51 406.
[58] Penrose, R. (1956) Math. Proc. Camb. Phil. Soc. 52 17.
[59] Adler, R. Introductio to General Relativity. McGraw-Hill, 1975.
[60] *On the Hypotheses which lie at the Bases of Geometry*. Bernhard Riemann. Translated by William Kingdon Clifford [Nature, Vol. VIII. Nos. 183, 184, pp. 14–17, 36, 37.]
[61] Barger, Vernon; Marfatia, Danny; Whisnant, Kerry Lewis (2012). *The Physics of Neutrinos*. Princeton University Press. ISBN 978-0-691-12853-5.
[62] R.P. Feynman and M. Gell-Mann. Theory of the Fermi Interaction, Phys. Rev. 109, 193 (1958).
[63] E.C.G. Sudarshan and R.K. Marshak. Chirality Invariance and the Universal Fermi Interaction. Phys. Rev. 109, 1860 (1958).
[64] M. Ruderman and R. Finkelstein. Note on the decay of the π-meson. Phys. Rev. 76, 1458 (1949).
[65] Noether, E.. Invariante Variations Probleme. Nachr.Ges. Wiss. Gottingen, Math-Phys. klasse (1918), 235.
[66] Tavid, M.A.. Transport theory and Statistical Mechanics 1 (3), 183 (1971).
[67] Atiyah, Michael F. (1988a), *Collected works. Vol. 1 Early papers: general papers*, Oxford Science Publications, The Clarendon Press Oxford University Press.
[68] Sorkin, R.D., Phys. Rev. Lett. 51, 87 (1983).
[69] Gross, D.J. and M. J. Perry, Nucl. Phys. B 226, 29 (1983)

[70] Winters-Hilt S. Topics in Quantum Gravity and Quantum field Theory in Curved Spacetime. UWM PhD Dissertation, 1997.
[71] Winters-Hilt S, I. H. Redmount, and L. Parker, "Physical distinction among alternative vacuum states in flat spacetime geometries," Phys. Rev. D 60, 124017 (1999).
[72] Friedman J. L., J. Louko, and S. Winters-Hilt, "Reduced Phase space formalism for spherically symmetric geometry with a massive dust shell," Phys. Rev. D 56, 7674-7691 (1997).
[73] Louko J and S. Winters-Hilt, "Hamiltonian thermodynamics of the Reissner-Nordstrom-anti de Sitter black hole," Phys. Rev. D 54, 2647-2663 (1996).
[74] Louko J, J. Z. Simon, and S. Winters-Hilt, "Hamiltonian thermodynamics of a Lovelock black hole," Phys. Rev. D 55, 3525-3535 (1997).
[75] Amari, S. and H. Nagaoka. Methods of Information Geometry. Oxford University Press. 2000.
[76] Winters-Hilt, S. Feynman-Cayley Path Integrals select Chiral Bi-Sedenions with 10-dimensional space-time propagation. Advanced Studies in Theoretical Physics, Vol. 9, 2015, no. 14, 667-683.
[77] Winters-Hilt, S. Unified Propagator Theory and a non-experimental derivation for the fine-structure constant. Advanced Studies in Theoretical Physics, Vol. 12, 2018, no. 5, 243-255.
[78] Winters-Hilt, S. The 22 letters of reality: chiral bisedenion properties for maximal information propagation. Advanced Studies in Theoretical Physics, Vol. 12, 2018, no. 7, 301-318.
[79] Winters-Hilt, S. Theory of Trigintaduonion Emanation and Origins of α and π. Researchgate 05/24/20.
[80] Winters-Hilt, S. Fiat Numero: Trigintaduonion Emanation Theory and its Relation to the Fine-Structure Constant α, the Feigenbaum Constant $C\infty$, and π. Advanced Studies in Theoretical Physics, Vol. 15, 2021, no. 2, 71-98.
[81] Winters-Hilt, S. Meromorphic precipitation of quantum matter with dimensionful action. May 2021. DOI:10.13140/RG.2.2.32294.24640.
[82] Winters-Hilt, S. Chiral Trigintaduonion Emanation Leads to the Standard Model of Particle Physics and to Quantum Matter. Advanced Studies in Theoretical Physics, Vol. 16, 2022, no. 3, 83-113.
[83] Winters-Hilt, S. Emanator Theory using split octonions is Manifestly Lorentz Invariant and reveals why the fundamental constant $\hbar$ should be so small. Advanced Studies in Theoretical Physics, 2023.

[84] Landau, Lev D.; Lifshitz, Evgeny M. (1969). Mechanics. Vol. 1 (2nd ed.). Pergamon Press.
[85] Goldstein, Herbert (1980). Classical Mechanics (2nd ed.). Addison-Wesley.
[86] Fetter, A.L and J.D Walecka, Theoretical Mechanics of Particles and Continua, Dover (2003).
[87] Percival, I.C. and D. Richards. Introduction to Dynamics. (1983) Cambridge University Press.
[88] Arnold, V.I. Ordinary Differential Equations. MIT Press. (1978).
[89] Arnold, Vladimir I. (1989). Mathematical Methods of Classical Mechanics (2nd ed.). New York: Springer.
[90] Robert L. Devaney. An Introduction to Chaotic Dynamical Systems. Addison -Wesley.
[91] Quigg, C. Gauge Theories of the Strong, Weak, and Electromagnetic Interactions. Princeton University Press. 2013.
[92] Hawking, Stephen & Ellis, G. F. R. (1973). The Large Scale Structure of Space-Time. Cambridge: Cambridge University Press.
[93] Peebles, P. J. E. (1980). Large-Scale Structure of the Universe. Princeton University Press.
[94] B. Abi et al. Measurement of the Positive Muon Anomalous Magnetic Moment to 0.46 ppm
Phys. Rev. Lett. 126, 141801 (2021).
[95] Balmer, J. J. (1885). "Notiz über die Spectrallinien des Wasserstoffs" [Note on the spectral lines of hydrogen]. Annalen der Physik und Chemie. 3rd series (in German). 25: 80–87.
[96] Werner Heisenberg (1925). "Über quantentheoretische Umdeutung kinematischer und mechanischer Beziehungen". Zeitschrift für Physik (in German). 33 (1): 879–893. ("Quantum theoretical re-interpretation of kinematic and mechanical relations")
[97] Schrödinger, E. (1926). "An Undulatory Theory of the Mechanics of Atoms and Molecules" (PDF). Physical Review. 28 (6): 1049–1070.
[98] Max Born; J. Robert Oppenheimer (1927). "Zur Quantentheorie der Molekeln" [On the Quantum Theory of Molecules]. Annalen der Physik (in German). 389 (20): 457–484.
[99] Dirac, P. A. M. (1928). "The Quantum Theory of the Electron" (PDF). Proceedings of the Royal Society A: Mathematical, Physical and Engineering Sciences. 117 (778): 610–624.
[100] Dirac, Paul Adrien Maurice (1930). The Principles of Quantum Mechanics. Oxford: Clarendon Press.
[101] Dirac, Paul A. M. (1933). "The Lagrangian in Quantum Mechanics" (PDF). Physikalische Zeitschrift der Sowjetunion. 3: 64–72.

[102] Feynman, Richard P. (1942). The Principle of Least Action in Quantum Mechanics (PhD). Princeton University.
[103] Feynman, Richard P. (1948). "Space-time approach to non-relativistic quantum mechanics". Reviews of Modern Physics. 20 (2): 367–387.
[104] Laplace, P S (1774), "Mémoires de Mathématique et de Physique, Tome Sixième", Statistical Science, 1 (3): 366–367.
[105] Erdeyli, A. Asymptotic Expansions. 1956 Dover.
[106] Erdeyli, A. Asymptotic Expansions of differential equations with turning points. Review of the Literature. Technical Report 1, Contract Nonr-220(11). Reference no. NR 043-121. Department of Mathematics, California Institute of Technology, 1953.
[107] Carrier, G.F, M. Crook and C.E. Pearson. Functions of a complex variable. 1983 Hod Books.
[108] Van Vleck, J. H. (1928). "The correspondence principle in the statistical interpretation of quantum mechanics". Proceedings of the National Academy of Sciences of the United States of America. 14 (2): 178–188.
[109] Chaichian, M.; Demichev, A. P. (2001). "Introduction". Path Integrals in Physics Volume 1: Stochastic Process & Quantum Mechanics. Taylor & Francis. p. 1ff. ISBN 978-0-7503-0801-4.
[110] Vinokur, V. M. (2015-02-27). "Dynamic Vortex Mott Transition".
[111] Hawking, S. W. (1974-03-01). "Black hole explosions?". Nature. 248 (5443): 30–31.
[112] Birrell, N.D. and Davies, P.C.W. (1982) Quantum Fields in Curved Space. Cambridge Monographs on Mathematical Physics. Cambridge University Press, Cambridge.
[113] Maldacena, Juan (1998). "The Large N limit of superconformal field theories and supergravity". Advances in Theoretical and Mathematical Physics. 2 (4): 231–252.
[114] Witten, Edward (1998). "Anti-de Sitter space and holography". Advances in Theoretical and Mathematical Physics. 2 (2): 253–291.
[115] Sommerfeld, Arnold (1916). "Zur Quantentheorie der Spektrallinien". Annalen der Physik. 4 (51): 51–52.
[116] Winters-Hilt, S. The Dynamics of Fields, Fluids, and Gauges. (Physics Series: "Physics from Maximal Information Emanation" #2.)
[117] R. N. Hall, J. Appl. Phys. 20, 925 (1949).
[118] D'Alembert, Jean Le Rond (1743). Traité de dynamique.
[119] Tolkien, J.R.R. (1990). *The Monsters and the Critics and Other Essays*. London: HarperCollinsPublishers.

Index

B

C

D

F

G

H

I

J

K

L

M

N

O

P

Q

R

S

T

U

V

W

X

Y

Z

www.ingramcontent.com/pod-product-compliance
Lightning Source LLC
Chambersburg PA
CBHW050452190326
41458CB00005B/1256
* 9 7 9 8 9 8 8 8 1 6 0 2 7 *